Research for Development

More information about this series at http://www.springer.com/series/13084

Carlo Andrea Castiglioni

Seismic Behavior of Steel Storage Pallet Racking Systems

Springer

Carlo Andrea Castiglioni
Department of Architecture, Built
Environment and Construction
Engineering
Politecnico di Milano
Milan
Italy

ISSN 2198-7300 ISSN 2198-7319 (electronic)
Research for Development
ISBN 978-3-319-28465-1 ISBN 978-3-319-28466-8 (eBook)
DOI 10.1007/978-3-319-28466-8

Library of Congress Control Number: 2015960817

Printed on acid-free paper

This Springer imprint is published by SpringerNature
The registered company is Springer International Publishing AG Switzerland

....

O frati, che per centomila
perigli siete giunti all'occidente;
a questa tanto picciola vigilia
de' nostri sensi ch'è del rimanente,
non vogliate negar l'esperienza,
dietro al sol, del mondo senza gente.
Considerate la vostra semenza:
fatti non foste a viver come bruti,
ma per seguir virtute e conoscenza.

....

Dante, *La Divina Commedia*
Inferno, XXVI, 112–120

To my family
and to all those, friends and colleagues,
who helped me during this research

Foreword by Luis Calado

Storage pallet racking systems are lightweight, high steel structures used to carry large loads. Compared to steel structures of rolled profiles, racking systems are more prone to buckling, requiring specific design rules to cover such phenomena. The numerical simulation of this type of structures is quite complex and depends on the stiffness of the beam-to-upright connections and on the base plate anchoring.

In the last decade, several collapses of storage racking systems were reported involving large economical losses. In some cases, collapses were due to earthquakes and to the lack of capacity of the structure to withstand horizontal forces. Also the high driving speeds inside the warehouse environment, due to competitive pressure in the logistic industry, may cause impact on racking systems originating dislodgment of loaded pallets and in some cases collapse of part of the racking with its loads. Sometimes, the partial collapse results in a global collapse due to a domino effect.

As storage racking systems cannot be considered as steel buildings, Eurocode 3 and usual structural standards cannot be directly applied. The peculiarities associated with geometrical dimensions, steel elements and loads need to be taken into account in the design. Besides these, the design of storage racking systems for earthquake resistance cannot apply directly the general design rules of steel buildings in seismic areas, because the response is strongly dependent on the peculiarities of such structural systems. Special attention should be paid to the potential movements of the pallets because they can fall down accidentally during earthquakes or horizontal impacts, hence requiring the design of appropriate accessories to reduce or prevent this risk of the storage goods.

The standards available for the seismic design of storage racking systems are few and have the limitation to consider static loads instead of mobile storage loads originated by the earthquake or impact loads. Besides this, experimental and analytical studies of the seismic behaviour of racking systems are few and the results are very often not available because they are property of companies. For that reason, these researches have had a little influence in the development of standards related to storage racking systems.

It is understandable that there are technical limitations in the field of safety and design of storage pallet racking systems in seismic areas due to lack of knowledge on actions, on structural behaviour and on sliding conditions of the pallets on the racks. In order to give a positive answer to these limitations, the EU sponsored through the Research Fund for Coal and Steel an RTD project titled "Storage Racks in Seismic Areas" (acronym SEISRACKS) which was coordinated by the author of the book. The scope of the project was to increment the knowledge on real service conditions of storage racks and on the real behaviour of elements, connections and structures and to assess design rules for storage racking systems in seismic zones.

This book is based on the research carried out under the project SEISRACKS, namely the extensive experimental tests in full scale of elements, connections and structures as well as the author's experience in the design of such structures. Following the background and the literature review on damages and standards, it is presented and explained in Chap. 2 the experimental campaign and results of the cyclic tests on full scale of components. Chapter 3 presents the problem of pallet sliding, while Chap. 4 deals with pushover tests in down-aisle and in cross-aisle direction. Chapter 5 presents the pseudo-dynamic tests performed, while in Chap. 6, the results of full-scale dynamic tests carried out using a shaking table are presented and discussed. In Chap. 7, some conclusive remarks are made and design recommendations are given.

This book sets out the seismic behavior of steel storage pallet racking systems in a logical manner giving helpful information to the reader which will allow designers to understand the cyclic behaviour in order to correctly conceive such type of structures.

Luis Calado
Professor at IST-UL

Foreword by Stefano Sesana

The industrial steel storage racks are a complex system of equipment of great importance for the modern logistic industry, nowadays also used in retail areas open to the public.

Designed mainly to support heavy vertical loads, the industrial racks are generally very light steel structures, flexible, sensible to the horizontal actions and for this reason apparently vulnerable when challenged by earthquake shakes.

Systematic studies on the behaviour of the steel storage racks challenged by seismic actions started in the USA in the early 80s (Chen, Scholl and Blume, Blume and Associates), promoted by the Rack Manufacturers Institute, the association of the rack manufacturers in the USA. These studies provided basic information to the RMI specifications that for a long time has been the only official document explicitly addressing to the topic of the design of pallet racks in seismic zone, now an official ANSI Norm since 2008.

The FEM Section X started in year 2000 studies on this argument using an "European" approach with the goal of developing a specific Norm; the FEM Section X—Racking and Shelving work Group, nowadays ERF (European Racking Federation), joins the European National Association of rack manufacturers; among them, the ACAI-CISI (now UNICSAL), the Association of the Italian manufacturers, has always played a proactive role promoting this project.

The starting points of this project have been the RMI standard and the Eurocode 8; during its development, several researches were carried out, mainly coordinated by the former Structural Engineering Department (DIS) of the Politecnico di Milano; starting from the observation and analysis of the dynamic behaviour of storage racks tested on a shaking table, the researches continued with experimental and numerical studies for the characterization of the components and of the structural systems; and finally, the development of advanced design methods explicitly considers phenomena typical of the racking systems such as the interaction between the structure and the supported loads.

The results of these researches are nowadays implemented in the standards and in the Norms developed by the European industry and used in the design practice that have been progressively extended worldwide and that are also attracting the interest of RMI.

In this book, the results of the first research on Seisracks are presented; the Seisracks was coordinated by ACAI in the role of research partner under the responsibility of Prof. Carlo Andrea Castiglioni of the Politecnico di Milano. This research was promoted by the European Industry of racks and shelving and was funded by the European Community: therefore, this is the result of a synergy between university and industry producing fundamental results of scientific and technical knowledge that are the basis of the first European guidelines for the design of storage racks in seismic areas, the FEM 10.2.08 issued in 2010, and of the first Italian Norm issued by UNI in 2009, the UNI/TS 11379.

These Norms are currently the specific reference for the design of steel storage racks in seismic zones based on the European approach, but they will be soon replaced by the new Norm EN16681, recently endorsed by the CEN Technical Committee TC-344, that is the evolution of the previous guidelines; this Norm is based, in addition to the results presented in this book, on the finding of the recent research project Seisracks2, also coordinated by Prof. Castiglioni, representing the continuation and completion of the first Seisracks.

The fundamental technical and scientific contents presented in this book are an evidence of a fruitful synergy between university and industry that has produced very important knowledge for a relevant branch of the logistic industry, in which the Politecnico di Milano and the Italian racking industry plaid and plays today a fundamental role.

Ing. Stefano Sesana
Convenor
CEN TC 344—WG5 "Seismic Design"
SCL Ingegneria Strutturale

Contents

List of Figures

List of Tables

Chapter 1
Introduction

1.1 Background

Despite their lightness, racking systems (see appendix A for detailed information about structural typologies, standard pallet types and sizes as well as nomenclature) carry very high live load (many times larger than the dead load, opposite of what happens for usual civil engineering structures) and can raise a considerable height. For these reasons they have to be properly designed (Fig. 1.1).

Many difficulties arise in the prediction of the structural behaviour of pallet racks, like instability (global, local and distortional) or modelling problems (beam-upright connections stiffness, base plate anchoring). The behaviour of these systems is affected by the particular geometry of their structural components, made by high slenderness elements, the non-linear behaviour of both the beam-to-column (Agatino et al. 2001) and the base-plate joints (Baldassino and Zandonini 2001).

Therefore, these structures cannot be considered as buildings, and reference cannot be made to usual Structural Design Recommendations and Standards.

The most recent Design Standards for steel storage racks (RMI 2002a, b; FEM 2001a, b; RAL 1990, A.S. 1993; FEM 2005) provide a combined numerical-experimental approach in which the design structural analysis is supported by specific tests to evaluate the performance of the key components (members and joints).

The design needs particular attention for storage racks installed in a seismic zone, where they must be able to withstand dynamic forces. Besides the usual global and local collapse mechanisms, an additional limit state for the system is represented by the fall of the pallets with subsequent damage to goods, people and to the structure itself. In Europe, no official document is currently available for the seismic design of pallet racks, and the designers are compelled to operate with a total lack of references and of commonly accepted design rules. Very often they make reference to the Rack Manufacturers Institute (R.M.I.) Specifications (R.M.I. 2002a, b), while the European Federation of Maintenance (F.E.M.) is presently working in order to produce an official document (FEM 2005).

C.A. Castiglioni, *Seismic Behavior of Steel Storage Pallet Racking Systems*, Research for Development, DOI 10.1007/978-3-319-28466-8_1

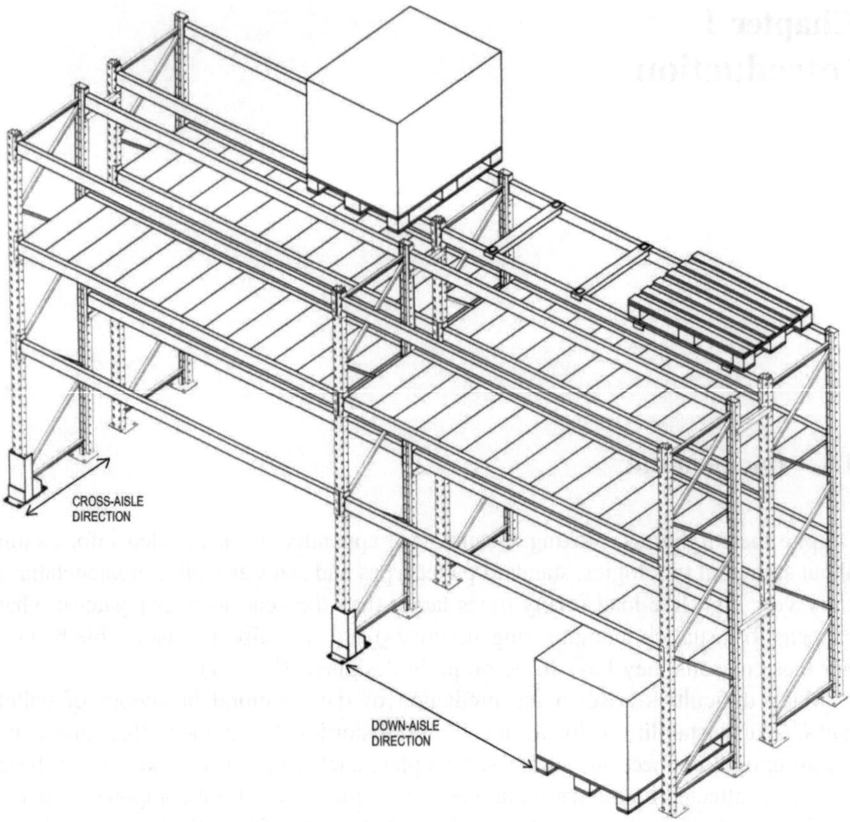

Fig. 1.1 Typical pallet rack configuration

It must be pointed out that the seismic behaviour of steel storage racks is not only a very interesting and challenging problem from a scientific point of view, but it has also a very large economic impact.

Racks, in fact, are widely adopted in warehouses where they are loaded with tons of (more or less) valuable goods. The loss of these goods during an earthquake may represent for the owner a very large economic loss, much larger than the cost of the whole rack on which the goods are stored, or of the cost for its seismic upgrade.

Racks are also more and more frequently adopted in supermarkets and shopping centres, in areas open to the public.

The falling of the pallets, in this case, may endanger the life of the clients as well as of the workmen and employees, involving not only Civil and Penal Right considerations about the liability of the owners, but also economic considerations related to the insurance coverage.

Sliding of the pallets on the racks and their consequent fall represents a situation that might occur during a seismic event also in the case of a well designed storage rack (the phenomenon depending only on the dynamic friction coefficient between

Fig. 1.2 Public warehouse store (from FEMA460)

the pallet and the steel beam of the rack) and that, as a limit state, must be prevented.

Many times, after an earthquake, loss of goods was reported, with or without contemporary failure of the steel rack structural system. Most probably, these structural failures are a consequence of the fall of the pallets and of the impact of the goods on the structure at the lower levels, triggering a progressive dynamic collapse.

The uncertainties associated with a clear assessment of the causes of such failures (due to structural design faults or caused by fall of the pallets) may result in long quarrels among constructors, users and insurance companies (Figs. 1.2 and 1.3).

Fig. 1.3 Damaged pallet (or non-allowed/non-standard pallet type)

This brief introduction shows that although these structures, made by thin-walled (and many times cold-formed) steel profiles, are very light and represent only a small percentage of the annual sales of steel profiles in the world, very large economic interests, as well as civil and penal Right liability problems might arise as a consequence of an earthquake event striking them.

1.2 The International Situation

During the past few decades, the number of large public warehouse stores (often referred to as big-box stores) has grown significantly, changing both consumer buying habits and the public's risk of injury during earthquakes. Whereas traditional retailers typically store goods and products outside the retail space in limited access storage rooms and warehouse facilities, big-box stores keep goods in close proximity to the consumer at all times. Typically, shoppers in these stores browse in aisles between steel storage racks, 5–6 m in height, that hold pallets of inventory goods, some of which can be very heavy.

During an earthquake, occupant safety in a big-box store depends on both the structural performance of the building and on the performance of the storage racks and their contents. Earthquake ground motions can cause storage racks to collapse or overturn if they are not properly designed, installed, maintained, and loaded. In addition, goods stored on the racks may spill or topple off. Both occurrences pose a life-safety risk to the exposed shopping public.

The seismic design for new warehouse stores, including both the buildings and storage racks, is governed by the building code in force in the jurisdiction where a store is built.

For example, the seismic requirements in building codes currently enforced in most U.S. jurisdictions are based on the *NEHRP Recommended Provisions for Seismic Regulations for New Buildings and Other Structures*, developed by the Building Seismic Safety Council (BSSC) of the National Institute of Building Sciences with National Earthquake Hazards Reduction Program (NEHRP) funding provided by the Federal Emergency Management Agency (FEMA).

The seismic requirements for new stores and storage racks, however, do not stipulate how goods are to be stored on the racks. For example, the state of California has enacted a California Labor Law that requires contents placed on storage racks 4 m or higher above ground level to be secured. While this law includes a few suggestions for securing goods, it does not include detailed procedures. Thus, in virtually all jurisdictions, requirements for securing storage rack contents are self-imposed by store owners and operators.

The situation is further complicated by the fact that these types of storage racks can be easily reconfigured (i.e., shelf level changed) to meet changing merchandising needs. The reconfiguration work, generally performed many times over the life of the structure, is done by store employees who may not always understand required procedures. Further, fork lifts are used to load goods on the racks and the racks can easily be damaged in the process. Finally, heavy merchandise stored on the floor near storage racks can topple during an earthquake and damage rack columns and braces, initiating rack collapse.

The stimulus for the development of the *NEHRP Recommended Provisions for Seismic Regulations for New Buildings and Other Structures* was a 2003 request from the State of Washington to FEMA for guidance concerning the life-safety risk posed by the storage racks in publicly accessible areas of retail stores, especially the risk of rack collapse and loss of stored goods during an earthquake. The goal was the possible development of state regulations in response to a fatality resulting from the breaking of a pallet on a storage rack in a commercial retail hardware facility. It should be pointed out that the accident was not associated with an earthquake. Considering the recent Nisqually earthquake, however, the State of Washington recognized that a significant life-safety hazard could be associated with storage racks behaviour during earthquakes.

FEMA was aware that current storage rack requirements were somewhat out of date. For example, the latest edition of the storage rack design specification developed in the U.S. by the Rack Manufacturers Institute (RMI) references the obsolete 1994 *NEHRP Recommended Provisions,* even though the *Provisions* document has been updated three times since then (i.e., in 1997, 2000, and 2003). Furthermore, during the 1994 Northridge earthquake (magnitude = 6.7), serious storage rack collapses occurred in several warehouse retail stores that would likely have resulted in injuries and possibly deaths if the earthquake had occurred during a time when the stores had significant public presence rather than at 4:30 a.m. on a federal holiday.

Many existing racks have been since voluntarily strengthened or replaced and stricter quality assurance programs for rack loading and reconfiguration have been implemented by some owners, in order to prevent a reoccurrence of the Northridge problems. In addition, the 1994 *NEHRP Recommended Provisions* included a 50 % increase of seismic loads for storage racks in areas accessible to the public. FEMA also recognized that the design process must take into account earthquakes larger than those recently experienced.

As already mentioned, in Europe, no official document is currently available for the seismic design of pallet racks and the designers are compelled to operate without references to commonly accepted European design rules.

Present Eurocodes 1, 3 and 8 give insufficient information on many design issues. Recently, rack manufacturers defined a set of conventional design criteria, based on engineering experience, and drafted a "code of good practice" for users, a strict application of which is intended to achieve a safe working environment (FEM 2001b). This Code gives insufficient information to some design aspects.

Very often designers make reference to the Rack Manufacturers Institute (R.M. I.) Specifications (R.M.I. 2002a, b). In the meanwhile, the Industry in Europe, under the guidance of the European Federation of Maintenance (F.E.M.) issued a Manufacturers' Design Recommendations called FEM 10.2.02. (FEM 2001a). These recommended the way in which components are brought together to provide the optimum strength and stability required to store specified pallet load size and maximum weight. An increasing number of European manufacturers are presently able to design according to these recommendations, if required by their customers. Furthermore, CEN has recently activated a Technical Committee (CEN-TC344), with the aim of developing a set of Eurocodes dedicated to racking and shelving.

1.3 Overview of Damage to Steel Pallet Storage Racks and Content Spillage

In 2003, estimated pan–European sale value for the racking industry exceeded 1.2 Billion Euro. Racking systems operated by industrial trucks represent approximately 70 % of the total yearly racking industry market. The current estimated yearly loss due to accidental impact is 600 million Euros. Moreover the losses due to consequent fires far exceed this value. Economical losses are expected to continue to rise due to competitive pressure in the logistic industry, resulting in higher driving speeds of industrial trucks within the racking environment.

The warehouse workplace is a potentially dangerous working environment. Careless driving of trucks can cause impact on racking and the dislodgment of loaded pallets onto operatives and even the collapse of part of the racking with its loads. In order to give an idea of the potential economic damage related to a collapse of one of these structures, it is enough to mention as an example that in the last two years, only in the Netherlands, at least two major collapses occurred, with a consequent fire. This fact made things of public domain (which is not usually the case). In these two collapses, there was more than 100 million Euro damage to goods and warehouse. Fortunately no person killed or injured. In the same period, in Europe, a conservative estimate gives more than 500 million Euro of goods lost due to racking system failures. Moreover, after a failure, the warehouse is usually out of service for a long period, increasing the economic damage.

The following images give a brief overview of possible causes and consequences of storage racks collapses, in daily practice, which include collapse of the pallet (e.g. Figure) that, falling down can cause the collapse of an entire bay of the rack due to the impact, or collision of a fork lift truck with the structure (e.g. Figures from Figs. 1.4, 1.5 and 1.6). Even in case of a more conditioned way of operation (as in case of Very Narrow Aisle (VNA) trucks) human failures can cause a very large accident (Fig. 1.7).

In case of a serious injuries or deadly accidents, also the police office is involved because of possible criminal offence (Fig. 1.8). The fire department can be involved in case of hazardous products fallen down (Fig. 1.9). But many times, information about a collapse is concealed to the outside world (Fig. 1.10).

Fig. 1.4 Damaged uprights due to industrial truck collisions

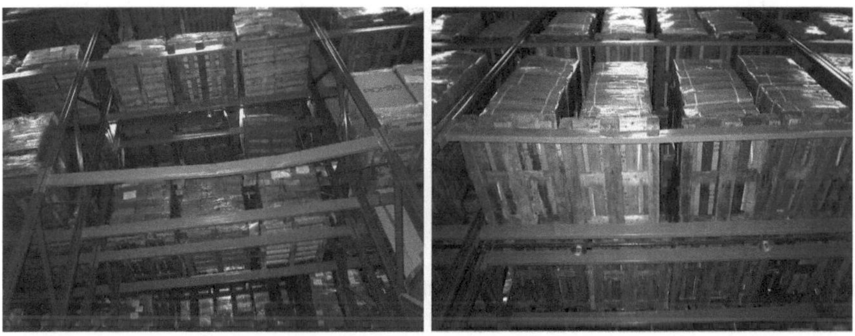

Fig. 1.5 Fork-lift truck collision on beams in a low passage way

In many cases, a partial collapse of part of a rack, results in a global collapse of the racking system in the warehouse, due to a "domino effect", with a structure falling on the adjacent one (Fig. 1.11).

Of course, failures can occur also in other types of racking systems. Drive-ins are a typology particularly prone to collapses due either to collisions with fork lift trucks, or to global and local stability problems (Fig. 1.12).

However, one shall never forget that tons of goods are stored above the persons who are daily working between racking, and a failure might cause injures or losses of human lives (Fig. 1.13).

Fig. 1.6 Serious calamity at the passage way. Fortunately, the roof of the truck fulfilled its protective function

Fig. 1.7 Collapse due to human error, despite the use of (VNA) trucks

Fig. 1.8 Police inspection after a serious damage

Fig. 1.9 Fire department officials inspection after a serious damage

Fig. 1.10 This collapse could not be hidden to the outside world

Fig. 1.11 Examples of "Domino-effect"

Fig. 1.12 Examples of collapse in "drive-in" structures

Fig. 1.13 Examples of collapses with goods spilled over the fork lift trucks

1.4 Damage to Steel Pallet Storage Racks and Content Spillage in Recent Earthquakes

In Europe, no official document is currently available for the seismic design of pallet racks, so the designers operate with a complete lack of references and of commonly accepted European design rules. Very often they make reference to the Rack Manufacturers Institute (R.M.I.) Specifications, while the European Federation of Maintenance (F.E.M./E.R.F.) is presently working on an official document (FEM 10.2.08 2005). The cause of this lack of Design Standards and Codes of Practice is the short knowledge of the actual behaviour of these structures under earthquake. Very little information is presently available related to the actual global ductility of the racks that is strongly influenced by the behaviour of the connections. Only a few experimental studies were carried out on this topic. Furthermore, only very limited research was carried out on the actual dynamic behaviour of pallet racks. Existing studies were performed mainly in the US where, after the Northridge Earthquake, the problem revealed all its economical impact, with enormous losses of stored goods. Only one study is presently available in Europe, carried out within the EC sponsored ECOLEADER program for Free Access to Large Scale Testing Facilities (Castiglioni 2003).

Many times, after an earthquake event, loss of goods was reported, with or without contemporary failure of the steel rack structural system. Most probably, the structural failures were a consequence of the fall of the pallets and of the impact of the goods on the beams at the lower levels, creating a progressive dynamic collapse.

Moreover no clearly established data and statistics exist related to the actual loading conditions of storage racks, in particular related to the "occupancy rate" of the rack during normal service, to different kinds of storage equipment, to different environments where racking systems are located and to different goods to be stored. This is due to the practically infinite possibilities of different applications occurring in the real practice.

In this context it shall also be mentioned that any observed damage to a rack component due to lift truck collisions, in general means loss of store capacity and flexibility over a certain period, so it requires an immediate repair.

1.4.1 Background

The use of steel single selective pallet storage racks in areas accessible to the public is a relatively recent development. Since being first introduced on a significant scale about 25 years ago, the number of big box stores utilizing storage racks in public areas has increased dramatically—especially during the past 15 years.

Although relatively few damaging earthquakes have occurred during this period of time, and vary little information is available, it is interesting to analyse steel storage rack damage and contents spillage during several earthquakes in California, (i.e., the 1987 Whittier, 1989 Loma Prieta, 1992 Landers, 1994 Northridge, and 2003 San Simeon earthquakes) and one in Washington (i.e., the 2001 Nisqually earthquake), as reported in (FEMA 2005).

To assess storage rack earthquake performance, it is important to determine the ground motions that occurred at the site, the design capacity of the rack, the actual loading of the rack, rack design details, and any other extenuating condition that would affect the rack performance. When an earthquake occurs, the actual ground motions at a given site are function of many factors. These factors include the magnitude of the earthquake, the depth of the earthquake, the distance of the site from the earthquake source and the soil conditions at the site. Of particular interest are sites where ground motion was equal to or exceeded the design level ground motion on which seismic code is based. Because of the nature of storage racks, another important issue is the actual pallet loading at the time of the earthquake relative to the rated capacity of the storage rack. Thus, during an earthquake, damage to a particular rack might not occur because the rack was lightly loaded or because the ground motion at the site was not severe. It is difficult to assess the performance of racks that have not really been subjected to their design limit.

The information publicly available concerns only a very small fraction of the facilities that have experienced earthquakes, in any case, according to (FEMA 2005), "less than 1 percent of the total reports and inspections made over almost 20 years in these stores describe unsatisfactory performance, failure or life safety hazards, while the other 99 % have performed without incident."

The two major failures described hereafter appear to be the result of overloading beyond any approved design limit. However, a full design-level earthquake has not yet occurred in an area where many of these stores are located.

The current database includes only one relevant event, the 1992 Landers earthquake in California, and the racks in the warehouse store performed adequately.

However, given the limited data available, it cannot be assumed that the past performances are indicative of what to expect in a design level earthquake in a heavily populated area.

Interesting information is reported in FEMA 460 (2005) and is summarized hereafter.

1.4.2 Damage Reported in the 1987 Whittier Earthquake

The Whittier earthquake, with a moment magnitude (*Mw*) of 5.9, occurred on October 1, 1987, at 7:42 a.m., its depth was 9.2 km. The epicenter was located in the Whittier Narrows region of southern California approximately 20 km southeast of downtown Los Angeles and approximately 48 km from Canoga Park. Several Home Depot and Home Club stores were inspected but no seismic damage was found except for damage associated with impact or reconfiguration. Of particular interest (for reasons that will be discussed later) were the Home Deport and Home

Club stores in Canoga Park, both at the time with racks in their originally installed configuration. The racks in these stores were not damaged, which is not surprising since the peak ground motions in the Canoga Park area were reported to be in the range of 0.05g, far below the design basis earthquake ground motion levels.

1.4.3 Damage Reported in the 1989 Loma Prieta Earthquake

The Loma Prieta earthquake, with a moment magnitude (Mw) of 6.9, occurred on October 17, 1989, at 5:04 p.m. at a depth of 18.4 km. The epicenter was located in the Santa Cruz Mountains about 96 km south of San Francisco. Many seismically vulnerable structures in the Bay Area were damaged. Because of the distance from the epicenter, measured ground motions in the Bay Area were significantly below the design level ground motions.

After the earthquake, 28 Bay Area Price Club stores were inspected and no damage was observed except for one line of racks in the bakery area of one store. These racks were heavily damaged but did not collapse. It was later determined that store personnel had removed the bottom beam level of these racks so they could put bread carts within the pallet racks to get them "out of the way." With the bottom beam level removed, the racks had less than a third of their rated capacity, which appears to be the primary reason for the damage.

1.4.4 Damage Reported in the 1992 Landers Earthquake

The Landers earthquake, with a moment magnitude (Mw) of 7.3, occurred on June 28, 1992, at 4:57 a.m. at a depth of 17.2 km. The epicenter was located near the town of Landers in the high desert area of California about 190 km east of downtown Los Angeles. Because of the remote location, there was only one store with storage racks accessible to the public located in the area of severe ground motion. A relatively new Builders Emporium (opened two to three years earlier) was located very close to the epicenter, less than 100 m from the fault rupture zone. The building was heavily damaged. Girders broke away from their anchorages in the wall columns and, in some places, battered their way through the walls. The ledgers were almost completely destroyed. Unanchored gondola shelving units were not seriously damaged and were placed back into service, but a great deal of small merchandise fell from the shelves and, in some cases, the units "walked" over 30 cm from their initial positions. Many small items also fell off storage racks, but items more than 60 cm in plan dimension (e.g., roof ventilators and HVAC pipes) stayed on the racks regardless of the height at which they were stored.

1.4.5 Damage Reported in the 1994 Northridge Earthquake

The Northridge earthquake, with a moment magnitude (*Mw*) of 6.7, occurred on January 17, 1994, at 4:31 a.m. at a depth of 17.6 km. The epicenter was located in Reseda in the San Fernando Valley. Because of the fault orientation and rupture direction, ground motions were at or above design levels in the north end of the valley and were just about code design levels in Santa Clarita, about 16 km north of the valley.

1.4.5.1 Price Club, Northridge, California

No damage was observed except for the "soda wall" racks (storage racks fully loaded with cases of soft drinks), which had failed longitudinally. It is believed that this failure resulted when floor-stacked merchandise toppled over and hit the racks, inflicting heavy damage, in fact identical racks standing nearby with the same or heavier loads were undamaged.

1.4.5.2 Home Depot, Santa Clarita, California

This store was opened a few days before the earthquake occurred. Several bays of racks failed completely in the cross-aisle direction, and there was also a considerable loss of contents (see from Fig. 1.14 through Fig. 1.20 courtesy of FEMA) (Figs. 1.15, 1.16, 1.17, 1.18 and 1.19).

A subsequent investigation revealed that the failed racks had been overloaded more than 50 % of their rated capacity (i.e., the ceramic tile racks were loaded three pallets wide rather than the assumed two pallets wide and building materials pallets, which weighed 16 kN each, had been improperly placed on general merchandise racks rated to carry only 6.8 kN per pallet position). Although no post-earthquake

Fig. 1.14 Cross-aisle collapse in the building material section of the Santa Clarita store

Fig. 1.15 Cross-aisle collapse in the tile section of the Santa Clarita store

Fig. 1.16 Cross-aisle collapse at Santa Clarita store, showing good performance of shrink wrapped merchandise

Fig. 1.17 Contents spillage in the door/window area of Santa Clarita store

Fig. 1.18 Contents spillage in the wood trim area of Santa Clarita store

Fig. 1.19 Contents spillage in the paint area of Santa Clarita store

Fig. 1.20 Contents spillage in a small-items section of Santa Clarita store

evaluation of the failed racks is publicly available, it was noted that properly loaded racks immediately adjacent to the overloaded bays were not damaged.

One row of building materials storage racks had a pronounced lean after the earthquake. This was apparently caused by material on the floor toppling onto the uprights and bracing during the earthquake. Once this material was removed, the racks self-centered, stood plumb again, and remained in service.

As a result of the failures at this new store, over 160 Home Depot stores were retrofitted with heavier, higher capacity racks in their ceramic tile areas. The cost of the retrofit is estimated to be approximately $15 million (in 1993 dollars).

1.4.5.3 Home Depot, Canoga Park, California

Home Depot replaced 32 % of their pallet positions in this store due to failed or seriously damaged racks. There is no question that serious injury and perhaps loss of life might have occurred had the store been open to the public at the time of the earthquake. It is believed that the cause of the failures was essentially significant overloading of racks, as in Santa Clarita.

Home Depots and Home Clubs initially did not sell large amounts of building materials, and their storage racks were designed for general merchandise rated at 6.8 kN per pallet position. As the stores grew, the merchandise mix changed and more building materials were kept in stock. Building materials storage racks at this store are rated to carry up to 16.8 kN per pallet position, or more than twice that of the general merchandise racks. Furthermore, the building materials storage racks usually are one level taller than the general merchandise racks in the older stores. The change in merchandise mix occurred in 1991 and, as noted above, the racks in these stores had been undamaged in the earlier 1987 Whittier earthquake.

This older store had been scheduled for a retrofit to heavier rack designs but it had a low ceiling that could not accommodate the taller rack designs. Thus, it appears that the retrofit was cancelled because of the low ceiling problem but the switch to heavier loads was made inadvertently without the concomitant stronger racks required. This oversight was largely responsible for the failures. The racks in this store subsequently were replaced with heavier, higher capacity 4 m high designs.

1.4.5.4 Home Depot, Glendale, California

Only minor damage was observed in this store. There were small amounts of fallen merchandise. One pallet did fall out of a storage rack, but the rack upright had been seriously damaged before the earthquake and was scheduled to be replaced the day the earthquake occurred.

1.4.5.5 Home Club, Canoga Park, California

This store was just down the street from the Home Depot and also suffered extensive damage, in some cases more severe than that in the Home Depot. As for the Home Depot, storage racks loaded beyond their rated capacity were deemed to cause the problems. The store was later fitted with heavier racks but subsequently closed. It is interesting to observe that a properly designed set of racks, fully loaded with heavy building materials, survived the earthquake without damage or incident. However, the racks were bootlegged and were sitting on asphalt with no anchorage.

1.4.6 Damage Reported in the 2001 Nisqually Earthquake

The Nisqually earthquake, with a moment magnitude (*Mw*) of 6.8, occurred on February 28, 2001, at 10.24 a.m. at a depth of 49.6 km. The epicenter was located in the Puget Sound area. Because of the great depth, the surface ground motions were considerably less than those of the Northridge earthquake even though the magnitude of the Nisqually earthquake was greater. The measured ground motions were significantly less than code level design ground motions. Stores in Seattle were investigated including one on reclaimed land, one adjacent to the fallen highway overpasses, and one south of Olympia.

1.4.6.1 Home Depot, Olympia, Washington

Of the stores examined, only the Olympia store located in the area nearest to the epicenter had storage rack damage, consisting in the buckling of some diagonal braces in the building materials racks. Very little merchandise fell out of the racks and almost all of what fell consisted of very small items that posed little apparent life safety hazards. The water heaters stored on the racks had recently been fitted with restraint. Many of them were leaning on these restraints, indicating the effectiveness of the maintenance.

1.4.7 Damage Reported in the 2003 San Simeon Earthquake

The San Simeon earthquake, with a moment magnitude (*Mw*) of 6.5, occurred in California on December 22, 2003 at 11:16 a.m. at depth of 7.2 km. The epicenter was located 9.6 km north of San Simeon and about 48 km north of Atascadero where the area's largest big-box store with pallet storage racks was located. Measured ground motions were far less than code level ground motions and no storage rack damage was observed at the store. Only minor damage occurred, a small lawn tractor on a steel pallet slid off the top of one rack and cans of roofing tar fell off another rack from a significant height. The fall of the lawn tractor on the steel pallets raises issues regarding the increasing use of pallets with a lower friction coefficient (e.g., steel and plastic) than the traditional wood pallets.

1.5 Codes and Standards for Storage Racks—Previous Researches

Racking systems are not "buildings" but a very peculiar steel construction work. They are different from buildings for the use, for the loads to be supported, for the geometrical dimensions and for the steel components, mainly made of thin gauge profiles and continuously perforated uprights, which ensure the typical functionality, adaptability and flexibility needed for the huge variability of requirements in storing goods. Only the clad warehouse, where racking systems support goods but also mezzanines, roof and walls, shall be considered as a very special "building".

For this reason it is necessary to explain how to consider the peculiarities of such kind of construction work when they are to be designed for seismic actions, because these peculiarities influence significantly the response to earthquakes and don't allow a designer to follow exactly the same approach for "ordinary steel structures", which is stated in the various Building Regulations.

While the basic technical description of earthquakes is obviously the same as for buildings, for racking systems it is of great importance to define whether or not it is possible to apply the "general design rules" which are enforced for ordinary steel structures, and how to correctly modify general principles and technical requirements, in order to take into account those peculiarities and to achieve the requested safety level.

Many specific physical phenomena affect the structural behavior of a racking system during an earthquake, such as the energy dissipation in the deformation of stored goods, or the sliding effect that can occur between pallets (or other unit-loads) and their directly supporting components, like beams, when seismic forces exceed certain limits, depending both on the acceleration values and also on the actual friction between the contact surfaces. Furthermore, the variable loads, like pallets or other unit-loads, can result in more than 95 % of the total mass, differently from buildings where dead weight and permanent loads sum generally in a significant percentage. Therefore the load presence and distribution on racking systems affect very much the response of the structure under seismic actions.

As far as the safety level is concerned, it is of great importance to consider the potential movements of the stored goods, which can fall down accidentally from the supporting beams, regardless of the strength of the racking systems against the earthquake. Therefore, proper designed accessories should be placed on the seismic resistant racks, in order to reduce as much as possible the risk of fall and the consequent risk of impacts, damages or even domino-collapse.

Methods of seismic isolation can be studied, to cut down the seismic forces and the rack oscillations, to prevent accidental movements of the stored goods.

At present, very few Codes are available, all over the world, dealing with the problem of the seismic design of racking systems.

In Europe, the Federation Europeenne de la Manutention (FEM) performed Standard development research activities for the European Union (EU). One result is the 2005 FEM seismic design standard, FEM10.2.08, *The Seismic Design of*

Static Steel Pallet Racks. Current FEM work includes analytical research, static and dynamic element testing as well as shake-table testing. Stub-column tests and beam-to-column connection tests for moment-rotation characteristics and properties have been conducted, using test facilities at the University of Trento and Politecnico di Milano in Italy. At the National Technical University in Athens, full-scale steel pallet racks have been tested at ground accelerations up to failure. This research indicates that movement of merchandise within packaged unit loads, movement of unit loads or packages on a pallet and movement of pallets on pallet beams within the rack occur even at relatively low ground accelerations. Specific sliding tests have been developed to improve the understanding of these phenomena and their influence on damping, period, and overall structural behaviour.

Since the early 1970s, in the U.S., RMI has sponsored many analytical and experimental storage rack research projects conducted at Cornell University. These studies have included full-scale, component, and element tests focusing on, hot-rolled and cold-formed structural elements, beams, columns, perforations, beam-to-column connectors and connections, base plates, flexural and torsional-flexural buckling, and testing and loading protocols.

During the late 1970s and early 1980s, major research projects were undertaken, including subassembly tests at Stanford University and full-scale shake-table testing at the University of California/Berkeley using El Centro 1940 records, by URS/Blume (see John A Blume and Associates 1973; Chen et al. 1980a, b, 1981), with funding from the RMI membership and a large grant from the National Science Foundation (NSF). The results of that testing, along with analytical studies, provided important baseline information about storage rack seismic performance, helped identify topics for further research, and articulated issues needing further study.

Storage racks are typically made of cold-formed steel members; therefore, their design depends on the thorough understanding and application of the American Iron and Steel Institute's (AISI) *Specification for the Design of Cold-Formed Steel Structural Members*. Other storage rack members are made with hot-rolled steel structural sections using the applicable seismic provisions of the American Institute of Steel Construction's (AISC) *Specification for Structural Steel Buildings, Allowable Stress Design,* and *Load and Resistance Factor Design Specification for Structural Steel Buildings*. AISI is currently developing a standard for the seismic design of structures using cold-formed steel members that may provide detailing requirements for the design of storage racks subjected to seismic loads.

The Canadian Standards Association has developed two storage racks standards: A344.1 *User Guide for Steel Storage Racks,* and A344.2, *Design and Construction of Steel Storage Racks.* RMI has supported these efforts.

Among the most important ongoing RMI initiatives there is the current testing program to determine the moment-rotation characteristics of the beam-to-column connectors of RMI members' products. The testing protocol will give information on the role of connector properties in the seismic performance of rack structures including information on damping, drift, base shear, and natural frequencies. The protocol is designed to mimic accepted testing provisions for building connections.

This testing program, being conducted for RMI by an independent testing laboratory, covers cold-formed and hot-rolled members as well as the linear elastic, nonlinear elastic, and inelastic behaviour of the connecting elements. The results of this beam-to-column connection testing program will yield proposals for changes in the RMI standard, the *NEHRP Recommended Provisions*, ASCE 7, the *IBC*, and *NFPA 5000* and should contribute to a convergence of the seismic requirements in those documents.

In the following the main codes for seismic design of racks are discussed.

1.5.1 Recommendations for the Design of Static Steel Pallet Racks Under Seismic Conditions (Pr FEM 10.2.08, European Racking Federation, the Racking and Shelving Product Group of the Federation Européenne de La Manutention)

Pr FEM 10.2.08 deals with all relevant and specific seismic design issues for racking systems, such as:

1. The seismic response could be significantly different in down-aisle direction and in cross-aisle direction, and could be considerably affected by the size and the distribution of the masses over the height. Reliable statistical evaluations are necessary to find the most probable mass distribution when the earthquake arises, depending on the racking system typology and dimensions. A most likely approach is given in this standard.
2. The natural damping of the "naked structure" is very small. But the real damping, measured in real condition, could be significantly more than that expected, because of the micro-movements in the stored goods and products and/or the sliding effects between pallets (or other unit-loads) and the supporting beams.
3. Cyclic forces due to earthquake can progressively damage connections and/or other components of a racking system. These changes could considerably affect the response of the structure and its reaction to seismic actions. A reliable modelling of the actual strength and stiffness is of fundamental importance in order to predict the structural behaviour of the structure.
4. In case of seismic isolation, the effectiveness of the isolation devices must be guaranteed for all the loading conditions and during the whole expected life of the racking system

The design procedures given in Pr FEM 10.2.08 apply to all types of static pallet racks fabricated from steel members for the seismic load case that are supported by floors lying on the ground. They do not apply to mobile storage systems.

The approach to the seismic design is based upon the philosophy of prEN 1998-1 (Eurocode 8), however the peculiar dynamic behavior of racking structures and of the stored unit loads carried by them is included in this document.

The reference to the design, tests and quality control of components and materials is expressly based on FEM 10.2.02 (FEM 2001a).

In case of clad racks, where the racking system is supporting roof and walls, this Code gives relevant information in addition to the National Building Regulation.

In the case of mechanical handling equipment connected to the racking (e.g. crane racking), the supporting forces and/or guide forces under seismic conditions shall be specified by the designer of that equipment, in co-operation with the rack designer, because of the interactivity of the dynamic systems.

Special analysis is also required for racks supported by other structures, such as platforms, to take properly into account the effects of the amplification of the supporting structures.

Making reference to prEN1998-1, it is stated that structures in seismic regions shall be designed and constructed in such a way that the following requirements are met, each with an adequate degree of reliability.

1. No collapse requirement

   The racking structure shall be designed and constructed to withstand the design seismic action without local or general collapse, thus retaining its structural integrity and a residual load bearing capacity after the seismic event

   Ultimate limit states are those associated with the collapse or with other forms of structural failure which may endanger the safety of people.
2. Damage limitation requirement

   In order to avoid damages and use limitation, whose costs would be disproportionately high in comparison with the cost of the structure itself, the racking structure shall be designed and constructed to withstand a seismic action having a larger probability of occurrence than the design seismic action, without local damage and the associate limitations of use

   Serviceability limit states are those associated with damage occurrence, corresponding to states beyond which specified service requirements are no longer met.

   SLS for racking systems is defined as elastic response under an earthquake with acceleration equal to 25 % of the design value.

   The movement of stored unit loads is not to be considered as a damage.
3. Movement of pallets

   The seismic accelerations can cause the sliding of the pallet on the beams, when the dynamic horizontal force on the pallet exceeds the static friction force between pallet and beam.

   Full scale tests showed that this effect occurs also for small ground accelerations values (low intensity earthquake) for ordinary situations (wooden pallets on painted steel beams), because of the amplification of the seismic forces at the highest storage levels.

The consequences of these phenomena are:

- the reduction of the seismic action on the rack, due to the energy dissipation, and the limitation of the horizontal action that can be transferred from the pallet to the rack's structure.
- the risk of fall of the pallet, that can produce the local or global collapse of the rack, when the pallet falls inside the rack between the pallet beams, or injury to people and damages to the equipment nearby the rack when the pallet falls out of the rack.

The modification of the seismic response is deemed in Pr FEM 10.2.08 by means of two coefficients that conventionally estimate the effects of typical phenomena of the racking systems, such as energy dissipation due to the pallet-beam friction, pallet damping due to the movement of the stored product, pallet flexibility and other phenomena:

E_{D1} = design spectrum modification factor
E_{D2} = mass modification factor.

1.5.1.1 Definition of the Intensity of the Seismic Action

With reference to prEN1998-1, two elastic spectra are defined in Pr FEM 10.2.08 to describe the earthquake motion (called Type 1 and Type 2).

1.5.1.2 Earthquake Design Return Period and Importance Factor γ_I

Reference importance factors defined in Pr FEM 10.2.08 are reported in Table 1.1.

For economical or strategic reasons the end-user may specify a higher importance factor; in any case the importance factor must be at least equal to the

Table 1.1 Importance factors for racks

Importance class	Description	Importance factor	
		Reference	Reduced [Note (2)]
I	Warehouses with fully automated storage operations low warehouse occupancy (1)	0.8	0.67
II	Normal warehouse conditions, including picking areas	1.0	0.84
III	Retail areas with public access	1.2	n.a.
IV	Hazardous product storage	1.4	n.a.

Notes

1. The reduced importance factor can only be used for racking systems not located on storey of a building and/or not used for retail areas or hazardous product storage
2. The reduced importance factor is based on design life of the rack of 30 years instead of 50

importance factor specified for the part of the building in which the racks are located.

According to Pr FEM 10.2.08 the vertical component of the seismic action shall only be taken into account in the following relevant cases:

1. Cantilever components
2. Beams supporting columns (for example in order picking tunnels).

The effect of the vertical seismic component must be taken into account only for the elements under consideration and their directly associated supporting elements or substructures.

For normal racking structures the vertical component of the seismic action generally is not considered.

1.5.1.3 Design Parameters for the Seismic Analysis

1.5.1.3.1 Design Spectrum Modification Factor

The design spectrum modification factor E_{D1} takes into account the effect of the energy dissipation and damping related to the interaction between the pallets and the racking structure, and can be assumed constant and equal to 0.8.

When pallets are blocked on the pallet beams by means of any special system (for example materials increasing the friction between pallet and beam) $E_{D1} = 1.0$ shall be assumed.

E_{D1} modifies the ordinate of the spectrum $S_d(T)$ (not the period).

1.5.1.3.2 Design Seismic Pallet Weight

The design pallet weight W_E to be considered in the seismic analysis for the evaluation of the horizontal seismic action, is determined as:

$$W_E = R_F \times E_{D2} \times Q_{P;rated}$$

where

R_F rack filling grade reduction factor
E_{D2} Pallet weight modification factor
$Q_{P;rated}$ rated (specified) pallet load for the compartment, upright frame or global down aisle design (see FEM 10.2.02)

The following values are proposed for the coefficients:

$R_F = 1.0$
$Q_{P;rated} = Q_P =$ design pallet load.

Table 1.2 Pallet weight modification factors

E_{D2}	Stored goods class	Example
0.8	Compact constrained	Frozen goods (cold storage) Steel sheet package Coils and paper rolls
0.65	Weak	Big number of pieces stored on the pallet whose size is small in comparison to the pallet size
0.5	Loose and unconstrained	
1.0	Liquid	Liquid in partially filled drums

For the analysis in the down aisle direction $R_F \geq 0.8$ must be used in any case, while for the analysis in the cross-aisle direction $R_F = 1.0$ and $Q_{P;rated}$ must be used in any case.

E_{D2} modifies the period and the horizontal action.

1.5.1.3.3 Pallet Weight Modification Factor E_{D2}

1. The pallet weight modification factor E_{D2} conventionally represents the global effects on the seismic behaviour of the rack of the particular nature of the load masses, such as:
   - the movement of the pallets on the pallet beams
   - the movement of the stored material
   - the damping of the stored material
   - the flexibility of the pallet
2. Unless otherwise specified, the values for E_D reported in Table 1.2 shall be considered, depending on the nature of the pallet and of the stored goods.

1.5.1.3.4 Pallet Sliding

1. The sliding of the pallets is expected to occur when the pallets are not restrained on the beams, and the horizontal seismic action on the pallets at one of the load levels evaluated by using the design spectrum with modification factor $E_{D1} = 1$ is greater than the horizontal reaction generated by the friction between the pallet and the beams:

$$F_i > \gamma_{\mu S} \times \mu_S \times Q_P$$

where:

F_i seismic force on pallet at level i

μ_S pallet-beam friction coefficient

Table 1.3 Reference values for the pallet-beam friction coefficient

Materials in contact	Environment	Friction coefficient μ_S
Painted steel beam wooden pallet	Warehouse conditions dry environment	0.15–0.25
Painted steel beam wooden pallet	Warehouse conditions wet environment	0.05–0.15
Steel beams, all finishing Any type of pallet	Cold store	0
Steel beams, all finishing Plastic pallet	Warehouse conditions	0.05–0.15
Steel containers on steel beams	Warehouse conditions	0.15–0.25

$\gamma_{\mu S}$ partial safety coefficient for pallet-beam friction coefficient (usually assumed $\gamma_{\mu S} = 1$)
Q_P design weight of the pallet

The above condition is equivalent to:

$$a_{g,i}/\ g\ > \gamma_{\mu S} \times \mu_S$$

where:
$a_{g,i}$ seismic horizontal acceleration at level i
g gravity acceleration

2. The pallet-beam friction coefficient to be considered is the static one; in the following Table 1.3 reference values are reported.
   Upper-bound values of the friction coefficient shall be used for assessment of the occurrence of sliding in the ULS analysis, while lower bound values shall be used for the assessment of the occurrence of sliding in the SLS analysis.
3. In case of sliding occurrence the pallets move and can reach improper equilibrium positions on the beam; therefore each single pallet beam must be designed considering the most severe possible position of the pallets in equilibrium in the load cell.
4. In any case the chance of falling down of the pallets must be limited by placing additional components or devices, such as depth beams, additional central beams etc.

1.5.1.4 Other Seismic Weights

All the permanent and live loads other than the pallet weight must be properly considered in the seismic analysis.

Actions (defined in FEM 10.2.02) to be considered acting simultaneously with the earthquake are:

1. Dead Loads (permanent actions)
   (a) weights of materials and constructions
   (b) weight of fixed equipment
2. Live loads (variable actions)
   (a) goods to be stored: the design pallet weight must be considered
3. Floor and walkways loads (variable actions)
   - 1.0 kN/m^2 on walkways and access floors not for storage, intended as global load
   - 3.5 kN/m^2 on floors for storage

   The following occupancy factors are suggested for the seismic analysis and the evaluation of the horizontal action:
   - 0.8 floors for storage
   - 0.3 walkways and access floors

The code characteristic vertical static load must be considered; no dynamic increment should be taken into account.

The following loads (defined in FEM 10.2.02) should not be considered concurrently with the seismic loads:

1. Wind loads
2. Vertical placement loads
3. Horizontal placement loads
4. Horizontal operational loads caused by rack-guided equipment
5. Thrusts on handrails
6. Temperature loads
7. Global imperfection loads
8. Impact loads.

1.5.1.5 Position of the Centre of Gravity of the Pallet

The elevation of the center of gravity of the pallet referred to the support beams (vertical eccentricity) is considered.

1. Cross-aisle direction:
   the masses of the pallets should be considered as positioned at the pertinent level over the support beam (center of gravity levels) for the evaluation of the period of vibration and of the seismic action (Fig. 1.21).

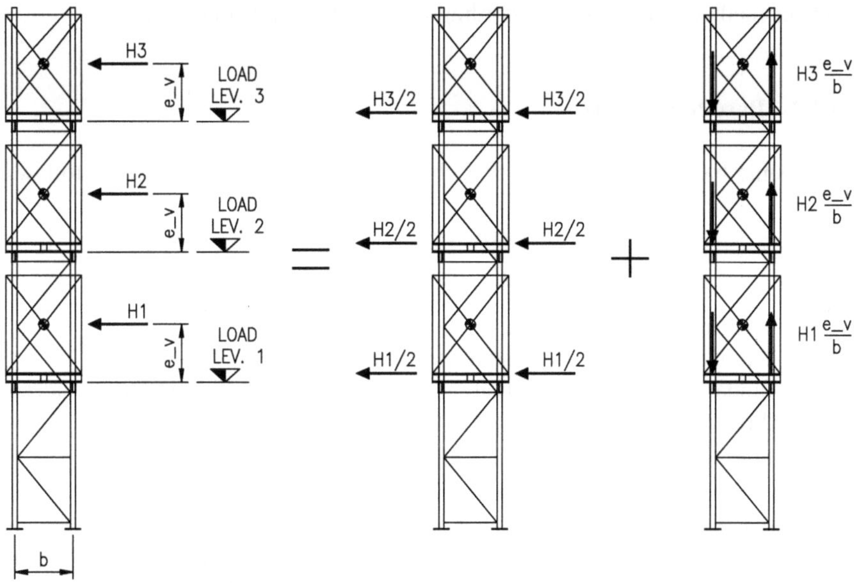

Fig. 1.21 Modelling the elevation of the center of gravity of the pallet in cross-aisle direction

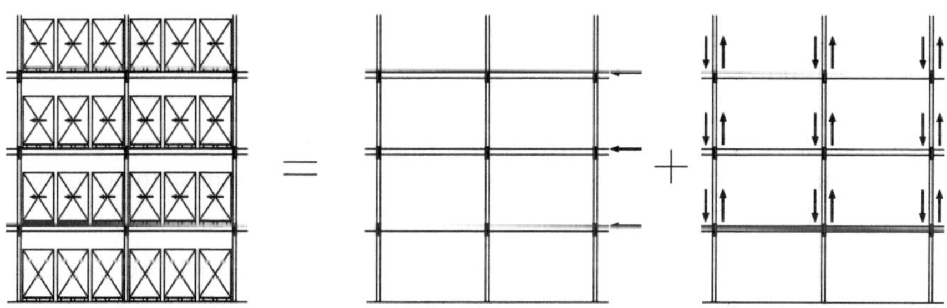

Fig. 1.22 Modelling the elevation of the center of gravity of the pallet in down-aisle direction

2. Down aisle direction:
   In the down aisle direction the eccentricity of the mass to the load level can in general be neglected for long racks, and it is not significant for the local check of the pallet beams and their connections (Fig. 1.22).

1.5.1.6 Methods of Analysis

The reference method for determining the seismic effects is the modal response spectrum analysis, using a linear-elastic model of the structure and the design spectrum multiplied by the modification factor E_{D1}.

Other methods of analysis, according to Eurocode 8, can be used.

1.5.1.7 Base Shear Force

The seismic base shear force F_b for each main direction is determined as follows:

$$F_b = E_{D1} \times S_d(T_1) \times W_{E,tot}$$

where:

E_{D1} spectrum modification factor
$S_d(T_1)$ ordinate of the elastic design spectrum
T_1 fundamental period of vibration for translational motion in the considered direction
$W_{E,tot}$ total seismic mass

The fundamental period of vibration of the rack must be evaluated by means of modal analysis, while simplified formulas proposed by various Codes and Standards for the evaluation of T_1, which are typical for buildings, are not allowed for racks.

1.5.1.8 Combination of the Horizontal Components of the Seismic Action

In general the components of the seismic action shall be considered as acting simultaneously, and their combination components may be accounted for as follows.

Method 1

- the structural response to each horizontal component shall be evaluated separately
- the maximum value of the effect of each action on the structure due to the two horizontal components of the seismic action is estimated by the square root of the squared responses to each horizontal component.

Method 2

As an alternative to the above method, the action effect on the structure due to the two horizontal components of the seismic action may be computed using the two following combinations:

(a) E_{Edx} "+" $0.30 \times E_{Edy}$

(b) $0.30 \times E_{Edx}$ "+" E_{Edy}

where

"+" implies "to be combined with"

E_{Edx} action effects due to the application of the seismic action along the horizontal axis x of the structure

E_{Edy} action effects due to the application of the same seismic action along the orthogonal horizontal axis y of the structure

The sign of each component in the above combinations shall be taken as the most unfavourable for the effect under consideration. When the regularity criteria are fulfilled and the lateral resisting systems are separated for the two principal directions, the seismic action may be assumed to act separately along the two main orthogonal horizontal axes of the structure.

1.5.1.9 Combination of the Vertical Component of the Seismic Action

In case the horizontal components of the seismic action are also relevant for those elements that are affected by the vertical seismic action, the following three combinations may be used for the computation of the action effects:

(a) $0.30 \times E_{Edx}$ "+" $0.30 \times E_{Edy}$ "+" E_{Edz}
(b) E_{Edx} "+" $0.30 \times E_{Edy}$ "+" $0.30 \times E_{Edz}$
(c) $0.30 \times E_{Edx}$ "+" E_{Edy} "+" $0.30 \times E_{Edz}$

where:

E_{Edx} and E_{Edy} have been defined previously;

E_{Edz} action effects due to the application of the vertical component of the design seismic action.

1.5.1.10 Displacements Analysis

The displacements induced by the design seismic action shall be calculated on the basis of the elastic deformation of the structure system obtained by the analysis by means of the following simplified expression:

$$d_s = q_d d_e$$

where

d_s displacement of a point of the structural system induced by the design seismic action

q_d displacement behavior factor, assumed equal to the behaviour factor q

d_e displacement of the same point of the structural system, as determined by a linear analysis based on the design response spectrum.

1.5.1.11 Regularity Criteria

1.5.1.11.1 Cross-Aisle Direction

Regularity in plan

- The upright frames not connected to the elements of a non-regular lateral resisting system (horizontal and vertical bracing) are considered as stiffness regular.

Regularity in elevation

- The upright frame can be regarded as regular if:
  the diagonal bracings are without interruptions from the floor to the top load level (stiffness regularity)
  and
  the ratio between the maximum and the minimum distance between the pallet beams in elevation is less than 2 (mass regularity).

1.5.1.11.2 Down Aisle Direction

Regularity in plan

- Racks not braced or symmetrically braced in the down aisle direction in the front and rear planes are stiffness regular in plan
- Racks braced on the rear plane only in the down aisle direction are not stiffness regular in plan

Regularity in elevation

- Position of the pallet beams

  The rack can be regarded as mass regular if:
  the pallet beams are on the same level for the entire length of the block
  and the ratio between the maximum and the minimum distance between the pallet beams in elevation is less than 2

- Type of vertical bracings
  Racks with vertical bracing continuous from the floor to the top load level are stiffness regular in elevation.
  Racks partially braced are not stiffness regular in elevation.

1.5.1.12 Rules for the Design of Low Dissipative Structures

In addition to general requirements related to materials, strength evaluation (for both members and connections) as well as for bolt tightening, the following specific rules are stated:

1. If ductility factor $q > 1.5$ is assumed, members which contribute to the seismic resistance of the structure by working in compression or bending must be of class 1, 2 and 3 according to EC-3.
2. If the structure is not regular in plan or elevation, q shall not exceed 1.50.
3. K bracings, in which the tension diagonals intersection lies on a column, should not be used in seismic zones.
4. For bolted shear connections the shear strength of the bolts should be higher than 1.20 times the bearing resistance.

1.5.1.13 Structural Systems Withstanding the Seismic Action

The seismic action can in general be studied separately in the down aisle direction and in the cross-aisle direction, and the lateral forces resisting systems can be uncoupled.

In FEM 10.2.08 only self-supporting racks are considered with regular layout in plan.

The relevant load condition for the seismic design of the rack is the fully loaded rack, as the horizontal seismic action is maximized. In case of a structural system other than the rack is provided to withstand the seismic action, it should be designed with the criteria of prEN 1998-1 (EC-8).

The rack's structural systems withstanding the seismic actions are:

1. in the cross-aisle direction, the upright frames
2. in the down-aisle direction, one of the following:

   (a) The unbraced frames
   The stability is provided by the beam—to—column joints of the rack, and no vertical bracings are provided; horizontal bracings are provided connecting the front and the rear frames.

   (b) The rear bracing.
   The bracing system, made of the following elements:

   - a rear bracing placed behind the rear frame, which can be an independent structure connected to the rack, or bracing elements connected to the elements of the rack itself
   - horizontal bracings connecting the front unbraced frame to the rear braced frame

   The vertical bracing withstand the horizontal seismic action.
   The horizontal bracings and the upright frames connected to the horizontal bracings withstand the eccentricity of the seismic action to the vertical

bracing, and transmit the action to the vertical bracing and to the connected horizontal bracing.
The system must be regarded as non-regular in plan. The vertical bracing should be continuous from the floor to the top load level; in this case the system may be regarded as regular in elevation, otherwise it must be regarded also as non regular in elevation.
Vertical bracings distributed among full rack height are strongly recommended. Interrupted vertical bracings should be avoided.

(c) Symmetrical vertical bracings (each row of the rack is braced).
Vertical bracings withstanding the seismic action are installed in a limited number of bays, in the front and in the rear frames.
Horizontal bracings connecting the front and the rear frames are also provided to share the seismic force between the two bracings.
The horizontal seismic forces can be equally divided between the front and the rear frames.
The rack can be regarded as regular in plan, and also in elevation when the vertical bracing is continuous from the floor to the top level of the rack.

For unbraced and symmetrically braced racks it is allowed to assume regular down-aisle behaviour if horizontal bracings are provided to ensure the uniform distribution of the seismic actions between the lateral resisting systems on the front and on the back frames.

Unbraced and symmetrically braced racks without horizontal bracings shall be regarded as non-regular in plan.

1.5.1.14 Specific Modelling Requirements for the Analysis

1. The modelling and analysis rules must be according to FEM 10.2.02.
2. For beam-to-column connections the stiffness obtained from tests according to FEM 10.2.02 for static load must be used. The pallet beam must be checked under pallet load with pinned ends and load factor $\gamma_L = 1.0$
3. The shear stiffness of the upright frame must be evaluated according to FEM 10.2.02.
4. When tension diagonals are provided, the bracing elements must be modelled in order to take into account the proper stiffness of the bracing considering the effect of the active component.
5. Bracings
    (a) the vertical bracings must be properly modelled with their eccentricity to the rack elements; also the elements connecting the vertical bracings to the rack must be modelled
    (b) the horizontal bracings must be properly modelled when present, with their proper stiffness, using the same criteria described in the previous clause.

1.5.1.15 Structural Types and Maximum Associated Behaviour Factors

1.5.1.15.1 Upright Frames

For frame types e_1 and d_1 the eccentricities of the scheme must be consistent with the rules of FEM 10.2.02 (Tables 1.4, 1.5).

1.5.1.15.2 Moment Resisting Frames

In moment resisting frames the dissipative zones are located in the beams or in the beam-to-column connections, and in the bottom of the columns.

Table 1.4 Upright frames types and maximum associated behaviour factors

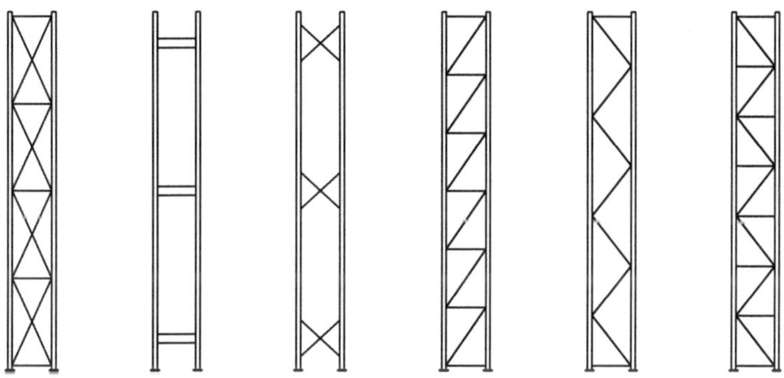

(a) Tension braced frame	(b) battened (vierendeel) frame	(c) partially braced frame	(d) Z- form braced frame	(e) D - form braced frame	(f) K - form braced frame

Frame type	Structural type	Detailing rules for dissipative elements	Maximum q factor	
			Regular rack	Not regular rack
a	Diagonal bracing with tension diagonals	Tension diagonal 5.2 diagonal connections 5.1.1	4	3.2
	diagonal bracing with tension and compression diagonals		2.5 (class DCH)	2.0
			2.0 (class DCM)	1.6
b	Dissipative battened frame can be used, provided that the requirements of moment resisting frames met; otherwise q = 1.5 must be assumed			
c			1.0	1.0
d-e-f	Low dissipative		1.5	1.5

Table 1.5 Upright frames types and maximum associated behaviour factors

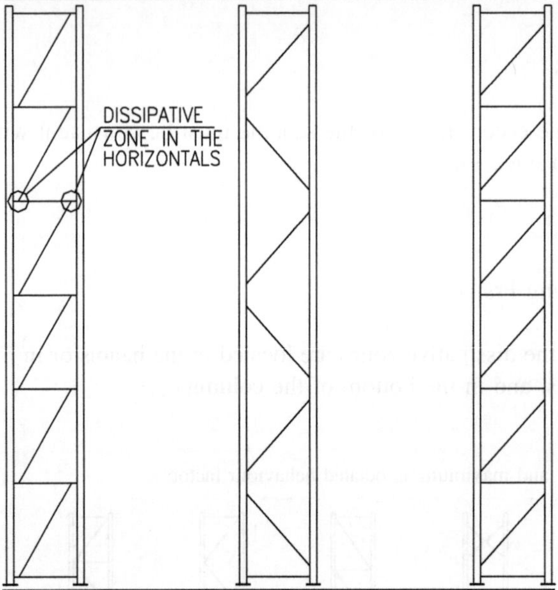

(d_1) Z-form dissipative braced frame	(e_1) D-form braced frame	(d_2) Z-form braced frame

Frame type	Structural type	Detailing rules for dissipative elements	Maximum q factor	
			Regular rack	Not regular rack
d_1	Eccentric braced frame (energy dissipation in the horizontals)	Design according to prEN 1998-1 Chapter 6.8	4	3.2
e_1-d_2			1.5	1.5

A behaviour factor q = 2 in general can be assumed for regular unbraced racks in down-aisle direction.

1.5.1.15.3 Vertical Bracings

The type of vertical bracings considered in FEM 10.2.08 is the "concentric diagonal bracing" with energy dissipation in the tension diagonals. Other types of vertical bracing can be used, that can be analyzed with the methods of prEN 1998-1.

The vertical bracing can be placed in the rear plane of the rack, or symmetrically positioned in the front and in the rear plane to allow structural regularity in plant.

In the first case the connection of the rack to the vertical bracing must be properly designed and considered in the structural analysis.

In particular the eccentricity of the vertical bracing to the rear uprights, that is necessary to allow the pallets overhang, must be properly modelled in the analysis, and the connection elements considered with their stiffness.

The connections to the bracings shall be designed in order to avoid unproper behaviour of the rack's elements; in particular the local torsion of the upright must be avoided.

The eccentricities of the bracings shall be modelled and properly considered according to FEM 10.2.02, while the elements connecting the rack to the bracings shall be designed with the overstrength requirements for dissipative connections.

1.5.1.15.4 Bracings with Tension Diagonals

For this typology the behavior factor q = 1.5 is applicable for ductility class Low (DCL) with no specific overstrength requirements for connections.

The behavior factor q = 4 is applicable for ductility class High (DCH) and Medium (DCM).

1.5.1.15.5 Bracings with Diagonals Working in Tension and Compression

For this typology the applicable behavior factors are:

q = 2.5 for ductility class DCH, and bracing elements belonging to class 1 and 2
q = 2.0 for ductility class DCM and bracing elements belonging to class 1, 2 and 3.

Diagonal bracings made with steel strips are allowed, provided that turnbuckles are used to put them in tension, and their tightness is periodically checked.

1.5.1.15.6 Horizontal Bracings

The horizontal bracings designed to transmit the horizontal actions to the lateral load resisting system shall fulfill overstrength requirements. No energy dissipation is allowed in the horizontal bracings.

1.5.1.16 Safety Verifications

1.5.1.16.1 Ultimate Limit State

The safety against collapse (ultimate limit state) under the seismic design situation is considered to be fulfilled if the following are met.

The design values of actions shall be combined using the following formula:

$$\sum \gamma_{GA} G_k \text{ “}+\text{” } \gamma_{QA} \sum \psi_{2,i} Q_{k,i} \text{ “}+\text{” } \gamma_I A_{E,d}$$

where
“+” implies “to be combined with”
Σ implies “the combined effects of”
G_k characteristic value of the permanent action (dead load)
$Q_{k,i}$ characteristic value of a typical variable action
$A_{E,d}$ design value of the seismic action for the reference return period
γ_{GA} partial safety factor for permanent actions = 1.0
γ_{QA} partial safety factor for variable actions = 1.0
$\psi_{2,i}$ partial reduction coefficient for variable actions:
$\psi_{2,1}$ 1.0 for the pallet load
$\psi_{2,2}$ 1.0 for floor loads on storage areas
$\psi_{2,3}$ 0.5 for passage floor and walkways
γ_I importance factor.

1.5.1.16.2 Serviceability Limit State

1.5.1.16.2.1 Damage Limitation Requirement

The requirement for limiting damage (serviceability limit state) is considered satisfied if the response of the rack remains in the elastic field under the seismic action having a larger probability of occurrence than the design seismic action corresponding to the “no collapse requirement” (See prEN 1998-1).

1.5.1.16.2.2 Pallet Sliding

The pallet sliding on the beams must be assessed under a seismic action calculated with the behavior factor q = 1 and ground acceleration reduced by factor E_{D1}.

The ratio between the horizontal seismic force and the pallet weight, or between the acceleration at rack level and the acceleration of gravity g, must be less than the friction coefficient of the pallet on the beams.

If the sliding of the pallet occurs, devices must be installed in order to prevent the falling of the pallets.

1.5.1.16.2.3 Pallet Falling

The sliding of pallets under the design condition must be controlled in order to assess the risk of the falling of pallets.

1.5.1.16.2.4 Falling of the Pallets Inside the Rack

The falling of the pallet inside the rack could cause the local or global collapse of the structure.

It must be prevented by providing special components such as front-to-rear supports connecting the pallet beams under the pallet supports, third beam, nets or others.

1.5.1.16.2.5 Falling of Pallets Outside the Rack

The falling of the pallets outside the rack can cause injury to persons, damage to the stored goods and can damage or cause the collapse of the equipment or racks in the proximity, especially in case of high racks and narrow aisles.

The designer must analyze this eventuality and take the necessary countermeasures.

In case of possible pallet falling outside the rack, the strength of the single pallet beam and its connections must be assessed considering the full weight of the pallets placed in normal condition on the couple of beams, with a material partial safety coefficient $\gamma_M = 1.0$.

The weight to be considered is the rated pallet weight.

1.5.1.16.2.6 Pallets Fixed on the Rack

When pallet are blocked on the pallet beams by means of any special system (for example materials increasing the friction between pallet and beam) the spectrum modification coefficient $E_{D,1}$ shall be assumed equal to 1.0.

1.5.1.16.2.7 Summary of the Effects Induced by Seismic Action on Rack's Elements

The effects induced by seismic action on rack's elements are summarized in Table 1.6.

1.5.1.17 Structural Systems Withstanding the Seismic Action

The following figures show possible schemes to be adopted in order to model the structural systems withstanding the seismic actions (Figs. 1.23 and 1.24).

The eccentricity of the rear bracing causes a global torsional behavior that affects the upright frames. In this case the rack must be considered not stiffness-regular in plane (Figs. 1.25 and 1.26).

In case of double entry rack the unsymmetry in plane is caused by the mass disposition as both the entries are in general not equally loaded (Figs. 1.27 and 1.28).

Table 1.6 Summary of the effects induced by seismic action on rack elements

Elements	Seismic action	
	Down-aisle direction	Cross-aisle direction
Uprights	Braced racks: (a) Compression and local bending (b) Tension and local bending	(a) Compression and local bending (b) Tension and local bending
	Unbraced racks: (a) Compression and bending b) Tension and bending	
Upright frame diagonals	Upright frames connected to the horizontal bracings: (a) Tension and compression or (b) Tension	(a) Tension and compression or (b) Tension
Pallet beams	Braced racks: (a) Tension and compression see note (1) below Unbraced racks: (a) Tension and compression (negligible) (b) Bending in the vertical plane (frame system)	(a) Bending and shear in the vertical plane due to the vertical eccentricity of the pallets (b) Bending and shear in the horizontal plane due to the Horizontal seismic action See note (1)
Base plates	Braced racks: Uprights connected to the longitudinal lateral resisting system (a) Compression (b) Uplift	(a) Compression (b) Uplift
	Unbraced racks: (a) Bending (frame system)	
Vertical bracing diagonals	(a) Tension and compression or (b) Tension	
Horizontal bracing diagonals	(a) Tension and compression or (b) Tension	
Beam-to-columns connections	Shear and bending	

Notes

(1) The pallet beam being also the horizontal element of the vertical bracing should be checked for bending due to pallet weight and compression due to seismic action

The stabilizing effect of friction between the pallets and the beams cannot be accounted for in the checks of the beams working in compression due to the seismic load

In the vertical plane the pallet beam of the rack braced in the down-aisle direction need not to be checked with pinned ends

The designer should assess the axial strength of the beam considering

- in the horizontal plane: the presence of horizontal bracings, if any, with beam-ends pinned in the horizontal plane
- in the vertical plane: a reduced value of the stiffness of the beam-to-upright connection.

Unless it is demonstrated by tests or by rational analysis that the connections maintain their efficiency under the design earthquake, the value of the beam-to-upright stiffness shall be reduced to 1/3 of the stiffness obtained from static tests performed according to FEM 10.2.02

To assume the beam buckling length equal to the distance between upright axes (pinned beam ends) is on the conservative side

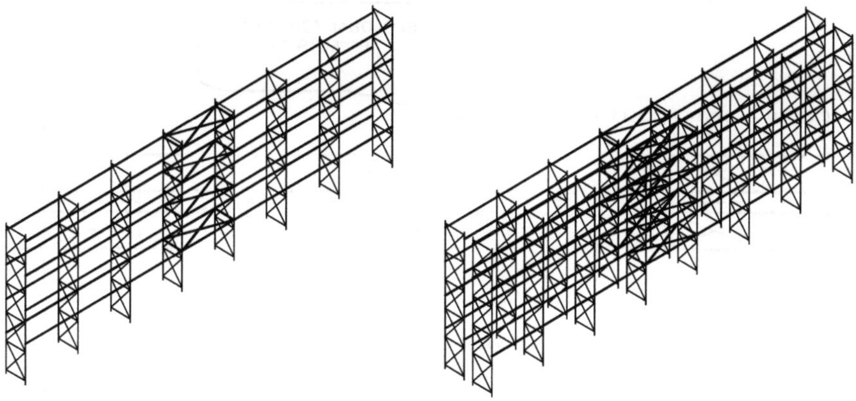

Fig. 1.23 Representation of unbraced frames

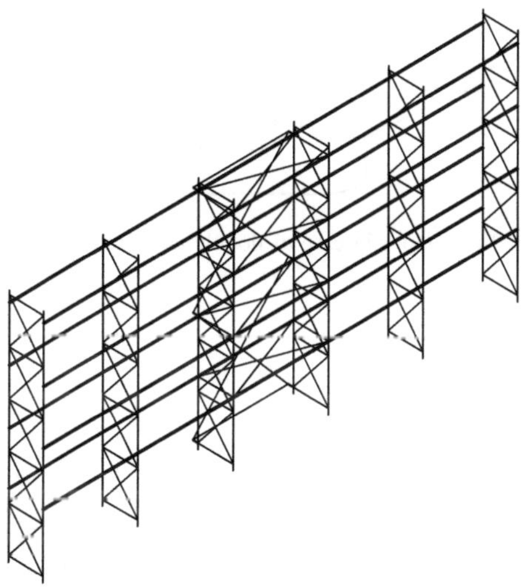

Fig. 1.24 Representation of the rear bracing and the bracing system

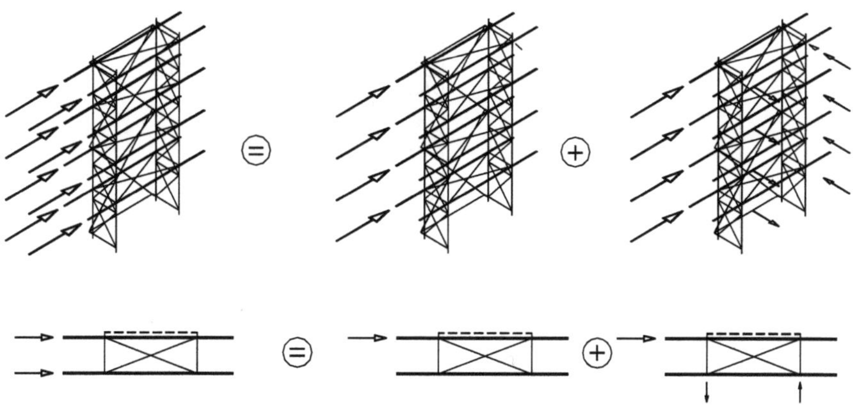

Fig. 1.25 Global torsional behaviour in case of eccentric rear bracing

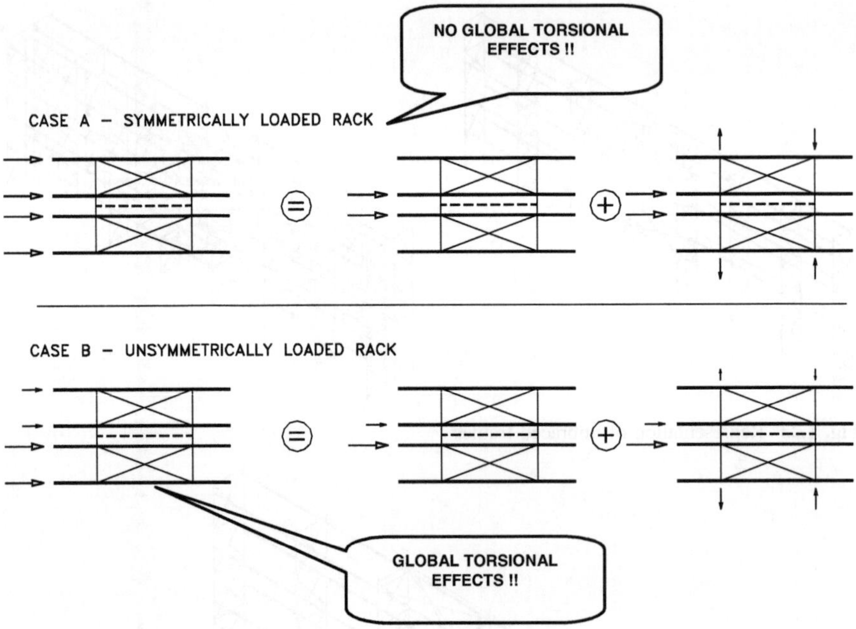

Fig. 1.26 Global torsional behaviour in case of double entry rack

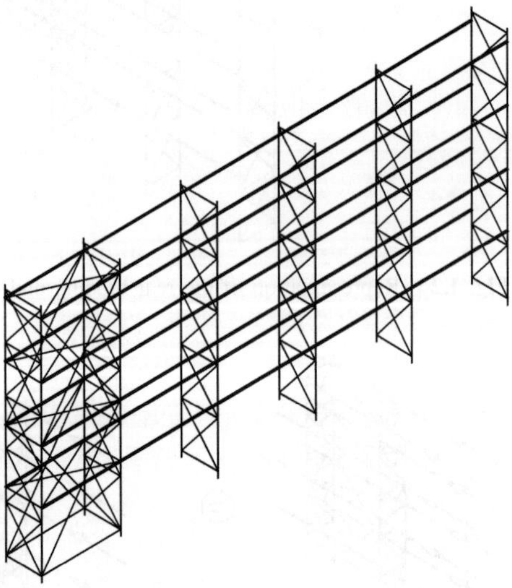

Fig. 1.27 Frame symmetrically braced in the down-aisle direction

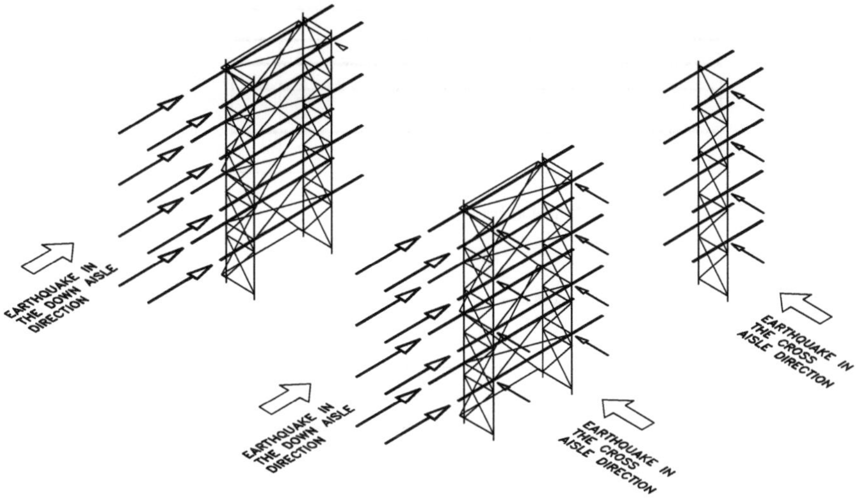

Fig. 1.28 Example of sub-models for the analysis of a spine-braced rack

1.5.1.18 Regularity Criteria—Down-Aisle Direction

The following Fig. 1.29 gives a representation of the regularity criteria in the down-aisle direction.

1.5.1.19 Typical Occurrences in the Seismic Analysis of Racks

Table 1.7 shows possible conditions occurring during the seismic analysis of racks.

1.5.2 Current Storage Rack Seismic Design Provisions and Practices in the U.S.

1.5.2.1 Development of Codes and Standards in the Past Decades

1.5.2.1.1 RMI Standards

Pallet storage racks were created to optimize warehouse and distribution center operations. As individual storage rack manufacturers developed new and competing products, the need for design and utilization standards and their implementation by the user and producer industries became obvious.

The Rack Manufacturers Institute (RMI) was established and incorporated in 1958 to deal with industry-wide issues. Among its initial activities was the development of the first edition of an RMI standard, *Minimum Engineering*

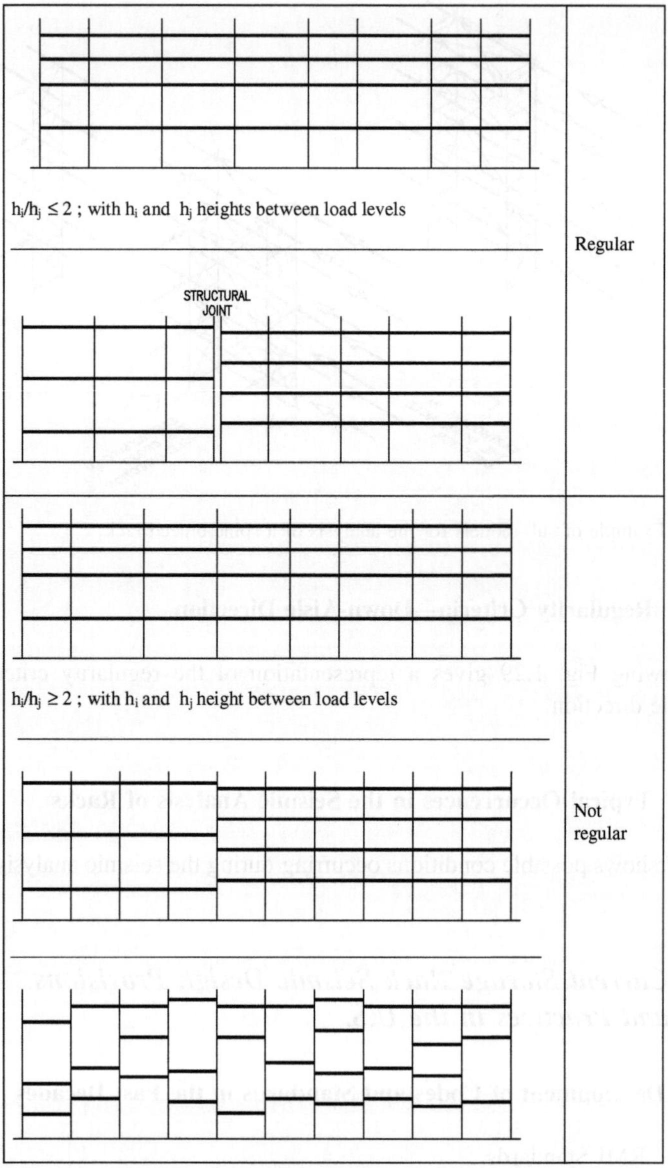

Fig. 1.29 The regularity criteria in the down-aisle direction

Standards for Industrial Steel Storage Racks, which was issued in 1964. This document was a short, simple, direct exposition of what had been developed and what was known by the members of the industry at that time. It represented the first step in developing specifications and other products designed to suit the needs of users, manufacturers, and the engineering and code-enforcement communities.

Table 1.7 Possible conditions occurring during the seismic analysis of racks

	Sub-structure *analysis*	Occurrence	Model	Method of analysis	Minimum loads configurations for seismic analysis *Notes*
1	Upright frame *cross-aisle direction*	(a) Single entry rack (b) Double entry rack—the connection between the upright frames is strong and allows the global cantilever scheme for both the upright frames	Planar	No modal simplified multimodal	All loading configurations, starting from fully loaded, unloading from bottom to top (1)
		(c) Double entry rack—the connection between the upright frames is weak and does not allows the global cantilever scheme for both the upright frames (d) Racks connected at top by crane racking top ties	Planar	Multimodal	(a) All upright frames equally loaded All loading configurations, starting from fully loaded, unloading from bottom to top (1) (b) One upright frame fully loaded, the others unloaded
2	Frame *down-aisle direction*	Rack unbraced in the down-aisle direction	Planar	No modal simplified multimodal	(a) Fully loaded rack must be considered (b) Other load configurations defined for the analysis under gravity loads on FEM 10.2.02 can be considered (2)
3	Vertical bracing of single entry rack symmetrically braced *down-aisle direction*	Rack symmetrically braced on front and rear planes	Planar	No modal simplified multimodal	Rack fully loaded
4	Vertical bracing of single entry rack braced on the back frame only *down-aisle and cross-aisle directions*	The horizontal bracings are placed in the bays where the vertical bracing is present, or in the adjacent ones	Spatial	No modal multimodal	Rack fully loaded for the analysis in the down-aisle direction Same loading criteria for cases (1a) and (1b) for the analysis in the cross-aisle direction
		The horizontal bracings are not placed in the bays where the vertical bracing is present, or in the adjacent ones	Planar (3)	No modal simplified multimodal	Rack fully loaded *Note: analysis of the vertical bracing only*
			Spatial	Simplified multimodal	Rack fully loaded for the analysis in the down-aisle direction Same loading criteria for cases (1a) and (1b) for the analysis in the cross-aisle direction *Note: the spatial model is necessary to consider the effect of the "twisting moment" and must include the upright frames connected by the horizontal bracings*

(continued)

Table 1.7 (continued)

	Sub-structure *analysis*	Occurrence	Model	Method of analysis	Minimum loads configurations for seismic analysis *Notes*
5	Vertical bracing of double entry rack symmetrically braced *down-aisle direction*	Rack symmetrically braced on front and back frames	Planar spatial	No modal simplified multimodal	Both racks fully loaded
			Spatial	No modal multimodal	One rack only fully loaded
6	Vertical bracing of double entry rack braced on the back frame only *down-aisle and cross-aisle directions*	The horizontal bracings are placed in the bays where the vertical bracing is present, or in the adjacent ones the horizontal bracings are not placed in the bays where the vertical bracing is present, or in the adjacent ones	Planar spatial	No modal simplified multimodal	Both racks fully loaded *Note: analysis of the vertical bracing*
			Spatial	Simplified multimodal	Rack fully loaded for the analysis in the down-aisle direction Same loading criteria for cases (1a) and (1b) for the analysis in the cross-aisle direction *Note: the spatial model is necessary to consider the effect of the "twisting moment" and must include the upright frames connected by the horizontal bracings*

Notes
(1) Generally the rack fully loaded is relevant for the upright compression
The uplift force must be checked considering the topmost level loaded, and, successively, the n upper levels loaded, with n = 2, 3, …, until the maximum uplift force is found
(2) With a conservative approach, the horizontal seismic action calculated for rack fully loaded can be considered, with the different load configurations defined for the check under gravity loads by FEM 10.2.02 that must be checked in any case also under the seismic action
(3) The matter should be analyzed with attention to consider the eccentricity of the bracing against the back frame, which affects the connected upright frames

In the late 1960s, RMI engaged Professor George Winter of Cornell University to undertake analysis and testing to provide a sound basis for the development of a more rigorous standard for the industry. Professor Winter was chosen because of his national and international reputation and his demonstrated expertise in the structural behaviour of cold-formed light-gage steel structures. The RMI membership and several other organizations provided financial and engineering support for the research effort that included analysis and testing related to the expanding range of products made by the growing number of industry members.

The results of the work conducted by Professor Winter and his team provided the basis for a new RMI standard, *Interim Specification for the Design, Testing, and Utilization of Industrial Steel Storage Racks*, which was issued in 1972 and which required earthquake loads to be considered in a manner mimicking the approach to building structures as stated in the *Uniform Building Code* (*UBC*).

In the *UBC*, design seismic forces for different types of building structures were based on *K* factors. The *K* factors for ordinary moment frame building structures and braced framed structures were 1.0 and 1.33, respectively. These were the factors used to define the seismic forces in 1972 edition of the RMI standard in the down-aisle and cross-aisle directions, respectively.

1.5.2.1.2 Model Building Code Requirements

As the use of storage rack structures increased across the nation, they were subjected to diverse loading conditions and received more scrutiny by the engineering and building code communities. Storage rack structures were mentioned for the first time in the 1973 edition of the *UBC* in the form of a footnote to a list of structures. The 1976 *UBC* referenced Standard 27-11 that addressed storage racks specifically and included design seismic forces for storage racks. The Building Officials Code Administrators International (BOCAI) and the Southern Building Code Congress International (SBCCI) soon added seismic provisions for storage rack structures to their codes—the *National Building Code* and *Standard Building Code*, respectively.

1.5.2.1.3 NEHRP Recommended Provisions

The first edition of the *NEHRP Recommended Provisions for Seismic Regulations for New Buildings* was issued in 1985. The *NEHRP Recommended Provisions* was intended to serve as a resource document for the organizations involved in developing seismic requirements to be included in codes and standards and to provide a venue for ongoing improvement of these requirements. A Provisions Update Committee (PUC) composed of structural engineers, industry and standards representatives, and building officials guides the triennial *Provisions* update process. RMI became a BSSC member organization in 1987 and, since that time, RMI representatives have served on the PUC Technical Subcommittees responsible for requirements for steel structures; for architectural, mechanical, and electrical components and systems; and for non-building structures.

The 1991 edition of the *NEHRP Recommended Provisions* introduced design values for storage racks in the chapter on architectural, mechanical, and electrical components. The design seismic forces were independent from the period, and the lateral force coefficient, based on allowable stress design (ASD), was 0.40 for the areas of highest seismicity. In the 1994 *Provisions*, design values for storage racks were significantly revised to be more consistent with the RMI seismic design criteria.

R values of 6 and 4 were assigned for storage racks in the down-aisle direction and the cross-aisle direction, respectively. Further, an importance factor of 1.5 was assigned for racks in areas accessible to the public. The R factor values of 6 and 4 were basically a translation of the *UBC* K values from the early 1970s.

Note that the first several editions of the *NEHRP Recommended Provisions* focused almost entirely on building structures; however, by the mid-1990s formal recognition was given to the fact that storage rack structures (and some other structures) are neither building structures nor architectural, mechanical, or electrical elements or components. Starting in the 1997 *NEHRP Recommended Provisions*, non-building structures including storage racks as well as cooling and storage towers, which had been treated in the chapter on architectural, mechanical, and electrical components, were now covered in a separate non-building structures chapter.

1.5.2.1.4 ASCE 7 Requirements

As the *NEHRP Recommended Provisions* became more influential with the various model code organizations, the American Society of Civil Engineers (ASCE) and developed the first of a series of updates to the ASCE 7 standard, *Minimum Design Loads for Buildings and Other Structures.* The 1993 edition of ASCE 7 adopted the *1991 NEHRP Recommended Provisions* as its seismic provisions and the 1995 edition of ASCE 7 reflected the 1994 edition of the *NEHRP Recommended Provisions* and covered storage rack structures under the category of architectural, mechanical, and electrical components and systems.

1.5.2.2 Current Seismic Requirements for Storage Racks

The 2003 edition of the *NEHRP Recommended Provisions* (FEMA 450) treats storage rack structures as non-building structures and references the seismic provisions of the RMI standard but also requires the use of mapped ground motions from the *NEHRP Recommended Provisions* for the design, and sets limitations on minimum base shear and seismic displacements. These requirements are "intended to assure comparable results from the use of the RMI standard, the *NEHRP Recommended Provisions*, and the *IBC* approaches to rack structural design." In addition, procedures are provided to "distinguish between the methods employed to design storage racks supported at grade (treated as non-building structures) from those supported above grade (treated as architectural, mechanical, and electrical components)." The treatment within the *NEHRP Recommended Provisions* "helps to clarify and coordinate the multiple references to rack structures in the *NEHRP Recommended Provisions* and the different means by which rack structures are analyzed and designed." Of particular interest to rack designers is the minimum base shear coefficient of about 10 % of the weight of the rack in areas of high seismicity. This minimum was originally imposed in the 2000 *IBC* to account for uncertainties of rack moment connection capacity to accommodate maximum considered earthquake (MCE) demands.

Based on the approach in the 2003 *NEHRP Recommended Provisions*, the 2002 and 2005 editions of ASCE 7 allow the use of the RMI standard subject to the *Provisions* requirements on ground motions, limitations on minimum base shear and seismic displacement or drift default values, again distinguishing between the approaches for storage racks supported at the base (which are to be designed as non-building structures) and those supported above the base (which are to be designed as architectural, mechanical, and electrical components and systems).

The 2003 edition of the *IBC* references the 2002 edition of ASCE 7 for its seismic requirements, thereby invoking the use of the RMI standard subject to the requirements and limitations imposed by ASCE 7. The 2006 edition of the *IBC* is expected to reference the 2005 edition of ASCE 7.

The 2003 edition of *NFPA 5000*, like the *IBC*, states that the design, testing, and utilization of industrial steel storage racks shall be in accordance with the RMI standard, and subject to the requirements and limitations imposed by the Sect. 9 (Earthquake Loads) of the 2002 edition of ASCE 7.

1.5.2.3 The Evolving RMI Standard

As noted previously, the 1972 edition of the RMI standard introduced seismic requirements for storage rack structures; it was updated and reissued in 1979, 1985, 1990, and 1997. Each new edition was an expanded version of the previous one and each represented an effort to reflect the seismic provisions articulated in the most current editions of the *NEHRP Recommended Provisions*, ASCE 7, and the model codes. To provide higher levels of safety in locations open to the general public, the 1997 edition of the RMI standard, which was based on the 1994 *NEHRP Recommended Provisions*, introduced a higher importance factor for storage rack installations in places such as retail warehouse stores that increased the magnitude of the design seismic base shear. As noted above, the 2002 edition of the RMI standard remains largely based on the 1994 edition of the *NEHRP Recommended Provisions* and the 1997 *UBC*, both of that are now considered out of date. In producing the 2002 edition, RMI followed the ANSI canvassing process and the document is designated ANSI Standard MH16.1-2004, *Specification for the Design, Testing, and Utilization of Industrial Steel Storage Racks*.

1.5.2.4 Current Storage Rack Seismic Design Practices in the U.S

Single selective steel pallet storage racks are typically designed for seismic forces using the equivalent lateral force procedures found in model building codes and in the RMI standard. Storage rack structural systems generally are moment frames in the down-aisle (longitudinal direction) and braced frames in the cross-aisle direction (transverse direction). Storage racks placed in the middle of a floor area usually are attached back to back, whereas single rack configurations are used near building walls. Storage racks in store areas accessible to the public typically are loaded with pallets; however, in some merchandising situations, merchandise is stored directly on the shelves. Intermediate shelf heights vary depending on merchandising needs. Storage racks are subjected to the greatest loads when pallets are fully loaded and all racks contain fully loaded pallets. Pallets are designed to carry the maximum pallet design load. Criteria for determining effective seismic weights of pallets used in rack seismic design procedures vary between the model building codes and the RMI standard.

Because storage racks normally do not have horizontal bracing at the shelf levels, they are treated as structures with flexible diaphragms and are evaluated analytically as two dimensional structural systems. For seismic loads, equivalent lateral forces are applied to the structural model, and the member forces are determined. These forces are then added to the other loads using a series of load combinations to obtain design member and connection forces. The design member loads are then checked against member capacities and/or allowable stresses. The capacities and allowable stresses take into account that the members generally are cold form steel members and that the columns are likely perforated. Moment connections typically are checked by comparing the computed moments against capacities (usually based on values obtained from monotonic tests with a factor of safety). These capacities typically are provided by the rack manufacturer. In the cross-aisle direction, base plates and anchor bolts are checked for the computed uplift forces. Other connections are part of a manufacturer's standardized components and are rarely included as part of the permit application process. Current storage rack design practice does not include increasing any of the connection forces or moments by the Ωo factor although that is inherent in current building code procedures for building structural systems.

Also, there are currently no ductility type prescriptive requirements for connection designs. P-Δ effects are typically considered by a moment magnifier for member design. The procedures currently used to compute rack seismic loads vary depending upon whether the prevailing requirements are from the model building codes, the *NEHRP Recommended Provisions* or ASCE 7, or the RMI standard. In some cases, there is more than one acceptable method of calculating seismic loads.

1.5.2.4.1 Pallet Loads and Effective Sesimic Weights

The pallet is the supporting structure for the basic unit that is stored on the typical single selective steel storage rack. There are many pallet sizes, shapes, materials, and types of construction; however, the vast majority of pallets used in the United States are what is termed the "GMA pallet." The GMA pallet may be a two-way pallet or a four-way pallet, depending upon whether the handling equipment can lift the pallet from two or four sides (Fig. 1.30).

The GMA pallet is 40 inches wide by 48 inches long and is constructed of hard wood with three stringers running front-to-back and many regularly spaced deck boards oriented side to side on both the top and the bottom of the pallet.

The unit load to be stored on the rack includes both the pallet and the product on the pallet and is the unit that is individually handled by the fork-lifts or other handling equipment. The weight, size, and pallet type is specified by the operator of the rack system.

The unit load weight (PL) that the rack components are designed for is usually determined by the end user.

The primary design weight is usually the maximum weight unit load ($PL_{Maximum}$) present in the warehouse. If there is more than one weight unit load and the weights

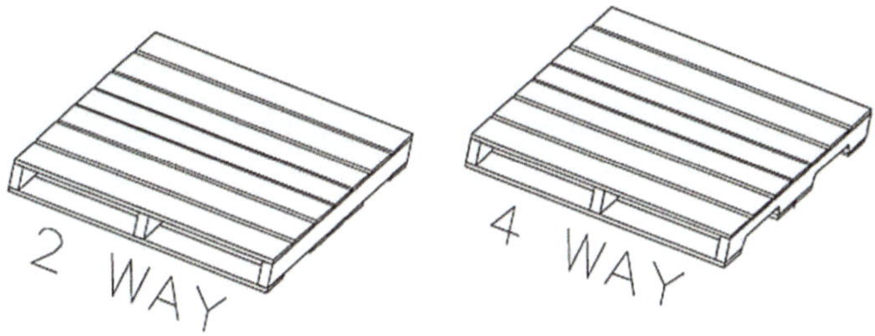

Fig. 1.30 Typical pallets used in retail operations

Fig. 1.31 Typical steel single selective pallet storage rack (from FEMA 460)

differ by a significant amount, then the second weight will also be specified and the locations of the storage of each weight load will be designated.

$PL_{Maximum}$ is used for the design of the load beams bending and deflection, the upright column axial capacity, and the cross-aisle frame bracing requirements (for vertical loads). The maximum pallet weight also is used to resist uplift when combined with cross-aisle seismic forces.

The average weight of the unit load ($PL_{Average}$) is the maximum total load expected in any individual rack row divided by the total number of storage positions in that row. This maximum total weight includes less-than-full-weight pallets and accounts for the number of storage positions normally expected to be empty. This is used for determining the down-aisle seismic force.

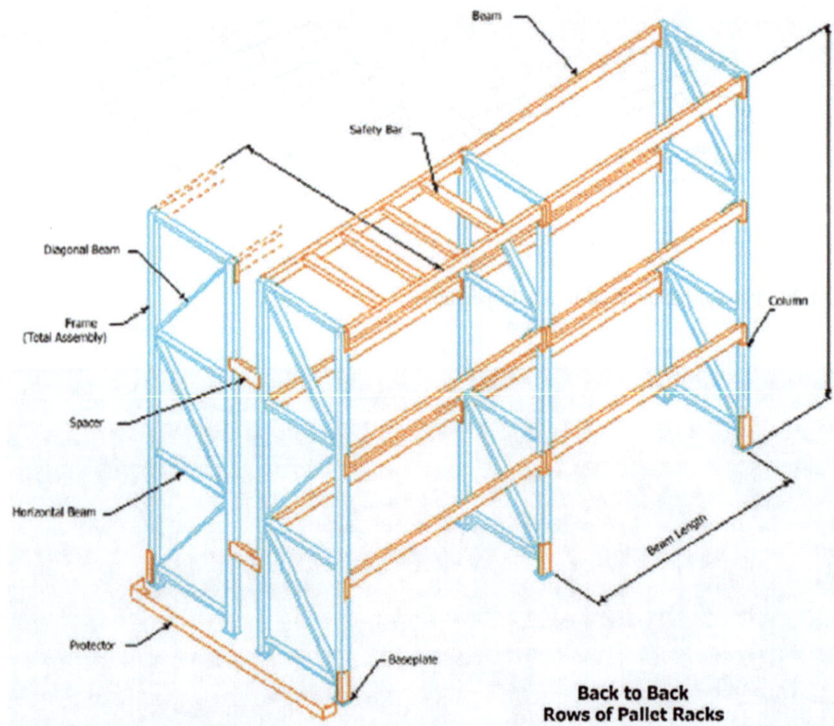

Fig. 1.32 Typical storage rack configuration (from FEMA 460)

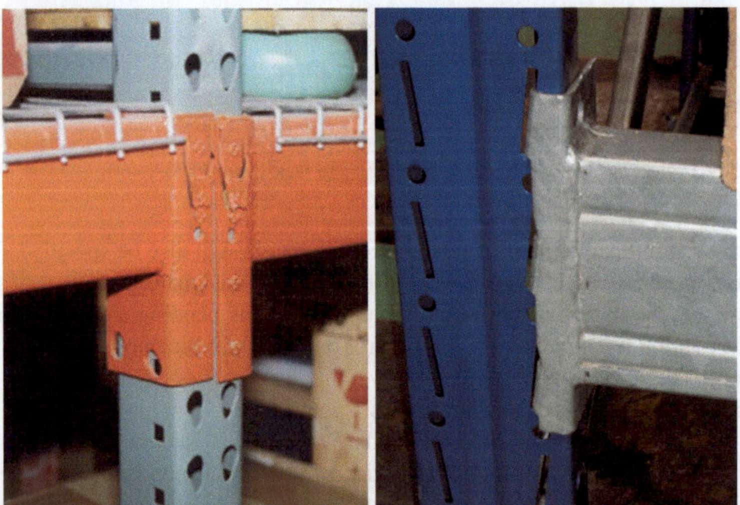

Fig. 1.33 Typical proprietary moment connection

In addition to the product load (*PL*), the rack structure is designed for the effects of the structure dead load (*DL*) and any live load (*LL*) that might be present. The dead load includes the frame weight, the shelf beam weight, and the weight of any accessory items on the rack such as pallet supports or wire mesh decks. The dead load is typically much less than 5 % of the weight of the product stored on the rack structure. The live load is typically 3.0 KN/m^2 or less.

The product load includes the pallet weight and the weight of the product on the pallet. The storage rack system is designed on the full, maximum indicated product load. For a warehouse store, all rack system positions are assumed with full weight pallets.

For most storage racks, the effective seismic weight (*Ws*) is a function of the ratio of the average pallet weight to the maximum pallet weight and is determined in accordance with the RMI standard, as:

$$Ws = (0.67 \times PL \times PL_{RF}) + DL + 0.25 \times LL$$

where

PL_{RF}	is the product load reduction factor
PL_{RF}	*1.0* when the seismic force is in cross-aisle direction
PL_{RF}	$PL_{Average}/PL_{Maximum}$ when the seismic force is in down-aisle direction
$PL_{Average}$	is the maximum total weight of product expected on the shelves in any row divided by the number of shelves in that row
$PL_{Maximum}$	is the maximum weight of product that will be placed on any one shelf in that row.

The 0.67 coefficient for the pallet weight comes not from the average load but from an evaluation of the amount of load that participates in developing the dynamic seismic force.

Experience has shown that the whole mass of the merchandise stored on the storage rack system does not participate entirely to the inertia generated from the ground motion. There is some friction inducing energy dissipation for the relative movement between the storage rack, the pallets stored on the rack, and the product on the pallets during the seismic motions. This 0.67 factor represents the fraction of the load on the fully loaded system contributing to the effective horizontal seismic weight. This effective seismic weight factor does not apply to vertical gravity loads.

It should be noted that the effective seismic weight ratios provided in the *UBC* are different from those above.

1.5.2.4.2 Rack Configuration

The number of pallet loads on each shelf between uprights generally is specified by the operator of the warehouse. In a typical storage rack system, there are usually two pallet loads between uprights. If there are cross-aisle tunnels in the system to allow movement between aisles, the bay will be three or four pallets wide and not have beams in the bottom two or three levels. At the tunnel bay, the columns will have shelves attached on one side in the bottom region and on both sides above the tunnel. Infrequently, the tunnel may be located at the end of a rack row and the last frame in the row will not have shelf beam support in the bottom region, in which case a special design for that frame is required.

The shelf elevations are determined by the operator of the warehouse based on the height of the loads and the shelf beam size and clearances required for storing and removing the load. The shelves may be spaced regularly for the full height of the rack system if all the loads are the same height or the spacing may be varied to accommodate different height loads with a minimal amount of lost space.

The overall storage height is based on the lifting height limitation of the handling equipment and the clear building height. Handling equipment characteristics also may necessitate a low bottom shelf beam at or near the floor. This shelf beam is typically 150 mm from the floor to the bottom of the beam and, considering a 100 mm beam, 200 mm to the centerline of the beam. When distributing the seismic forces vertically, this bottom beam does not participate in the determination of seismic forces to levels above as would be indicated by the traditional vertical distribution. The seismic force including this beam will only govern the design on the horizontal force on the anchors. This situation is addressed in the RMI standard.

1.5.2.4.3 Seismic Loads

The seismic loading requirements are determined by the building authority having jurisdiction. The *International Building Code* (*IBC*) currently is used by most building authorities. There is, however, one notable exception, the State of California. The *California Building Code* (*CBC*) is still based on the 1997 edition of the *Uniform Building Code*, although it appears that the state will adopt the *IBC*. While no longer supported by any code development organization, the *UBC* will continue to serve as the basis for the *CBC* until the state adopts a new model building code.

1.6 Review of Past Seismic Research on Storage Racks

Experimental and analytical studies of the seismic performance of storage racks are scarce and the results are often proprietary; and consequently, they have not significantly influenced the development of codes and regulations related to storage rack systems.

Available results from experimental and analytical investigations on the seismic response of storage racks are briefly reviewed hereafter, and gaps in knowledge requiring further research studies are identified.

1.6.1 Review of Experimental Research

Experimental research related to the seismic behaviour of storage racks can be categorized into different types of testing procedures:

- *Cantilever* testing of subassemblies in which quasi-static cyclic loads are applied to beam-to-upright connections.
- *Portal* testing of subassemblies in which beam and uprights portal subassemblies are loaded laterally to simulate seismic loading.
- Quasi-static testing of storage rack systems for which completely loaded storage racks are loaded laterally to simulate seismic loading.
- In situ dynamic testing of storage rack systems with small shakers or under ambient vibrations, in order to obtain their dynamic characteristics (e.g., natural periods and damping).
- Shake-table testing of storage rack systems, with completely loaded storage racks, excited by recorded or artificially generated ground motions.
- Testing of cold-formed steel members and structures, from which most storage racks are built.
- Testing of merchandise.

Table 1.8 lists the experimental investigations documented in the public literature, and the various testing techniques. The major conclusions obtained from these experimental investigations are reviewed briefly below.

1.6.1.1 Cantilever Testing of Storage Rack Subassemblies

The lateral stiffness of storage rack systems in the down-aisle (moment-resisting frame) direction is greatly influenced by the distortions that occur at the

Table 1.8 Investigations of storage rack systems and subassemblies

Authors (year)	Testing types (number of specimens)
Blume and Associates (1973, 1974)	In situ dynamic tests (19)
Krawinkler et al. (1979)	Cantilever tests (29), portal tests (6), quasi-static tests of storage rack systems (4), dynamic tests (2)
Chen et al. (1980a, b)	Shake-table tests (4), merchandise tests (2)
Bernuzzi and Castiglioni (2001)	Cantilever tests (22)
Filiatrault (2001)	Shake-table tests with real merchandise (5)
Castiglioni et al. (2003)	Shake-table tests (4)
Higgins (2004)	Cantilever tests (22)
Filiatrault (2004)	Shake-table tests (4)
Castiglioni (2004)	Shake table tests with base isolators (2)

beam-to-upright connections. For analytical modelling purposes, these distortions often are represented by simple rotational spring elements inserted between the beam ends and the upright center line. The rotational spring constant to be used in a numerical model can be obtained from moment-rotation relationships between a beam end and an upright using the so-called cantilever test method (RMI 2002a, b).

Krawinkler et al. (1979) performed cantilever tests on 20 different beam-to-upright subassemblies of standard pallet racks. In all connections, the beam ends were welded to angle connectors that, in turn, permitted connection to the perforations on the upright through either hooks (Type A) or button grips (Type B). The experimental results indicated that, because of local deformations at the beam-to-upright connections, moment-rotation hysteretic loops have a pinched shape similar to that obtained for reinforced concrete elements with high shear load. Low cycle fatigue phenomena also may affect the strength and ductility of beam-to-upright and upright-to-floor (base plate) connections. The strength of the Type A subassemblies was limited by the capacity of the hook-type grips that started to pull out of the upright perforations. In Type B subassemblies, fractures of the fillet weld between the beam and the connection angle limited the moment capacity.

More recently, Bernuzzi and Castiglioni (2001) performed a series of 11 monotonic and 11 cyclic tests on two different types of beam-to-upright connections used in Europe. The experimental results obtained from the monotonic tests indicated that the connections were characterized by significant ductile behaviour. None of the test specimens failed before the maximum rotation achievable by the testing equipment was reached. This maximum rotation was way beyond practical design values. The results of the cyclic tests exhibited, with increasing number of response cycles, pronounced pinching behaviour associated with slippage and plastic deformations of the connectors leading to significant reduction of energy dissipation capacity.

Quasi-static testing was recently conducted on 22 different types of interior beam-to-upright subassemblies (Higgins 2004). The test data indicate that beam-to-upright connections exhibit very ductile and stable behaviour, with rotational capacities beyond the values observed during shake-table tests and expected from a design seismic event. The hysteretic responses of some of the tested beam-to-upright connections, however, exhibited significant pinching similar to those tested by Bernuzzi and Castiglioni (2001).

Sarawit and Peköz (2003) recently proposed a new beam-to-upright connection test to replace the cantilever test method. The actual frames bending moment-to-shear force ratio is better represented in this proposed test method than the current cantilever test.

1.6.1.2 Portal Testing of Storage Rack Subassemblies

The moment-rotation characteristics obtained from the cantilever test method summarized in the previous section is highly dependent on the bending moment-to-shear force ratio applied to the beam-to-upright connection. This ratio varies continuously during the application of seismic lateral force to a vertically loaded storage rack in its down-aisle direction. In order to better represent the bending moment-to-shear force ratio in a beam-to-upright connection, the portal test method has been proposed. In this testing procedure, a portal assembly of generally one beam connected to two uprights is loaded by constant static gravity loads on the beam and variable lateral loads on one of the upright at the elevation of the beam. Moment-rotation at both beam-to-upright connections can be monitored during the tests.

Krawinkler et al. (1979) performed six portal tests on three different beam-to-upright subassemblies of standard pallet racks. Four types of beam-to-upright connections were investigated. In all connections, the beam ends were welded to angle connectors which in turn permitted connection to perforations on the upright through either hooks (Type A) or button grips (Types B and C). In Type D connection, additional devices were used to join the connector angles to the uprights. The global hysteretic response of the subassemblies resulted relatively insensitive to the gravity loads induced by the merchandise. When the moment-rotation hysteretic loops of the beam-to-upright connections were compared to that of the cantilever tests described in the previous section, it was observed that the loops from the portal tests exhibited a significantly higher initial stiffness. This result confirmed that moment-rotation characteristics of beam-to-upright connections depend on the bending moment-to-shear force ratio that is significantly higher in the portal tests due to the presence of the vertical merchandise loads.

1.6.1.3 Quasi-Static Cyclic Testing of Complete Storage Rack Systems

Quasi-static cyclic testing of complete storage racks represents an efficient experimental procedure to study the interaction between beams, uprights and connections under gravity (merchandise) loads and simulated seismic lateral loads. The beams of the racks are loaded by either concrete blocks on pallets or real merchandise, and hydraulic actuators apply lateral loads to the uprights at the various beam levels. With this testing procedure, racks can be tested separately in their down-aisle or cross-aisle directions. Krawinkler et al. (1979) performed four quasi-static tests of complete three-story storage racks.

Two tests were performed in the down-aisle direction and two others were performed in the cross-aisle direction. The first two rack specimens contained hook beam-to-upright connectors, while the two others incorporated button grip connectors. The lateral load was applied only at the top level of the racks. The global hysteretic response of the racks along the down-aisle direction was characterized by a nonlinear pinched hysteretic shape. It was observed that for constant lateral displacement amplitudes the second load cycle led to a significant decrease in energy dissipation capacity while the third cycle was practically identical to the second one. Failure in the down-aisle direction typically initiated by weld cracking between the beam ends and the connector angles. This weld cracking in the cyclically loaded racks occurred at smaller lateral displacements than in the monotonically loaded racks. Up to the point when weld cracking was adverted, the ductility of storage racks in their down-aisle direction depended strongly on the axial load ratios in the uprights. For small axial load ratios, a very ductile behaviour characterized by flexural plastic hinging in the uprights was achieved. It also was found that second order (P-Δ) effects greatly affected the lateral strength and stiffness of storage racks in the down-aisle direction. Finally, the ductility and energy dissipation capacity of storage racks resulted much larger in the down-aisle moment-resisting direction than in the cross-aisle braced frame direction.

1.6.1.4 Dynamic in Situ Testing of Storage Rack Systems

The first published in situ dynamic investigation of storage racks was performed in the mid 1970s at various distribution centers in the San Francisco Bay Area (Blume and Associates 1973). Ambient and man-made vibration measurements were applied to representative steel industrial storage racks of standard pallets, drive-in and drive-through, cantilever, and stacker crane types to obtain a range of natural frequencies and damping ratios. The ambient vibration and man-made excitation measurements generated average response accelerations at the top of the racks on the order of 0.005 and 0.015 g, respectively. The experimental results indicated that the fundamental translational period of storage racks obtained by means of the empirical formula for building periods of the 1973 edition of the *Uniform Building Code*, was not reliable

Measured fundamental periods over a range of actual merchandise loading conditions averaged 0.6 s in the down-aisle direction, and 0.2 s in the cross-aisle direction. Torsional periods averaging 0.4 s were identified in many of the rack configurations. It was noted that these period values would increase at least by 20 % under response amplitudes representative of a strong earthquake. Measured structural damping ratios for storage racks averaged 2–3 % of critical at root-mean-square response acceleration levels of 0.01–0.02g. It was noted that these damping values would increase at least by a factor of 2 under response amplitude representative of a strong earthquake for which significant energy dissipation would occur due to rocking, slippage and interaction of stored merchandise. Thus, it was concluded that a damping ratio of 5 % of critical would be a reasonable value for storage racks under seismic excitations.

Krawinkler et al. (1979) subjected two full pallet rack assemblies with gravity loads to forced and free vibration tests to obtain information on natural frequencies, mode shapes and damping characteristics in the down-aisle and cross-aisle direction. Measured fundamental periods averaged 0.7 s in the down-aisle direction, and 0.5 s in the cross-aisle direction. The vibration decay obtained from the free-vibration tests in the down-aisle direction exhibited a textbook example of Coulomb-type friction decay. At large amplitudes, the friction between the grip-type connectors and the perforations in the uprights caused significant damping (on the order of 2.5–3.5 % of critical). Once the connectors locked up at smaller amplitudes, the damping dropped drastically to a very small value (on the order of

0.7 % of critical). The damping characteristics in the cross-aisle direction were more constant with vibration amplitudes (on the order of 1–2 % of critical).

1.6.1.5 Shake-Table Testing of Storage Rack Systems

Shake-table testing complete storage rack systems loaded with real merchandise represents the most direct experimental procedure to assess their seismic behaviour. Unfortunately, this type of testing is expensive compared to other testing procedures and only a very limited number of shake-table studies on storage racks have been performed to date.

The first published shake-table studies on storage racks in the United States, was performed in the late seventies on the 20-ft-square shake-table at the University of California, Berkeley (Chen et al. 1980a, b, 1981). Four types of full-scale industrial steel storage racks were subjected to scaled ground motions actually recorded during two California earthquakes (1940 El Centro and 1966 Parkfield earthquakes). The ground motions were scaled so that the resulting base shear coefficients approximately equaled the design base shear coefficients of the 1979 edition of the *Uniform Building Code* for ordinary moment frame buildings (with $K = 1.0$) in the down-aisle directions and ordinary brace frame buildings (with $K = 1.33$) in the cross-aisle direction.

The types of storage racks tested were: single standard pallet racks, back-to-back pallet racks, drive-in racks, and stacker racks. In general, the storage racks performed well during the tests, with the exception of the drive-in stacker racks in the cross-aisle direction, for which considerable buckling was observed in the first story diagonal members. The fundamental periods of vibration ranged from 2 to 3 s for the standard pallet and drive-in racks in the down-aisle direction and 0.5–1.0 s for the standard pallet, drive-in, and stacker racks in the cross-aisle direction. The first mode damping values were much larger in the down-aisle direction (on the order of 3–9 % of critical) than in the cross-aisle direction (0.5–3 % of critical). It was also observed that the ductility and energy dissipation capacity of the racks were much larger in the down-aisle, moment resisting frame direction than in the cross-aisle, braced frame direction. Therefore, the racks could undergo significant inelastic deformations without suffering major damage in the down-aisle direction, but could only develop limited amount of inelastic deformations in the cross-aisle direction. Second order (P-Δ) effects contributed significantly to the response of the racks in the down-aisle direction.

More recently (Filiatrault 2001), five different back-to-back pallet racks loaded with real merchandise were tested on a uniaxial shake-table under a single component, scaled at various amplitudes, of the ground motion recorded at Canoga Park during the 1994 Northridge earthquake in California. Three of the tests were conducted in the cross-aisle direction, while the two other tests were conducted in the down-aisle direction. In general, the racks performed well.

Significantly more flexibility, ductility, and energy dissipation capacity were observed in the down-aisle direction than in the cross-aisle direction. The fundamental periods of vibration averaged 1.4 s in the down-aisle direction and 0.6 in the cross-aisle direction. No structural damage occurred in any of the rack configurations for peak ground motion amplitudes less than 0.42*g*.

Castiglioni et al. (2003) performed shake-table tests on four full-scale steel storage pallet racks loaded by concrete blocks mounted on pallets simulating content merchandise. The four specimens were chosen among six structures designed by two different European manufacturers based on Eurocode 8. The experimental results indicated that sliding of pallets occurred for ground motion intensities less than the considered design levels. Also, the diagonal bracing configuration in the down-aisle and cross-aisle directions has a significant influence on the seismic response of steel storage pallet racks. In particular eccentric bracing configurations can lead to significant torsional response. The authors stressed out the importance of a regular configuration of bracing systems.

A shake-table study was recently conducted at the University of Buffalo on four different pallet rack configurations, incorporating bolted beam-to-upright connections (Filiatrault and Wanitkorkul 2004).

All racks were tested in the down-aisle direction. The main objectives of the tests were:

- to determine the variations of in-plane dynamic characteristics of the industrial storage racks during service life.
- to deduce average beam-to-upright rotational stiffness from measured in-plane fundamental period of racks.
- Evaluate the response of storage racks under strong ground motions.

The test results indicated that the rotational stiffness of beam-to-upright connections is the main factor influencing the down-aisle seismic response of pallet racks. The initial values obtained for the rotational stiffness of beam-to-upright connections are two to three times higher than values used in current design. These values reduce with the amplitude of vibration but generally return to their initial values once the structure returns to its undeformed position. Furthermore, very ductile seismic behavior was observed in the down-aisle direction with peak interstorey drifts exceeding 7 % without any sign of incipient collapse.

1.6.1.6 Experimental Research on Cold-Formed Steel Members

The lateral load-resisting systems of storage racks often include cold-formed steel bracing members. Therefore, research information related to the behaviour of cold-formed steel structural members and systems have influenced the design of storage racks.

Cheng (1973) performed axial load tests on cold-formed steel open sections used as primarily load carrying structural members in storage racks. It was observed that local flexural-torsional buckling is the primary mode of failure for axially loaded perforated open section segments. An analytical expression was proposed to predict the axial load carrying capacity of these members.

Kotha and Peköz (2000) studied the behavior of cold-formed pallet storage racks with semi-rigid beam-to-upright connections and with flexible upright bases through finite element analyses. A general moment-rotation relationship was established to model the beam-to-upright connection stiffness of pallet storage racks. Also, the upright base flexibility caused by base plate bending was quantified. Guidelines were provided to carry out nonlinear finite element analysis of storage racks accounting for these influencing parameters.

1.6.1.7 Testing of Merchandise

As part of the shake-table testing conducted by Chen et al. (1980a), the seismic response of a two-story, one bay wide, one bay deep rack was obtained for two different cases: (a) the case in which the merchandise was tied with metal straps to the rack beams, and (b) the case in which the merchandise was not fastened to the rack. For the purpose of this investigation, eight types of merchandise with different geometry and weight were considered. The tests were conducted with scaled ground motions that resulted in base shear coefficients about the same as code level base shear coefficients for buildings. The results of the tests indicated that neither merchandise nor wood pallets fell off the racks during the tests. Only some uppermost paper products and canned goods moved slightly during the tests in the down-aisle direction. It was found, however, that the seismic response of the racks without the merchandise tied up to the beams was smaller compared to that of the same rack with the contents rigidly attached to the beams. This response reduction was particularly obvious during the strongest portion of the response. Low amplitude free vibration tests also indicated that the damping was slightly higher for the case where the merchandise was not tied up to the rack.

The shaking table tests conducted recently by Filiatrault (2001) included also real merchandise. It was clearly demonstrated that the use of plastic wrap was a very efficient mean of restraining the merchandise from falling off the racks. No fallen merchandise was observed for ground motion amplitude less than 0.30*g*. Significant slippage of heavy merchandise on pallets was observed but no pallet overturned during any of the tests. However, tall merchandise items (e.g., water heaters) resulted vulnerable to overturning, and need a restraint.

1.6.2 Review of Analytical and Numerical Researches

As evidenced by the experimental research previously reviewed, the seismic response of storage racks in the down-aisle direction is strongly affected by the nonlinear moment-rotation response of the beam-to-upright connections. In the cross-aisle direction, on the other hand, the seismic response of storage racks depends on the characteristics of the bracing members used in the truss configuration. Therefore, numerical models that have been used to predict the seismic response of storage racks incorporate these different lateral load-resisting systems to various degrees. The analytical and numerical research related to the seismic behaviour of storage racks can be divided into two different types of models:

- *Linear models* for which the moment-rotation response of beam-to-upright connections is linearized by simple linear rotational springs representing secant properties at the anticipated response level of the storage racks. For dynamic analysis, an equivalent linear viscous damping model is also used to represent the energy dissipation of these connections during inelastic actions.
- *Nonlinear models* for which the nonlinear response of beam-to-upright connections is followed over the time-history response of storage racks by the use of nonlinear moment-rotation hysteretic rules. This nonlinear modeling is used mainly for research purposes and rarely in design situation.

1.6.2.1 Linear Modelling of Storage Rack Systems

Blume and Associates (1973) developed and analyzed equivalent lumped mass numerical models representing selected storage racks in order to compare their predicted fundamental periods to in situ measured values. Pinned upright bases were assumed for all rack configurations except for the cross-aisle direction of cantilever racks. Rigid beam-to-upright connections were assumed in the down-aisle direction.

Reasonable agreement was achieved between measured and computed storage fundamental periods.

Chen et al. (1980a) conducted frequency analyses of linear mathematical models in order to compare calculated periods of vibration and mode shapes with those observed during low amplitude shaking table tests and pull-release free-vibration tests that they had previously conducted. These calculated periods and mode shapes were then used to perform response spectrum analyses. The calculated fundamental periods of vibration were also used to determine the base shear coefficients according to the 1973 edition of the *Uniform Building Code* and to the ATC-3 procedure (ATC 1978). It was found that two-dimensional models with minimum net section properties and centreline dimensions were adequate for practical purposes.

Modeling parameters such a semi-rigid beam-to-upright and base connections should be considered in theoretical predictions of rack response. It was also found that in the down-aisle direction, the lateral forces determined by the 1973 edition of the *Uniform Building Code* were roughly equivalent to those obtained from response spectrum analyses with intensity levels slightly more than 50 % of the 1940 El Centro and 1966 Parkfield earthquake records. In the cross-aisle direction, however, the code lateral forces were approximately equivalent to 25–50 % of the El Centro and Parkfield records. In the cross-aisle direction the lateral forces predicted by the *UBC* were higher than those predicted by the ATC-3 (ATC 1978) procedure. Opposite results were obtained in the down-aisle direction.

Blume and Associates (1987) performed static and response spectrum analyses to investigate the applicability of the eccentric braced frame concept (Roeder and Popov 1978) to storage racks in order to improve their seismic behavior in the cross-aisle direction. The results of the study indicated that, aside from a considerable savings in steel material, the eccentric bracing system could undergo significantly more inelastic deformations without structural instability than conventional bracing systems. Although the analytical results were promising, the authors recommended also that experimental investigations needed to be conducted before implementing the eccentric bracing system in storage racks. Such experimental results are not available to date.

1.6.2.2 Nonlinear Modelling of Storage Rack Systems

Chen et al. (1980b) developed also two-dimensional non-linear numerical models for standard pallet racks in both down-aisle and cross-aisle directions and compared their predictions to the results obtained from shake-table tests. In the down-aisle direction, semi-rigid beam-to-upright connections and semi-fixed upright bases were assumed in evaluating their initial stiffness. Bilinear moment-rotation hysteretic rules were considered for the semi-rigid beam-to-upright connections. Second order (P-Δ) effects also were considered in the time-history dynamic analyses. The stiffness of the connections between the beams and the uprights was adjusted in the model to simulate the observed local deformations. The responses predicted by the model were in good agreement with the experimental results.

Baldassino and Bernuzzi (2000) conducted a numerical study on the lateral-load response of the steel storage pallet rack systems commonly used in Europe. The results confirmed that the nonlinear rotational behaviour of beam-to-upright connections influenced significantly the response of storage rack systems in the down-aisle direction. Also, the numerical investigation confirmed the significant influence of the base plate connections on the overall rack response in both directions. The authors pointed out the need for test data on the non-linear moment-rotation behaviour of base upright connections (Fig. 1.34).

Fig. 1.34 Typical rack bracing members and connections

1.6.3 Research Needs

As presented in this brief review, the current engineering knowledge base concerning the earthquake safety and vulnerability of storage racks is 20–30 years old and is limited to contents and racks unlike many modern applications. The retail industry and the state-of-the art of the design of storage racks have changed considerably in the interim. Large chains of stores now routinely invite the public to shop in a physical environment that formerly was found only in a warehouse, racks have more complex configurations and are taller, and their contents have become heavier. These facts clearly pinpoint to urgent research needs related to the seismic behaviour of storage rack systems. In this section, experimental and analytical research that is perceived to be the most urgently needed is briefly listed.

1.6.3.1 Experimental Research Needs

Only two full-scale shake-table testing investigations of storage racks fully loaded have been performed in Europe (Castiglioni et al. 2003) and other three in the United States (Chen et al. 1980a, b, 1981; Filiatrault 2001). There is an urgent need to increase the experimental database of the seismic response of complete storage rack systems through shake-table testing. The main variables that need to be investigated in such experimental programs are:

- The layout and types of storage racks representing current construction practices and innovative systems such as eccentric bracing.
- The layout and types of merchandise contents.
- The types of seismic restraints (e.g., plastic wraps, screens, ledges, etc.) for contents.
- The structural interaction between neighbouring racks.
- The direction of the horizontal seismic input, relative to the rack's orientation (transverse, longitudinal, or non-orthogonal).
- The characteristics of the input ground motions, including consideration of whether vertical accelerations must be characterized and near-field motions, and relating these input motions to seismic hazard mapping and codes.

As demonstrated by available experimental and analytical results, the seismic response of storage racks in their down-aisle direction is strongly affected by the non-linear response of the beam-to-upright and base plate connections. Since numerous variables enter in the design of these connections, an experimental parametric study on the cyclic response of beam-to-upright and base plate connections is urgently needed.

While the needs of the down-aisle direction are urgent, testing needs in the cross-aisle are even more urgent since the understanding of this directions behaviour is even less understood. Failures of racks in earthquakes are most commonly reported as cross-aisle failures.

The information on the seismic response of merchandise contents installed in storage racks is very limited. There is an urgent need to conduct shake-table studies of merchandise. For this purpose, shake-table testing could be used to simulate the motions experienced by various levels of storage racks during earthquakes. A robust numerical model would be required to develop these input motions. Various merchandise items could be mounted on the shake-table via a rigid assembly representative of the level on which they are mounted. Various types of merchandise contents would be investigated experimentally under a large number of input motions representative of several seismic hazard levels. Furthermore, these results could be compared with the ones obtained when various types of seismic restraints are introduced. With this information, clear recommendations could be provided on the types of seismic restraint to be used for a particular type of merchandise content.

1.6.3.2 Analytical Research Needs

There is a need to develop a general purpose computer-based numerical model for the prediction of the seismic response of storage racks and contents.

The development of such a general-purpose model requires close coordination and interaction with the experimental work.

1.7 Scope and Aim of the Research

As clearly understandable from the previous overview, at present, there are technical limitations in the field of safety and design of storage racks in seismic areas: lack of knowledge on actions challenging the structures, lack of knowledge on structural behaviour in terms of ductility and of sliding conditions of the pallets on the racks and lack of Standard Design Codes in Europe.

To solve some of these limitations, the EU sponsored through the Research Fund for Coal and Steel an RTD project titled "Storage Racks in Seismic Areas" (acronym SEISRACKS, Contract Number: RFS-PR-03114).

The objectives of this project, initiated in December 2004 and terminated in June 2007, are:

- to increase knowledge on actual service conditions of storage racks,
- to increase knowledge on racks' actual structural behaviour
- to assess design rules for racks under earthquake conditions.

The research team is composed by the following units: ACAI the Italian Association of Steel Constructors (Coordinator), Instituto Superior Tecnico of Lisbon (P), National Technical University of Athens (GR), Politecnico di Milano (I), University of Liege (B) and the European Laboratory for Structural Assessment

(ELSA) of the Joint Research Center of Ispra (Subcontractor of Politecnico di Milano).

The scientific objectives of this project can be summarised as follows:

Increase knowledge on actual service conditions of storage racks collecting data by continuous monitoring of a structure located in a warehouse in seismic area.

These data refer, in particular to:

(a) actions (actual live load distribution on the rack, occupancy ratio, vertical loads, accidental actions due to impacts, loading cycles, etc.)
(b) structural response (vibrations, frequencies, settlements, permanent deformations, etc.)

Increase knowledge on actual structural behaviour of storage racks

(a) by definition of the sliding properties of pallets on the racks, as a function of: (i) type of pallet, (ii) stored material, (iii) acceleration, (iv) frequency of the excitation
(b) by identification of base isolation devices with characteristics suitable to storage racks in seismic areas in order to minimise the pallet sliding phenomenon, and verification by full scale testing of one full-scale base-isolated storage rack
(c) by assessment of the actual lateral load carrying capacity and ductility of storage racks by means of pseudo-dynamic tests carried out up to failure of full-scale structures.

Assessment of design rules for racks under earthquake conditions

(a) by definition of a set of design actions for serviceability and ultimate state design for racks in seismic areas.
(b) by definition of q-factors to be adopted in seismic design of racks.
(c) carrying out a revision of the most updated draft of FEM 10.2.08 Design Standard on the basis of the previous work and collected data, in order to incorporate into the document all the information relevant for a safe, although competitive, design of storage racks in seismic areas.

The research activities to be carried out in co-operation among the partners, in order to achieve the aforementioned objectives within this project, are subdivided in the following Work Packages:

WP 1—Full scale dynamic tests of storage racks
WP 2—Full scale pseudo-dynamic and pushover tests of storage racks
WP 3—In situ monitoring of storage racks
WP 4—Cyclic testing of components
WP 5—Assessment of seismic design rules for storage racks.

The project focuses on steel selective pallet storage racks (Fig. 1.31) located in retail warehouse stores and other facilities, eventually accessible to the general public.

The structural elements in a typical pallet storage rack are shown in Fig. 1.32. Storage racks are composed of specially designed steel elements that permit easy installation and reconfiguration consistent with the merchandising needs of a warehouse retail store. Except where adjacent to walls, storage racks normally are configured as two rows of racks that are interconnected. Pallets typically can have plan areas of approximately one square meter (see appendix A for detailed information about structural typologies, standard pallet types and sizes as well as nomenclature) and can have a maximum loaded weight of approximately 10–15 kN. Storage rack bays are typically 1.0–1.1 m deep and 1.8–2.7 m wide and can accommodate two or three pallets. The overall height of pallet rack structural frames found in retail warehouse stores varies between 5 and 6 m. In industrial warehouse facilities, racking system can reach considerable heights, such as 12–15 m or more.

The rack industry calls the longitudinal direction the down-aisle direction and the transverse direction, the cross-aisle direction. Proprietary moment connections are typically used as the structural system in the down-aisle direction and braced frames are typically used in the structural system in the cross-aisle direction. Photographs of typical down-aisle moment frame connections, cross-aisle braced frame connections, and column base plate connections are presented in Fig. 1.33 through Fig. 1.35.

This work was conceived during the development of the SEISRACKS Project, as the author, "team-leader" for the "Politecnico di Milano" unit, was also acting as "Scientific Co-ordinator" on behalf of ACAI.

The results of WP 4 ("Cyclic tests of rack components"), of WP2 ("Full scale, Pseudo-dynamic and push-over tests of storage racks") as well as those of sliding tests and full scale shaking table tests carried out within WP1 ("Full scale dynamic tests of storage racks) will be presented.

Hereafter, in Chap. 2 the results of the cyclic tests on full scale rack components (namely beam-to-upright connections and column bases) carried out at the Department of Civil Engineering and Architecture of Instituto Superior Tecnico of

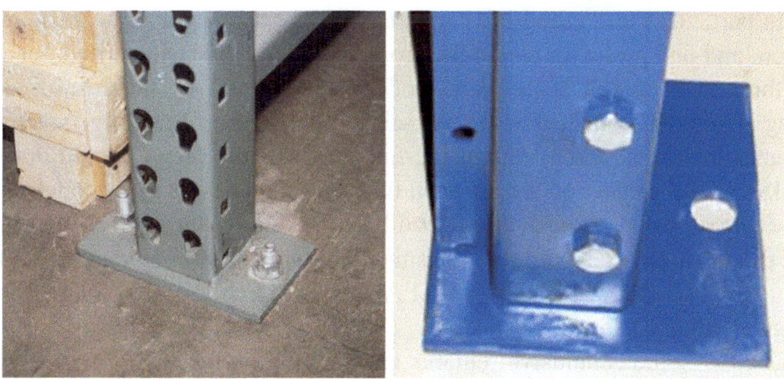

Fig. 1.35 Typical column base plate connections

Lisbon (with the cooperation of the author) are presented and discussed. In particular, an innovative cyclic testing procedure for structural elements and components will be presented and discussed, alternative to the classic ones recommended by ECCS (1986) and ATC (1992). Results of 30 tests carried out on two different types of beams, connected to the same type of upright will be presented and discussed. Furthermore, results of 40 tests on column bases will be presented. Tests were carried out on the same type of column base, but different loading directions and axial load in the column, as well as different type of steel base-to-foundation connections were considered (concrete foundation or steel foundation).

Chapter 3 deals with the assessment of the friction factor between pallets and rack beams, which is governing the "pallet sliding" phenomenon. This turned out to be a very important effect governing the dynamic behavior of racking systems.

About 1260 "static" sliding tests were carried out, for the assessment of the "static" pallet-beam friction factor. Six different types of beams, with different types of surface finishes (produced by different manufactures, from 3 different European countries) were considered in combination with different types of pallet characteristics (wooden, plastic, Europallets, American pallets, new, old, dry, wet).

Re-analysis of the results allowed the assessment of the "static" friction factor for the different combinations. Furthermore, nearly 200 dynamic tests were carried out (under the co-ordination of the author) on the shaking table of the National Technical University of Athens, in order to assess the "dynamic" pallet-beam friction factor. Tests were carried out both in the cross-aisle and in the down-aisle direction. The response in the cross-aisle direction was influenced by the out of plane bending and torsional response of the beam.

Test results confirm that "sliding" is, under severe dynamic conditions, the main factor influencing the rack response. Hysteresis loops were obtained showing both the presence of energy dissipation through sliding as well as the existence of an upper limit to the acceleration at the center-of-gravity of the masses (i.e. on the horizontal actions applied to the structure).

Chapters 4 and 5 deal respectively with the full scale pushover and pseudo-dynamic tests that were carried out (under the coordination of the author) at the European Laboratory for Structural Assessment of the Joint Research Center of Ispra. In particular, one pseudo-dynamic test and a pushover test in the down-aisle direction, and one pushover test in the cross-aisle direction were carried out on two bays, three storeys full scale rack models. Re-analysis of the results allowed drawing significant conclusions on the seismic behaviour of racking systems.

Chapter 6 presents the results of the full scale dynamic tests carried out within the SEISRACKS project on six different two-bays three-levels rack models, as well as those carried out previously, on four different two-bays three-levels rack models within the research project on the "Seismic behaviour of steel storage pallet racks" sponsored by the EU within the ECOLEADER Research Program for Free Access to Large Testing Facilities.

In Chap. 7 some conclusive remarks are finally presented.

References

A.S. 4084. (1993). *Steel Storage Racking*, Australian Standards.

Agatino, M. R., Bernuzzi, C., & Castiglioni, C. A. (2001). Joints under cyclic reversal loading in steel storage pallet racks. In *Proceedings of XVIII Conference C.T.A.*, Venezia, September 2001, pp. 105–114.

ATC-3. (1978). *Tentative provisions for the development of seismic regulations for buildings*. Redwood City, CA: Applied Technology Council.

ATC-24. (1992). *Guidelines for cyclic testing of components of steel structures*. Redwood City, CA: Applied Technology Council.

Baldassino, N., & Bernuzzi, C. (2000). Analysis and behaviour of steel storage pallet racks. *Thin-Walled Structures, 37*(4), 277–304.

Baldassino, N., & Zandonini, R. (2001). Numerical and experimental analysis of base-plate connections of steel storage pallet racks. In *Proceedings of XVIII conference C.T.A.*, Venezia, pp. 127–136.

Bernuzzi, C., & Castiglioni, C. A. (2001). Experimental analysis on the cyclic behaviour of beam-to-column joints in steel storage pallet racks. *Thin-Walled Structures, 39*, 841–859.

Castiglioni, C. A., Panzeri, N., Brescianini, J. C., & Carydis, P. (2003). Shaking table tests on steel pallet racks. In *Proceedings of the conference on behaviour of steel structures in seismic areas-stessa 2003*, Naples, Italy, pp. 775–781.

Castiglioni, C. A. (2003). Dynamic tests on steel pallet racks. *Costruzioni Metalliche, 55*(3), 35–44.

Chen, C. K., Scholl, R. E., & Blume, J. A. (1980a). *Seismic study of industrial storage racks, report prepared for the national science foundation and for the rack manufacturers institute and automated storage and retrieval systems (sections of the material handling institute)*. San Francisco, CA: John A. Blume & Associates.

Chen, C. K., Scholl, R. E., & Blume J. A. (1980b). Earthquake simulation tests of industrial steel storage racks. In *Proceedings of the seventh world conference on earthquake engineering*, Istanbul, Turkey, pp. 379–386.

Chen, C. K., Scholl, R. E., & Blume, J. A. (1981). Seismic-resistant design of industrial storage racks In *Proceedings of the second specialty conference on dynamic response of structures: experimentation, observation and control,* Atlanta, GA, pp. 745–759.

Cheng, P. H. (1973). Effect of semi-rigid connection to steel column of cold-formed perforated singly symmetric section. In *Proceedings of the second specialty conference on cold formed steel structures*, St Louis, Missouri, pp. 641–670.

ECCS. (1986). *Recommended testing procedure for assessing the behaviour of structural elements under cyclic loads*. European Convention for Constructional Steelworks, Publication No. 45.

FEM 10.2.02. (2001a). *The design of static steel pallet racks*. Federation Europeen de la Manutention, Vers. 1.02.

FEM 10.2.04. (2001b). *Guidelines fot the safe use of static steel racking and shelving*. Federation Europeen de la Manutention, Vers. 1.02.

FEM 10.2.08. (2005). *The seismic design of static steel pallet racks*. Federation Europeen de la Manutention, final draft, December 2005.

FEMA. (2005). *Seismic considerations for steel storage racks located in areas accessible to the public, FEMA 460,* September 2005.

Filiatrault, A. (2001). Shake table tests of Storage Racks and Contents. *Presentation material at the April 12*, 2001 Seismic Safety Commission Hearing on Industrial Storage Racks, San Francisco, CA (available from author).

Filiatrault, A., & Wanitkorkul, A. (2004). *Shake-table testing of frazier industrial storage racks*, Report No. CSEE-SEESL-2005-02, Structural Engineering and Earthquake Simulation Laboratory, Departmental of Civil Structural and Environmental Engineering, SUNY Buffalo.

Higgins, P. (2004). *Personal Communication*. Peter Higgings and Associates, Malibu, CA.

John A. Blume & Associates. (1973). *Seismic investigation of steel industrial storage racks*. San Francisco, CA: Report prepared for the Rack Manufacturer's Institute.

John A. Blume & Associates. (1974). *Seismic design examples of industrial storage racks*. San Francisco, CA: Report prepared for the Rack Manufacturer's Institute.

John A. Blume & Associates. (1987). *Application of eccentric bracing to rack structures*. San Francisco, CA: Report prepared for the Rack Manufacturer's Institute.

Kotha, R., & Peköz, T. (2000). *Cold-formed steel frame and beam-column design*. Ithaca: School of Civil and Environmental Engineering, Cornell University.

Krawinkler, H., Cofie, N. G., Astiz, M. A., & Kircher, C. A. (1979). *Experimental study on the seismic behavior of industrial storage racks*. Report No. 41, The John A. Blume Earthquake Engineering Center, Department of Civil Engineering, Stanford University, Stanford, CA.

RAL. (1990). *Storage and associated equipment,* Deutsches Institut für Gutersichcherung und Kennzeichnung (German Institute for Quality Assurance and Marketing).

RMI. (2002a). *Specification for the design testing and utilization of industrial steel storage racks*. Charlotte, NC: Rack Manufacturers Institute, 2002 edition.

RMI. (2002b). *Commentary to specification for the design testing and utilization of industrial steel storage racks*. Charlotte NC: Rack Manufacturers Institute.

Roeder, C. W., & Popov, E. P. (1978). Eccentrically braced steel frames for earthquakes. *Journal of the Structural Division, ASCE, 104*(3), 391–412.

Sarawit, A. T., & Peköz, T. (2003). *Cold-formed steel frame and beam-column design, Report 03-03*. Ithaca, NY: School of Civil and Environmental Engineering, Cornell University.

Further Readings

AISC 341. (2002). *Seismic provisions for structural steel buildings*. Chicago, Illinois: American Iron and Steel Institute, 2002 edition.

AISC 341. (2005). *Seismic provisions for structural steel buildings*. Chicago, Illinois: American Iron and Steel Institute, 2005 edition.

AISI. (1996). *Cold-formed steel design manual*. Washington DC: American Iron and Steel Institute, 1996 edition.

AISI. (2002). *Cold-formed steel design manual*. Washington DC: American Iron and Steel Institute, 2002 edition.

ANSI MH16.2-1984. (1996). *American national standard for the use of industrial and commercial steel storage racks: Manual of safety practices/a code of safety practices*. American National Standards Institute Inc., 10 p.

ASCE-7-2002. (2002). *Standard-minimum design loads for buildings and other structures*. Reston, VA: American Society of Civil Engineers, 2002 edition.

ASCE-7-2005. (2005). *Standard-minimum design loads for buildings and other structures*. Reston, VA: American Society of Civil Engineers, 2005 edition.

ATC-20. (1989). Procedures for the post earthquake safety evaluation of buildings. Redwood City, CA: Applied Technology Council.

Aydin, K., & Tung, C. C. (2001). Energy balance equation for estimating overturning potential of an unanchored rocking body subject to earthquake excitation. *Earthquake Spectra, 17* (2), 209–220.

Baldassino, N., Bernuzzi, C., & Zandonini, R. (1999). Structural analysis of steel storage pallet racks. In *Proceedings of the 6th international colloquium on stability and ductility of steel structures (SDSS'99),* Timisoara, Romania, pp. 465–475.

Baldassino, N., Berardi, C., & Bernuzzi, C. (2001). Influence of the bracing systems on the performance of steel storage pallet racks. In *Proceedings of XVIII conference C.T.A*. Venezia, pp. 115–125.

Ballio, G., Bernuzzi, C., & Castiglioni, C. A. (2000). An approach for the seismic design of steel storage pallet racks. *Stahlbau, 68*(11), 919–928.

Brescianini, J. C., Castiglioni, C. A., & Panzeri, N. (2003). Dynamic experimental tests on steel pallet racks. In *Proceedings of XIX CTA conference*, Genova, Italy, pp. 107–116.

Brown, B. J. (1983). Seismic design of pallet racking systems. *Bulletin of the New Zealand National Society for Earthquake Engineering, 16*(4), 291–305.

CEN/TC 344 N 57. (2006). Adoption of WI 00344006 'steel static storage systems—racking and shelving systems—terms and definitions', Draft resolution a/2006.

Chan, D. H., & Raymond, K. Y. (2005). *Structural behaviour of storage racks under seismic ground motion*. Washington D.C.: FEMA.

Choi, B., & Tung, C. C. (2002). Estimating sliding displacement of and unanchored body subjected to earthquake excitation. *Earthquake Spectra, 18*(4), 601–613.

Davies, J. M. (1992). Down-aisle stability of rack structures. In *Eleventh international specialty conference on cold-formed steel structures*, St-Louis, Missouri, pp. 417–434.

DIN-EN 13199-1. (2000a). *Packaging. Small load carrier systems. Part 1: Common requirements and test methods.*

DIN-EN 13199-2. (2000b). *Packaging. Small load carrier systems. Part 2: Column stackable system (CSS).*

DIN-EN 13199-3. (2000c). *Packaging. Small load carrier systems. Part 3: Bond stackable system (BSS).*

DIN-EN 13626. (2003). *Packaging. Box pallets. Common requirements and test methods.*

DIN-EN 13698-1. (2004). *Pallet production specification. Part 1: Construction specification for 800 mm × 1200 mm flat wooden.*

DIN-EN 13698-2. (2004). *Pallet production specification. Part 2: Construction specification for 1000 mm × 1200 mm flat wooden.*

DIN-Pr EN 13382–1. (1998). *Flat pallets for materials handling. Principal dimensions.*

Filiatrault, A., Bachman, R. E., & Mahoney, M. G. (2006). Performance-based seismic design of pallet-type steel storage racks. *Earthquake Spectra, 22*(1), 47–64.

Gallagher, R. P., & Steinberg, J. (2003). *Seismic retrofit of storage racks and shelving at warehouse for high-value equipment and parts, ATC-29-2 seminar on seismic design retrofit and performance of non-structural components* (pp. 227–241). Redwood City, CA: Applied Technology Council.

Godley, M. H. R. (2002). The behaviour of drive-in storage structures. In *Sixteenth international specialty conference on cold-formed steel structures*, Orlando, Fl, pp. 340–352.

ISO 445. (1996). *Pallets for materials handling. Vocabulary.*

Lewis, G. M. (1991). Stability of rack structures. *Thin-Walled Structures, 12*, 163–174.

Mahoney, M., & Lusk, J. (2004). *FEMA mitigation investigation and opportunities report for the san simeon earthquake*, Federal Emergency Management Agency.

Pekoz, T., & Winter, G. (1973). Cold-formed steel rack structures. In *Second specialty conference on cold formed steel structures,* St. Louis, Missouri, pp. 603–615.

Piepiekarz, M., Koza, H., Ellis, M., & Boardman, P. (1995). Seismic performance of commercial building support equipment. In *Pacific conference on earthquake engineering*, Melbourne, Australia, pp. 187–196.

Rihal, S. S., & Gates, W. E. (1998). *Performance and behavior of library shelving and storage rack systems during the 1994 Northridge earthquake, ATC-29-1 seminar on seismic design retrofit and performance of non-structural components* (pp. 121–135). Redwood City, CA: Applied Technology Council.

RMI. (1990). *Specification for the design testing and utilization of industrial steel storage racks*. Charlotte, NC: Rack Manufacturers Institute, 1990 edition.

RMI. (2004). *Testing guidelines for pallet stacking frames*. Charlotte, NC: Rack Manufacturing Institute.

Salmon, M. A., Welch, R. E., & Longinow, A. M. (1973). Analysis of drive-in and drive-thru storage racks. In *Second specialty conference on cold formed steel structures*, St. Louis, Missouri, pp. 617–639.

SEAOC. (1999). *Recommended lateral force requirements and commentary* (7th ed.). Sacramento, CA: Seismology Committee of the Structural Engineers Association of California.

Shao, Y., & Tung, C. C. (1999). Seismic response of unanchored bodies. *Earthquake Spectra, 15* (3), 523–536.

Shenton, I. I. I. (1996). Criteria for initiation of slide rock and slide-rock rigid-body modes. *Journal of Engineering Mechanics, ASCE, 122*(7), 690–693.
Tung, C. C. (2005). Criteria for initiation of motion of rigid bodies to base excitation. *Earthquake Engineering and Structural Dynamics, 34*(10), 1343–1350.
Yiu, F., & Pekoz, T. (2001). *Design of cold-formed steel plain channels progress report: Project No CF9501*. Ithaca, NY: School of Engineering, Cornell University.

Chapter 2
Component Tests

2.1 Overview

Component tests were performed with the aim of characterizing the behaviour of both beam-to-upright and base connections, in order to allow a correct interpretation of full scale tests as well as in order to calibrate numerical models. It has to be noticed that the behaviour of both components is strongly influenced by the following factors:

- Nature and geometry of the profiles (unsymmetrical cross section of the upright, thin walled sections of both beams and uprights, see Fig. 2.1)
- Asymmetry of the connections. In the case of the beam-to-upright connection, the asymmetry is caused by the inclined hooks and by the presence of the safety-bolt on the upper side of the beam end-plate connector only (Fig. 2.2a). In the case of the base connections, the asymmetry is due to the eccentric position of the upright on the base-plate, and by the asymmetrical disposition of the bolts (Fig. 2.2b).

Special attention was paid to component and connections testing procedures presented in Chap. 5 ("Tests") of FEM 10.2.02 (2001): *The Design of Static Steel Pallet Racks*, namely:

- 5.5—Bending tests on beam end connections
- 5.8—Tests on floor connections

Based on such specifications the following tests were performed at the Laboratory for Testing Materials of the Civil Engineering and Architecture Department of Instituto Superior Tecnico of Lisbon, by the team leaded by prof. L. Calado, and in co-ordination with this author, who attended the tests in two occasions, in November 2005 and in January 2006.

C.A. Castiglioni, *Seismic Behavior of Steel Storage Pallet Racking Systems*,
Research for Development, DOI 10.1007/978-3-319-28466-8_2

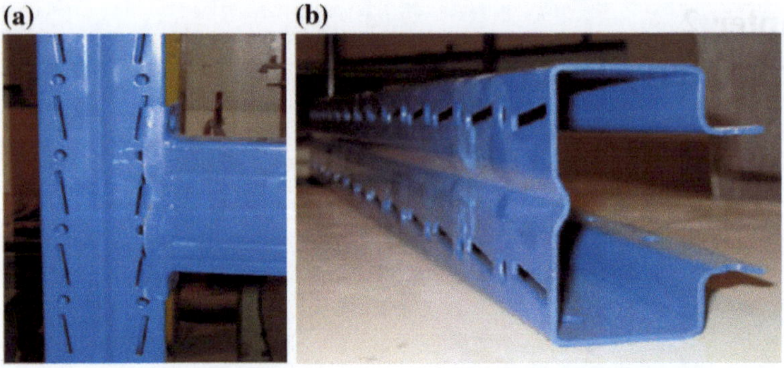

Fig. 2.1 Specimens for beam-to-upright connection tests: **a** connection, **b** upright (100/20b)

Fig. 2.2 Asymmetries in the connections: **a** beam-to-upright, **b** bases

2.2 Beam-to-Upright Connections

Proprietary moment connections are typically used as beam-to-upright connections for steel selective pallet storage racks. This study focused on the beam-to-upright connection type shown in Fig. 2.3. A "hooked" end plate connector is welded to the beam at both ends. Connection is attained by introducing the hooks in the openings (punched during fabrication) on the uprights, and by adding a safety bolt connecting the upper part of the extended end-plate to the upright.

As a general remark, it should be noticed that this proprietary beam-to-upright connection is strongly non-symmetric in both vertical and horizontal planes.

In the vertical plane, non-symmetry is due to the presence of the safety bolt on the upper side of the beam only (Fig. 2.3) and by the fact that the beam is fillet welded to the end-plate on three sides only, leaving the lower flange un-welded.

In the horizontal plane, non-symmetry is due to the shape of the end-plate connector that has hooks on one side only, and is obtained by cold forming of a thin plate, bent in shape of an L, so with a stiffened edge (the same edge where hooks are present).

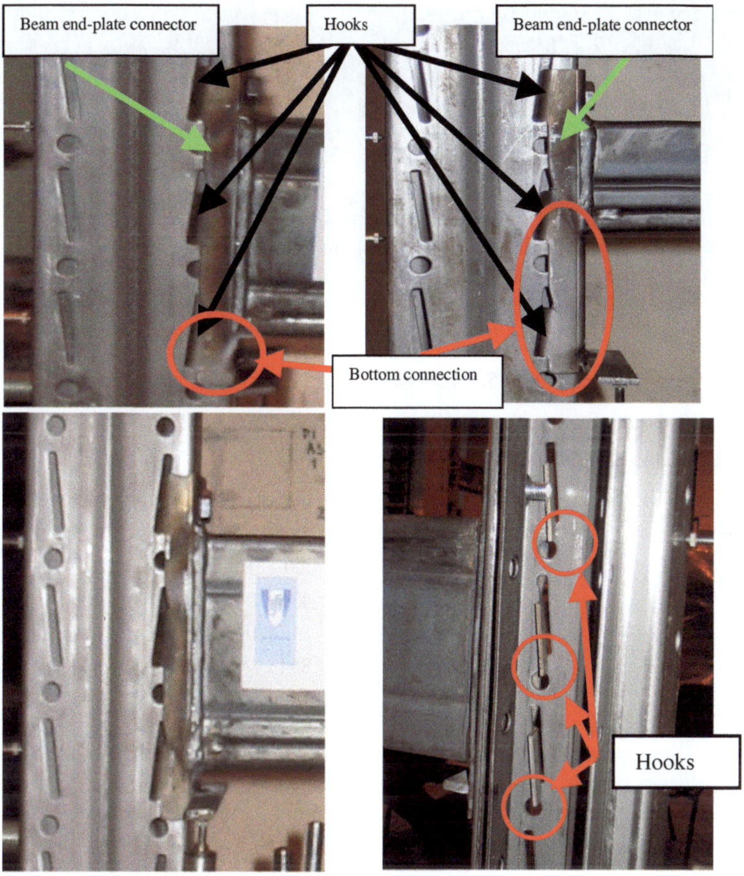

Fig. 2.3 Typical beam-to-upright connection considered in the study

A non-symmetric response is hence to be expected under hogging and sagging bending.

Objectives of the tests were:

- Assessment of the moment-rotation curves.
- Assessment of the collapse modes of these connections under monotonic and cyclic loads.

Two different cross-section for beams (TG 70 × 45 × 1.5 and TG 130 × 45 × 1.5 mm) were adopted (Fig. 2.4), with upright of identical cross-section (100/20b) shown in Fig. 2.1b. Consequently, beam height varied between 70 and 130 mm. The member geometrical properties are shown in Table 2.1.

The material used for beams and upright is S275 steel, with actual values of yield and ultimate stress showed in Table 2.2.

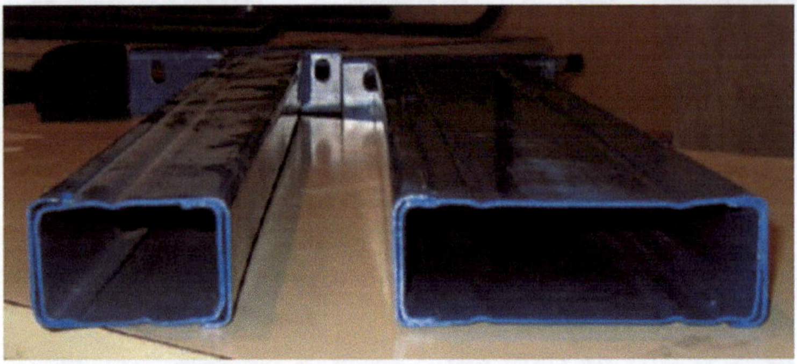

Fig. 2.4 Types of beams (70 × 45 × 1.5 and 130 × 45 × 1.5)

Table 2.1 Geometrical properties of the members

Member	Properties	Gross section	Net section
100/20b	A [mm^2]	588.8	525.7
	t [mm]	2.0	2.0
	J_x [mm^4]	436020	406100
	J_y [mm^4]	812280	694680
	W_x [mm^3]	8543	8330
	W_y [mm^3]	16747	14323
TG 70 × 45 × 1.5 mm	A [mm^2]	514.06	514.06
	t [mm]	1.5	1.5
	J_x [mm^4]	407500	407500
	J_y [mm^4]	136272	136272
	W_x [mm^3]	11754	11754
	W_y [mm^3]	4764	4764
TG 130 × 45 × 1.5 mm	A [mm^2]	697.66	697.66
	t [mm]	1.5	1.5
	J_x [mm^4]	1742220	1742220
	J_y [mm^4]	168662	168662
	W_x [mm^3]	26792	26792
	W_y [mm^3]	5171	5171

Table 2.2 Material characteristics

Member	f_y [MPa]	f_u [MPa]	ε_u [%]
Beam 70 × 45 × 1.5	353.6	446.1	26.90
Beam 130 × 45 × 1.5	357.0	458.0	27.00
Upright 100 × 82 × 2.0	348.0	493.0	25.50
Beam end connector	263.0	385.0	41.10

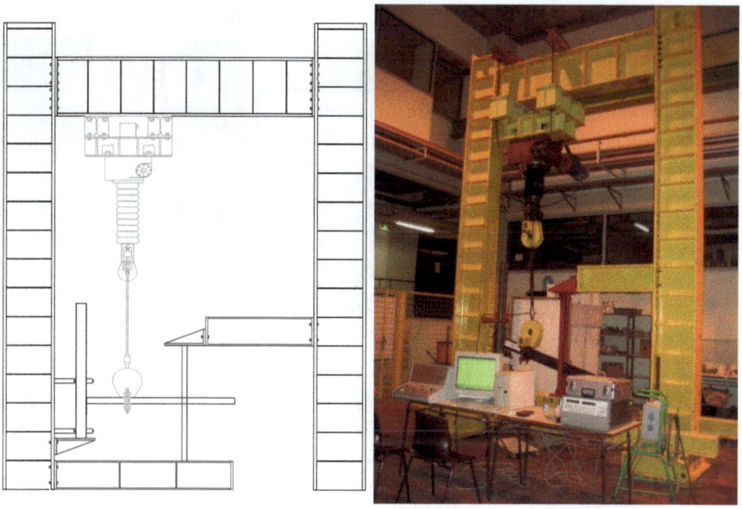

Fig. 2.5 Test set up

The test set-up and instrumentation are shown in the following Figs. 2.5 and 2.6. Specimen size, configuration and instrumentation (Figs. 2.6 and 2.7) were adopted according to FEM 10.2.02 (2001) Recommendations.

The moment-rotation curves were plotted for each test. The bending moment was defined as ($M = F * a$) and the rotation of the connection by Eq. (2.1).

$$\phi = \frac{V}{a} - \left(\frac{\delta_1 - \delta_2}{d}\right) \tag{2.1}$$

where (see Fig. 2.7):

V displacement due to load F;
a lever arm for the load F;
δ_1 deflection measured by transducer T1;
δ_2 deflection measured by transducer T2;
d beam height.

Table 2.3 summarizes and identifies the 30 tests carried out on beam-to-upright specimens. In particular, monotonic tests were performed under both hogging (MB) and sagging (MT) bending moments.

As the Standard cyclic testing procedure proposed by ECCS (1986), in particular in the case of unsimmetric behaviour (as in the case under exam), leads to a correct evaluation of the cyclic behaviour of components only for the condition of unloaded structure (vertical load F = 0), an innovative cyclic testing procedure has been identified and applied.

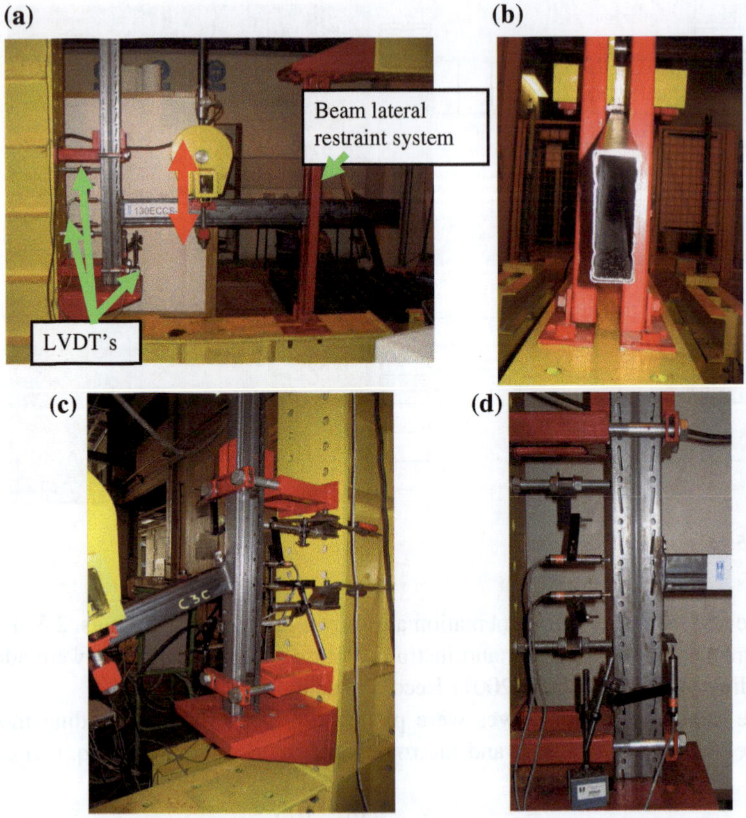

Fig. 2.6 Test set up, **a** test sut-up, **b** lateral restraint system, **c**, **d** instrumentation

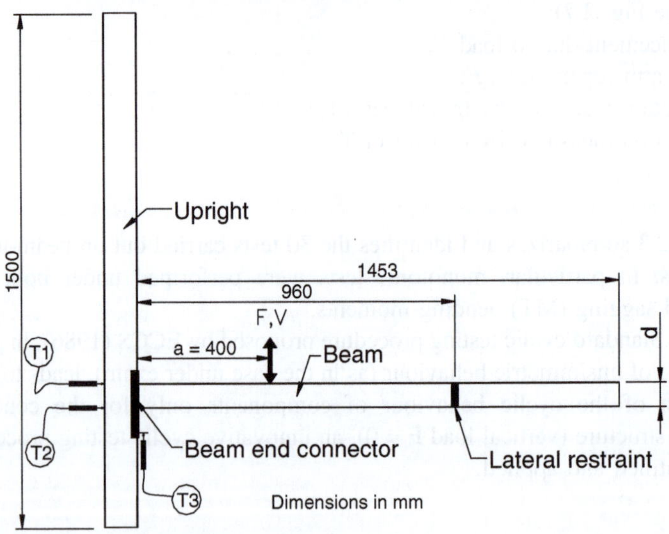

Fig. 2.7 Specimen dimensions and instrumentation

Table 2.3 Tests on beam-to-upright connections

TEST	Beam width—70 mm	Beam width—130 mm
Monotonic—below (hogging bending)	70MB-1 70MB-3 70MB-4	130MB-2 130MB-3 130MB-4
Monotonic—top (sagging bending)	70MT-3 70MT-5 70MT-6	130MT-2 130MT-3 130MT-5
Cyclic—ECCS	70ECCS-1 70ECCS-2	130ECCS-2 130ECCS-3
Cyclic—Fy25	70Fy25-3 70Fy25-4	130Fy25-1 130Fy25-2
Cyclic—Fy50	70Fy50-1 70Fy50-2	130Fy50-1 130Fy50-2
Cyclic—Fy66	70Fy66-1 70Fy66-2	130Fy66-1 130Fy66-2
Cyclic—Fy75	70Fy75-1	130Fy75-1

Hence, cyclic tests were carried out under different levels of vertical load (namely 0, 25, 50, 66 and 75 of the yield load F_y of the connection) in order to simulate the presence of a service gravity load on the beam.

The yield strength F_y, as well as other relevant parameters, such as the yield displacement v_y, the ultimate strength F_u, as well as the ultimate displacement v_u, and the initial elastic stiffness $S_{j,ini}$, can be conventionally defined according to ECCS (1986), with reference to the results of a monotonic test (Fig. 2.8).

2.2.1 Monotonic Tests

2.2.1.1 Beam 70 × 45 × 1.5

The monotonic tests listed in Table 2.4 were carried out on beam-to-upright connections specimens with beam 70 × 45 × 1.5 mm.

Table 2.4 reports also the values of the yield (M_y) and ultimate bending moment (M_u), of the yield (ϕ_y) and ultimate rotation (ϕ_u) and of the initial elastic stiffness ($S_{j,ini}$) measured for each test, as well as the mean values of the same parameters.

Figure 2.9 compares the moment-rotation curves for both hogging (MB) and sagging (MT) bending, in the case of beam 70 × 45 × 1.5. It can be noticed that the specimens under hogging bending show a slightly larger ultimate strength than the specimens under sagging bending moments, but about the same yield strength. The ductile behaviour is quite different: while the initial elastic stiffness is similar, the average ultimate rotation of the specimens under hogging bending is approximately 2.3 times larger than that of the specimens under sagging bending.

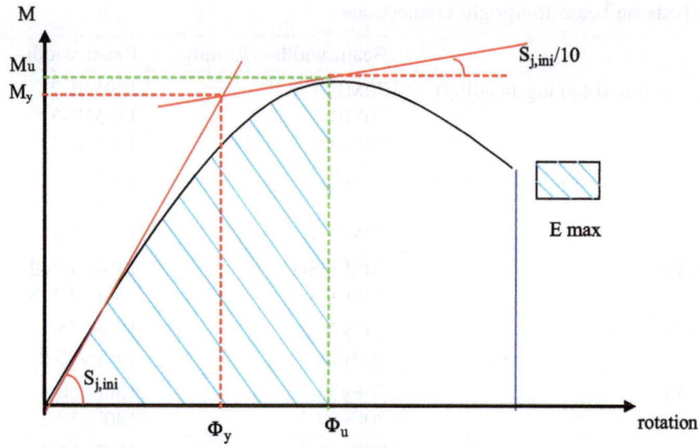

Fig. 2.8 Conventional definitions according to ECCS (1986)

Table 2.4 Monotonic tests on beam-to-upright connections with 70 × 45 × 1.5 mm beam

Loading	Specimen	M_y [kNm]	ϕ_y [mrad]	$S_{j,ini}$ [kNm/rad]	M_u [kNm]	ϕ_u [mrad]
Hogging bending	70MB-1	1.30	34.7	42.7	1.58	143
	70MB-3	1.22	26.0	48.7	1.68	191
	70MB-4	1.18	22.7	52.4	1.53	202
	Average	**1.23**	**27.8**	**47.9**	**1.60**	**178.7**
Sagging bending	70MT-3	1.27	27.3	46.0	1.44	86
	70MT-5	1.20	24.0	49.4	1.37	76
	70MT-6	1.26	26.7	51.3	1.44	71
	Average	**1.24**	**26.0**	**48.9**	**1.42**	**77.7**

Fig. 2.9 Comparison of the moment-rotation curves for beam 70 × 45 × 1.5

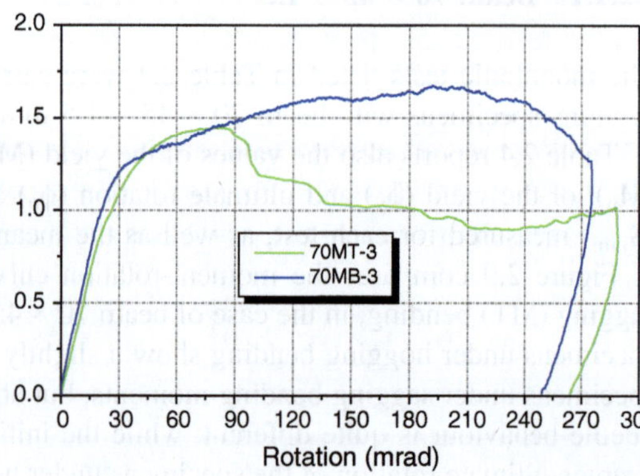

This large difference can be explained (Figs. 2.10 and 2.11) by the presence of the extended end-plate connector (which has a standard size, and is the same for 70 mm as well as for 130 mm deep beams), by its unsymmetrical connection to the beam and by the presence of the safety bolt on the upper side of the beam only. All these factors cause an unsymmetrical behaviour of the connection for hogging and sagging bending moments. The failure mode of MB specimens, subjected to hogging bending, consisted of large deformations in the top zone of the beam end connector, leading to loosening of the hooked connection, with the hooks exiting from the holes in the upright.

The failure mode of MT specimens subjected to sagging bending was the fracture of the beam close to the weld between the beam and the end connector. Typical failure modes observed during the monotonic tests are presented in Figs. 2.10 and 2.11. For all tests the connection behaviour was not influenced by bolt deformation.

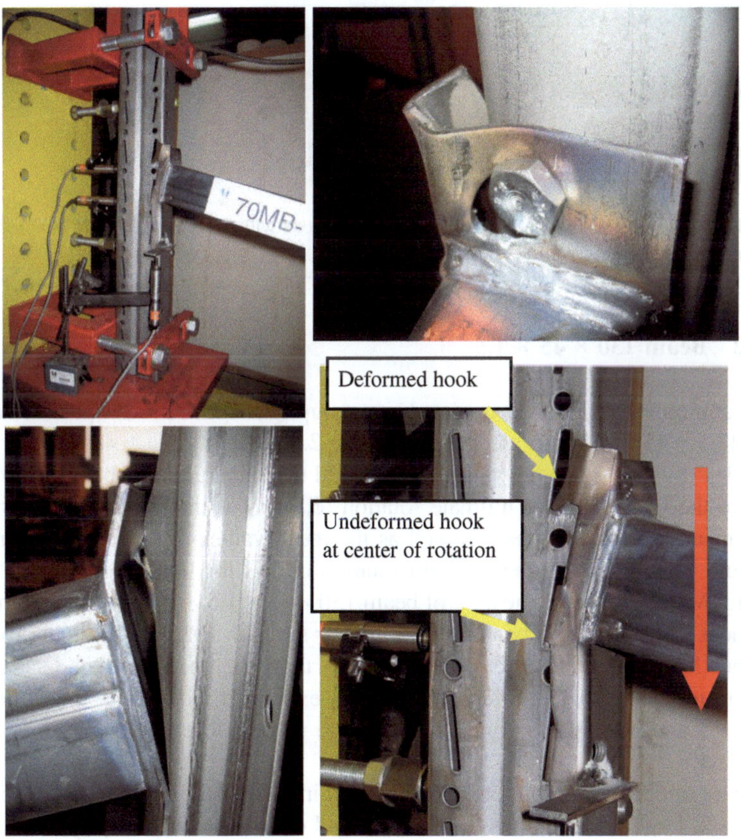

Fig. 2.10 Monotonic tests on beam-to-upright connections with 70 × 45 × 1.5 mm beam. Typical collapse mode under hogging bending moments

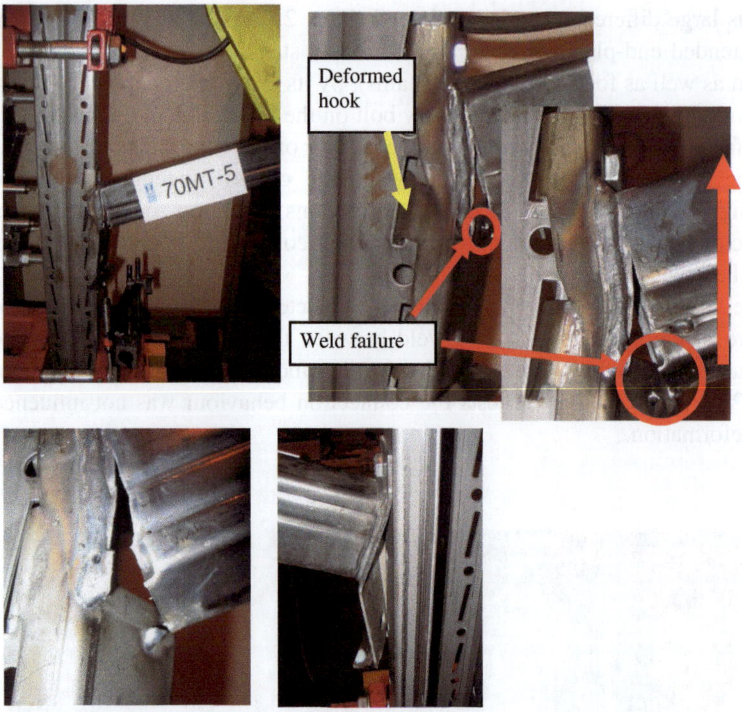

Fig. 2.11 Monotonic tests on beam-to-upright connections with 70 × 45 × 1.5 mm beam. Typical collapse mode under sagging bending moments

2.2.1.2 Beam 130 × 45 × 1.5

The monotonic tests carried out on beam-to-upright connections specimens with beam 130 × 45 × 1.5 mm are listed in Table 2.5.

Table 2.5 reports also the values of the yield (M_y) and ultimate bending moment (M_u), of the yield (ϕ_y) and ultimate rotation (ϕ_u) and of the initial elastic stiffness ($S_{j,ini}$) measured for each test, as well as the mean values of these parameters.

Figure 2.12 compares the moment-rotation curves for both hogging (MB) and sagging (MT) bending, in the case of beam 130 × 45 × 1.5. It can be noticed that the specimens under hogging bending show larger strength (M_u) and ultimate rotation (ϕ_u), but lower initial elastic stiffness ($S_{j,ini}$) than the specimens under sagging bending moments. This difference can be explained observing the following Figs. 2.13 and 2.14, presenting typical failure modes observed during the monotonic tests.

Failure of both specimens subjected to hogging (MB) and sagging (MT) bending was due to large deformations in the beam end connector, respectively resulting in the top and bottom hooks coming out of the openings in the upright.

Table 2.5 Monotonic tests on 130 × 45 × 1.5 mm beam specimens

Loading	Specimen	M_y [kNm]	ϕ_y [mrad]	$S_{j,ini}$ [kNm/rad]	M_u [kNm]	ϕ_u [mrad]
Hogging bending	130MB-2	2.52	23.0	101.9	3.18	122
	130MB-3	2.59	27.9	94.6	3.19	127
	130MB-4	2.49	27.1	90.9	3.05	140
	Average	**2.53**	**26.0**	**95.8**	**3.14**	**129.7**
Sagging bending	130MT-2	1.99	18.2	109.7	2.26	111
	130MT-3	2.22	18.6	122.2	2.63	97
	130MT-5	2.11	19.7	125.0	2.46	114
	Average	**2.11**	**18.8**	**119.0**	**2.44**	**107.3**

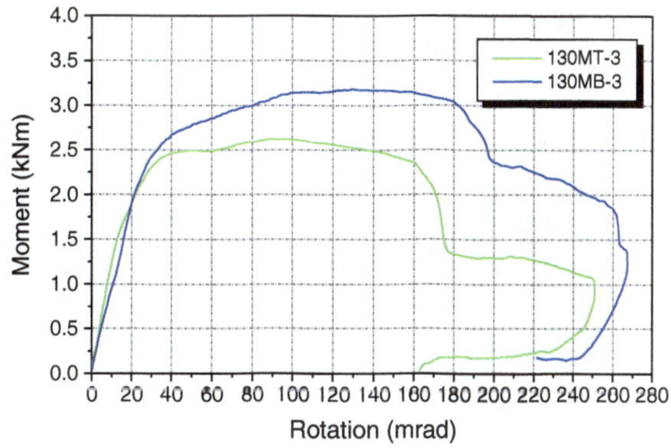

Fig. 2.12 Comparison of the moment-rotation curves for beam 130 × 45 × 1.5

Under hogging bending (Fig. 2.13) the safety bolt is in tension. Until it is completely punched through the hole in the end-plate connector, the presence of the bolt restrains the deformation of the upper part of the end-plate. Hence, only the upper hook can come out of the opening. On the contrary, under sagging bending, the collapse mechanism is governed by the failure of the hooks in the lower part of the end-plate connector. The openings on the uprights are inclined. This causes a different hook-opening interaction, in the case of hogging bending and of sagging bending. In this last case, in fact, due to a larger deformation sustained by the hooks, failure is attained with evident cracks in the hooks (Fig. 2.15).

2.2.2 Cyclic Tests

2.2.2.1 Cyclic Loading History

The seismic behaviour of beam-to-upright connections of racking systems is similar to that of more common beam-to-column connections of conventional reinforced

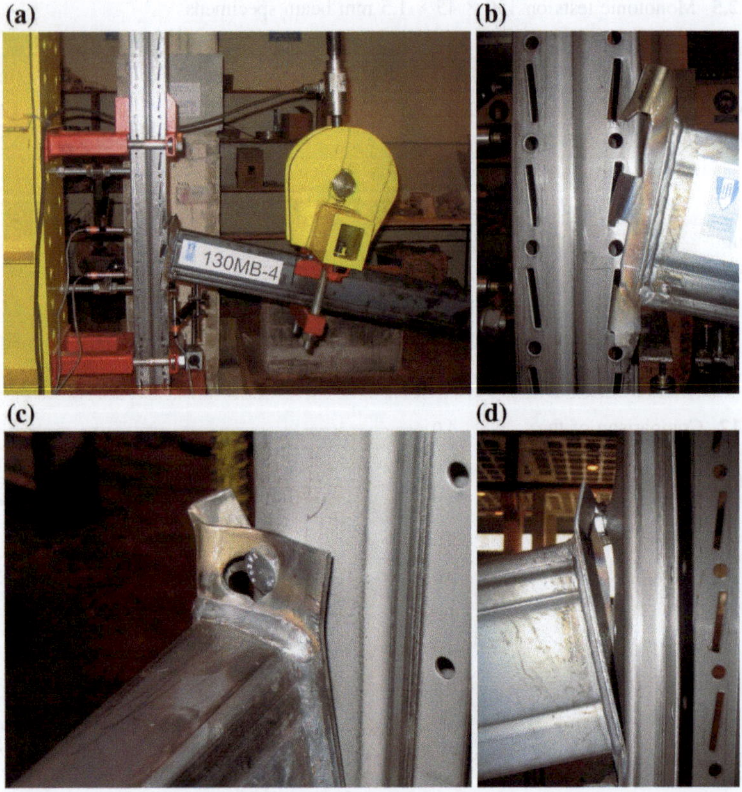

Fig. 2.13 Monotonic tests on beam-to-upright connections with 130 × 45 × 1.5 mm beam. Typical collapse mode under hogging bending moments

concrete or steel buildings. These more common structural sub-assemblages are generally tested with reference to recommendations available in the literature, such as those produced by ECCS (1986), ACI (1999) and ATC (1992). All these recommendations share the same basic characteristics listed hereafter:

Tests are displacement-controlled;

Specimens are subjected to a symmetrical reversed cyclic displacement history;

Displacement history is expressed in terms of imposed displacement ductility, making reference to the yield displacement, generally evaluated through monotonic tests;

Loading cycles are repeated at every displacement amplitude in the post-elastic range;

Failure is conventionally defined.

The positive and negative cycle amplitudes are based upon the following series:

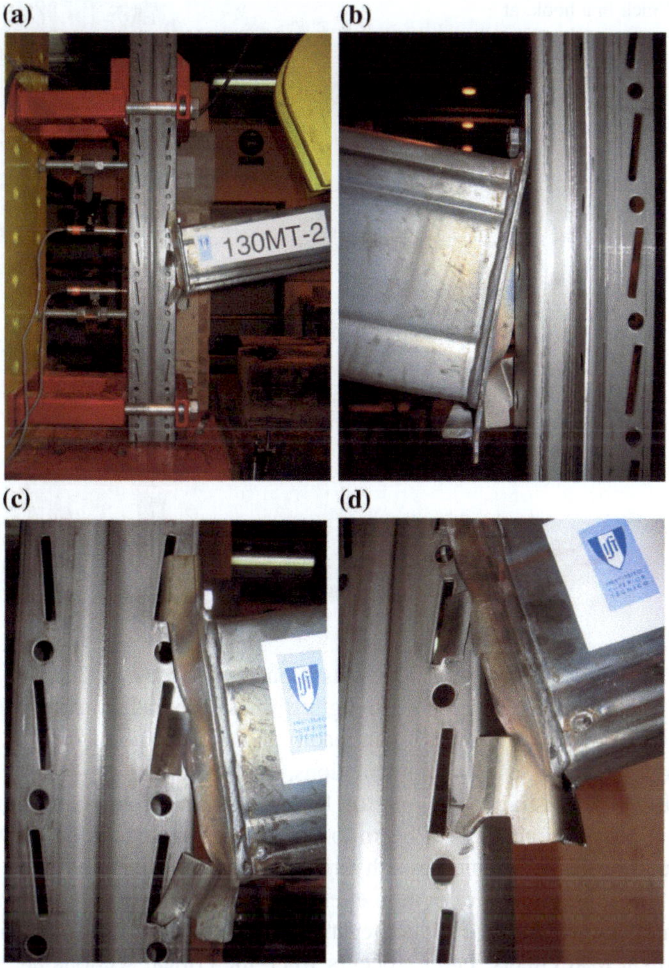

Fig. 2.14 Monotonic tests on beam-to-upright connections with 130 × 45 × 1.5 mm beam. Typical collapse mode under sagging bending moments

one cycle in the $d_y^+/4$, $d_y^-/4$ interval;
one cycle in the $2d_y^+/4$, $2d_y^-/4$ interval;
one cycle in the $3d_y^+/4$, $3d_y^-/4$ interval;
one cycle in the d_y^+, d_y^- interval;

two or three cycle in the $(2+n)d_y^+$, $(2+n)d_y^-$ interval with n = 0, 1, 2, 3, … until failure occurs.

These generally accepted recommendations served as a common ground for the comparison of different testing programmes (e.g. see Proença et al. 1994), as well as for the calibration of nonlinear numerical models and derivation of detailing rules

Fig. 2.15 Crack in a hook, at failure under sagging bending

for codes and other comparable documents. General performance indexes—such as available displacement ductility, strength/stiffness degradation and hysteretic energy-dissipation characteristics—can readily be inferred from cyclic tests carried out with reference to these recommendations.

In reality, the seismic behaviour of beam-to-upright connections of racking systems and, more in general, of beam-to-column connections in conventional r.c. or steel buildings is determined by a hybrid loading consisting of vertical (gravity) loads and horizontal (seismic) motion effects, and the beam-to-column connections bending moments are shifted in the negative, hogging, direction as a consequence of the vertical load effects. Failure occurs when the critical sections are no longer able to withstand vertical loads as a consequence of accumulated damage (induced by horizontal motion and vertical forces acting together).

However, general recommended testing procedures, encompassing only displacement controlled conditions, fail to address the unsymmetrical displacement histories experienced by critical beam-to-column connections when subjected to earthquake motion acting simultaneously with vertical (live and dead) loads.

2.2.2.1.1 Innovative Cyclic Testing Procedure

Considering the former limitations of the commonly accepted testing procedures (particularly evident when testing is performed in the inverted T configuration, with the beam standing vertically and the column horizontally), an innovative testing procedure was developed, which intends to capture the hybrid nature of loading imposed to beam-to-column connections when subjected to vertical and horizontal load effects.

This innovative testing procedure can be considered a development of the commonly accepted testing procedures as it inherits some of their characteristics, such as the cycle repetition in the post-elastic range and the fact that the controlled-displacement part of the testing cycles is indexed to the yield displacement (determined through monotonic tests).

Gravitational load effects are expressed through F_g which can in turn be expressed as a fraction of the force leading to yielding (F_y) at the critical cross section of the connection. Values of F_g corresponding to 25, 50, 66 and 75 % of the yield force (F_y) were considered in the beam-to-upright testing programme. F_y can be determined by a monotonic test.

The testing procedure is composed by single fully reversed cycles at displacement ductility of 1/4, 1/2, 3/4 and 1, followed by a sequence of groups of two (or three) cycles at multiples (2, 3, 4, etc.) of the yield displacement until a conventional failure criterion is met.

Each cycle has an initial force-controlled part and a final displacement-controlled part.

In what follows, the gravity force is considered to be positive. As a consequence of this conventional choice, the "positive" bending moments are those associated with "hogging" bending.

Figure 2.16 depicts typical positive and negative succeeding cycles.

A typical *positive* cycle is composed of the following two parts:

Application of the force correspondent to vertical (gravitational) load effects F_g on the beam-to-column connection (force-controlled part of the cycle);

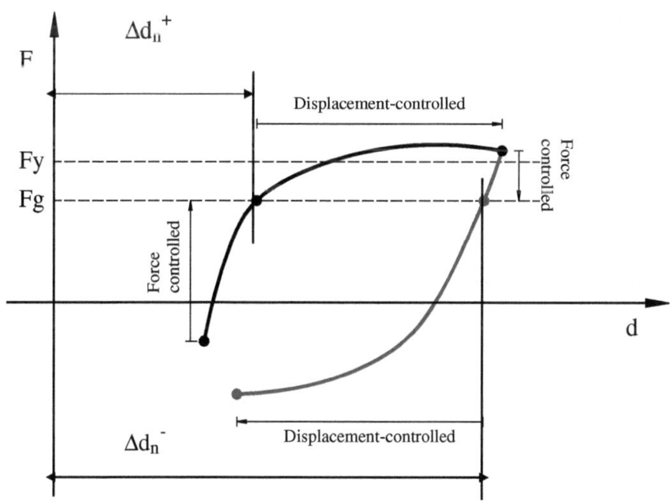

Fig. 2.16 Typical positive and subsequent negative cycles (positive forces induce hogging bending moments in the beam-to-column connection)

Starting from the displacement attained at the end of the force-controlled part of the cycle (Δd_n^+) the displacement controlled part of the cycle is imposed. The displacement amplitude is considered a multiple of the yield displacement.

Also the following *negative* cycle is composed of two different parts:

Force-controlled unloading until attainment of the force F_g associated with the presence of the vertical (gravity) loads alone;

Starting from the displacement (Δd_n^-) reached at the end of the force-controlled part of the cycle, the displacement-controlled part of the cycle is imposed to the specimen, until the intended displacement amplitude is reached.

The positive and negative cycles (in the post-elastic range) can be derived from the following equation:

$$\left((2+n)d_y^+ + \Delta d_n^+\right), \left((2+n)d_y^- + \Delta d_n^-\right) \quad (\mathrm{n} = 0, 1, 2, 3, \ldots.)$$

where:

Δd_n displacement amplitude value obtained when the force-controlled part of the cycle reaches the force correspondent to gravitational load (F_g).

Adopting this innovative cyclic testing procedure, failure can be identified in one of the following situations:

When, in the force-controlled part of the positive cycle, the specimen fails to develop the force correspondent to gravitational loads (Fig. 2.17a).

When, in the displacement-controlled part of the positive cycle, the restoring force decreases to values below those corresponding to gravitational loads (Fig. 2.17b).

Obviously, failure can also occur in any of the positive or negative cycles when cross-sectional collapse occurs.

2.2.2.1.2 Comparison Between Testing Procedures

A comparison between the results obtainable by means of the ECCS and of the "innovative" testing procedure is presented hereafter, in the case of tests on 130 × 45 × 1.5 mm beams, without compromising the generality of the conclusions.

ECCS-type cyclic tests results can be exemplified in Fig. 2.18, which shows the force-displacement chart for specimen 130ECCS-2. The force-displacement chart is almost symmetrical with some slight differences in the positive and negative strength, which result from unsymmetrical detailing of the beam-to-upright connection. Low values of stiffness during most of the duration of these tests (as a consequence of the connection remaining fully opened) led to pronounced pinching effects.

Application of the innovative cyclic testing procedure led to a displacement history that could not be predicted *a priori* as a consequence of the force-controlled part of the cycles. Loading histories applied in tests with different values of F_g are different, as depicted in Fig. 2.19.

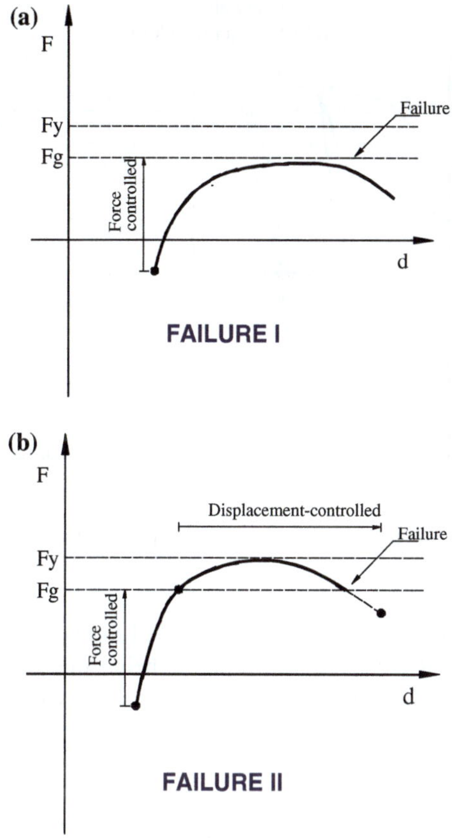

Fig. 2.17 Possible failure of specimen

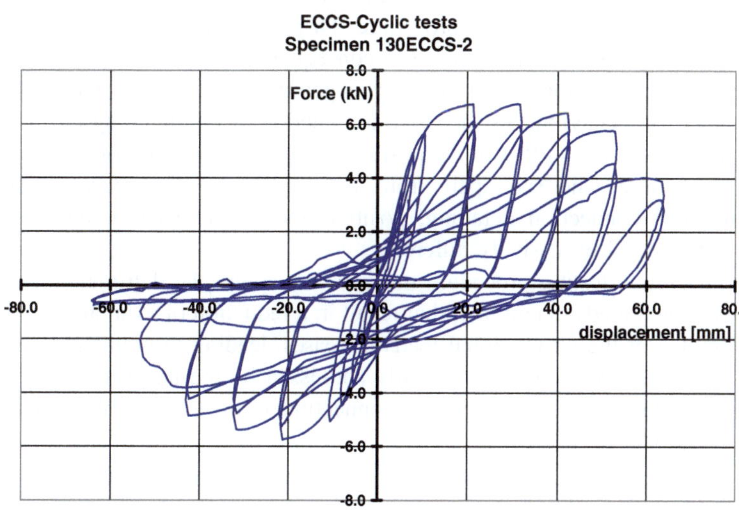

Fig. 2.18 ECCS cyclic test results for 130 mm height beam specimen

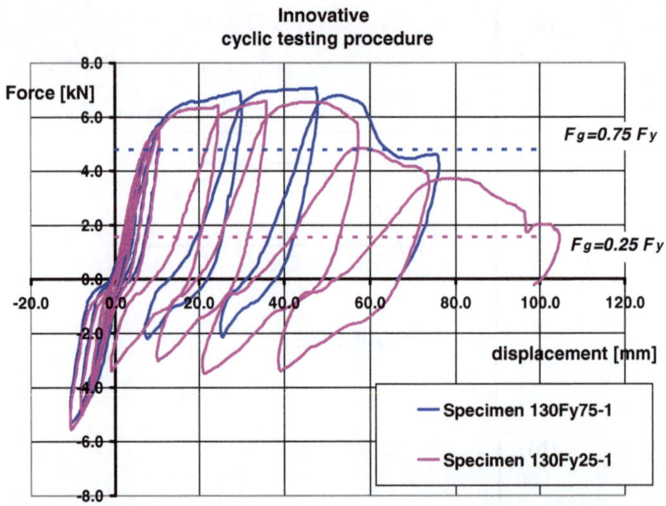

Fig. 2.19 Innovative cyclic testing procedure results for tests 130Fy25-1 and 130Fy75-1

In test 130Fy25-1 the specimen was subjected to elastic cycles, after which it endured two cycles of ±2 and 4 times d_y (d_y being the yield displacement) and failure occurred in the first positive 6 d_y displacement cycle. In test 130Fy75-1 the specimen failed in the first positive cycle with 4 d_y amplitude. The first, elastic, cycles are similar in both tests.

These two test results (as all the remaining tests conducted with the innovative cyclic testing procedure) are fundamentally different from those obtained through the application of the ECCS recommended testing procedure. Among the others, the following differences could be identified:

Imposed displacement history is unsymmetrical in the tests performed according to the innovative cyclic testing procedure. Displacements tend to systematically accumulate in the positive direction as a consequence of absence of closure of the top part of the connection (the bottom part is generally closed throughout the tests). This connection behaviour also leads to a reduction of the pinching effects.

Imposed forces are shifted in the positive (hogging) direction for the innovative testing procedure whereas, apart from the asymmetry that may result from unsymmetrical connection detailing, positive and negative force amplitudes for ECCS tests are not excessively different.

The innovative testing procedure manages to capture the detrimental effects of the gravitational loads magnitude, that result in a reduction of available ductility as the magnitude of the gravitational loads increases, leading to premature collapse.

Failure of the connection is explicitly addressed by the innovative testing procedure since failure occurs when the connection is no longer able to withstand vertical load effects.

2.2.2.2 Beam 70 × 45 × 1.5

Figure 2.20 shows the hysteresis loops in terms of moment and rotation, for the cyclic tests performed on beam-to-upright connections specimens with beam 70 × 45 × 1.5 mm.

It can be observed that in the case of the tests performed according to the innovative cyclic testing procedure, the resulting imposed displacement history is unsymmetrical. In fact, displacements tend to systematically accumulate in the positive direction as a consequence of the absence of closure of the top part of the connection (the bottom part is generally closed throughout the tests). This connection behaviour also leads to a reduction of the pinching effects.

The response parameters for all cyclic tests are presented in Table 2.6 where, in addition to the elastic parameters (i.e. initial stiffness $S_{j,ini}$, yield moment M_y and rotation ϕ_y), the maximum positive (hogging) moment (M_u^+) and corresponding rotation (ϕ_u^+), as well as the minimum negative (sagging) moment (M_u^-) and corresponding rotation (ϕ_u^-) are reported, together with the number of cycles at failure N_c, the failure rotation ϕ_c and the bending moment for gravity loads (M_g).

The cyclic tests show practically the same initial stiffness of the monotonic tests (as shown in Fig. 2.21, in the case of an ECCS test). Strength deterioration can be observed in the cyclic test (70ECCS-1) for rotations larger than 40 mrad, in the case of sagging bending (top loading).

Strength is shifted in the positive (hogging) direction for the innovative testing procedure while, in the case of ECCS tests, the load carrying capacities under positive and negative bending are not excessively different, apart from the small asymmetry that may result from unsymmetrical connection detailing.

The innovative testing procedure allows a clear assessment of the detrimental effects of the magnitude of the gravitational loads that result in a reduction of available ductility as the magnitude of the gravitational loads increases, leading to premature collapse.

Figure 2.22 shows a re-analysis (in terms of Resistance, Absorbed energy and Rigidity ratios) of the cyclic test results carried out under different values of the gravity loads, while the typical observed failure modes are shown in Fig. 2.23.

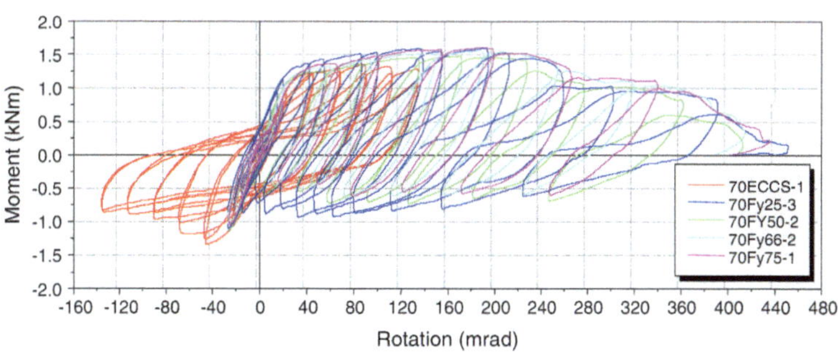

Fig. 2.20 Cyclic tests on beam 70 × 45 × 1.5 mm specimens

Table 2.6 Response parameters of the cyclic tests on beam 70 × 45 × 1.5 mm

Test	$S_{j,ini}$ [kNm/rad]	M_y [kNm]	ϕ_y [mrad]	M_u^+ [kNm]	ϕ_u^+ [mrad]	M_u^- [kNm]	ϕ_u^- [mrad]	N_c	ϕ_c [mrad]	M_g [kNm]
70ECCS-1	41.2	1.17	28.1	1.36	89.9	−1.34	−45.0	10	134.0	0.00
70ECCS-2	43.3	1.21	28.1	1.36	113.0	−1.40	−45.0	13	178.6	0.00
70Fy25-3	47.6	1.36	29.7	1.59	195.4	−0.94	65.0	11	393.1	0.30
70Fy25-4	47.7	1.22	26.0	1.55	191.2	−0.87	145.2	9	338.5	0.30
70Fy50-1	50.8	1.19	28.2	1.50	212.3	−0.81	125.1	8	427.6	0.60
70Fy50-2	50.9	1.17	28.2	1.48	197.4	−0.74	129.1	7	364.0	0.60
70Fy66-1	51.1	1.19	27.4	1.55	194.7	−0.61	74.6	6	331.5	0.79
70Fy66-2	43.8	1.20	26.4	1.57	193.0	−0.66	75.7	6	321.5	0.79
70Fy75-1	47.5	1.25	25.7	1.61	196.4	−0.64	81.7	6	342.0	0.90
Average All	**47.1**	**1.22**	**27.5**							
Average ECCS	**42.3**	**1.19**	**28.1**	**1.36**	**101.5**	**−1.37**	**−45**		**156.3**	
Average cyclic	**48.5**	**1.23**	**27.4**	**1.55**	**197.2**	**−0.75**	99.5		**359.7**	

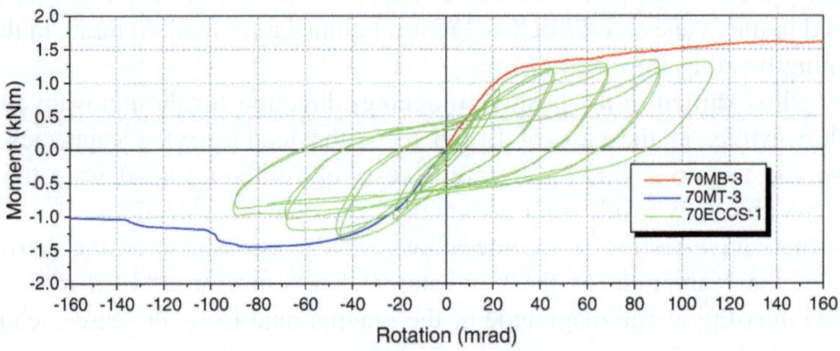

Fig. 2.21 Comparisons of the ECCS cyclic and monotonic tests for 70 × 45 × 1.5 mm beam specimen

In the case of ECCS cyclic testing procedure, the degradation of resistance and stiffness of the connection under hogging bending were lower than in sagging bending.

In the case of tests performed according to the innovative procedure, the amplitude of the displacements in the post-elastic phase (depending on different values of gravitational force F_g) did not change significantly the resistance, stiffness and ductility of the connection. According to the previous definitions (Fig. 2.17), connection 70Fy50-2 exhibited type I failure mode, while all other (70 × 45 × 1.5 mm beam) specimens exhibited failure mode type II.

Premature collapse occurred as the magnitude of the gravity loads increased.

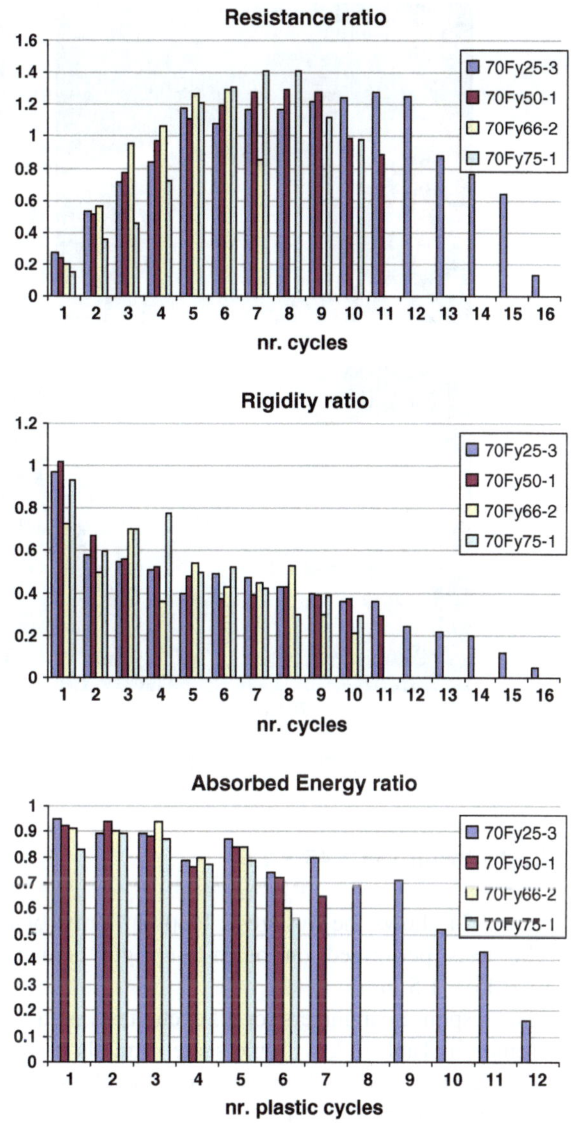

Fig. 2.22 Cyclic tests on 70 × 45 × 1.5 mm beam specimens

In all cyclic tests, the observed collapse mechanisms observed were the same as in monotonic tests under hogging bending, but with amplified deformations. At the end of some tests with a gravitational force of 50, 66 and 75 % of the yield strength, fracture of the beam end-plate connector occurred around the upper hole (Fig. 2.23).

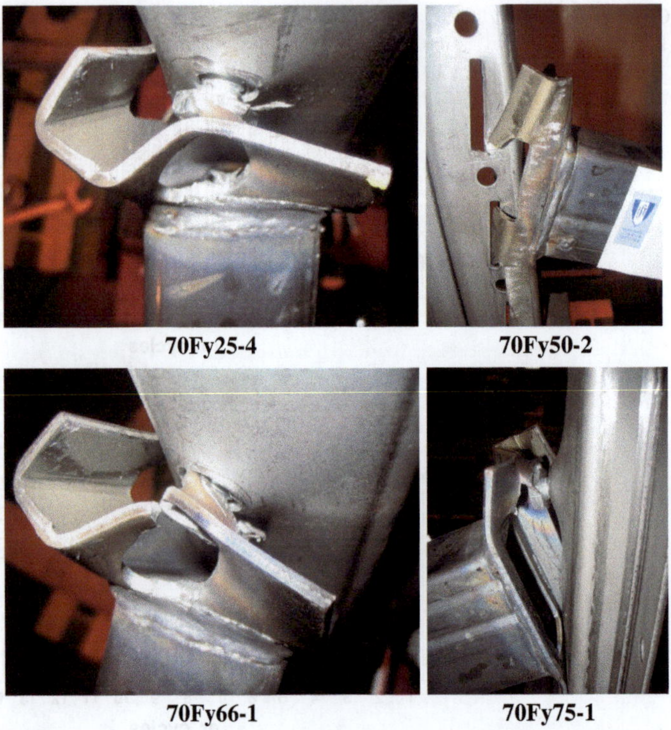

Fig. 2.23 Cyclic tests on 70 × 45 × 1.5 mm beam specimens—failure modes

2.2.2.3 Beam 130 × 45 × 1.5

Figure 2.24 shows the hysteresis loops in terms of moment and rotation, for the cyclic tests performed on beam-to-upright connections specimens with beam 130 × 45 × 1.5 mm.

The response parameters for all cyclic tests with a beam 130 × 45 × 1.5 mm are presented in Table 2.7, with the same symbols used (and described) in Table 2.6.

As shown in Fig. 2.24, the number of cycles to failure is strongly influenced by the intensity of the vertical (gravity) load.

The cyclic tests show practically the same initial stiffness when compared to the monotonic tests (as shown in Fig. 2.25, in the case of an ECCS test). The degradation levels of the resistance during the cyclic test (130ECCS-3) were nearly the same for both ways of loading (hogging and sagging bending).

Figure 2.26 shows a comparison (in terms of Resistance, Absorbed energy and Rigidity ratios) of the cyclic test results carried out under different values of the gravity loads.

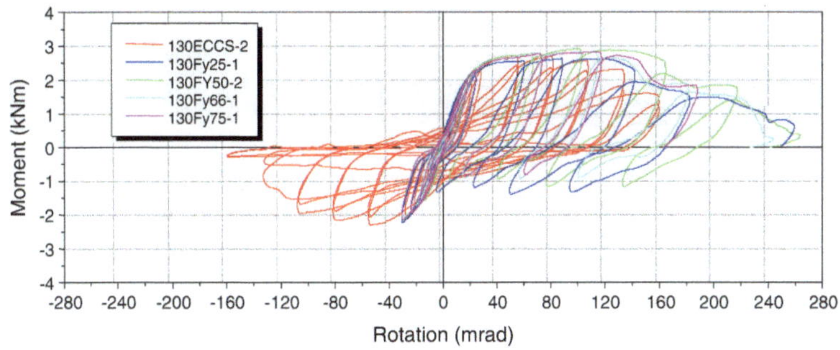

Fig. 2.24 Cyclic tests on beam 130 × 45 × 1.5 mm specimens

In the case of ECCS cyclic testing procedure, the resistance and stiffness ratios of the connection under hogging bending (bottom loading) were higher than under sagging bending (top loading).

In the case of tests performed according to the innovative procedure, a gradual deterioration of the energy dissipation capacity was observed. The resistance rate was satisfactory until the sixth cycle, after which a reduction occurred.

Despite of the high degradation of stiffness in the third cycle for the model 130Fy75-1, the ratio remained stable throughout the subsequent part of the test. In the other specimens, the stiffness degradation was satisfactory since no sudden decrease from one cycle to another occurred.

Figure 2.27 shows some images of the collapse mechanisms during the cyclic tests.

Premature collapse as well as reduction of the rotation capacity was observed as the magnitude of the gravity loads increased. At the end of the test the beam end connector was connected to the upright only by means of the bolt which sometimes could punch through the end-plate (Fig. 2.27).

According to the previous definitions (Fig. 2.17), the collapse mode for the connections 130Fy25-1 and 130Fy75-1 was type I while for 130Fy50-2 and 130Fy66-1 was type II.

2.2.3 Comparison and Analysis of Test Results

2.2.3.1 Monotonic Tests

Effect of the beam size on the main parameters (yield and ultimate strength, initial stiffness, yield and ultimate rotation) is shown in Table 2.8, with reference to monotonic tests.

Figure 2.28 shows the comparison between the moment—rotation curves for both beam sizes, and loading conditions.

Table 2.7 Response parameters of the cyclic tests for beam 130 × 45 × 1.5 mm

Test	$S_{j,ini}$ [kNm/rad]	M_y [kNm]	ϕ_y [mrad]	M_u^+ [kNm]	ϕ_u^+ [mrad]	M_u^- [kNm]	ϕ_u^- [mrad]	N_c	ϕ_c [mrad]	M_g [kNm]
130ECCS-2	85.0	2.51	25.9	2.70	79.3	−2.30	−53.2	10	157.5	0.00
130ECCS-3	104.3	2.47	22.2	2.80	79.5	−2.23	−52.9	9	157.5	0.00
130Fy25-1	107.2	2.38	25.5	2.64	117.9	−1.39	52.6	5	259.6	0.63
130Fy25-2	80.1	2.51	24.5	2.83	123.1	−1.27	49.5	4	175.6	0.63
130Fy50-1	115.4	2.62	29.6	2.97	130.9	−1.13	74.0	4	210.8	1.26
130Fy50-2	95.3	2.54	28.3	2.89	126.2	−1.19	136.4	4	218.3	1.26
130Fy66-1	84.8	2.46	23.1	2.84	116.1	−0.95	103	3	182.8	1.66
130Fy66-2	92.2	2.40	22.8	2.73	111.7	−1.19	136.4	3	182.8	1.66
130Fy75-1	107.0	2.47	24.2	2.83	118.7	−0.87	63	2	118.3	1.89
Average All	**96.8**	**2.48**	**25.1**							
Average ECCS	**94.7**	**2.49**	**24.05**	**2.75**	**79.4**	**−2.27**	**−53.1**		**157.5**	
Average cyclic	**97.4**	**2.48**	**25.4**	**2.82**	**120.7**	**−1.14**	**87.8**		**192.6**	

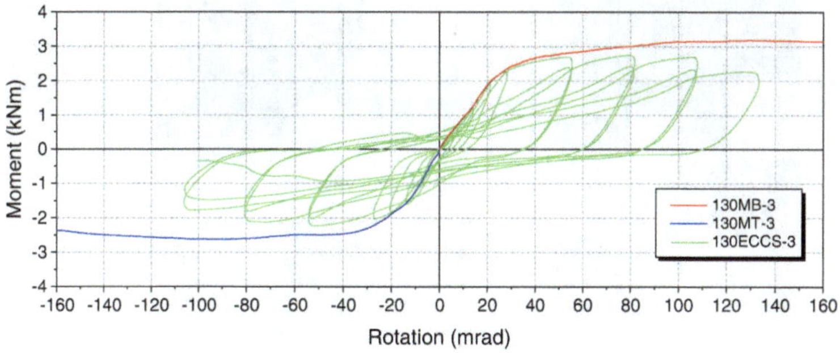

Fig. 2.25 Comparisons of the cyclic and monotonic tests for 130 × 45 × 1.5 mm beam

Fig. 2.26 Cyclic tests on 130 × 45 × 1.5 mm beam specimens

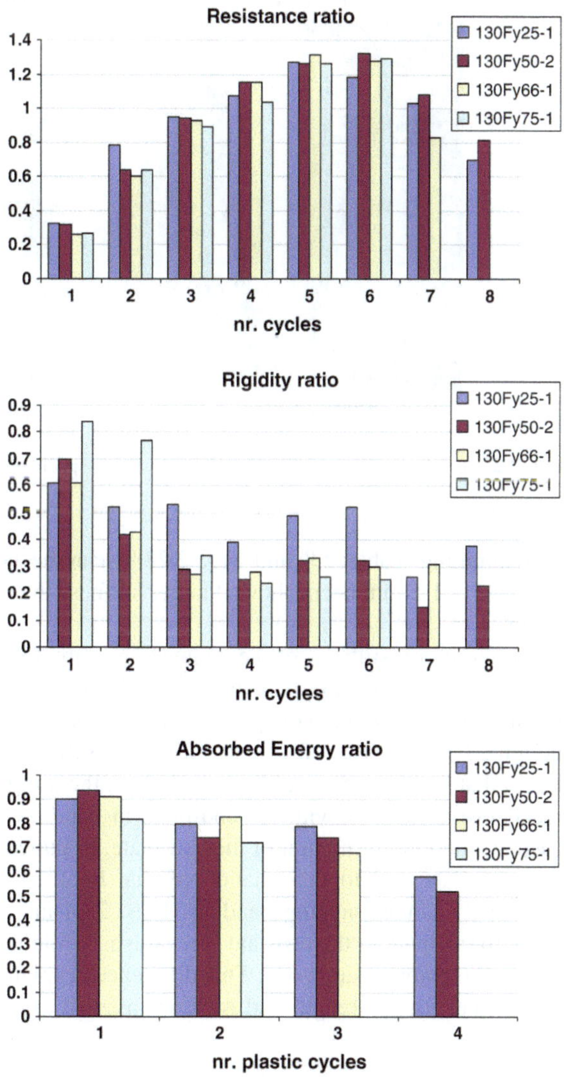

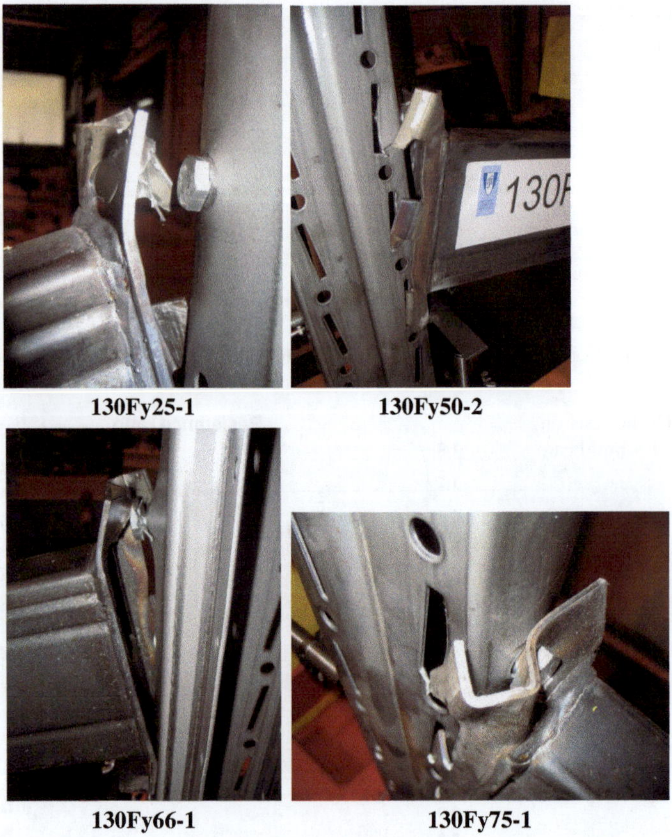

Fig. 2.27 Cyclic tests on 130 × 45 × 1.5 mm beam specimens—failure modes

Examining Table 2.8 and Fig. 2.28 it can be noticed that connections with 130 × 45 × 1.5 mm beam showed larger stiffness than those with 70 × 45 × 1.5 mm beam.

Furthermore, stiffness of the connections with 130 × 45 × 1.5 mm beams under sagging bending is approximately 25 % larger than that under hogging bending, because of the different collapse mechanisms, shown in Figs. 2.13 and 2.14. Under hogging bending (Figs. 2.13 and 2.29c), the safety bolt is in tension, and only one hook (that finally comes out from the hole in the upright) participates to the resistant mechanism, together with the end-plate in bending. Collapse mechanism, in this case, is due to deformation of the end-plate, to punching of the safety bolt through the end-plate and to deformation of one hook (a second one is only partially deformed). Under sagging bending (Figs. 2.14 and 2.29b), on the contrary, two hooks participate to the resistant mechanism, while the safety bolt doesn't, because it is in the compression zone. This differences explain both the higher stiffness and the smaller yield strength and rotation (as well as ultimate strength and rotation)

Table 2.8 Effect of the beam size—monotonic tests

Loading	Beam	M_y [kNm]	ϕ_y [mrad]	$S_{j,ini}$ [kNm/rad]	M_u [kNm]	ϕ_u [mrad]
Hogging bending	130 × 45 × 1.5	2.53	26.0	95.8	3.14	129.7
	70 × 45 × 1.5	1.23	27.8	47.9	1.60	178.7
Ratio	**130/70**	**2.06**	**0.94**	**2.00**	**1.96**	**0.73**
Sagging bending	130 × 45 × 1.5	2.11	18.8	119.0	2.44	107.3
	70 × 45 × 1.5	1.24	26.0	48.9	1.42	77.7
Ratio	**130/70**	**1.70**	**0.72**	**2.43**	**1.72**	**1.38**

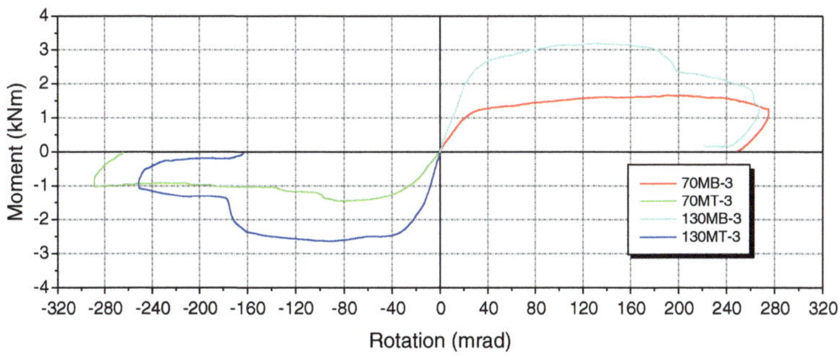

Fig. 2.28 Monotonic moment—rotation curves for both beam sizes and loading directions

exhibited by the 130 × 45 × 1.5 mm beam-to-upright connection under sagging bending with respect to the same connection under hogging bending.

The non symmetric geometry (and stiffness) of the end plate would cause a rotation of the beam in the horizontal plane (as shown in Fig. 2.29b). Such a rotation is however prevented by the test set up. Hence, all deformation concentrates in the end-plate connector and in the hooks.

The elastic behaviour of the 70 × 45 × 1.5 mm. beam-to-upright connection was practically symmetric, under hogging and sagging bending, as shown in Table 2.8. On the contrary, the ultimate behaviour of this connection under sagging bending was very different from the one exhibited under hogging bending, as evidenced in the same table, as well as in Figs. 2.10 and 2.11.

Under hogging bending, the safety bolt is in tension, and the center of rotation is located at the lower flange of the beam (in compression) as sketched in Fig. 2.29c. The end plate is extended, below the lower flange of the beam, in order to allow presence of a third hook. As evidenced in Fig. 2.10, the central hook is located approximately in correspondence of the lower flange of the 70 × 45 × 1.5 mm beam (i.e. at the center of rotation), so it doesn't participate to the resistant mechanism. Collapse mechanism, in this case, is due to deformation of the end-plate, punching

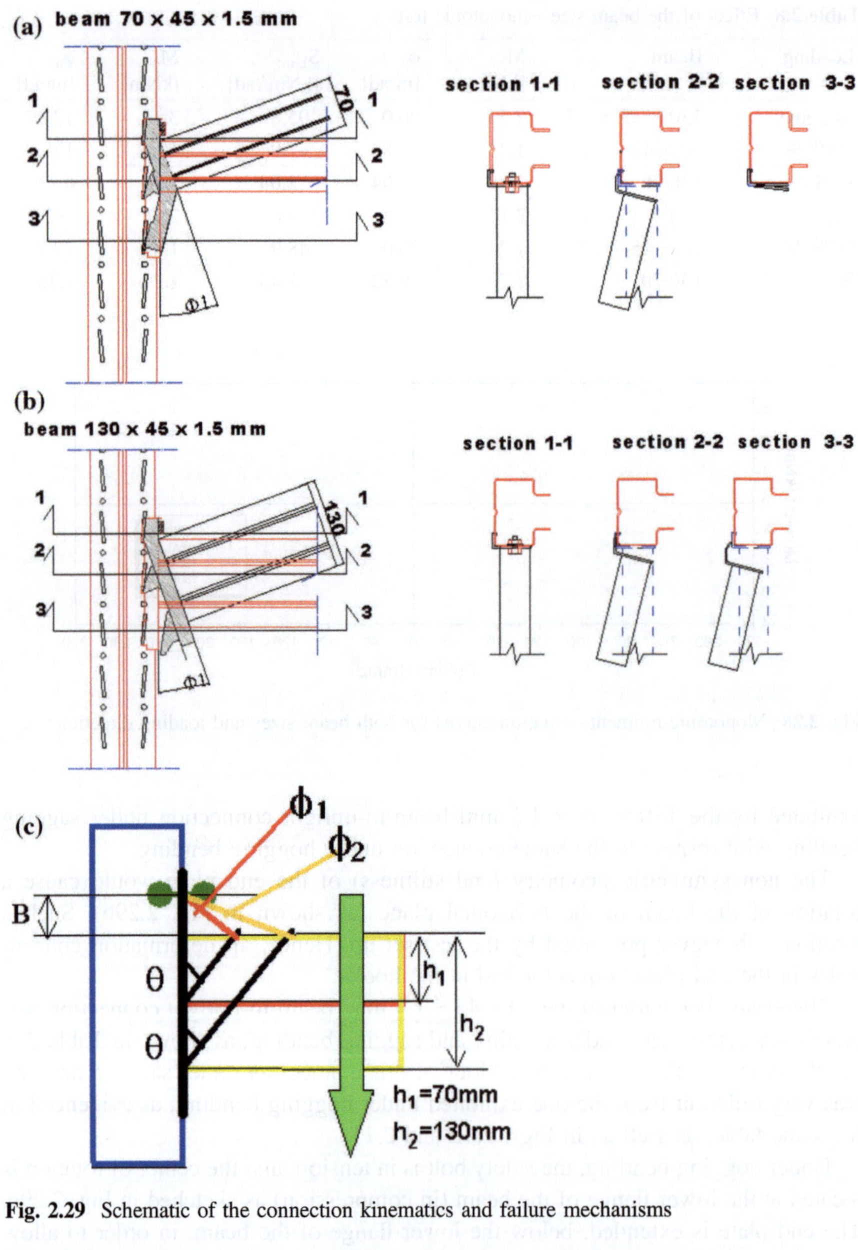

Fig. 2.29 Schematic of the connection kinematics and failure mechanisms

of the safety bolt through the end-plate and to deformation of one hook (the upper one, which can be seen as partially deformed in Fig. 2.10).

Under sagging bending (Fig. 2.11), on the contrary, the center of rotation is close to the upper flange of the beam, while the safety bolt is in the compression zone. The portion of the end-plate, extended below the lower flange, is restrained, on one side, by the lowest hook. Furthermore, the whole end-plate is stiffened on the "hooked" side, by the presence of the edge stiffener represented by the bent portion of the L shaped plate. When the beam is bent upward, the deformation of the lower part of the end plate, should comply with the restraints represented by the lower flange of the beam, by the lowest hook and by the stiffened edge. Its deformed shape is hence the one shown in Fig. 2.11 and sketched in Fig. 2.29a. The presence of the hooks and of the stiffened edge, forces the end-plate to bend in both the vertical and the horizontal plane. Accordingly should do the beam, which is however restrained to out-of-plane bending by the test set-up (as shown in Fig. 2.6).

This results in a concentration of deformations at the fillet weld connecting the beam to the end plate, on the side of the stiffened (hooked) edge, that fractures with consequent collapse of the connection (Fig. 2.11). Crack initiates at the end of the vertical leg of the fillet weld connecting the beam to the end-plate. This is the most stressed point of the connection, also because of the stress concentration due to the fact that the beam is fillet welded to the end plate on three sides only, and the fillet is not returned around the corner, despite of what is recommended by clause 4.3.2.1 (4) of EN 1993-1-8:2005 (*"Fillet welds finishing at the ends or sides of parts should be returned continuously, full size, around the corner for a distance of at least twice the leg length of the weld, unless access or the configuration of the joint renders this impracticable."*).

Figure 2.30 show a classification of both the connections, according to Eurocode 3 (2005)—Part 1.8, adopting respectively the length of the beam $L_b = 1.8$ m, as for the structures tested full scale within this research, and $L_b = 2.7$ m, as in the most common real applications of racks where, at each level and for each span, up to three pallets can be placed side-by-side on a couple of beams.

In Fig. 2.30, according to Eurocode 3 (2005)—Part 1.8, $m = M/M_{y,b}$ and $\phi = \phi/\phi_{y,b}$, where M and ϕ are the actual moment and rotation experimentally measured, and $M_{y,b}$ and $\phi_{y,b}$ are the conventional yield moment and rotation of the beam. $M_{y,b} = f_y{*}W_b$ is the product of the yield stress of the material (f_y) times the section modulus (W_b) of the beam cross section, while $\phi_{y,b} = M_{y,b}/k_b$ is the ratio of $M_{y,b}$ (previously defined) to the beam conventional elastic stiffness $k_b = (EI_b)/L_b$, I_b being the moment of inertia of the beam and E the Young's Modulus.

It can be noticed that for a beam length of 1800 mm (Fig. 2.30a) the connection with beam 70 × 45 × 1.5 can be considered as semi-rigid while in the case of beam 130 × 45 × 1.5, the same end-plate connection behaves as flexible. When the beam length is increased to 2700 mm (Fig. 2.30b), both connections can be considered semi-rigid.

2.2.3.2 Cyclic Tests

The effect of the beam size and of the gravity loads in terms of hysteresis loops, is shown in Figs. 2.31 and 2.32, respectively with reference to ECCS and to cyclic tests, as well as in Table 2.9 that summarizes the mean values of the response

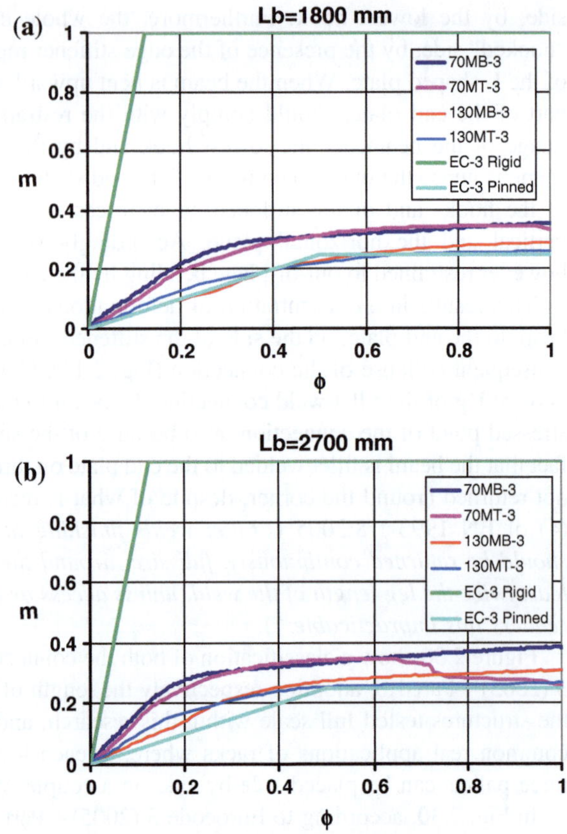

Fig. 2.30 Classification of the beam-to-upright connections according to EN 1993-1-8:2005

parameters for both 70 × 45 × 1.5 mm (a) and 130 × 45 × 1.5 mm (b) beam, and for different values of the gravity loads.

In Table 2.9, in addition to the elastic parameters (i.e. initial stiffness $S_{j,ini}$, yield moment M_y and rotation ϕ_y), the maximum positive (hogging) moment (M_u^+) and corresponding rotation (ϕ_u^+), as well as the minimum negative (sagging) moment (M_u^-) and corresponding rotation (ϕ_u^-) are reported together with the number of cycles at failure N_c, the failure rotation ϕ_c and the bending moment for gravity loads (M_g). It is evident that the response of the connections under symmetric cycles and in absence of gravity loads (ECCS tests, Fig. 2.31) is very different from that of the same connections tested in presence of gravity loads (Fig. 2.32).

In particular, the response of the specimens tested in presence of gravity loads is characterized by cyclic creep (ratcheting), with a progressive accumulation of plastic deformation in the same direction of the mean applied load. The measured ratcheting rate increases with the applied gravity load.

Collapse of the connection is attained approximately under the same rotation ϕ_c (360 mrad for 70 × 45 × 1.5 mm beams and 192 mrad for 130 × 45 × 1.5 mm beams), although some dependence of ϕ_c on the applied gravity load can be

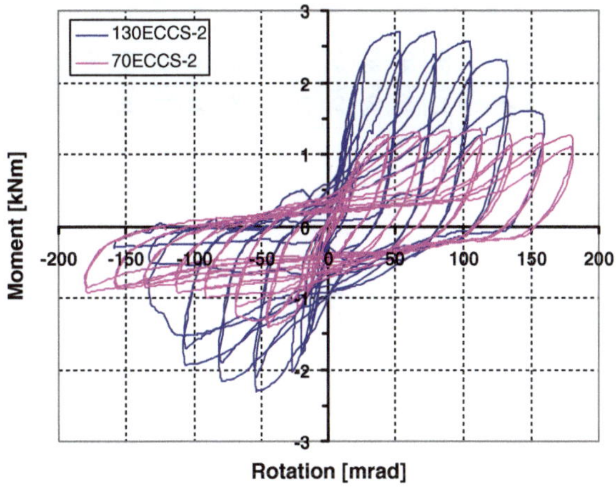

Fig. 2.31 Effect of the beam size—ECCS cyclic tests

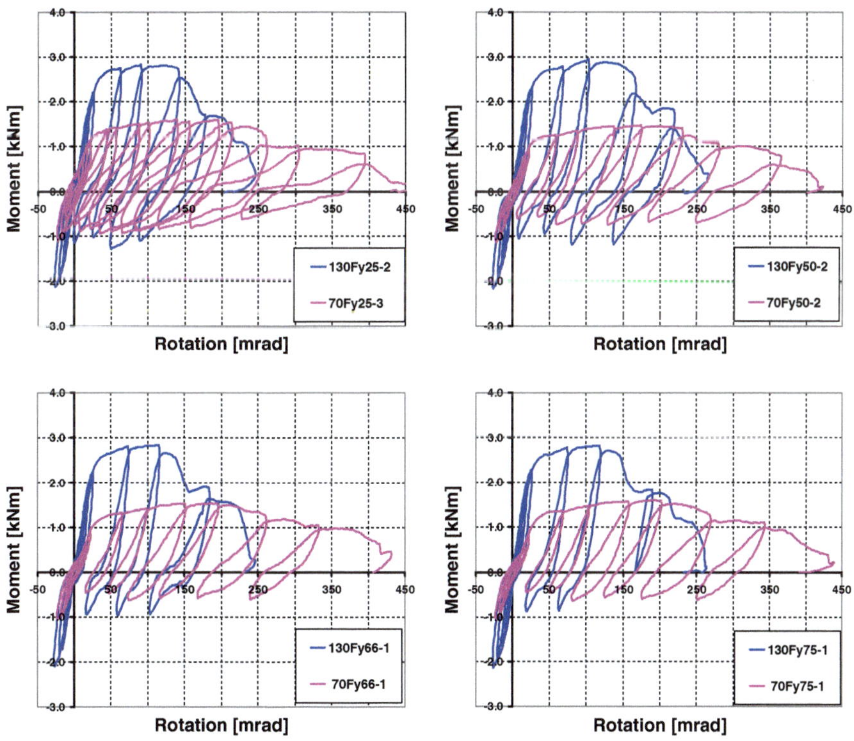

Fig. 2.32 Effect of the beam size and of the gravity loads—cyclic tests

Table 2.9 Average values of response parameters for 70 × 45 × 1.5 mm (a) and 130 × 45 × 1.5 mm (b) beams

	M_g [kNm]	$S_{j,ini}$ [kNm/rad]	M_y [kNm]	ϕ_y [mrad]	M_u^+ [kNm]	ϕ_u^+ [mrad]	M_u^- [kNm]	ϕ_u^- [mrad]	N_c	ϕ_c [mrad]
(a)										
70-ECCS	0.00	42.25	1.19	28.10	1.36	101.45	−1.37	−45.00	11.5	156.30
70-Fy25	0.30	47.65	1.29	27.85	1.57	193.30	−0.91	105.10	10	365.80
70-Fy50	0.60	50.85	1.18	28.20	1.49	204.85	−0.78	127.10	7.5	395.80
70-Fy66	0.79	47.45	1.20	26.90	1.56	193.85	−0.64	75.15	6	326.50
70-Fy75	0.90	47.50	1.25	25.70	1.61	196.40	−0.64	81.70	6	342.00
(b)										
130-ECCS	0.00	94.65	2.49	24.05	2.75	79.40	−2.27	−53.05	9.5	157.50
130-Fy25	0.63	93.65	2.45	25.00	2.74	120.50	−1.33	51.05	4.5	217.60
130-Fy50	1.26	105.35	2.58	28.95	2.93	128.55	−1.16	105.20	4	214.55
130-Fy66	1.66	88.50	2.43	22.95	2.79	113.90	−1.07	119.70	3	182.80
130-Fy75	1.89	107.00	2.47	24.20	2.83	118.70	−0.87	63.00	2	118.30

observed, at least for large values of the gravity load (66 and 75 % of the yield load). Hence, connections of beams supporting a high gravity load collapse at a number of cycles (N_c) smaller than those supporting a smaller gravity load, due to the different creep rate.

Hysteresis loops are clearly non symmetric, with similar maximum values of the hogging moment (M_u^+) and of its corresponding rotation (ϕ_u^+), practically independent on the applied gravity load, but with the minimum values of the sagging moment (M_u^-) and of its corresponding rotation (ϕ_u^-) becoming (in absolute value) smaller and smaller when increasing the gravity load. Of course, the elastic parameters (i.e. initial stiffness $S_{j,ini}$, yield moment M_y and rotation ϕ_y) are not influenced by the value of the applied gravity load, as shown in Fig. 2.33.

Dependence of the various inelastic response parameters on the applied gravity load (M_g) is shown in Fig. 2.34.

In absence of gravity load (ECCS tests), the response of the specimens is nearly symmetric; little non symmetry is due to the different response of the connection under hogging and sagging bending, already highlighted in the case of monotonic tests. In particular, it should be noticed that under sagging bending the connections show a lower ductility than under hogging bending ($\phi_u^- < \phi_u^+$). The values of M_u^+, ϕ_u^+, M_u^- and ϕ_u^- (Table 2.9) are (in absolute value) smaller than those measured in monotonic tests (reported in Table 2.8), because obtained under cyclic loading.

Differences can also be noticed in the failure modes of the different specimens.

In absence of gravity loads (ECCS tests), failure occurred always under sagging bending, i.e. under the loading condition for which the connection showed the

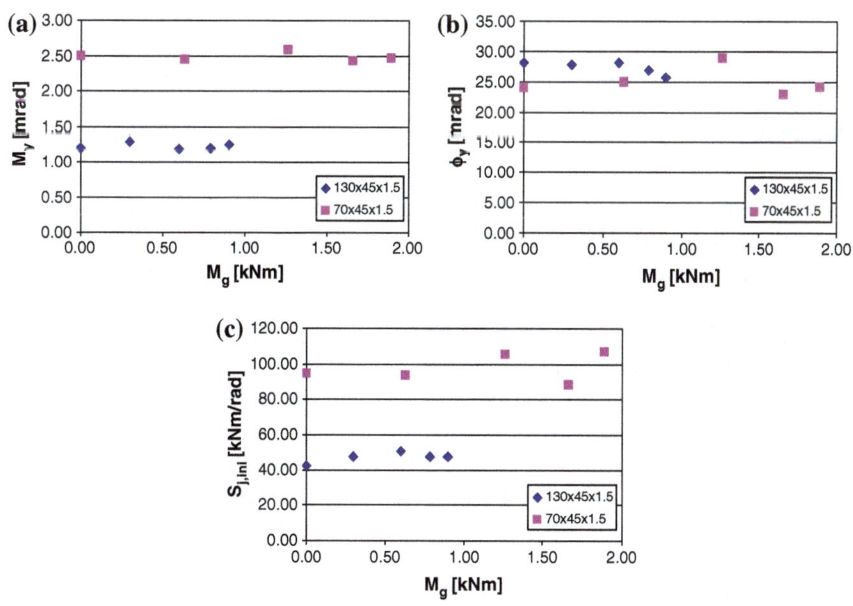

Fig. 2.33 Comparison of elastic response parameters for ECCS and cyclic tests

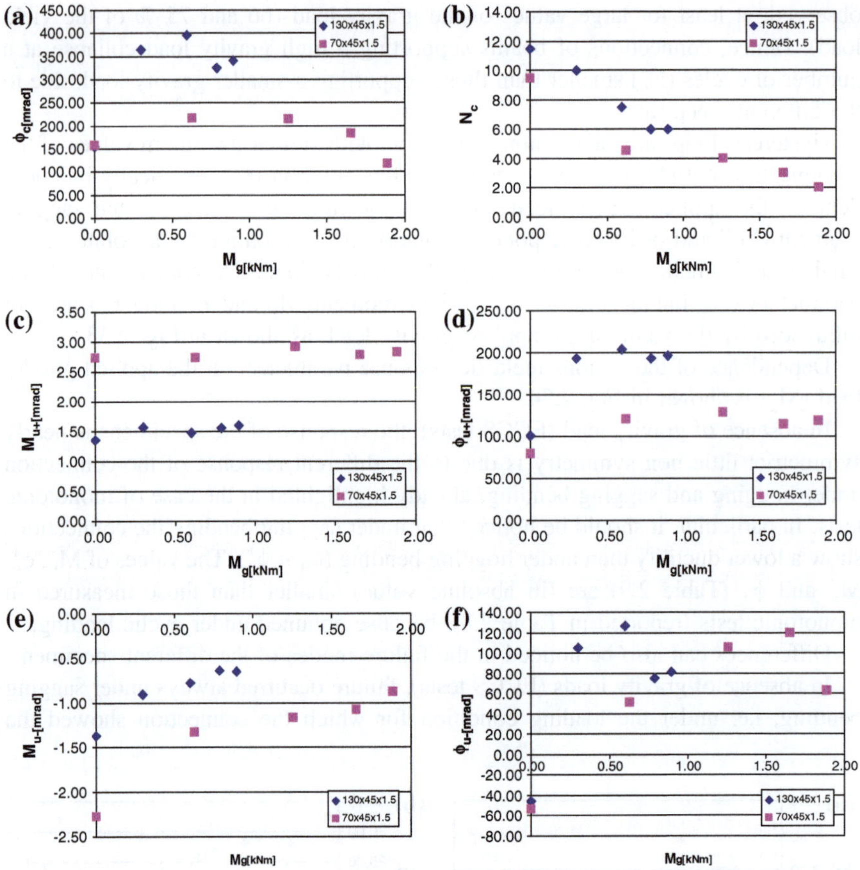

Fig. 2.34 Comparison of ultimate response parameters for ECCS and cyclic tests

lower strength and ductility in monotonic tests. In the case of 70 × 45 × 1.5 mm beams the collapse mechanism involved cracking of the fillet weld connecting the beam to the end-plate, deformation of the end plate and of the central hook and a small deformation of the lowest hook. In this case, however, due to the cyclic load reversals, deformation of the upper hook as well as partial punching of the safety bolt through the end-plate could be observed. Also in the case of 130 × 45 × 1.5 mm beams, the collapse mechanism for cyclic (ECCS) tests is similar to the one described for monotonic tests under sagging bending, with deformation of the two lowest hooks. Furthermore, also in this case, due to the cyclic load reversals, large deformation of the upper hook as well as partial punching of the safety bolt through the end-plate could be observed. In presence of gravity loads, on the contrary, failure occurred always under hogging bending. No difference could be noticed in the failure mechanisms under increasing values of the gravity loads, but only a smaller number of cycles to failure. Collapse mechanism, for both 70 × 45 × 1.5 mm and 130 × 45 × 1.5 mm beams, involved complete punching of the safety bolt through the end-plate, with partial deformation of the central and of the lowest hooks, because of the cyclic reversal loading. Cracking of the hooks was also observed.

2.2.4 Conclusions

- The failure mode for 70 mm (hogging bending) and 130 mm (hogging and sagging bending) deep beams consisted of large deformations in the top zone of the beam end connector.
- In the 70 mm deep beam subject to sagging bending, the failure mode was the fracture of the fillet weld between the beam and the end-plate connector.
- The connection behaviour for all tests was not influenced by safety bolt deformation, whose axial stiffness is much larger than the bending stiffness of the end-plate connector.
- The rotation capacity difference between the loading directions was larger for the 70 mm deep beam due to different collapse occurred in the sagging loading.
- According to Eurocode 3 (2005) the connection with a 70 mm deep beam can be considered as semi-rigid, for both beam lengths of 1.8 and 2.7 m.
- According to Eurocode 3 (2005) the connection with a 130 mm deep beam can be considered as semi-rigid for a beam lengths of 2.7 m, and as flexible for a beam lengths of 1.8 m (as in the case of the structures tested full scale within this research); hence, in this last case, the influence of the connection behaviour may be ignored in structural analysis.
- The results of test performed with the innovative cyclic testing procedure were fundamentally different from those obtained through the application of the ECCS recommended testing procedure.
- In the tests performed according to the innovative cyclic testing procedure, the imposed displacement history is unsymmetrical. Displacements tend to systematically accumulate in the positive direction as a consequence of the absence of closure of the top part of the connection (the bottom part is generally closed throughout the tests). This connection behaviour also leads to a reduction of the pinching effects.
- In the innovative testing procedures, the imposed forces are shifted in the positive (hogging) direction whereas, apart from the asymmetry that may result from unsymmetrical connection detailing, positive and negative force amplitudes for ECCS tests are not excessively different.
- Failure of the connection is explicitly addressed by the innovative testing procedure since failure occurs when the connection is no longer able to withstand vertical load effects.

2.3 Column Base Connections

Proprietary moment connections are typically used as column-base connections for steel selective pallet storage racks. This study focuses on the base connection type shown in Fig. 2.35.

The column bases consist of two vertical steel gusset plates fillet-welded to the base plate. The upright is connected to the base by bolting through the slotted holes in the gusset plates using two M10 × 25 (grade 8.8) bolts for each vertical plate.

The base plate is connected to the foundation surface (in this case either in concrete or in steel) by means of two M16 (grade 8.8) bolts. All bolts are preloaded.

As a general remark, it should be noticed that this proprietary column-base connection is strongly non-symmetric in cross-aisle direction, where an eccentricity of 66.5 mm exists between the bolt line and the c.o.g. of the up-right profile.

A non-symmetric response is hence to be expected in cross-aisle direction under transverse load reversals, when the bolts may be either in tension or in compression.

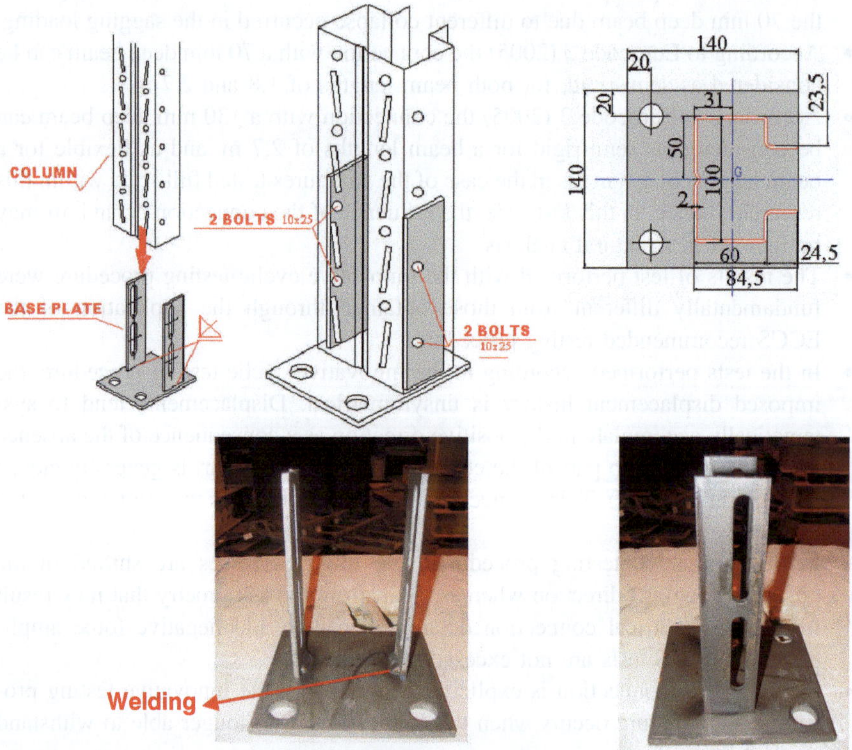

Fig. 2.35 Details of base plate and base-to-upright connections

Objectives of the tests on column base connections were:

- Assessment of the moment-rotation curves
- Assessment of the collapse modes

under either monotonic or cyclic transversal loads, for different levels of axial load applied to the upright.

The specimens were tested in the cross-aisle and down-aisle directions. In the cross-aisle direction the specimens were subjected to a monotonic load in two different directions, in order to allow respectively tension and compression in the anchor bolts of the column base connection.

One type of upright cross-section (100/20b), whose geometrical properties are shown in Table 2.10 was studied under four values of axial load: 0, 25, 50 and 75 kN. These correspond respectively to 0, 17, 34 and 52 % of the yield resistance of the net cross section under axial load (Af_y), f_y being the nominal yield strength of 275 MPa.

Column bases connected to steel and concrete surfaces were examined.

In the full scale dynamic tests carried out at NTUA the base plates were welded to the steel deck of the shaking table; hence some tests were carried out with the steel base welded and bolted to a steel surface. These tests are indicated hereafter as "Athens base" tests.

Table 2.11 summarizes the performed tests. The size and shape of the test specimen are shown in Fig. 2.36, while the test set up and instrumentation are shown in Figs. 2.37 and 2.38.

The tested specimens comprise a portion of upright 100/20b, 650 mm long, connected to a special steel plate at the top and to the ordinary base plate at the bottom. The steel plate at the top, welded to the upright, allowed the coupling of the specimen to the actuator, by means of four M16 (grade 8.8) preloaded bolts.

The specimens were tested in a system composed by a main reaction frame, a mechanical actuator and a secondary frame used to induce the axial force in the specimens.

In order to simulate different "floor conditions", the specimens were connected either to a steel member or to a concrete block. These, in turn, where connected to the main frame using bolts. The lateral displacements at the top of the specimens

Table 2.10 Geometrical properties of the upright 100/20b

Properties	Gross section	Net section
A [mm^2]	588.8	525.7
t [mm]	2.0	2.0
J_x [mm^4]	436020	406100
J_y [mm^4]	812280	694680
W_x [mm^3]	8543	8330
W_y [mm^3]	16747	14323

Table 2.11 Column-base tests

Type	Test	Steel base (bolted)			Athens base (bolted & welded)
		No axial load	Axial load 25 kN	Axial load 50 kN	Axial load 25 kN
(a)					
	Monotonic	SBCB-1 SBCB-2 SBCB-3 SBCB-4 SBCB-5	SBCBF25-1 SBCBF25-2	SBCBF50-1 SBCBF50-2	ASBCBF25-1
	Monotonic	SBCT-1 SBCT-2 SBCT-3 SBCT-4 SBCT-5	SBCTF25-1 SBCTF25-2	SBCTF50-1 SBCTF50-2	ASBCTF25-1
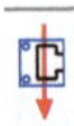	Monotonic	SBDB-1 SBDB-2 SBDB-3 SBDB-4 SBDB-5	SBDBF25-1 SBDBF25-2 SBDBF25-3	SBDBF50-1 SBDBF50-2	ASBDBF25-1
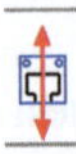	Cyclic	——–	SBCCF25-1	———	ASBCCF25-1
	Cyclic	–	SBDCF25-1	–	ASBDCF25-1

Type	Test	Concrete base—Hilti			
		No axial load	Axial load 25 kN	Axial load 50 kN	Axial load 75 kN
(b)					
	Monotonic		CBCBF25-1 CBCBF25-2	CBCBF50-1 CBCBF50-2	CBCBF75-1
	Monotonic		CBCTF25-1	CBCTF50-1	CBCTF75-1
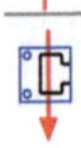	Monotonic	CBDB-1	CBDBF25-1 CBDBF25-2	CBDBF50-1 CBDBF50-2 CBDBF50-3	CBDBF75-1
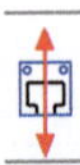	Cyclic		CBCCF25-1	CBCCF50-1	CBCCF75-1
	Cyclic		CBDCF25-1	CBDCF50-1	CBDCF75-1

Connected to (a) a steel surface, (b) a concrete surface

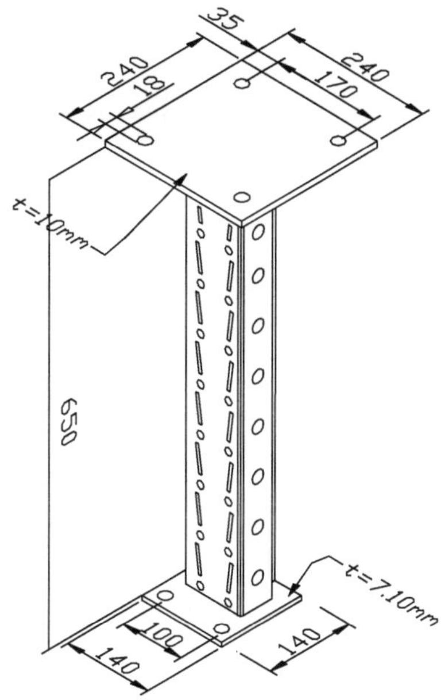

Fig. 2.36 Specimen shape and size

were restrained using double angles welded to a metallic member that was connected to the main frame, as shown in Fig. 2.37.

The axial force was imposed in the centre of gravity of the upright using two 36 mm Dywidag bars. The system of application of the axial force comprised a secondary reaction frame, a beam, two hydraulic actuators and two loading cells, as shown in Fig. 2.38.

The Dywidag bars, 8 m long, were connected to the head of the actuator, which was bolted to the plate at the top of the upright.

Tests were carried out under displacement controlled conditions. The actuator applied the transversal displacement at a distance of 1 m from the column base, while the imposed load was measured through a load cell. Four electric displacement transducers (LVDTs) were positioned in the column base to evaluate the relative bending of the base, as shown in Fig. 2.37.

The moment-rotation curves were plotted for each test; the moment at the column base is $M = F * a$ and the rotation of the connection is evaluated according to Eq. 2.2.

$$\phi = \frac{V}{a} \tag{2.2}$$

where V is the displacement imposed at the top of the specimen by the transversal load F applied with a lever arm a = 1.0 m with respect to the specimen base.

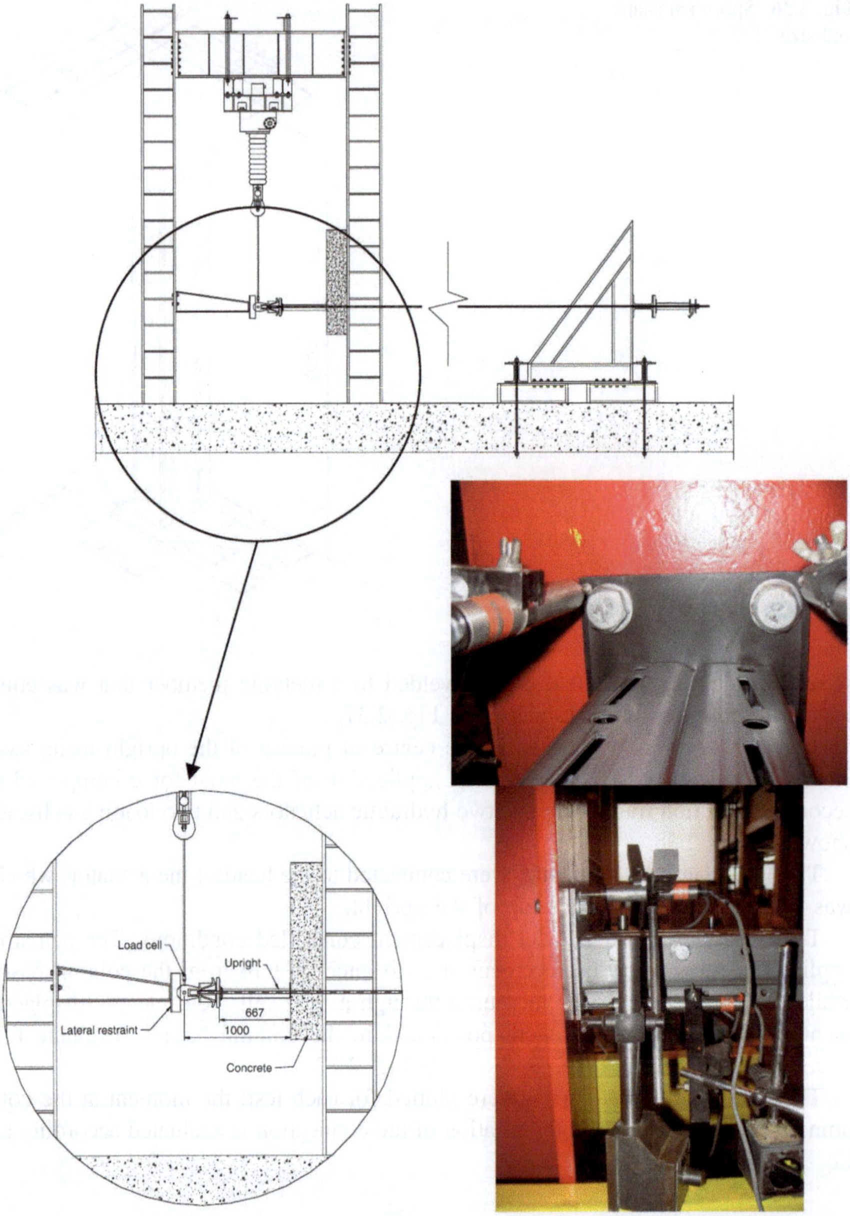

Fig. 2.37 Test set-up and instrumentation

Fig. 2.38 Test set-up: **a** main frame, **b** secondary reaction frame, **c** axial load system connected to the secondary reaction frame, **d** details of jacks and Dywidag bars connected to the secondary reaction frame, **e** details of Dywidag bars, their connection to the head of the actuator, as well as of the lateral displacements restraint

2.3.1 Monotonic Tests

2.3.1.1 Cross-Aisle

2.3.1.1.1 Bolts in Tension

Table 2.12 summarizes the results of monotonic tests carried out in cross-aisle direction on column-base connections, with anchor bolts in tension. Different values of the axial load were considered (0, 25, 50 and 75 kN). Different types of connection of the upright steel base to the floor were also considered: bolted to a steel or to a concrete floor, as well as bolted-and-welded to a steel deck (as in the

Athens dynamic shaking table tests). Table 2.12 summarizes, for the different types of connections, and for each test, the yield bending moment M_y and rotation ϕ_y, the initial elastic stiffness $S_{j,ini}$, the maximum bending moment M_{max} and associated rotation ϕ_{max}, the energy absorbed up to the attainment of the maximum rotation ϕ_{max}, as well as the failure mode exhibited by the specimen.

Repetitions were carried out under similar testing conditions, in order to verify the reliability of the results.

In some cases, test repetition resulted in different failure modes. For this reason, Table 2.12 report the average values of tests carried out under similar conditions, as well as the average values for those tests, performed under similar conditions, but characterized by the same failure mode.

Figures 2.39 and 2.40 show a comparison of the moment-rotation curves for column-base connections with anchor bolts in tension, under different values of the axial load, connected respectively to a steel deck (Fig. 2.39) and to a concrete deck (Fig. 2.40).

It can be noticed that the specimens with an axial force in the upright of 25 and 50 kN exhibited a similar initial stiffness, as shown in Figs. 2.39 and 2.40. The ductility of the connection decreases when the axial force is increased. Nevertheless, those specimens, like SBCB-1 and ASBCBF25-1 or CBCBF25-1, that suffered a premature collapse due to weld fracture, showed a lower ductility than the specimens with an axial force of 50 kN.

The increase of the axial force in the upright resulted in a decrease of resistance of the specimens.

Specimen ASBCBF25-1, bolted-and-welded to a steel frame, had a higher initial stiffness and resistance. This was due to the lack of bending in the base of the column.

The collapse modes in the cross-aisle resulted mostly of the weld failure, base bending and distortion buckling of the upright cross section. In Table 2.12 it can be observed that, for the specimens without axial force or with an axial force of 25 kN, the collapse modes were weld failure and base bending. Rotation capacity of those specimens suffering weld failure is heavily reduced.

In no one of the specimens with an axial force of 50 and 75 kN failure due to weld failure or base bending could be observed. The typical collapse mode was distortion buckling of the upright cross section. Figure 2.41 shows each collapse mode mentioned in Table 2.12.

2.3.1.1.2 Bolts in the Compression Zone

Table 2.13 summarizes the results of monotonic tests carried out in cross-aisle direction on column-base connections, with anchor bolts in the compression zone. Also in this case different values of the axial load were considered (0, 25, 50 and 75 kN) as well as different types of connection of the upright steel base to the floor: bolted to a steel or to a concrete floor, as well as bolted-and-welded to a steel deck (as in the Athens dynamic shaking table tests).

Table 2.12 Monotonic tests on column-bases bent in cross-aisle direction with bolts in tension: (a) bolted to a steel surface, (b) bolted to a concrete surface (c) bolted-and-welded to a steel surface

	Axial load [kN]	M_y [kNm]	ϕ_y [mrad]	$S_{j,ini}$ [kNm/rad]	M_{max} [kNm]	ϕ_{max} [mrad]	E_{max} [kNm*mrad]	Failure mode
(a) Steel base								
sbcb-1	0	1.73	24.5	70.4	1.83	41.7	50.5	Weld failure
sbcb-2	0	2.15	41.4	51.9	2.59	178.0	344.4	Base bending
sbcb-3	0	2.34	27.0	86.6	2.44	46.7	59.8	Weld failure
sbcb-4	0	2.25	36.9	60.9	2.86	175.2	388.5	Base bending
sbcb-5	0	2.55	44.1	57.9	3.24	179.6	446.8	Base bending
Average	**0**	**2.20**	**34.8**	**65.5**	**2.59**	**124.2**		**All**
Average	**0**	**2.04**	**25.8**	**78.5**	**2.14**	**44.2**	**55.2**	**Weld failure**
Average	**0**	**2.32**	**40.8**	**56.9**	**2.89**	**177.6**	**393.2**	**Base bending**
sbcbf25-1	25	1.90	21.6	87.9	2.03	37.3	51.7	Weld failure + Base bending
sbcbf25-2	25	2.06	26.4	77.9	2.26	62.5	89.3	Weld failure + Base bending
Average	**25**	**1.98**	**24.0**	**82.9**	**2.14**	**49.9**	**70.5**	**Weld failure + Base bending**
sbcbf50-1	50	1.98	22.9	86.4	2.09	42.0	58.0	Distorsional buckling
sbcbf50-2	50	1.52	14.4	106	1.69	34.6	40.8	Distorsional buckling
Average	**50**	**1.75**	**18.7**	**96.2**	**1.89**	**38.3**	**49.4**	**Distorsional buckling**
(b) Concrete base								
cbcbf25-1	25	1.74	24.9	70.1	1.87	42.6	54.0	Weld failure
cbcbf25-2	25	1.55	22.5	69.0	1.75	56.8	70.8	Weld failure + Base bending
Average	**25**	**1.65**	**23.7**	**69.6**	**1.81**	**49.7**	**62.4**	**All**

(continued)

Table 2.12 (continued)

	Axial load [kN]	M_y [kNm]	ϕ_y [mrad]	$S_{j,ini}$ [kNm/rad]	M_{max} [kNm]	ϕ_{max} [mrad]	E_{max} [kNm*mrad]	Failure mode
cbcbf50-1	50	1.42	17.2	82.6	1.55	36.4	39.0	Distorsional buckling
cbcbf50-2	50	1.89	32.7	57.8	1.96	45.8	54.4	Distorsional buckling
Average	**50**	**1.66**	**25.0**	**70.2**	**1.76**	**41.1**	**46.7**	**Distorsional buckling**
cbcbf75-1	75	0.54	12.4	43.6	0.58	21.4	8.3	Distorsional buckling
	75	**0.54**	**12.4**	**43.6**	**0.58**	**21.4**	**8.3**	
(c) Athen base (bolted + welded)								
asbcbf25-1	25	2.37	29.0	81.8	2.46	46.5	64.2	Weld failure
	25	**2.37**	**29.0**	**81.8**	**2.46**	**46.5**	**64.2**	

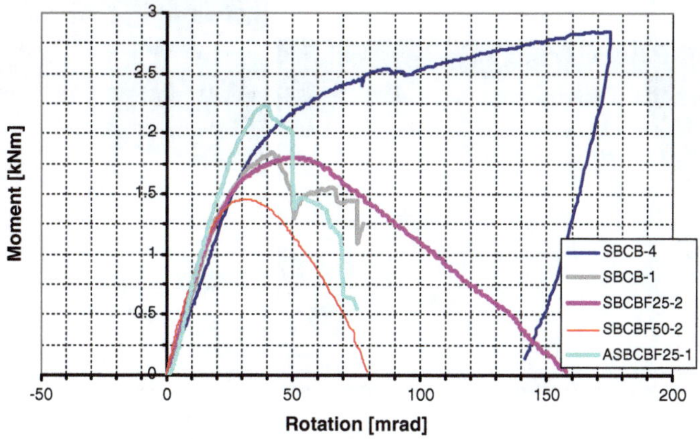

Fig. 2.39 Comparison of the moment-rotation curves for column-base connections bent in cross-aisle direction, with anchor bolts in tension, connected to a steel deck, under different values of the axial load

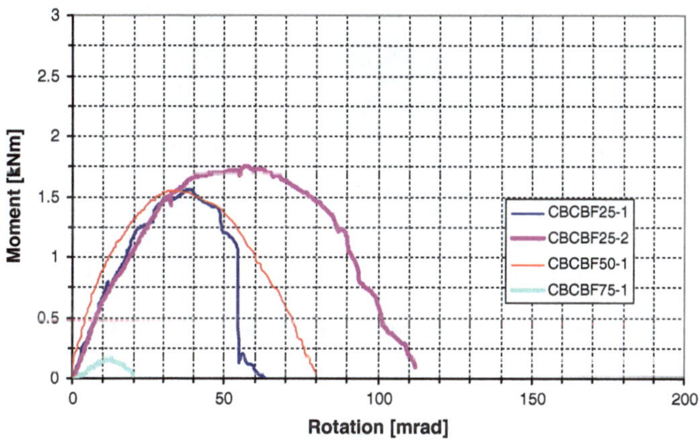

Fig. 2.40 Comparison of the moment-rotation curves for column-base connections bent in cross-aisle direction, with anchor bolts in tension, connected to a concrete deck, under different values of the axial load

Table 2.13 summarizes, for the different types of connections, and for each test, the yield bending moment M_y and rotation ϕ_y, the initial elastic stiffness $S_{j,ini}$, the maximum bending moment M_{max} and associated rotation ϕ_{max}, the energy absorbed up to the attainment of the maximum rotation ϕ_{max}, as well as the failure mode exhibited by the specimen.

Figures 2.42 and 2.43 show a comparison of the moment-rotation curves for column-base connections, with anchor bolts in tension, under different values of the

Weld failure | Base bending and weld failure

Distortion buckling | Base bending

Fig. 2.41 Collapse modes for specimens bent in cross-aisle direction with bolts in tension

axial load, connected respectively to a steel deck (Fig. 2.42) and to a concrete deck (Fig. 2.43).

In the tests in cross-aisle with bolts in the compression zone, increasing the axial force resulted in an increase of the initial stiffness $S_{j,ini}$, and resistance of the connection, but in a reduction of its ductility.

In the specimen bolted-and-welded to a steel plate (ASBCTF25-1) an increase in the resistance and rotation capacity of the connection was observed, as expected.

Distortional buckling of the upright cross section was not exhibited by any test. The collapse mode for all specimens, except for the ones with bolted-and-welded base, was bending of the base plate, as shown in Fig. 2.44. The magnitude of the uplift of the edges of the base plate, under different values of the axial force is shown in Fig. 2.45. It is possible to observe that higher axial forces correspond to lower base bending.

Table 2.13 Monotonic tests on column-bases bent in cross-aisle direction with bolts in compression zone: (a) bolted to a steel surface, (b) bolted to a concrete surface (c) bolted-and-welded to a steel surface

	Axial load [kN]	M_y [kNm]	ϕ_y [mrad]	$S_{j,ini}$ [kNm/rad]	M_{max} [kNm]	ϕ_{max} [mrad]	E_{max} [kNm mrad]	Failure mode
(a) Steel base								
sbct-1	0	−0.74	−44.7	16.6	−0.91	−199.6	145.8	Base bending
sbct-2	0	−0.62	−68.4	9.1	−0.74	−191.9	104.9	Base bending
sbct-3	0	−0.67	16.2	41.4	−1.03	−199.2	173.9	Base bending
sbct-4	0	−0.65	−25.4	25.6	−0.92	−184.9	138.1	Base bending
sbct-5	0	−0.64	−31.5	20.3	−0.85	183.7	126.2	Base bending
Average	**0**	**−0.66**	**−30.8**	**22.6**	**−0.89**	**−118.4**	**137.8**	**Base bending**
sbctf25-1	25	−0.72	−21.1	34	−0.75	−32.7	16.4	Base bending
sbctf25-2	25	−0.91	−20.6	44	−0.97	−37.8	25.4	Base bending
Average	**25**	**−0.82**	**−20.9**	**39**	**−0.86**	**−35.2**	**20.9**	**Base bending**
sbctf50-1	50	−1.07	−17.7	60.7	−1.13	−27.3	21.4	Base bending
sbctf50-2	50	−1.07	−29.5	36.1	−1.09	−37.5	24.2	Base bending
Average	50	−1.07	−23.6	48.4	−1.11	−32.4	22.8	Base bending
(b) Concrete base								
cbctf25-1	25	−0.80	−10.1	79.0	−0.90	−29.1	20.3	Base bending
	25	**−0.80**	**−10.1**	**79.0**	**−0.90**	**−29.1**	**20.3**	
cbctf50-1	50	−0.95	−12.4	77.0	−1.08	−32.0	24.6	Base bending + Local buckling
	50	**−0.95**	**−12.4**	**77.0**	**−1.08**	**−32.0**	**24.6**	
cbctf75-1	75	−1.04	−9.9	105.2	−1.23	−29.8	26.1	Base bending + Local buckling

(continued)

Table 2.13 (continued)

	Axial load [kN]	M_y [kNm]	ϕ_y [mrad]	$S_{j,ini}$ [kNm/rad]	M_{max} [kNm]	ϕ_{max} [mrad]	E_{max} [kNm mrad]	Failure mode
	75	**−1.04**	**−9.9**	**105.2**	**−1.23**	**−29.8**	**26.1**	
(c) Athen base (bolted + welded)								
asbctf25-1	25	−0.96	−26.8	35.9	−1.14	−76.1	60.0	Weld failure + Local buckling
	25	**−0.96**	**−26.8**	**35.9**	**−1.14**	**−76.1**	**60.0**	

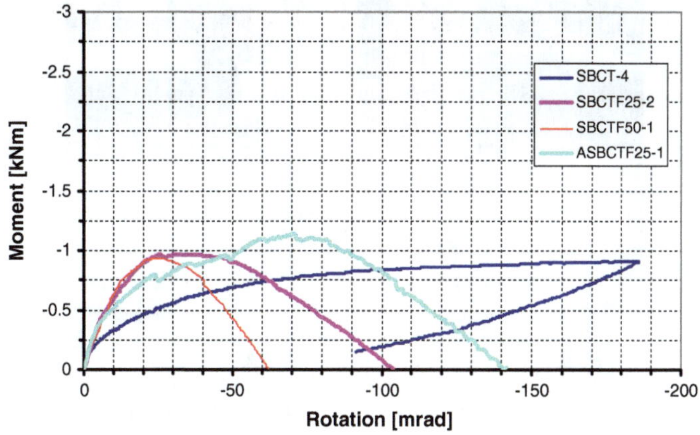

Fig. 2.42 Comparison of the moment-rotation curves for column-base connections bent in cross-aisle direction, with anchor bolts in compression zone, connected to a steel deck, under different values of the axial load

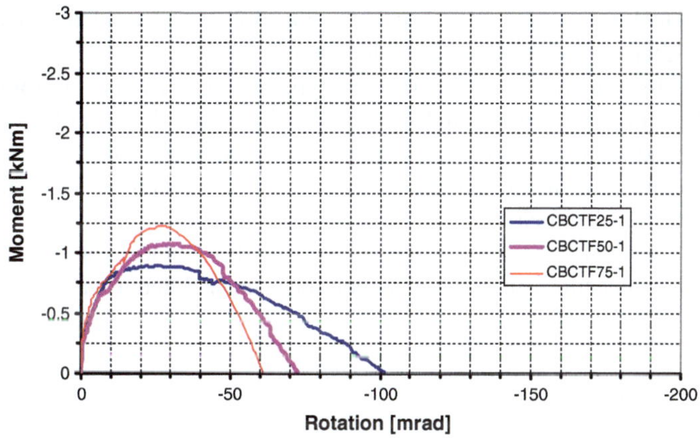

Fig. 2.43 Comparison of the moment-rotation curves for column-base connections bent in cross-aisle direction, with anchor bolts in compression zone, connected to a concrete deck, under different values of the axial load

No base plate bending was observed in the case of bolted-and-welded connections.

2.3.1.2 Down-Aisle

Table 2.14 summarizes the results of monotonic tests carried out on column-base connections in down-aisle direction.

Fig. 2.44 Typical collapse modes for specimens bent in cross-aisle direction with bolts in the compression zone

Also in this case different values of the axial load were considered (0, 25, 50 and 75 kN) as well as different types of connection of the upright steel base to the floor: bolted to a steel or to a concrete floor, as well as bolted-and-welded to a steel deck (as in the Athens dynamic shaking table tests).

Table 2.14 summarizes, for the different types of connections, and for each test, the yield bending moment M_y and rotation ϕ_y, the initial elastic stiffness $S_{j,ini}$, the maximum bending moment M_{max} and associated rotation ϕ_{max}, the energy absorbed up to the attainment of the maximum rotation ϕ_{max}, as well as the failure mode exhibited by the specimen.

Figures 2.46 and 2.47 show a comparison of the moment-rotation curves for column-base connections in down-aisle bending, under different values of the axial

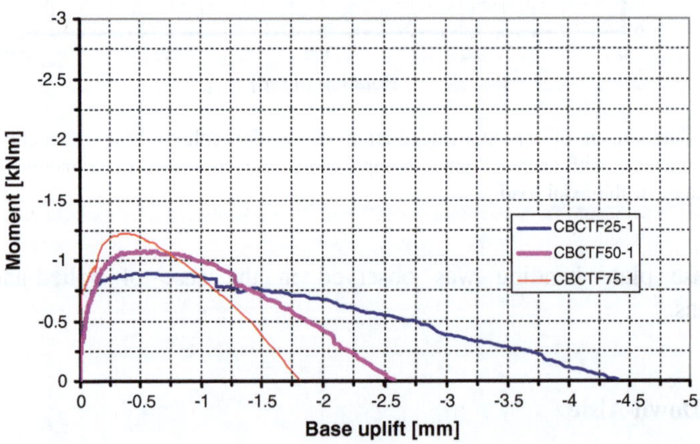

Fig. 2.45 Comparison of the moment-uplift curves for column-base connections bent in cross-aisle direction, with anchor bolts in compression zone, connected to a concrete deck, under different values of the axial load

Table 2.14 Monotonic tests on column-bases bent in down-aisle direction: (a) bolted to a steel surface, (b) bolted to a concrete surface (c) bolted-and-welded to a steel surface

	Axial load [kN]	M_y [kNm]	ϕ_y [mrad]	$S_{j,ini}$ [kNm/rad]	M_{max} [kNm]	ϕ_{max} [mrad]	E_{max} [kNm mrad]	Failure mode
(a) Steel base								
sbdb-1	0	1.66	28.3	58.7	1.70	30.2	31.9	Weld failure
sbdb-2	0	1.46	17.8	81.8	2.04	79.1	118.7	Weld failure
sbdb-3	0	1.95	40.2	48.5	2.36	140.9	254.7	Base bending
sbdb-4	0	1.06	19.3	55.0	1.16	46.1	41.0	Weld failure
sbdb-5	0	1.90	42.1	45.2	2.44	167.4	294.6	Base bending
Average	**0**	**1.61**	**29.5**	**57.8**	**1.94**	**92.7**	**148.2**	**All**
Average	**0**	**1.39**	**21.8**	**65.2**	**1.63**	**51.8**	**63.9**	**Weld failure**
Average	**0**	**1.93**	**41.2**	**46.9**	**2.40**	**154.2**	**274.7**	**Base bending**
sbdbf25-1	25	1.76	13.0	135.7	1.83	25.1	34.7	Base bending
sbdbf25-2	25	1.82	18.4	98.7	2.00	37.3	52.4	Base bending
sbdbf25-3	25	1.30	13.0	99.8	1.42	25.2	23.3	Weld failure
Average	**25**	**1.63**	**14.8**	**111.4**	**1.75**	**29.2**	**36.8**	**All**
Average	**25**	**1.30**	**13.0**	**99.8**	**1.42**	**25.2**	**23.3**	**Weld failure**
Average	**25**	**1.79**	**15.7**	**117.2**	**1.92**	**31.2**	**43.5**	**Base bending**
sbdbf50-1	50	1.65	14.4	114.4	1.82	31.5	40.8	
sbdbf50-2	50	2.49	14.5	172.0	2.61	26.7	51.8	Distorsional buckling + base bending
Average	**50**	**2.07**	**14.5**	**143.2**	**2.21**	**29.1**	**46.3**	
(b) Concrete base								
cbdb-1	0	1.57	27.7	56.6	2.37	173.9	305.1	Base bending
	0	**1.57**	**27.7**	**56.6**	**2.37**	**173.9**	**305.1**	

(continued)

Table 2.14 (continued)

	Axial load [kN]	M_y [kNm]	ϕ_y [mrad]	$S_{j,ini}$ [kNm/rad]	M_{max} [kNm]	ϕ_{max} [mrad]	E_{max} [kNm mrad]	Failure mode
cbdbf25-1	25	1.50	12.9	116.2	1.69	31.5	38.1	Base bending
cbdbf25-2	25	1.47	17.9	81.9	1.58	43.7	49.6	Base bending
Average	**25**	**1.49**	**15.4**	**99.1**	**1.63**	**37.6**	**43.8**	**Base bending**
cbdbf50-1	50	2.09	11.5	181.7	2.26	25.1	41.7	Distorsional buckling + base bending
cbdbf50-2	50	1.94	11.2	173.6	2.15	23.6	37.5	Distorsional buckling + base bending
cbdbf50-3		1.87	11.2	167.3	2.04	25.5	39.5	Distorsional buckling + base bending
Average	**50**	**1.97**	**11.3**	**174.2**	**2.15**	**24.7**	**39.5**	**Distorsional buckling + base bending**
cbdbf75-1	75	1.62	8.4	191.9	1.76	16.5	19.6	Distorsional buckling
	75	**1.62**	**8.4**	**191.9**	**1.76**	**16.5**	**19.6**	
(c) Athen base (bolted + welded)								
asbdbf25-1	25	2.43	21.0	115.6	2.37	60.8	117.0	Bolt slip
	25	**2.43**	**21.0**	**115.6**	**2.37**	**60.8**	**117.0**	

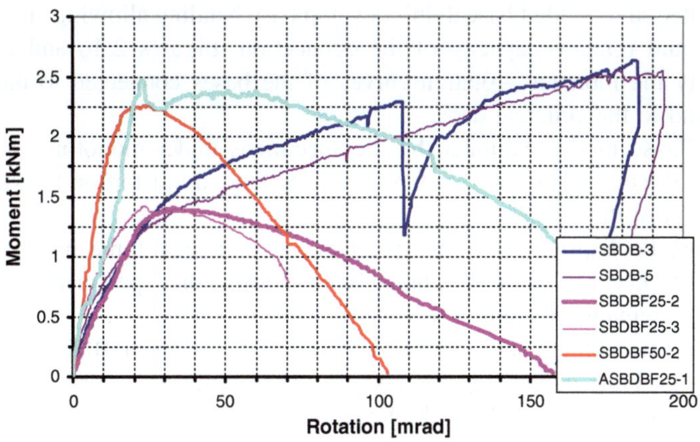

Fig. 2.46 Comparison of the moment-rotation curves for column-base connections bent in down-aisle direction, connected to a steel deck, under different values of the axial load

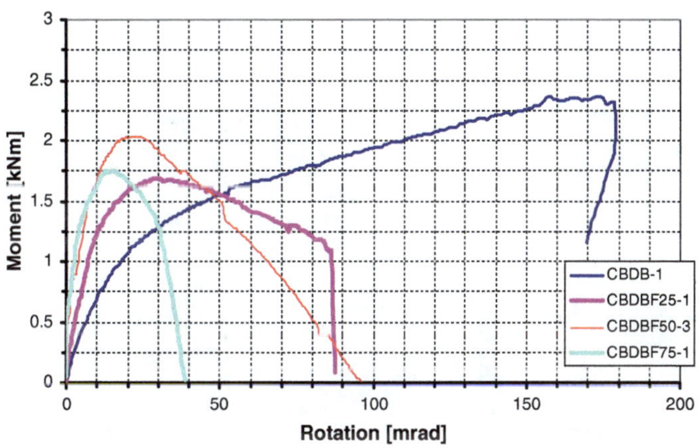

Fig. 2.47 Comparison of the moment-rotation curves for column-base connections bent in down-aisle direction, connected to a concrete deck, under different values of the axial load

load, connected respectively to a steel deck (Fig. 2.46) and to a concrete deck (Fig. 2.47).

The tests performed on column-base connections bent in down-aisle show that the application of an axial force of 25 and 50 kN increases the initial stiffness and resistance, but decreases the rotation capacity of the connection. With an axial force of 75 kN the connection does not exhibit great difference in terms of initial stiffness when compared to the specimen under 50 kN, but both resistance and ductility are lower.

The bolted-and-welded base didn't develop base bending allowing an increase of resistance and rotation capacity of the connection. Figures 2.46 and 2.47 show respectively the moment—rotation curves of the bases connected to the metallic member and to the concrete deck.

The collapse modes exhibited by the specimens under monotonic bending in down-aisle direction were weld failure, base bending and distortion buckling in the cross section of the upright.

The typical collapse mode of the specimens without axial force and with an axial force of 25 kN was base bending. Nevertheless, the specimen SBDBF25-3 exhibited a premature collapse due to weld fracture.

The specimen bolted-and-welded to the steel deck (ASBDBF25-1) evidenced sliding of the bolts connecting the base to the upright, as a consequence of the restraint imposed by the weld to the bending of the base plate.

Specimens with an axial force of 50 kN exhibited distortion buckling of the cross section of the upright associated to base bending. On the other hand, in the tests with an axial force of 75 kN, there was only distortion buckling of the cross section. The typical failure modes of the specimens tested under monotonic bending in down-aisle direction are shown in Fig. 2.48.

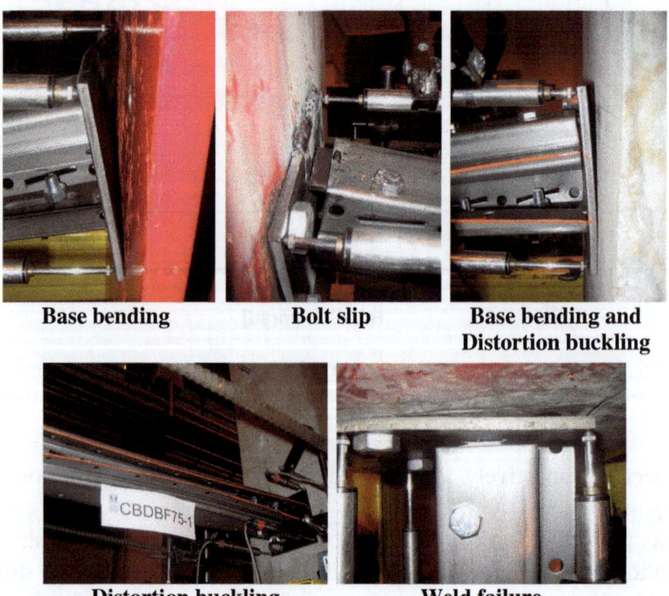

Fig. 2.48 Collapse modes for specimens under monotonic bending in down-aisle direction

2.3.2 *Cyclic Tests*

2.3.2.1 Cross-Aisle

The cyclic tests on column-base connections under bending in cross-aisle direction were performed considering, as in the monotonic tests, the base plates connected either to the concrete block or to the metallic member (in this case either bolted or bolted-and-welded).

The adopted testing procedure is the one recommended by ECCS (1986).

The tests on the column-base specimens connected to the concrete block (this being the most frequent application in real practice) were performed considering three different values of the compressive axial force, namely 25, 50 and 75 kN (corresponding respectively to 17, 34 and 52 % of the yield strength of the net section of the upright) as shown in Fig. 2.49. The column-base specimens connected to the steel deck were tested considering only a single axial force value of 25 kN, in order to simulate the testing conditions of the dynamic tests carried out on the shaking table at NTUA.

Figures 2.50 and 2.51 show the hysteresis loops in terms of moment and rotation respectively for the specimen simply bolted (Fig. 2.50) and bolted-and-welded (Fig. 2.51) to the steel surface, considering an axial force of 25 kN.

The hysteresis loops related to specimens under cyclic cross-aisle bending presented hereafter are plotted in such a way that positive bending moments and rotation correspond to tensile forces in the bolts, while negative bending moments and rotations correspond to situations in which bolts are in the compression zone.

The test results are summarized in Table 2.15, where in addition to the elastic parameters (yield moment M_y, yield rotation ϕ_y and initial stiffness $S_{j,ini}$) the number of plastic cycles to failure N_u and the total absorbed energy E_{tot} are also

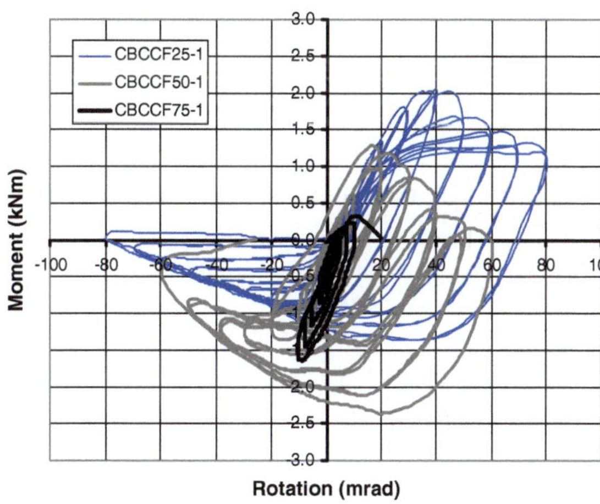

Fig. 2.49 Hysteresis loops for specimen connected to a concrete deck, under cyclic bending in cross-aisle direction and different values of the axial load

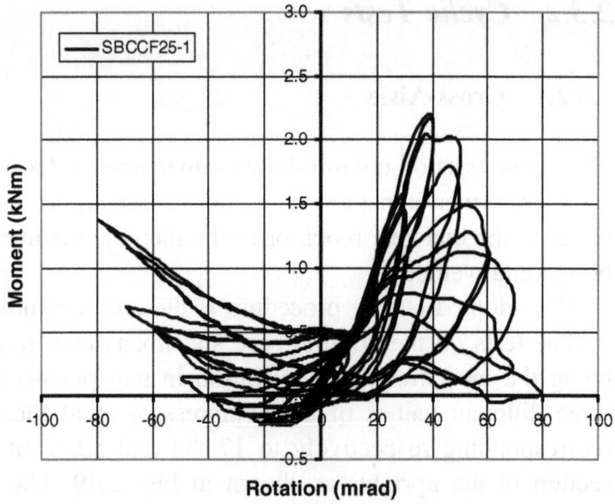

Fig. 2.50 Hysteresis loops for the specimen connected to a steel deck, under cyclic bending in cross-aisle direction

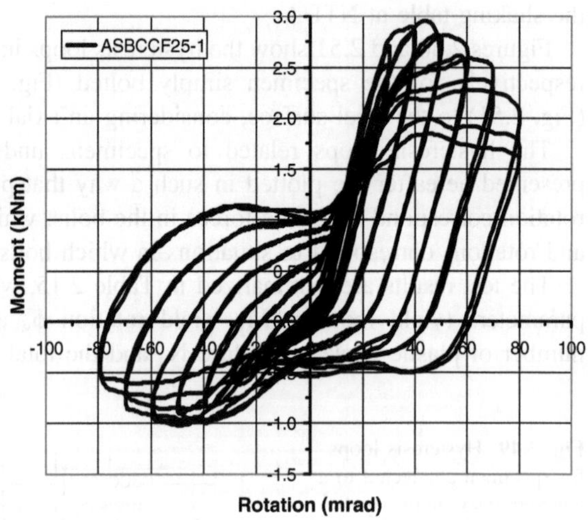

Fig. 2.51 Hysteresis loops for the specimen bolted-and-welded to a steel deck, under cyclic bending in cross-aisle direction

Table 2.15 Cyclic tests on column-bases bent in cross-aisle direction

	Base	Axial load [kN]	M_y [kNm]	ϕ_y [mrad]	$S_{j,ini}$ [kNm/rad]	N_u	E_{tot} [kNm mrad]
asbccf25-1	Athens	25	2.83	33.0	85.9	10	1756
sbccf25-1	Steel	25	1.45	20.0	72.4	8	496
cbccf25-1	Concrete	25	2.00	26.3	75.8	12	1305
cbccf50-1	Concrete	50	0.92	10.3	89.7	9	634
cbccf75-1	Concrete	75	0.31	3.5	89.2	1	27.8

presented. Assessment of the number of cycles to failure was performed according to (Bernuzzi et al. 1997a, 2000).

Collapse of specimens with axial force of 25 kN was due to base bending under negative bending moments (when bolts are in the compression zone).

For specimens with an axial force of 50 and 75 kN (Fig. 2.49), collapse occurred as a consequence of distortion buckling of the cross section of the upright. As in the monotonic tests, the higher the axial force, the lower the strength of the specimen under positive bending moments (bolts in tension). On the other hand, presence of the axial force is beneficial to the resistance of the specimens, reducing the base bending deformation that develops when the base is under negative bending moments, with bolts in the compression zone.

Specimen CBCCF75-1 exhibited a very low load carrying capacity when the base was under negative bending with bolts in the compression zone, due to evident distortion buckling. Consequently, the test was stopped at the first plastic cycle.

Welding between the base and the steel deck increased the resistance of specimen ASBCCF25-1 (Fig. 2.51). The shape of the half cycle with negative bending moment is different when compared to those of specimens CBCCF25-1 (Fig. 2.49) and SBCCF25-1 (Fig. 2.50). This difference is due to the lack of bending deformation of the base plate, that is restrained by the fillet welds, and allows an increase of dissipated energy.

2.3.2.2 Down-Aisle

As in the case of cross-aisle testing, also the cyclic tests on column-base connections under bending in down-aisle direction were performed considering the base connected either to a concrete block or to a metallic member (in this case either bolted or bolted-and-welded).

The adopted testing procedure is the one recommended by ECCS (1986).

The tests on the column-base specimens connected to the concrete block (being the most frequent application in real practice) were performed considering three values of the compressive axial force: 25, 50 and 75 kN, as shown in Fig. 2.53. The column-base specimens connected to the steel deck were tested considering only an axial force of 25 kN, in order to simulate the testing conditions of the dynamic tests carried out on the shaking table at NTUA.

Figures 2.52 and 2.54 show the hysteresis loops in terms of moment and rotation respectively for the specimen simply bolted (Fig. 2.52) and bolted-and-welded (Fig. 2.54) to the steel surface, considering an axial force of 25 kN.

The test results are summarized in Table 2.16, where in addition to the elastic parameters (yield moment M_y, yield rotation ϕ_y and initial stiffness $S_{j,ini}$) the number of plastic cycles to failure N_u and the total absorbed energy E_{tot} are also presented.

Despite of the geometrical symmetry of the column base in this loading configuration, the load carrying capacity of the specimens under negative bending moment is slightly higher than that exhibited under positive bending.

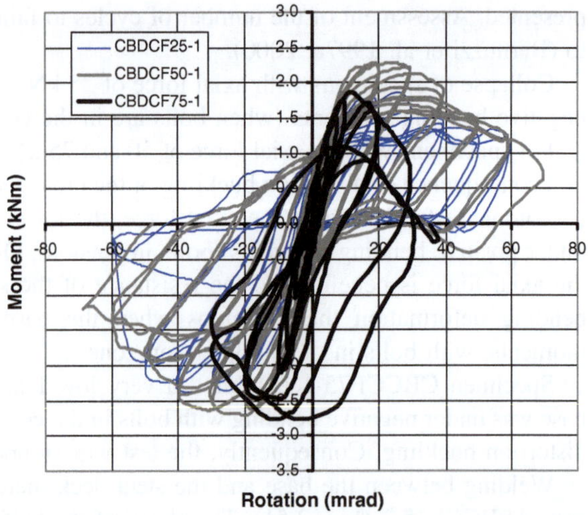

Fig. 2.52 Hysteresis loops for specimens connected to a concrete deck, under cyclic bending in down-aisle direction and different values of the axial load

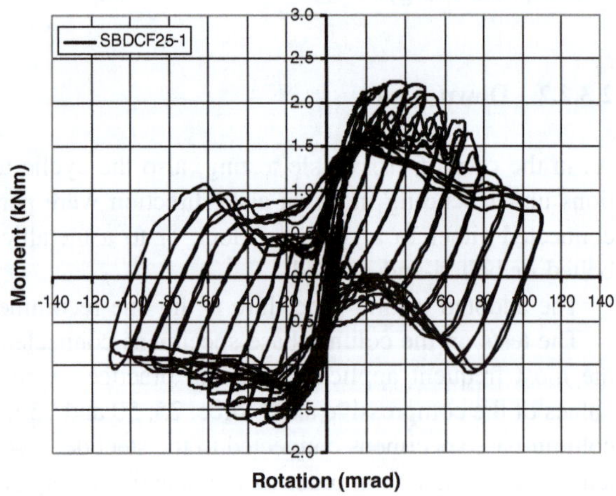

Fig. 2.53 Hysteresis loops for the specimen connected to a steel deck, under cyclic bending in down-aisle direction

This is probably due to the combined effect of the axial load superimposed to an initial out-of-straightness of the specimens, caused by the gravity load of the test apparatus (approximately estimated in 0.3 kN), that resulted in second order effects, leading to an un-symmetric response of the specimens, that enhanced when increasing the axial force.

In any case, the axial load improved the load carrying capacity of the column base under both positive and negative bending. This was true for all specimens, with the exception of CBDCF75-1 (Fig. 2.53) due to the premature distortional buckling of the upright cross section.

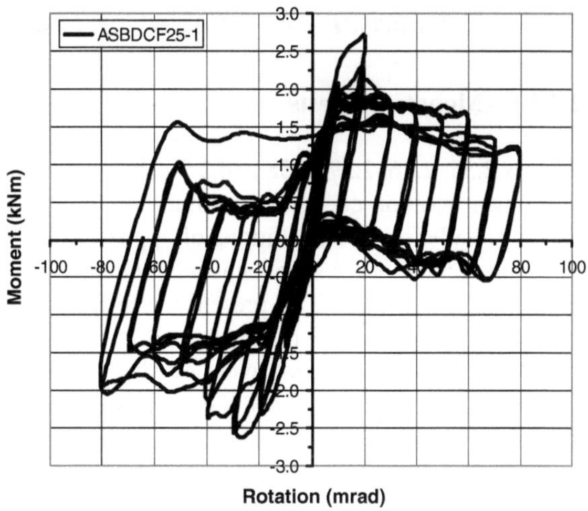

Fig. 2.54 Hysteresis loops for the specimen bolted-and-welded to a steel deck, under cyclic bending in down-aisle direction

Table 2.16 Cyclic tests on column-bases bent in down-aisle direction

	Base	Axial load [kN]	M_y [kNm]	ϕ_y [mrad]	$S_{j,ini}$ [kNm/rad]	N_u	E_{tot} [kNm mrad]
asbdcf25-1	Athens	25	2.57	15.4	167.4	14	2123
sbdcf25-1	Steel	25	1.94	15.0	129.7	20	3460
cbdcf25-1	Concrete	25	1.75	20.7	84.8	8	806
cbdcf50-1	Concrete	50	2.17	12.8	169.4	11	1196
cbdcf75-1	Concrete	75	1.80	9.5	190.7	6	389

The base plate welded and bolted to the metallic deck (specimen ASBDCF25-1, Fig. 2.54) improved the resistance, stiffness and energy dissipation capacity of the connection, when compared to the specimen simply bolted to the steel deck (specimen SBDCF25-1, Fig. 2.52). Weld failure, however occurred in specimen ASBDCF25-1 after 14 plastic cycles, under a maximum rotation of 80 mrad, while specimen SBDCF25-1, simply bolted to the steel deck, sustained 20 cycles in the plastic range, with a maximum rotation capacity of approximately 110 mrad, and consequently a larger total absorbed energy E_{tot}.

Specimen CBDCF25-1 collapsed due to weld failure, while specimen CBDCF50-1 collapsed due to plate bending and onset of distortional buckling in the upright cross section.

2.3.3 Comparison and Analysis of Test Results

2.3.3.1 Monotonic Tests

Figures 2.55, 2.56, 2.57, 2.58 and 2.59 show a comparison of the response, in terms of moment-rotation curves, of column-base connections for different directions of bending, under similar conditions of applied axial load and "deck connection".

Figure 2.55 refers to specimens without axial load and connected to a steel deck. Figures 2.56, 2.57 and 2.58 present the results for specimens connected to a concrete deck, respectively with 25, 50 and 75 kN axial load, while Fig. 2.59 refers to specimens with 25 kN axial load bolted-and-welded to a steel deck.

Tables 2.17 and 2.18 show a comparison of the mean values of the results of monotonic tests on column-base connections bent in cross-aisle direction, respectively with bolts in tension (Table 2.17) and in compression (Tables 2.18), while Table 2.19 refers to the case of down-aisle bending. Each table shows the comparison of the results obtained for different types of deck connection (namely bolted to a concrete deck (b) and either bolted (a) or bolted-and-welded (c) to a steel deck).

Table 2.20 shows a comparison of the mean values of the results obtained in monotonic tests on column-bases bent in cross-aisle direction with bolts in tension versus bolts in compression zone. Similarly, Table 2.21 compares the results of cross-aisle bending with bolts in tension with those of down-aisle bending, while Table 2.22 compares the results of down-aisle bending with those of cross-aisle bending with bolts in compression.

Examining Table 2.17 it can be noticed that in monotonic bending tests with bolts in tension, increasing the axial load in the column always results in an increment of stiffness but in a reduction of strength, ductility and energy absorption capacity of the specimens bolted either to a steel or a concrete deck. Bolting and welding the specimens to a steel surface is always beneficial with respect to simply bolting to either a concrete or a steel deck, under the same axial load.

In monotonic bending tests with bolts in the compression zone (Table 2.18), increasing the axial load in the column results in an increment of stiffness, strength,

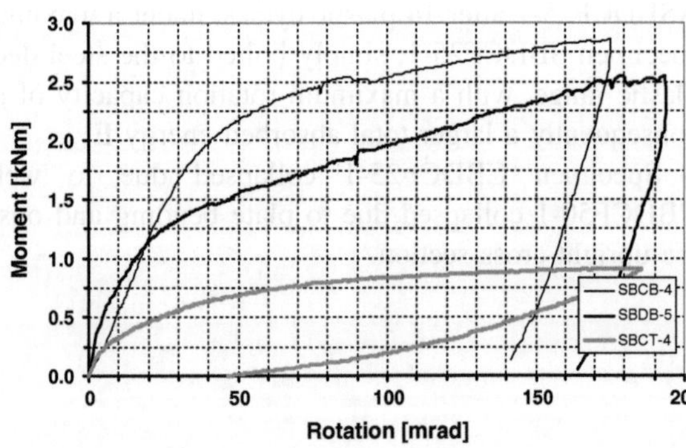

Fig. 2.55 Comparison of the moment-rotation curves for column-base connections, without axial load and connected to a steel deck, for different directions of bending

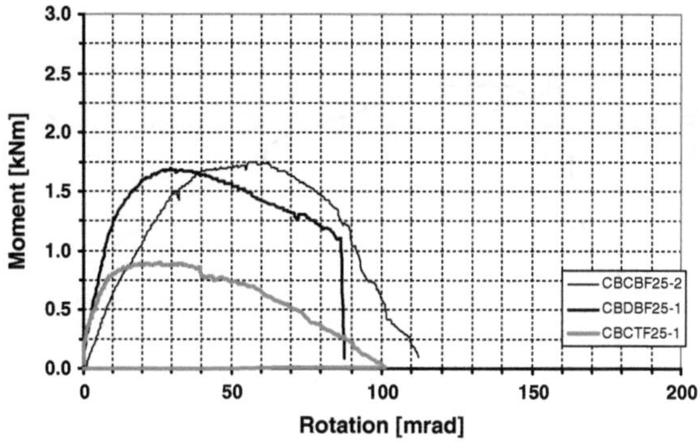

Fig. 2.56 Comparison of the moment-rotation curves for column-base connections, with 25 kN axial load and connected to a concrete deck, for different directions of bending

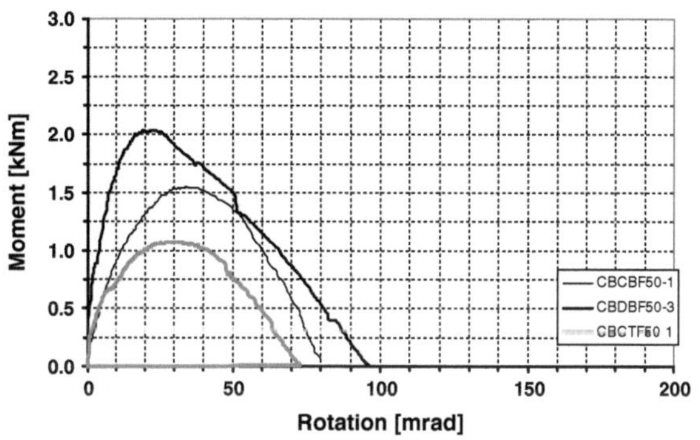

Fig. 2.57 Comparison of the moment-rotation curves for column-base connections, with 50 kN axial load and connected to a concrete deck, for different directions of bending

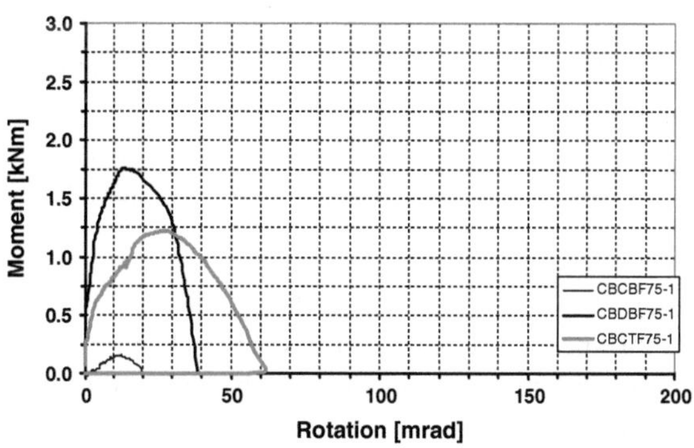

Fig. 2.58 Comparison of the moment-rotation curves for column-base connections, with 75 kN axial load and connected to a concrete deck, for different directions of bending

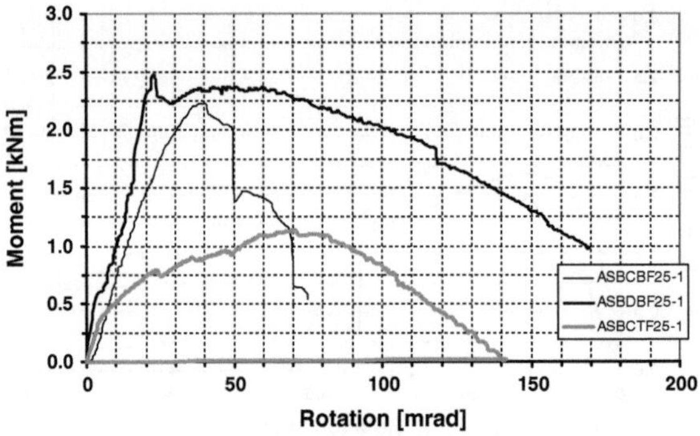

Fig. 2.59 Comparison of the moment-rotation curves for column-base connections, with 25 kN axial load bolted-and-welded to a steel deck, for different directions of bending

Table 2.17 Comparison of the results of monotonic tests on column-bases bent in cross-aisle direction, with bolts in tension

M_y	ϕ_y	$S_{j,ini}$	M_{max}	ϕ_{max}	E_{max}	
(a) Steel base						
1.14	1.58	0.72	1.35	4.02	7.12	sbcb Base bending/weld failure
0.95	0.85	1.15	0.90	0.70	0.66	sbcb All/base bending
1.08	1.35	0.83	1.21	2.81	4.67	sbcb All/weld failure
0.90	0.69	1.27	0.83	0.40	0.27	sbcbf25 /sbcb
0.88	0.78	1.16	0.88	0.77	0.70	sbcbf50 /sbcbf25
0.80	0.54	1.47	0.73	0.31	0.19	sbcbf50 /sbcb
(b) Concrete base						
1.01	1.05	1.01	0.97	0.83	0.75	cbcbf50 /cbcbf25
0.33	0.50	0.62	0.33	0.52	0.18	cbcbf75 /cbcbf50
0.33	0.52	0.63	0.32	0.43	0.13	cbcbf75 /cbcbf25
(c) Athen base (bolted + welded)						
1.20	1.21	0.99	1.15	0.93	0.91	asbcbf25/sbcbf25
1.44	1.22	1.18	1.36	0.94	1.03	asbcbf25/cbcbf25

ductility and energy absorption capacity of the specimens bolted a concrete deck. On the contrary, for specimens bolted to a steel deck, increasing the axial load results in an increment of stiffness and strength, but in a reduction of ductility and energy absorption capacity. Bolting and welding the specimens to a steel surface is always beneficial with respect to strength, ductility and energy absorption capacity with respect to simply bolting either to a concrete or to a steel deck, under the same axial load.

Table 2.18 Comparison of the results of monotonic tests on column-bases bent in cross-aisle direction, with bolts in compression zone

M_y	ϕ_y	$S_{j,ini}$	M_{max}	ϕ_{max}	E_{max}	
(a) Steel base						
1.24	0.68	1.73	0.97	0.30	0.15	sbctf25 /sbct
1.30	1.13	1.24	1.29	0.92	1.09	sbctf50 /sbctf25
1.62	0.77	2.14	1.25	0.27	0.17	sbctf50 /sbct
(b) Concrete base						
1.19	1.23	0.97	1.20	1.10	1.21	cbctf50 /cbctf25
1.09	0.80	1.37	1.14	0.93	1.06	cbctf75 /cbctf50
1.30	0.98	1.33	1.37	1.02	1.29	cbctf75 /cbctf25
(c) Athen base (bolted + welded)						
1.17	1.28	0.92	1.33	2.16	2.87	asbctf25/sbctf25
1.20	2.65	0.45	1.27	2.62	2.96	asbctf25/cbctf25

In the case of monotonic bending in down-aisle direction (Table 2.19) it can be noticed that increasing the axial load results in an increment of strength and stiffness, but in a reduction of ductility and energy absorption capacity for specimens bolted to a steel deck. When specimens are bolted to a concrete surface, increasing the axial load results in an increment of stiffness, a reduction of ductility and energy absorption capacity, while strength increases for low and intermediate values of the applied axial force (up to 50 kN), but reduces for higher values of the axial load (75 kN).

Bolting and welding the specimens to a steel surface is, in this loading direction, always beneficial for stiffness, strength, ductility and energy absorption capacity with respect to simply bolting either to a concrete or to a steel deck, under the same axial load.

When comparing the main response parameters of monotonic tests in cross-aisle bending with bolts in tension with the corresponding parameters related to tests performed with bolts in compression zone (Table 2.20), it can be noticed that the yield and maximum bending moments as well as the maximum absorbed energy of tests performed with bolts in tension are always larger than those of tests performed with bolts in compression zone, when the axial load is less than 75 kN.

The yield and maximum rotation of specimens tested with bolts in tension are always larger than those of tests performed with bolts in compression zone, with the exception of specimens bolted to a steel deck, under a 50 kN axial load (for which is $\phi_{y,CT} > \phi_{y,CB}$) and of those bolted to a concrete surface with an axial load of 75 kN (for which is $\phi_{max,CT} > \phi_{max,CB}$). Stiffness of specimens connected to a steel deck (simply bolted or bolted-and-welded) tested with bolts in tension was always higher than that of specimens tested with bolts in compression zone. The opposite is true for specimens bolted to a concrete deck, for which $S_{j,ini,CT} > S_{j,ini,CB}$. This is probably due to problems in the bond between the anchor bolts and the concrete or eventually to cracking of the concrete under tensile loads.

Table 2.19 Comparison of the results of monotonic tests on column-bases bent in down-aisle direction

M_y	ϕ_y	$S_{j,ini}$	M_{max}	ϕ_{max}	E_{max}	
(a) Steel base						
1.39	1.89	0.72	1.47	2.98	4.30	sbdb Base bending/weld failure
0.83	0.72	1.23	0.81	0.60	0.54	sbdb All/base bending
1.16	1.35	0.89	1.19	1.79	2.32	sbdb All/weld failure
1.38	1.21	1.17	1.35	1.24	1.87	sbdbf25 Base bending/weld failure
0.91	0.94	0.95	0.91	0.94	0.85	sbdbf25 All/base bending
1.25	1.14	1.12	1.23	1.16	1.58	sbdbf25 All/weld failure
1.01	0.50	1.93	0.90	0.31	0.25	sbdbf25/sbdb All
0.94	0.60	1.53	0.87	0.49	0.36	sbdbf25/sbdb Weld failure
0.93	0.38	2.50	0.80	0.20	0.16	sbdbf25/sbdb Base bending
1.27	0.98	1.29	1.26	1.00	1.26	sbdbf50/sbdbf25
1.29	0.49	2.48	1.14	0.31	0.31	sbdbf50/sbdb
(b) Concrete base						
0.95	0.56	1.75	0.69	0.22	0.14	cbdbf25/cbdb
1.32	0.73	1.76	1.32	0.66	0.90	cbdbf50/cbdbf25
1.25	0.41	3.08	0.91	0.14	0.13	cbdbf50/cbdb
0.82	0.74	1.10	0.82	0.67	0.50	cbdbf75/cbdbf50
1.09	0.55	1.94	1.08	0.44	0.45	cbdbf75/cbdbf25
1.03	0.30	3.39	0.74	0.09	0.06	cbdbf75/cbdb
(c) Athens base (bolted + welded)						
1.49	1.42	1.04	1.35	2.08	3.18	asbdbf25/sbdbf25
1.63	1.36	1.17	1.45	1.62	2.67	asbdbf25/cbdbf25

When comparing the main response parameters of monotonic tests in cross-aisle bending with bolts in tension with the corresponding parameters related to tests performed in down-aisle direction, (Table 2.21) the following considerations can be drawn: stiffness $S_{j,ini}$ in down-aisle direction is always higher than in cross-aisle direction, in presence of axial load in the column. Both the yield M_y and maximum M_{max} bending moments in down-aisle are larger than in cross-aisle, for axial load larger than 25 kN. The yield rotation ϕ_y in cross-aisle direction, with bolts in tension is always larger than that in down-aisle direction. The same consideration is

Table 2.20 Comparison of the results of monotonic tests on column-bases bent in cross-aisle direction: bolts in tension versus bolts in compression

M_y	ϕ_y	$S_{j,ini}$	M_{max}	ϕ_{max}	E_{max}	
(a) Steel base						
3.33	1.13	2.90	2.91	1.05	1.87	sbcb/sbct All
3.52	1.32	2.52	3.25	1.50	2.85	sbcb/sbct Base bending
2.41	1.15	2.13	2.49	1.42	3.37	sbcb25/sbct25
1.64	0.79	1.99	1.70	1.18	2.17	sbcb50/sbct50
(b) Concrete base						
2.06	2.35	0.88	2.01	1.71	3.07	cbcb25/cbct25
1.75	2.02	0.91	1.63	1.28	1.90	cbcb50/cbct50
0.52	1.25	0.41	0.47	0.72	0.32	cbcb75/cbct75
(c) Athen base (bolted + welded)						
2.47	1.08	2.28	2.16	0.61	1.07	asbcb25/asbct25

Table 2.21 Comparison of the results of monotonic tests on column-bases in cross-aisle bending with bolts in tension versus down-aisle bending

M_y	ϕ_y	$S_{j,ini}$	M_{max}	ϕ_{max}	E_{max}	
(a) Steel base						
1.37	1.18	1.13	1.34	1.34	1.74	sbcb/sbdb All
1.47	1.18	1.20	1.31	0.85	0.86	sbcb/sbdb Weld failure
1.20	0.99	1.21	1.20	1.15	1.43	sbcb/sbdb Base bending
1.21	1.62	0.74	1.22	1.71	1.92	sbcb25/sbdb25
0.85	1.29	0.67	0.86	1.32	1.07	sbcb50/sbdb50
(b) Concrete base						
1.10	1.54	0.70	1.11	1.32	1.42	cbcb25/cbdb25
0.84	2.21	0.40	0.82	1.66	1.18	cbcb50/cbdb50
0.33	1.48	0.23	0.33	1.30	0.42	cbcb75/cbdb75
(c) Athen base (bolted + welded)						
0.98	1.38	0.71	1.04	0.76	0.55	asbcb25/asbdb25

true also for the maximum rotation ϕ_{max}, with the exception of the case when the base is bolted-and-welded to the steel deck.

When comparing the main response parameters of monotonic tests in cross-aisle bending with bolts in compression zone, with the corresponding parameters related to tests performed in down-aisle direction (Table 2.22), the following considerations can be drawn: stiffness $S_{j,ini}$ as well as both the yield M_y and maximum M_{max}

Table 2.22 Comparison of the results of monotonic tests on column-bases in down-aisle bending versus cross-aisle bending with bolts in compression

M_y	ϕ_y	$S_{j,ini}$	M_{max}	ϕ_{max}	E_{max}	
(a) Steel base						
2.44	0.96	2.56	2.18	0.78	1.08	sbdb/sbct All
2.92	1.34	2.08	2.70	1.30	1.99	sbdb/sbct Base bending
1.99	0.71	2.86	2.03	0.83	1.76	sbdb25/sbct25
1.93	0.61	2.96	1.99	0.90	2.03	sbdb50/sbct50
(b) Concrete base						
1.86	1.52	1.25	1.81	1.29	2.16	cbdb25/cbct25
2.07	0.91	2.26	1.99	0.77	1.61	cdb50/cbct50
1.56	0.85	1.82	1.43	0.55	0.75	cbdb75/cbct75
(c) Athen base (bolted + welded)						
2.53	0.78	3.22	2.08	0.80	1.95	asbdb25/asbct25

bending moments in down-aisle direction are always higher than in cross-aisle direction. The yield rotation ϕ_y as well as the maximum rotation ϕ_{max}, in cross-aisle direction, with bolts in compression zone, are always larger than those in down-aisle direction, in presence of an axial load in the column. The maximum absorbed energy in down-aisle direction is always larger than in cross-aisle direction, for axial loads lower than 75 kN.

Figures 2.60, 2.61, 2.62, 2.63, 2.64 and 2.65 summarize the previous considerations, plotting the trend of the mean value of the main response parameters of all tests, under increasing the axial load in the column.

As a general consideration, examining Figs. 2.60 to 2.65 it is evident that there is a rather good agreement in the trends of tests performed with specimens bolted to a concrete or to a steel deck. Some scatter is evident, in particular in the case of the initial stiffness (Fig. 2.62), although this is to be expected due to the different behaviour of the anchor bolts, when connected to a steel plate or chemically "bonded" into a concrete block.

It can be observed that the yield moment M_y (Fig. 2.60), in the case of tests in cross-aisle direction with bolts in tension (cbcb and sbcb test series) decreases steadily up to an axial load of 50 kN, and drops abruptly for further increments of the axial load. In the case of tests in cross-aisle direction with bolts in the compression zone (cbct and sbct test series) M_y steadily increases with the axial load, which is beneficial in reducing the base-plate bending effect. In the case of tests in cross-aisle direction (cbdb and sbdb test series) M_y attains a maximum value for an axial load of 50 kN.

Both the yield rotation ϕ_y (Fig. 2.61) and the maximum rotation ϕ_{max} (Fig. 2.64) are always heavily reduced by the application of the axial load, although the value of ϕ_{max} doesn't seem to be much influenced by increment of the axial load beyond

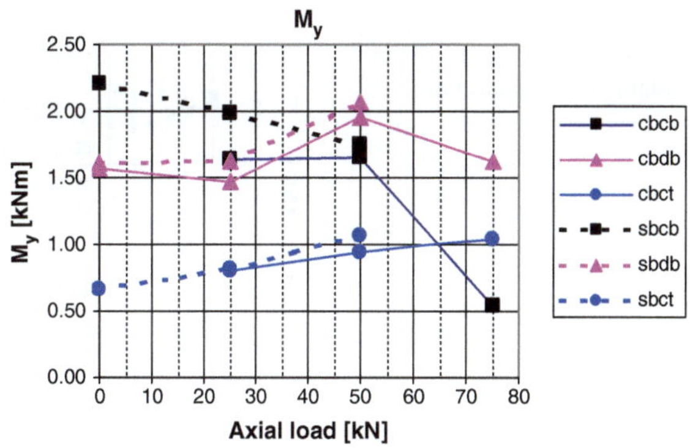

Fig. 2.60 Yield-moment versus axial load for column-base connections

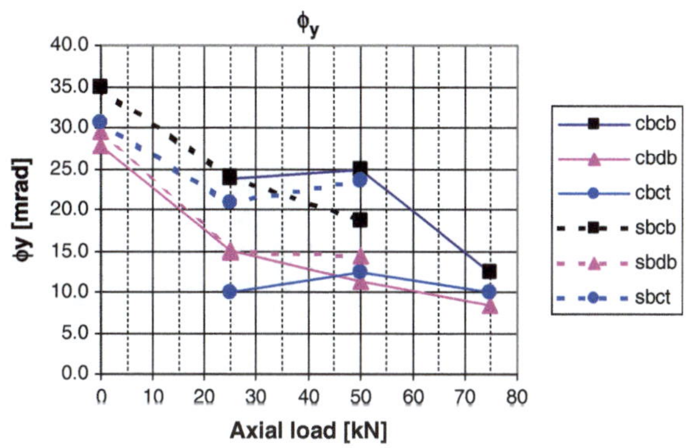

Fig. 2.61 Yield-rotation versus axial load for column-base connections

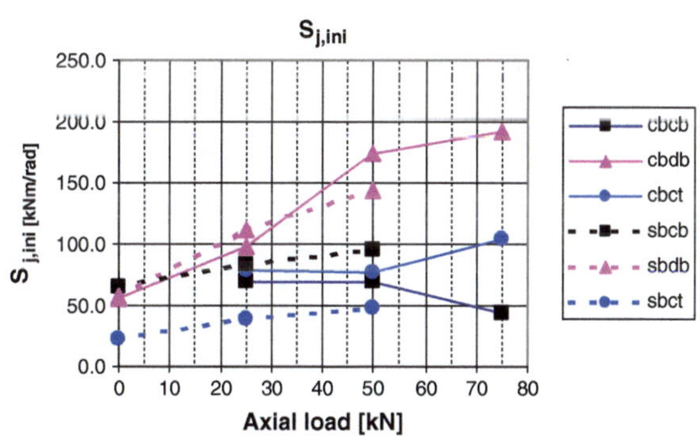

Fig. 2.62 Initial elastic stiffness versus axial load for column-base connections

25 kN. The initial elastic stiffness $S_{j,ini}$ (Fig. 2.62) increases with the axial load in the case of tests in down-aisle bending (cbdb and sbdb test series) and in cross-aisle bending with bolts in the compression zone (cbct and sbct test series). In the case of tests performed in cross-aisle bending with bolts in tension (cbcb and sbcb test series), $S_{j,ini}$ initially moderately increases with the axial load, but diminishes when the applied axial load is increased up to 75 kN.

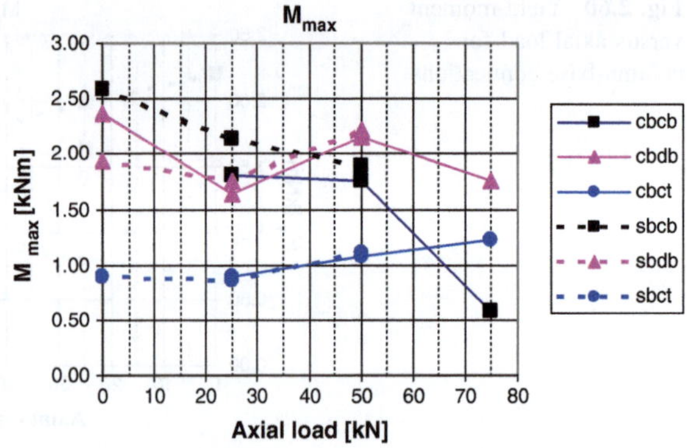

Fig. 2.63 Maximum bending strength versus axial load for column-base connections

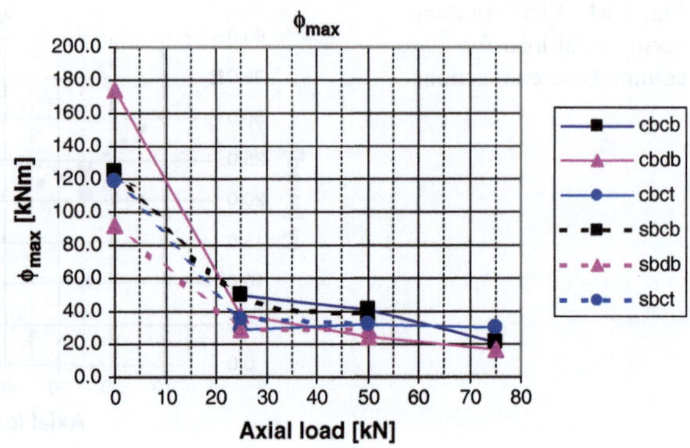

Fig. 2.64 Rotation corresponding to maximum bending strength versus axial load for column-base connections

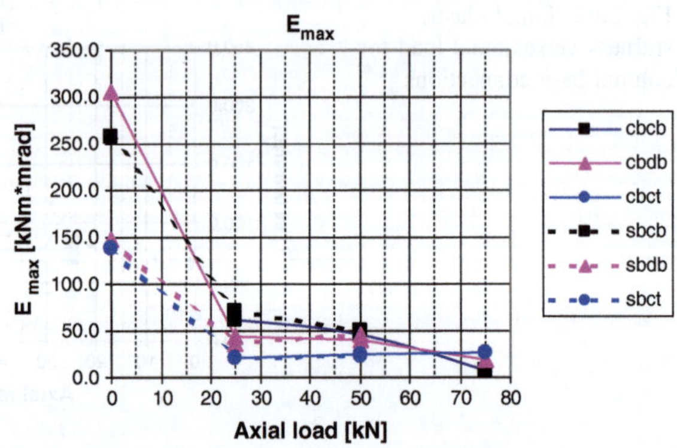

Fig. 2.65 Absorbed energy corresponding to maximum bending strength versus axial load for column-base connections

The maximum bending strength M_{max} (Fig. 2.63), in the case of tests in cross-aisle direction with bolts in tension (cbcb and sbcb test series) decreases steadily up to an axial load of 50 kN, and drops abruptly for further increments of the axial load. This is due to onset of distortion buckling of the cross section of the column. In the case of tests in cross-aisle direction with bolts in compression zone (cbct and sbct test series)

M_{max} steadily increases with the axial load. In the case of tests in cross-aisle direction (cbdb and sbdb test series) M_{max} oscillates around a practically constant value. The same considerations drawn for the maximum rotation ϕ_{max} apply also to the case of the maximum absorbed energy E_{tot} (Fig. 2.65). This parameter is however more sensible to increments of the axial load beyond 50 kN.

2.3.3.2 Cyclic Tests

The following Figs. 2.66, 2.67, 2.68, 2.69, 2.70, 2.71, 2.72, 2.73, 2.74, 2.75 and 2.76 show comparisons of the hysteretic behaviour of the specimens, tested under different loading and/or restraint conditions, in terms of applied bending moment versus rotation.

It should be pointed out that the hysteresis loops related to specimens under cyclic cross-aisle bending presented hereafter are plotted in such a way that positive bending moments and rotation correspond to tensile forces in the bolts, while negative bending moments and rotations correspond to situations in which bolts are in compression zone.

Figures 2.66 to 2.70 present a comparison of the specimens response, under the same axial load, but bent in different directions (namely cross and down-aisle).

Figure 2.66 refers to the specimens simply bolted to the steel deck, with a 25 kN axial load, and shows the poor response in cross-aisle direction of this type of connection, due to the low load carrying capacity when bolts are in the compression zone.

Figures 2.67, 2.68 and 2.69 refer to specimens simply bolted to the concrete deck, respectively with a 25, 50 and 75 kN axial load.

Response in cross-aisle direction of this type of connection, when bolts are in the compression zone, is better than the one evidenced by specimens connected to a

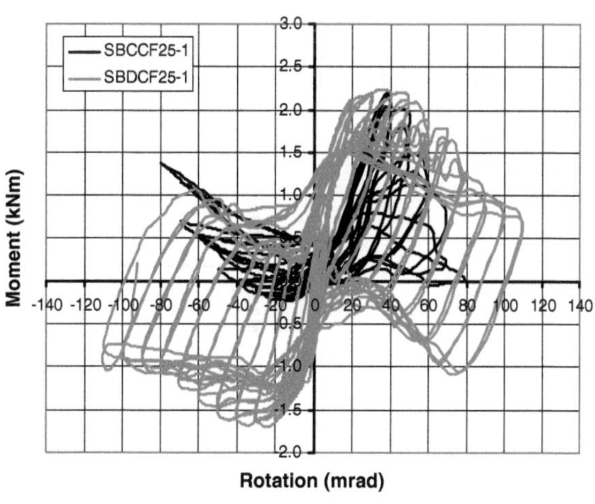

Fig. 2.66 Comparison of the hysteretic behaviour of column-bases connected to a steel deck, with 25 kN axial load, for different bending directions

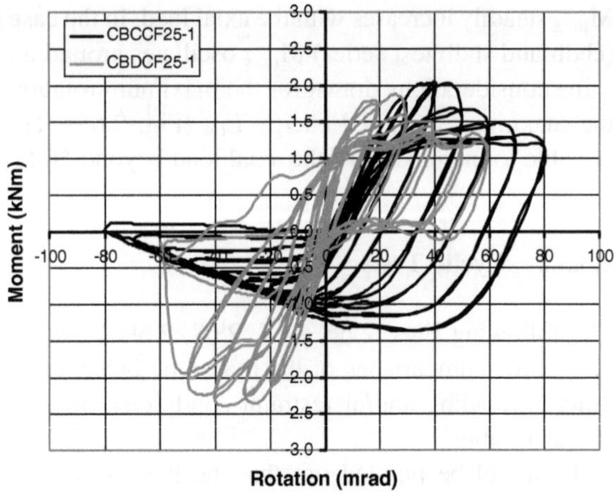

Fig. 2.67 Comparison of the hysteretic behaviour of column-bases connected to a concrete deck, with 25 kN axial load, for different bending directions

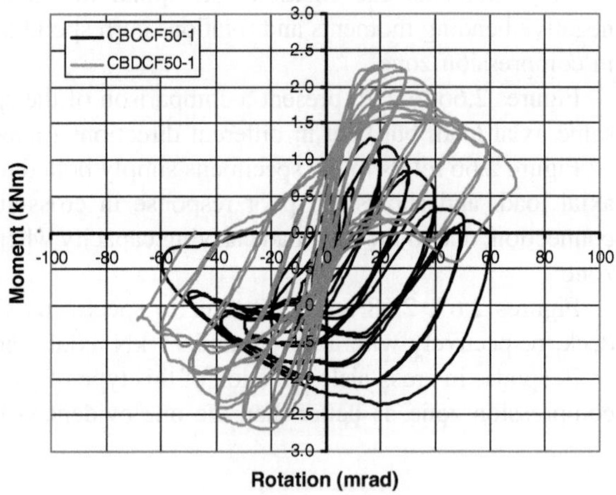

Fig. 2.68 Comparison of the hysteretic behaviour of column-bases connected to a concrete deck, with 50 kN axial load, for different bending directions

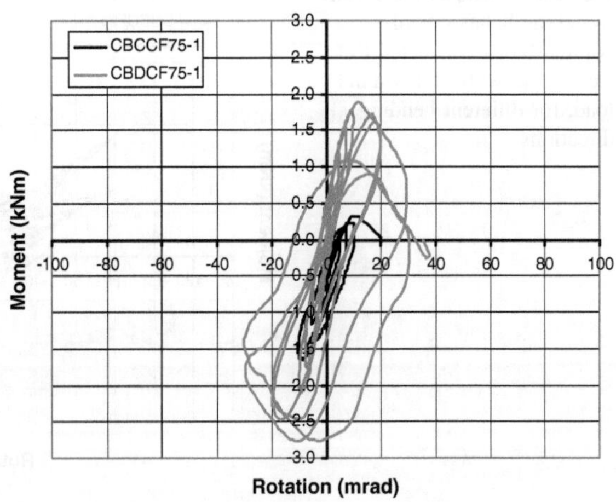

Fig. 2.69 Comparison of the hysteretic behaviour of column-bases connected to a concrete deck, with 75 kN axial load, for different bending directions

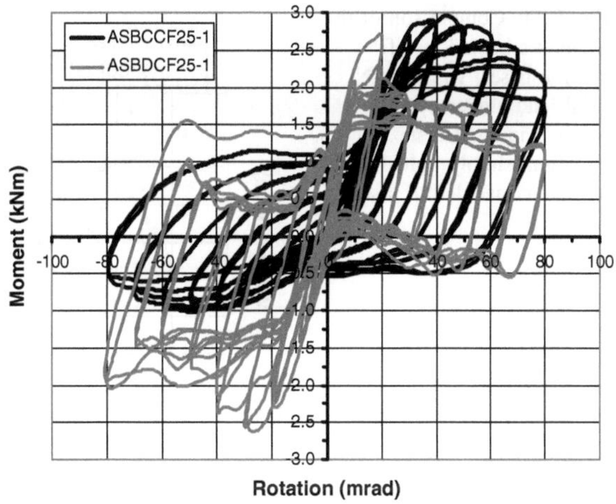

Fig. 2.70 Comparison of the hysteretic behaviour of column-bases bolted-and-welded to a steel deck, with 25 kN axial load, for different bending directions

steel deck, most probably due to the local deformability of the concrete surface, or in any case to the different behaviour at the base-deck interface.

It is also evident that increasing the axial load up to 50 kN is beneficial for the specimen response in down-aisle direction. In cross-aisle direction, increasing the axial load up to 50 kN results in a reduction of rotation capacity. The load carrying capacity is also reduced when bolts are in tension (due to occurrence of buckling at the free edges of the cross section of the column profile which are in compression). On the contrary, the bending strength is improved by increasing the axial load for those conditions in which the bolts are in the compression zone (as, in this case, the axial load reduces the bending deformation of the steel base plate).

Further increasing of the axial load to 75 kN causes a loss of load-carrying as well as rotation capacity of the specimens in both bending directions (as shown in Fig. 2.69).

Figure 2.70 refers to the specimens bolted-and-welded to the steel deck, with a 25 kN axial load, and shows the good response in both cross-aisle and down-aisle direction of this type of connection in terms of both load-carrying and rotation capacity.

Figures 2.71 and 2.72 compare the response of specimens simply bolted and bolted-and-welded to a steel deck, under 25 kN axial load, respectively under bending in cross-aisle and in down-aisle direction. Similarly, Figs. 2.73 and 2.74 compare the response of specimens simply bolted to a concrete floor and bolted-and-welded to a steel deck, under 25 kN axial load, respectively under bending in cross-aisle and in down-aisle direction. The beneficial effect of the bolted-and-welded type of connection is evident, when comparing the behaviour of specimens connected to a steel floor, particularly for cross-aisle bending. When compared to specimens simply bolted to a concrete floor, the response of bolted-and-welded base connections shows a small enhancement of both load-carrying and rotation capacity in down-aisle direction. In cross-aisle direction,

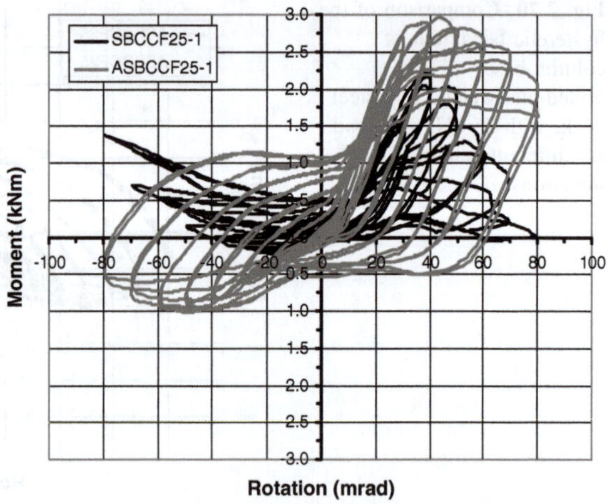

Fig. 2.71 Comparison of the hysteretic behaviour of column-bases, with 25 kN axial load, under cyclic bending in cross-aisle direction, simply bolted and bolted-and-welded to a steel deck

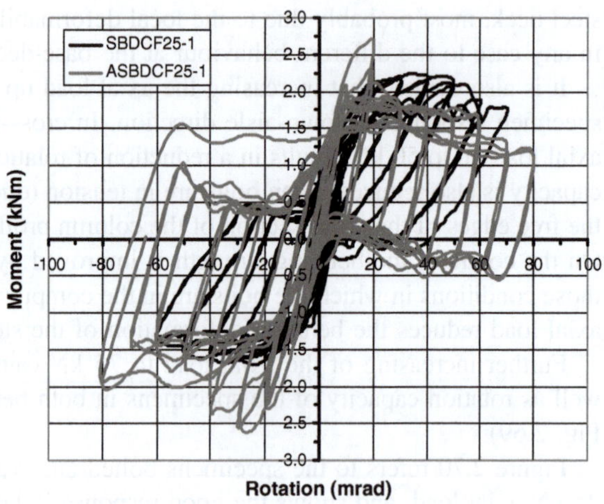

Fig. 2.72 Comparison of the hysteretic behaviour of column-bases, with 25 kN axial load, under cyclic bending in down-aisle direction, simply bolted and bolted-and-welded to a steel deck

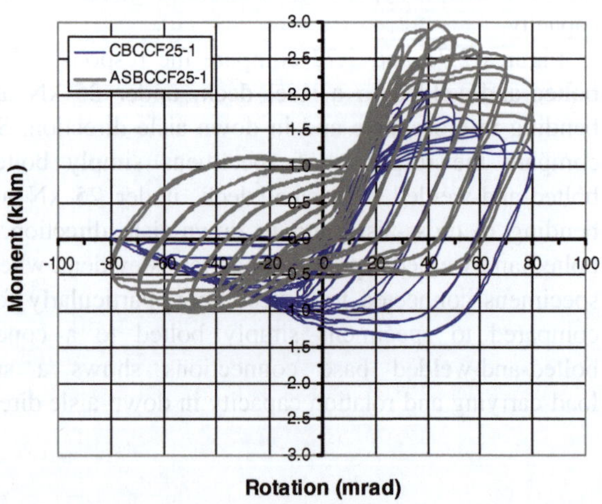

Fig. 2.73 Comparison of the hysteretic behaviour of column-bases, with 25 kN axial load, under cyclic bending in cross-aisle direction, connected to a concrete deck or bolted-and-welded to a steel deck

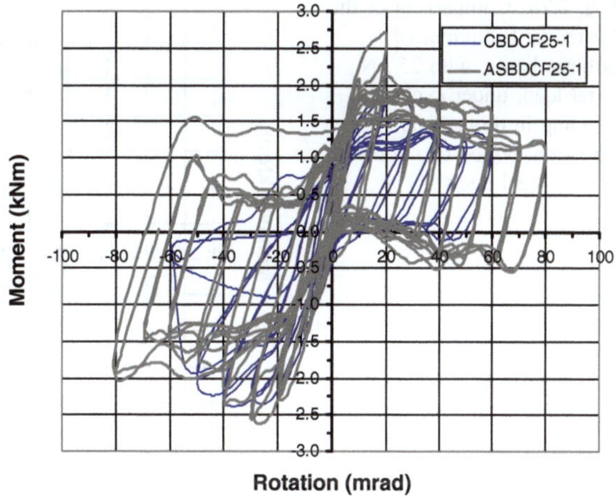

Fig. 2.74 Comparison of the hysteretic behaviour of column-bases, with 25 kN axial load, under cyclic bending in down-aisle direction, connected to a concrete deck or bolted-and-welded to a steel deck

the response of the bolted-and-welded specimens is superior when bolts are in tension, but is comparable in terms of load carrying and rotation capacity with that of specimens connected to the concrete floor when bolts are in the compression zone, despite a much higher energy dissipation capacity evidenced by the shape of the hysteresis loops.

Figures 2.75 and 2.76 compare the response of specimens simply bolted to a steel and to a concrete deck, under 25 kN axial load, respectively under bending in cross-aisle and in down-aisle direction.

The response of the specimens bolted to a steel deck is similar to that of the specimens bolted to a concrete floor, in the case of cross-aisle bending, with bolts in tension. When the bending direction is reversed, and bolts are in the compression zone, the specimens connected to the concrete deck show higher load-carrying and rotation capacity (Fig. 2.75).

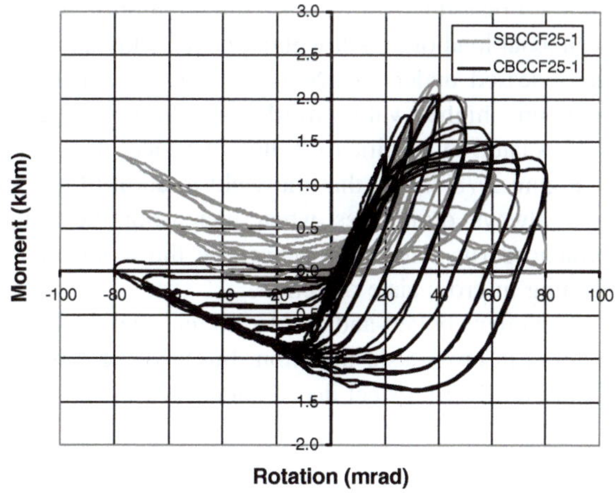

Fig. 2.75 Comparison of the hysteretic behaviour of column-bases, with 25 kN axial load, under cyclic bending in cross-aisle direction, bolted to a concrete or to a steel deck

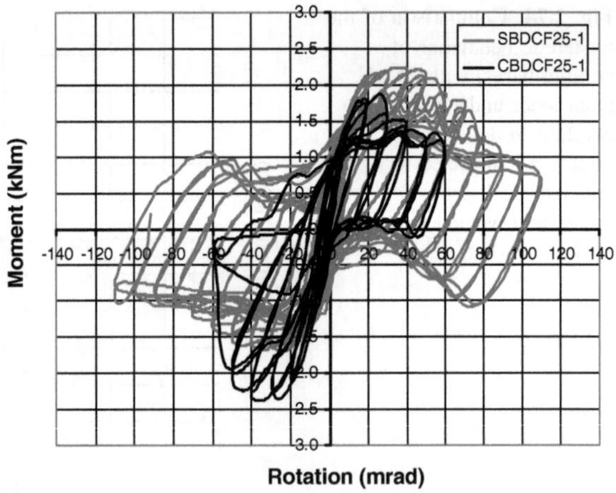

Fig. 2.76 Comparison of the hysteretic behaviour of column-bases, with 25 kN axial load, under cyclic bending in down-aisle direction, bolted to a concrete or to a steel deck

In down-aisle direction, specimens connected to a steel deck show a larger rotation capacity than those connected to a concrete floor. However, under bending with bolts in the compression zone, the load-carrying capacity of specimens connected to a concrete floor is higher.

Table 2.23 summarizes these type of comparisons, in terms of ratios of the main response parameters (yield moment M_y and rotation ϕ_y, initial elastic stiffness $S_{j,ini}$ number of plastic cycles to failure N_u and total absorbed energy E_{tot}) of the various specimens.

Figures 2.77, 2.78, 2.79 and 2.80 compare the main response parameters versus the applied axial load for specimens tested under cyclic bending. Information is available for all six combinations of floor connection and bending directions only in the case of 25 kN axial load. In fact, tests with higher values of the applied axial load were carried out only on specimens connected to a concrete floor.

Figure 2.77 shows that the yield moment of specimens connected to a concrete floor diminishes when increasing the axial load, for bending in cross-aisle direction, while for bending in down-aisle direction presents a maximum value corresponding to an axial load of 50 kN while, under 75 kN axial load, M_y is comparable with the one assessed under 25 kN axial load. Specimens bolted-and-welded to the steel floor show highest values of M_y for both bending directions, while the lowest value is associated to specimens bolted to a steel deck, under cross-aisle bending.

Figure 2.78 shows that the yield rotation of specimens connected to a concrete floor always diminishes when increasing the axial load. ϕ_y is however more influenced (in a negative sense) by the axial load when the specimen is under bending in cross-aisle direction.

Specimens bolted-and-welded to the steel floor show the highest values of ϕ_y for bending in cross-aisle direction, but also the lowest (together with the specimens simply bolted to a steel deck) under down-aisle bending.

Table 2.23 Comparison of results of cyclic tests on column-bases

M_y	ϕ_y	$S_{j,ini}$	N_u	E_{tot}	
(a) Steel base					
1.34	0.75	1.79	2.50	6.98	sbdcf25 /sbccf25
0.73	0.76	0.96	0.67	0.38	sbccf25 /cbccf25
1.11	0.72	1.53	2.50	4.29	sbdcf25 /cbdcf25
(b) Concrete base					
0.88	0.79	1.12	0.67	0.62	cbdcf25 /cbccf25
2.36	1.24	1.89	1.22	1.89	cbdcf50 /cbccf50
0.17	0.37	0.47	3.00	13.99	cbdcf75 /cbccf75
0.46	0.39	1.18	0.75	0.49	cbccf50 /cbccf25
1.24	0.62	2.00	1.38	1.48	cbdcf50 /cbdcf25
0.34	0.34	0.99	0.22	0.04	cbccf75 /cbccf50
0.83	0.74	1.13	0.55	0.33	cbdcf75/cbdcf50
0.16	0.13	1.18	0.17	0.02	cbccf75 /cbccf25
1.03	0.46	2.25	0.75	0.48	cbdcf75 /cbdcf25
(c) Athens base (bolted + welded)					
0.91	0.47	1.95	1.40	1.21	asbdcf25 /asbccf25
1.95	1.65	1.19	1.25	3.54	asbccf25 /sbccf25
1.32	1.03	1.29	0.70	0.61	asbdcf25 /sbdcf25
1.42	1.25	1.13	0.83	1.35	asbccf25/cbccf25
1.47	0.74	1.97	1.75	2.63	asbdcf25/cbdcf25

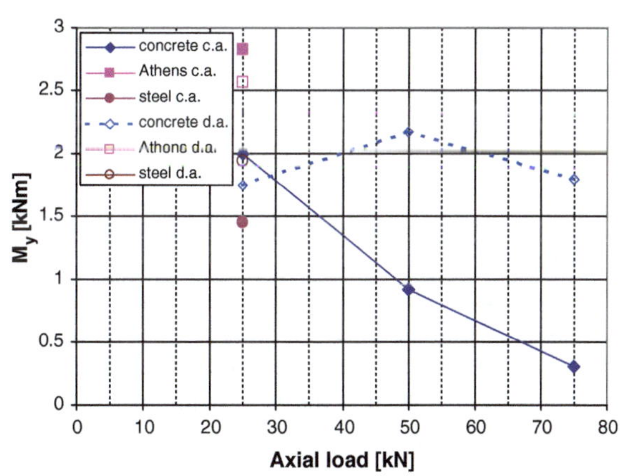

Fig. 2.77 Yield moment versus axial load

Figure 2.79 shows that the initial elastic bending stiffness $S_{j,ini}$ of specimens connected to a concrete floor always increases when increasing the axial load.

Specimens bolted-and-welded to the steel floor show the highest values of $S_{j,ini}$ for bending in down-aisle direction. Stiffness $S_{j,ini}$ of specimens connected to a concrete floor (under both bending directions) is similar and comparable with those

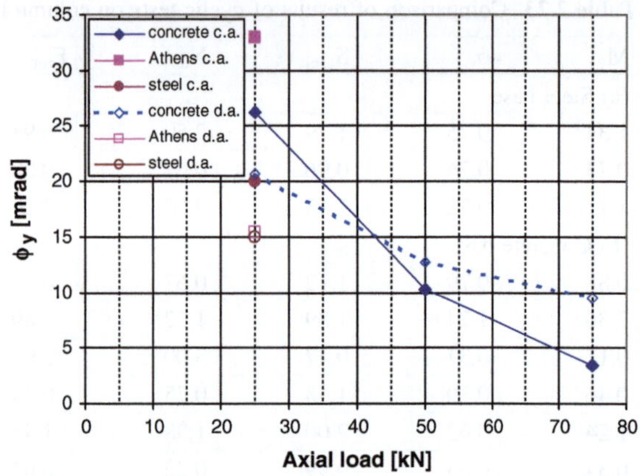

Fig. 2.78 Yield rotation versus axial load

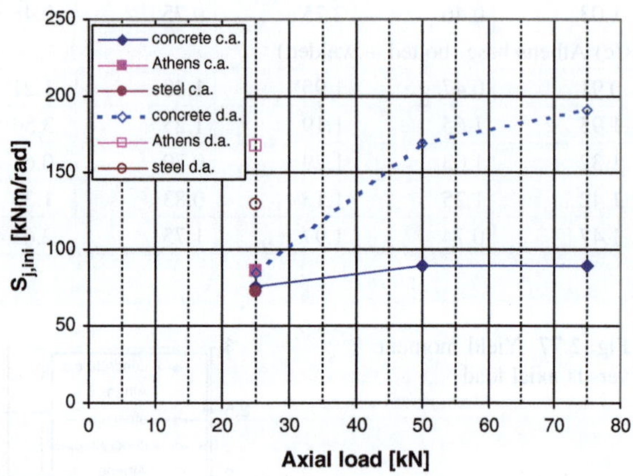

Fig. 2.79 Initial elastic stiffness versus axial load

of specimens simply bolted and bolted-and-welded under bending in cross-aisle direction.

Figure 2.80 shows that the number of cycles to failure for specimens under cross-aisle bending always diminishes when increasing the axial load while, under down-aisle bending, attains a maximum for an applied axial load of 50 kN. The specimen that sustained the largest number of plastic cycles was that bolted to a steel floor under down-aisle bending. Specimen simply bolted to a steel deck under cross-aisle bending and simply bolted to a concrete slab under down-aisle bending exhibited the same number of plastic cycles to failure.

The trend of the total absorbed energy E_{tot} is similar to the one of N_u as shown in Fig. 2.81. E_{tot} for specimens under cross-aisle bending always diminishes when increasing the axial load while, under down-aisle bending, attains a maximum for an applied axial load of 50 kN. The specimen that could absorb the largest amount

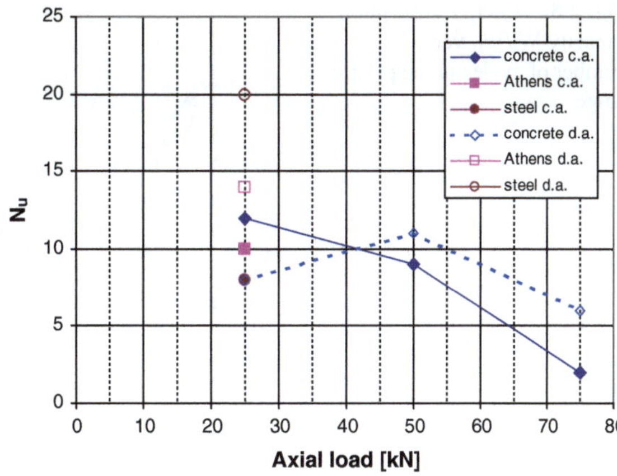

Fig. 2.80 Number of plastic cycles to failure versus axial load

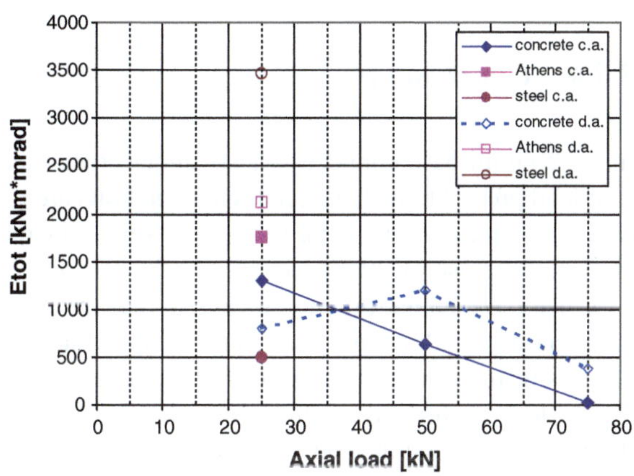

Fig. 2.81 Total absorbed energy versus axial load

of energy was the one bolted to a steel floor under down-aisle bending. Specimen simply bolted to a steel deck under cross-aisle bending exhibited the lowest energy absorption capacity.

Figure 2.82 shows the trend of the energy absorption capacity in terms of the ratio W_i/W_0 versus the number of cycles in the plastic range applied to the specimen, where W_i is the ratio of the energy absorbed in *i*th cycle to the energy absorbable in the same cycle by an elastic perfectly-plastic specimen, while W_0 is the ratio of the energy absorbed in the 1st cycle in the plastic range to the energy absorbable in the same cycle by an elastic perfectly-plastic specimen.

It can be noticed that all specimens show a similar trend of W_i/W_0, that abruptly drops (loosing approximately 40-60 % of energy dissipation capacity) in the first three or four plastic cycles, and then stabilizes until failure.

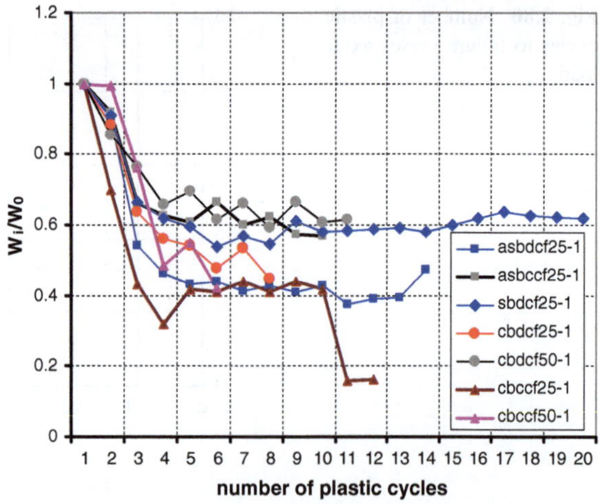

Fig. 2.82 Energy absorption capacity W_i/W_0 versus number of cycles in the plastic range

2.3.4 *Conclusions*

In the column base connection tests under cross-aisle bending, the axial compression load proved to be beneficial when the bolts are in the compression zone. In this case, an increase of the axial force results in an increase of resistance and stiffness, but in a decrease of rotation capacity of the specimens.

When the loading direction results in tension in the bolts, the axial compression load causes a reduction of resistance and stiffness of the column bases, because induces distortional buckling of the free edges of the cross-section profile.

The tests performed on the column base in the down-aisle direction proved that the axial compression load (25 and 50 kN) increases the initial stiffness and the resistance but decreases the rotation capacity of the connection.

In the cyclic tests, the higher the axial compression force, the bigger the difference between the resistance under positive and negative bending moments.

The collapse modes exhibited by the specimens were weld failure, base bending and distortion buckling of the cross section of the upright. In some tests premature weld failure caused a reduction in the rotation capacity of the connection. Therefore, it is extremely important to control the welding process during the manufacturing of the components.

Distortion buckling of the cross section of the upright occurred in the tests with an axial compression force of 50 and 75 kN, except for the monotonic test when the loading direction is such that bolts are in the compression zone. In the tests in cross-aisle direction with axial force of 75 kN distortion buckling occurred prematurely, drastically reducing the mechanical properties of the connection.

The base welded-and-bolted to a steel deck exhibited an increase of resistance and rotation capacity of the connection since there was no base plate bending. In the cyclic tests a higher capacity of energy dissipation was observed, when compared to simply bolted connections.

The results prove that connecting the column base to a concrete slab or to a steel deck does not substantially change the mechanical properties or the failure mode of the connection.

In both cases, however, pre-tension in the bolts should be provided.

References

ACI. (1999). *Acceptance criteria for moment frames based on structural testing*. Farmington Hills: American Concrete Institute, ACI, 374.

ATC-24. (1992). *Guidelines for cyclic testing of components of steel structures*. Redwood City, CA: Applied Technology Council.

Bernuzzi, C., Calado, L., & Castiglioni, C. A. (1997a). Ductility and load carrying capacity prediction of steel beam-to-column connections under cyclic reversal loading. *Journal of Earthquake Engineering, 1*(2), 401–432.

Bernuzzi, C., Calado, L., & Castiglioni, C. A. (2000). Low cycle fatigue of structural steel components: a method for re-analysis of test data and a design approach based on ductility. *Indian Society of Earthquake Technology (ISET), Special Issue on Experimental Techniques* (pp. 47–63), December.

ECCS. (1986). *Recommended testing procedure for assessing the behaviour of structural elements under cyclic loads*. European Convention for Constructional Steelworks, Publication No. 45.

Proença, J., Calado, L., & Ferreira, J. (1994). Some case studies of cyclic tests in RC and steel structural subassemblages. In S. Gomes et al. (Eds.), *Recent advances in experimental mechanics*. Balkema. *Proceedings of the 10th International Conference on Experimental Mechanics* (pp. 495–500), Lisbon.

Further Readings

Agatino M. R., Bernuzzi, C., & Castiglioni C. A. (2001). Joints under cyclic reversal loading in steel storage pallet racks. In *Proceedings. XVIII C.T.A. Conference, Venezia* (Vol. 2, pp. 105–114).

AS4084. (1993). *Steel storage racking*. Standards Australia.

Baldassino, N., Berardi, C., & Bernuzzi, C. (2001). Influence of the bracing systems on the performance of steel storage pallet racks. In *Proceedings of XVIII Conference C.T.A.* Venezia (pp. 115–125).

Baldassino, N., & Bernuzzi, C. (2000). Analysis and behaviour of steel storage pallet racks. *Thin-Walled Structures, 37*(4), 277–304.

Baldassino, N., Bernuzzi, C., & Zandonini R. (1999). Structural analysis of steel storage pallet racks. In *Proceedings of the 6th International Colloquium on Stability and Ductility of Steel Structures (SDSS'99)*, Timisoara, Romania (pp. 465–475).

Baldassino, N., Bernuzzi, C., & Zandonini R. (2000). Performance of joints in steel storage pallet racks. In *Proceedings of Connections in Steel Structures IV: Steel Connections in the New Millennium, Roanoke, Virginia, USA.*

Baldassino, N., Bernuzzi, C., Zandonini, R., & Hancock, G. (1998). Overall, local and distortional buckling in pallet racks. In *Proceedings of the Structural Stability Research Council Conference (S.S.R.C.)*, September, Atlanta U.S.A.

Baldassino, N., & Hancock, G. (1999). Distortional buckling of cold-formed steel storage rack sections including perforations. In P. Makelainen, P. Hassinen (Eds.), *Proceedings Fourth*

International Conference on Steel and Aluminium Structures (ICSAS'99), Helsinky (pp. 131–138). Oxford: Elsevier.

Baldassino, N., & Zandonini R. (2001). Numerical and experimental analysis of base-plate connections of steel storage pallet racks. In *Proceedings of XVIII Conference C.T.A.* (127–136), Venezia.

Baldassino, N., & Zandonini, R. (2002). Design by testing of steel storage pallet racks. In *Proceedings Of EUROSTEEL—3rd European Conference on Steel Structures*, Coimbra, Portugal.

Baldassino N., & Zandonini R. (2003). Industrial steel racks: tests, design and Codes. In *Proceedings ASSCCA'03 Advanced in Structures Steel, Concrete, Composite and Aluminium,* Sidney, Australia.

Baldassino N., & Zandonini R. (2004). Performance of base-palte connections of steel storage pallet racks. In D. Dubina & D. Grecea (Eds.), *Recent advances and new trends in structural design* (pp. 11–20). Timisoara: Editura Orizonturi Universitare. *Proceedings Of International Colloquium Dedicated to the 70th Anniversary of Professor Victor Gioncu,* Timisoara, May 7–8, 2004.

Ballio, G., Bernuzzi, C., & Castiglioni, C. A. (1999). An approach for the seismic design of steel storage pallet racks, spec. In *Stahlbau*, November.

Ballio, G., Calado, L., & Castiglioni, C. A. (1997). Low cycle fatigue and fracture of structural steel members and connections. *Fatigue and Fracture of Engineering Materials and Structures, 20*(8), 1129–1146.

Ballio, G., & Castiglioni, C. A. (1994). Seismic behavior of steel sections. *Journal of Constructional Steel Research, 29*, 21–54.

Ballio, G., & Castiglioni, C. A. (1995). A unified approach for the design of steel structures under low/or high cycle fatigue. *Journal of Constructional Steel Research, 43*(1995), 75–101.

Bernuzzi, C., Calado, L., & Castiglioni C. A. (1997b). Steel beam-to-column connections: failure criteria and cumulative damage models. In *Proceedings of the STESSA Conference, Kyoto* (pp. 538–545).

Bernuzzi, C., Calado, L., & Castiglioni, C. A. (1997c). Behaviour of steel beam-to-column joints under cyclic reversal loading: an experimental study. In *Proceedings of the 5th International Colloquium on Stability and Ductility of Steel Structures, SDSS, Nagoya* (Vol. 1, pp. 335–342).

Bernuzzi, C., Calado, L., & Castiglioni C. A. (1998). Structural steel components under low-cycle fatigue: Design assisted by testing. In *Proceedings of the SEWC, San Francisco.*

Bernuzzi, C., & Castiglioni, C. A. (2001). Experimental analysis on the cyclic behaviour of beam-to-column joints in steel storage pallet racks. *Thin-Walled Structures, 39*, 841–859.

Bernuzzi, C., & Zandonini, R. (1993). *Serviceability and analysis models of steel buildings.* Goteborg, Sweden: International Colloquium on Structural Serviceability of Buildings.

Bernuzzi, C., Zandonini, R., & Zanon, P. (1991). Rotational behaviour of extended end plate. *Costruzioni Metalliche, 43*(2), 74–103.

Brescianini, J. C., Castiglioni, C. A., & Panzeri, N. (2003). Dynamic experimental tests on steel pallet racks. In *Proceedings of CTA, Genova* (pp. 107–116).

Calado, L., & Castiglioni, C. A. (1995). Low-cycle fatigue testing of semi-rigid beam-to-column connections. In *Proceedings Of 3rd International Workshop on Connections in Steel Structures, IABSE/AISC, Trento.*

Calado, L., & Castiglioni, C. A. (1996). Analise experimental do comportamento ciclico de ligacoes metalicas viga-pilar. *Mecanica Experimental, 1*, 61–72.

Calado, L., Castiglioni, C. A., & Bernuzzi, C. (1998). Behaviour of steel beam-to-column joints under cyclic reversal loading: An experimental study. In T. Usami & Y. Itoh (Eds.), *Stability and ductility of steel structures* (pp. 279–292). Amsterdam: Elsevier Science Ltd.

Calado, L., Castiglioni, C. A., Mele, E., & Ferreira, C. (1999). Damage accumulation design of steel beam-to-column connections. In *Proceedings of the 2nd Encontro Nacional de Construção Metalica e Mista, Coimbra* (pp. 551–563).

Castiglioni, C. A. (1994). Effects of local buckling on the cyclic behavior of steel members. In *S.S. R.C. 50th Anniversary Meeting, Bethlehem, PA, June 1994, Proceedings of the Technical Session* (pp. 381–395).

Castiglioni, C. A. (1995a). Seismic damage assessment of steel members and joints. In *SSRC, Theme Conference on Stability Problems related to aging, damaged and deteriorated structures, Kansas City* (pp. 55–66).

Castiglioni, C. A. (1995b). Seismic behavior of steel beam-column members. In *SSRC, Theme Conference on Stability Problems related to aging, damaged and deteriorated structures, Kansas City* (pp. 77–88).

Castiglioni, C. A. (1999). Failure criteria and cumulative damage models for steel components under low-cycle fatigue. In *Proceedings XVII C.T.A. Conference, Napoli.*

Castiglioni, C. A. (2003a). Dynamic tests on steel pallet racks. *Costruzioni Metalliche, 55*(3), 35–44.

Castiglioni, C. A. (2003b). Seismic behaviour of steel storage racks. In *Proceedings of the IV Congresso de Construcao Metalica e Mista, Lisbon* (pp. 41–62), December 4–5, 2003.

Castiglioni, C. A. (2005). Effects of the loading history on the local buckling behavior and failure mode of welded beam-to-column joints in moment-resisting steel frames. *Journal of Engineering Mechanics, ASCE, 131*(6), 568–585.

Castiglioni, C. A., Bernuzzi, C., Calado, L., & Agatino, M. R. (1997). Experimental study on steel beam-to-column joints under cyclic reversal loading. In *Proceedings Of the Northridge Earthquake Research Conference, Los Angeles* (526–533).

Castiglioni, C. A., Brescianini, J. C., & Panzeri, N. (2003a). Variable Amplitude, low-cycle fatigue testing of welded steel beam-to-column joints. *Costruzioni Metalliche, 55*(2), 35–51.

Castiglioni, C. A., & Calado, L. (1996). Seismic damage assessment and failure criteria for steel members and connections. *Proceedings of the International Conference on Advances in Steel Structures, Hong Kong, II*, 1021–1026.

Castiglioni, C. A., Mouzakis, H., & Carydis, P. (2007). Constant and variable amplitude cyclic behaviour of welded steel beam-to-column connections. *Journal of Earthquake Engineering, 11*(6), 876–902.

Castiglioni, C. A., Panzeri, N., Brescianini, J. C., & Carydis, P. (2003b). Shaking table tests on steel pallet racks. In *Proceedings STESSA 2003, Napoli* (pp. 775–781).

CEB. (1996). *RC frames under earthquake loading—State of the art report.* Comité Euro-International du Beton, Thomas Telford Ed.

CEN. (2005). *Eurocode 3: Design of Steel Structures—Part 1.1 general rules and rules for buildings*. European Committee for Standardization.

Chen, C. K. (1980). *Seismic study on industrial steel storage racks*. Prepared for the National Science Foundation, URS/ John A. Blume & Associate Engineers, June.

Davies, M., & Jiang, C. (1998). Design for distorsional buckling. *Journal of Constructional Steel Research, 46*(1–3), 174–175.

ECCS. (1992). *Analysis and design of steel frames with semi-rigid joints*. European Convention for Constructional Steelworks, Publication n. 67.

FEM 10.2.02. (2000). *The design of static steel pallet racks*. Federation Europeen de la Manutention, Vers. 1.01.

FEM 10.2.08. (2005). *The seismic design of static steel pallet racks*. Federation Europeen de la Manutention, Final draft, December 2005.

FEMA. (2005). *Seismic Considerations for Steel Storage Racks Located in Areas Accessible to the Public. FEMA 460,* September 2005.

Godley, M. H. R. (1991). Storage racking, Chapter 1. In Rhodes (Ed.), *Design of cold formed steel members* (pp. 361–399). London, UK: Elsevier Applied Science.

Hancock, G. J. (1985). Distortional buckling of steel storage rack columns. *Journal of Structural Engineering, ASCE, 111*(12), 2770–2783.

Markazi, F. D., Beale, R. G., & Godley, M. H. R. (1997). Experimental analysis of semi-rigid boltless connectors. *Thin-Walled Structures, 28*(1), 57–87.

Mazzolani, F. M., & Piluso V. (1996). *Theory and design of seismic resistant steel frames*. London: Champmann and Hall, E&FN Spon.

Proença, J., Calado, L., Castiglioni, C. A., & Tristão G. (2006). Cyclic testing on steel storage beam-to-upright subassemblages. An innovative cyclic testing procedure. In *1st European Conference on Earthquake and Seismology, Geneva*, paper n. 1152.

RAL. (1990). *Storage and associated equipment*. RAL Deutsches Institut fur Gutersicherung und Kennzeichnung (German Institute for Quality Assurance and Marketing).

RMI. (1990). *Specification for the design testing and utilization of industrial steel storage racks*. Charlotte, NC: Rack Manufacturers Institute, 1990 edition.

RMI. (1997). *Specification for the design, testing and utilization of industrial steel storageracks*. Charlotte, NC: Rack Manufactures Institute.

RMI. (2002a). *Specification for the design testing and utilization of industrial steel storage racks*. Charlotte, NC: Rack Manufacturers Institute, 2002 edition.

RMI. (2002b). *Commentary to specification for the design testing and utilization of industrial steel storage racks*. Charlotte NC: Rack Manufacturers Institute.

RMI. (2004). *Testing guidelines for pallet stacking frames*. Charlotte, NC: Rack Manufacturing Institute.

Zandonini, R., Baldassino, N., & Bernuzzi, C. (1998). Experimental and numerical studies on pallet racks. In *Proceedings of the Conference "Professor Otto Halase-Memorial Session"*, Budapest TU, Budapest, Rumania.

Zandonini, R., Baldassino, N., & Bernuzzi, C. (1999a). Structural analysis of steel storage pallet racks. In D. Dubina, & M. Ivanyi (Eds.), *Proceedings Of Stability and Ductility of Steel Structures (SDSS'99)* (pp. 465–475), Timisoara, Rumania. Oxford: Elsevier.

Zandonini, R., Baldassino, N., & Bernuzzi, C. (1999b). Influence of beam-to-column joint modelling on the performance of steel pallet racks. In G. Huber, T. Michl (Ed.), *Festschrift Commemorative Publication Prof. Dr. Ferdinand Tschemmernegg* (pp. 351–363). Innsbruck: Institut fur Stahlbau, Holzbau und Mischbautechnologie.

Zandonini, R., Baldassino, N., Bernuzzi, C., & Hancock, G. (1998). Overall, local and distortional buckling in pallet racks. In *Proceedings of the Structural Stability Research Council,(S.S.R.C.)*, Atlanta, GA, September 1998.

Chapter 3
Pallet Sliding

3.1 Overview

3.1.1 Friction Models

Friction is the tangential reaction force between two surfaces in contact. Physically these reaction forces are the results of many different mechanisms, which depend on contact geometry and typology, properties of the bulk and surface materials of the bodies, displacement and relative velocity of the bodies and presence of lubrication.

In dry sliding contacts between flat surfaces, friction can be modelled as elastic and plastic deformation forces of microscopical asperities in contact. For each asperity contact the tangential deformation is elastic until the applied shear pressure exceeds the shear strength τ_y of the surface materials, when it becomes plastic.

There are different models of friction that consider stationary condition, e.g. constant velocity of the contact surfaces, and other, developed in the last century, that consider friction with a dynamic model.

In the Coulomb model (1776), the main idea is that friction opposes motion and that its magnitude is independent of velocity and contact area (Fig. 3.1). It can therefore be described as

$$F = F_C \mathrm{sgn}(v) \tag{3.1}$$

where the friction force F_C is proportional to the normal load, i.e. $F_C = \mu \cdot F_N$.

The Coulomb (1776) friction model does not specify the friction force for zero velocity. It may be zero or it can take any value in the interval between $-F_C$ and F_C, depending on how the sign function is defined. This very simple model is often modified with the introduction of viscosity parameters in order to take into account a dependence on velocity, as shown in Eq. 3.2, and in Fig. 3.2.

C.A. Castiglioni, *Seismic Behavior of Steel Storage Pallet Racking Systems*, Research for Development, DOI 10.1007/978-3-319-28466-8_3

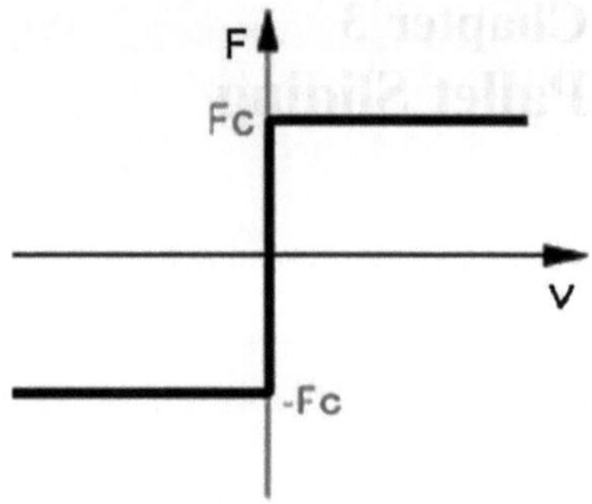

Fig. 3.1 Coulomb friction model

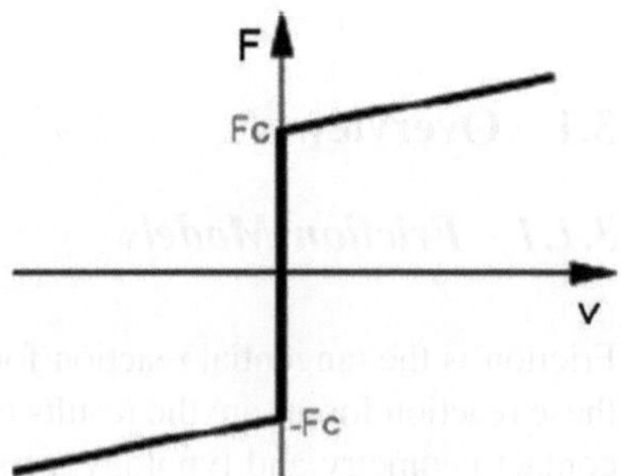

Fig. 3.2 Coulomb friction model with the adding of viscosity

$$F = (F_c + F_v v)\mathrm{sgn}(v) \tag{3.2}$$

Stiction is short for static friction as opposed to dynamic friction. It describes the friction force at rest. Morin (1833) introduced the idea of a friction force at rest that is higher than the Coulomb friction level. Static friction counteracts external forces below a certain level and thus keeps an object from moving. It is hence clear that friction at rest cannot be described as a function of velocity only. Instead it has to be modelled using the external force F_E in the following way.

$$F = \begin{cases} F_E & \text{if } v = 0 \text{ and } |F_E| < F_S \\ F_S\mathrm{sgn}(F_E) & \text{if } v = 0 \text{ and } |F_E| \geq F_S \end{cases}. \tag{3.3}$$

The friction force for zero velocity is a function of the external force and not of the velocity. Stribeck (1902) observed that the friction force does not decrease discontinuously as in Fig. 3.3, but that the velocity dependence is continuous as shown in Fig. 3.4. This is called Stribeck friction. A more general description of friction than the classical models is, therefore:

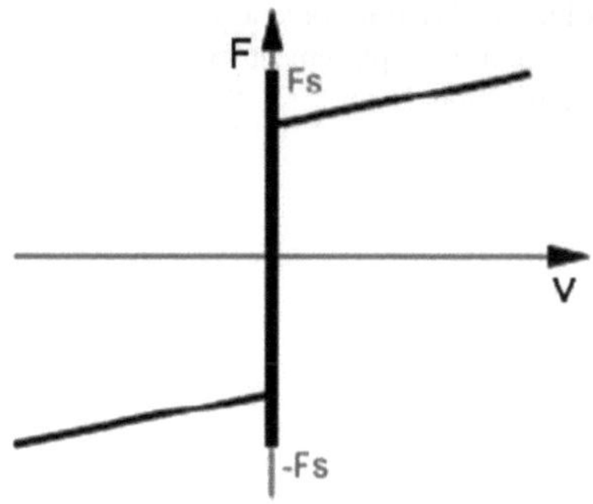

Fig. 3.3 Stiction plus Coulomb friction model

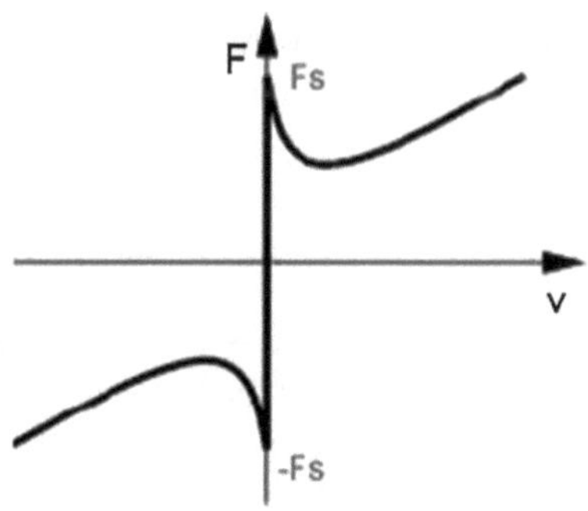

Fig. 3.4 Coulomb friction model with a continuous decrease of the friction force

$$F = \begin{cases} F(v) & \mathit{if}\ v \neq 0 \\ F_E & \mathit{if}\ v = 0\ \mathit{and}\ |F_E| < F_S\ . \\ F_S \mathrm{sgn}(F_E) & \mathit{otherwise} \end{cases} \tag{3.4}$$

where $F(v)$ is an arbitrary function, which may look in Fig. 3.4. A common form to interpret in a general way the nonlinearity of the friction force is

$$F(v) = F_C + (F_S - F_C) \cdot e^{-|v/v_S|^{\delta_S}} + F_v \cdot v \tag{3.5}$$

where v_S is called the Stribeck velocity. Function F is easily obtained by measuring the friction force for motions with constant velocity. The curve schematically shown in Fig. 3.4, is often asymmetrical.

Other static models of friction, as Karnopp model (1985) and Armstrong's model (1994) are described by Olsson et al. (1998).

Lately there has been a significant interest in dynamic friction models that describe friction, as a dynamic system, with differential equations. In the Dahl model (1968, 1975 and 1976) it is assumed that the friction force is only position dependent, i.e. it depends only on the relative displacement x of the two surfaces. The starting point of this model is the stress–strain curve of the classical solid-mechanics theory. When subjected to stress the friction force increases

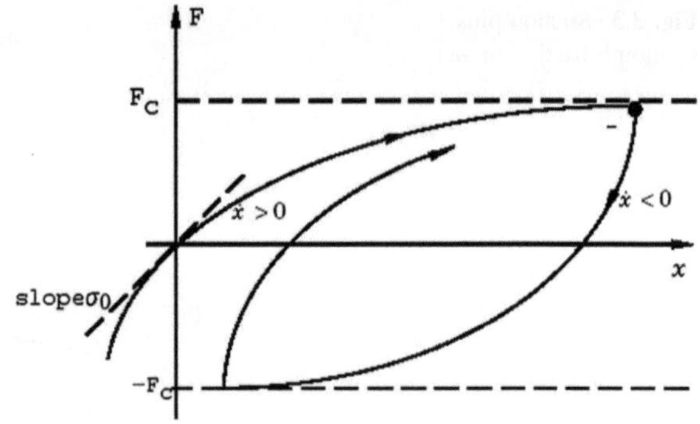

Fig. 3.5 Friction force as a function of displacement for Dahl's model

gradually until rupture occurs. Let x be the displacement, F the friction force and F_c the Coulomb friction force. Then, Dahl's model has the form:

$$\frac{dF}{dx} = \sigma \cdot \left(1 - \frac{F}{F_C} \cdot \operatorname{sgn}(v)\right)^{\alpha} \tag{3.6}$$

where σ is the stiffness coefficient and α is a parameter that determines the shape of the stress–strain curve. The value α = 1 is commonly used, while higher values give a stress–strain curve with a sharper bend. Figure 3.5 shows a graphical representation of this model.

To obtain a time domain model, Dahl observed that:

$$\frac{dF}{dt} = \frac{dF}{dx} \cdot \frac{dx}{dt} = \frac{dF}{dx} \cdot v = \sigma \cdot \left(1 - \frac{F}{F_C} \operatorname{sgn}(v)\right)^{\alpha} \cdot v \tag{3.7}$$

This model is hence a generalization of the Coulomb model, but it doesn't include the Stribeck effect, which is a rate dependent phenomenon, and doesn't capture stiction.

Bliman and Sorine (1991, 1993 and 1995) developed various dynamic models based on the experimental investigations of Rabinowicz (1951).

Models by Bliman and Sorine assume that the magnitude of the friction force depends only on sgn(v) and on the space variable s defined by

$$s = \int_0^t |v(\tau)| d\tau \tag{3.8}$$

In the Bliman and Sorine models friction is then a function depending on the path only, but is rate independent. The models can be expressed as linear systems in the space variable s.

$$\begin{aligned} \frac{dx_S}{ds} &= Ax_S + Bv_S \\ F &= C \cdot x_S \end{aligned} \tag{3.9}$$

Bliman and Sorine proposed models of different complexity, depending on the definition of the coefficients A, B and C in Eq. (3.9).

The simplest, first order model is given by:

$$\mathrm{A} = -1/\varepsilon_{\mathrm{f}} \quad \mathrm{B} = \mathrm{f}_1/\varepsilon_{\mathrm{f}} \quad \text{and} \quad \mathrm{C} = 1$$

which is identical to Dahl's model with $F_c = f_1$, $\sigma = f_1/\varepsilon_f$ and $\alpha = 1$. When $\varepsilon_f \rightarrow 0$, the model tends to behave as a classical Coulomb friction model. However higher order models can be defined, capable of interpreting phenomena like Stiction, and the presence of friction peaks at specific break-away distance.

The LuGre Model (Canudas de Wit et al. 1995) is a dynamic friction model related to the bristle interpretation of friction. Friction is modelled as the average deflection force of elastic springs. When a tangential force is applied the bristles will deflect like springs. If the deflection is sufficiently large the bristles start to slip. The average bristle deflection for a steady state motion is determined by the velocity. It is lower at low velocities, which implies that the steady state deflection decreases with increasing velocity. This models the phenomenon that the surfaces are pushed apart by the lubricant, and models the Stribeck effect. The model also includes rate dependent friction phenomena such as varying break-away force and frictional lag. The model has the form:

$$\begin{aligned} \frac{dz}{dt} &= v - \sigma_0 \cdot \frac{|v|}{g(v)} z \\ F &= \sigma_0 \cdot z + \sigma_1(v) \frac{dz}{dt} + f(v) \end{aligned} \tag{3.10}$$

where z is the average bristle deflection, σ_0 is the stiffness of the bristles, $\sigma_1(v)$ is the damping, function $g(v)$ models the Stribeck effect and $f(v)$ is the viscous friction. The model behaves like a spring for small displacements. For constant velocity, the model gives the steady state friction force as:

$$F = g(v)\mathrm{sgn}(v) + f(v)$$

3.1.2 Numerical Models for Sliding of Pallets

In a study developed within the SEISRACKS project, Denoël and Degée (2005) developed an analytical model for a SDOF sliding system subjected to a sinusoidal ground motion (Fig. 3.6).

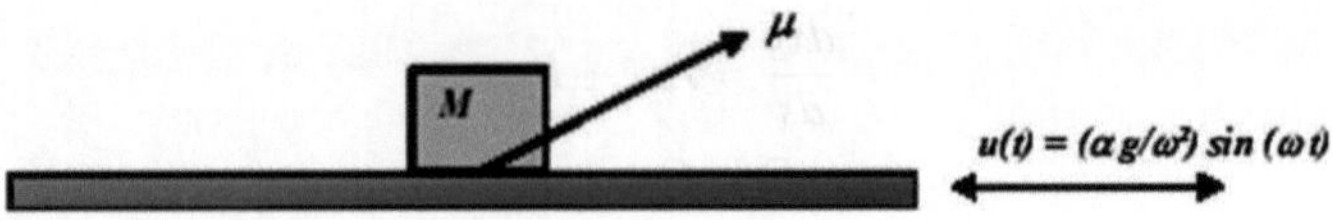

Fig. 3.6 SDOF system analysed in the research

The behaviour of the system can be separated into two phases.

While the inertial force on the mass doesn't exceed the friction force at the interface, the mass is stuck to the support. The governing equations are thus:

$$\begin{aligned}\ddot{x} &= \ddot{u}(t) = a \cdot g \cdot \sin(\omega \cdot t)\\ \dot{x} &= \dot{x}(t_0) + \frac{a \cdot g}{\omega} \cdot (\cos \omega t - \cos \omega t_0)\\ x(t) &= x(t_0) + (\dot{x}(t_0) - a \cdot g \cdot \cos \omega t_0) \cdot (t - t_0) + \frac{a \cdot g}{\omega^2} \cdot (\sin \omega t - \sin \omega t_0)\end{aligned} \quad (3.11)$$

$$\text{while}: \quad F(t) = M \cdot \ddot{x}(t) < \mu \cdot M \cdot g \quad (3.12)$$

No sliding occurs if the non dimensional maximum acceleration of the support $\alpha = a/g$ is lower than the friction factor μ ($\alpha < \mu$).

As soon as the inertial force exceeds the friction resistance, the mass starts sliding on the support and is subjected to a constant horizontal force equal to the friction force.

$$\begin{aligned}\ddot{x}(t) &= \frac{\pm \mu \cdot M \cdot g}{M}\\ \dot{x}(t) &= \dot{x}(t_0) \pm \mu \cdot g \cdot (t - t_0)\\ x(t) &= x(t_0) + \dot{x}(t_0)(t - t_0) \pm \mu \cdot g \cdot \frac{(t - t_0)^2}{2}\end{aligned} \quad (3.13)$$

The mass starts again sticking to the support as soon as the relative velocity between the mass and the support become equal to zero.

$$\frac{\mu}{a} \arcsin\left(\frac{\mu}{a}\right) + \sqrt{1 - \left(\frac{\mu}{a}\right)^2} \quad (3.14)$$

An advanced numerical tool has also been developed (Degée and Denoël 2007) in order to be able to evaluate accurately the behaviour of racks subjected to seismic action with a due account for possible sliding of supported pallets. The tool is included in the non linear finite element software FineLg developed in University of Liège (1996).

The starting point of the development of the sliding-mass model is the use of the concept of "mathematical deck" already available in FineLg since its development

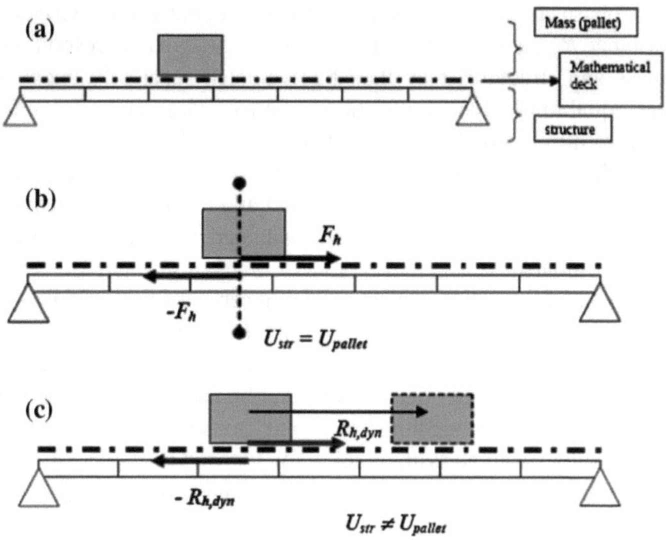

Fig. 3.7 **a** Evolution of the mathematical deck for the sliding mass model (basic scheme); **b** Sliding mass model in "stick" phase; **c** Sliding mass model in "slip" phase

by Yang (1996). The mathematical deck was elaborated to study the dynamic behaviour of structures subjected to moving loads or vehicles and particularly to study the bridge-vehicles interaction.

According to this concept, the interactive behaviour is obtained by solving two uncoupled sets of equations, respectively for the structure and for the vehicles, and then by ensuring compatibility and equilibrium at the contact points between the structure and the vehicles with an iterative procedure. In this scheme, the mathematical deck acts as an interface element to evaluate the position of the vehicles with respect to the physical deck and to perform the iterative compatibility process.

When elaborating the "sliding mass" model, the idea has been to start from a "moving mass" vehicle without any user-imposed speed and to obtain the horizontal behaviour of the mass as the result of a dynamic computation according to a stick/slip model (Fig. 3.7a).

During the "stick" phase, the displacement, velocity and acceleration of both the mass and the underlying structure are identical. The mathematical deck computes thus the horizontal friction force F_h ensuring simultaneously this equality of displacement and the general equations of dynamics for both the mass and the structure, including also the earthquake action (Fig. 3.7b).

For structures like racks, the supported mass is much heavier than the mass of the structure itself (M up to more than 100 times the structural mass). In this case, convergence of the iterative procedure to ensure equilibrium and compatibility of the coupled system is very difficult to be achieved unless specific methods are used. In the present case, an Aitken acceleration procedure was developed.

The "slip" phase starts as soon as the horizontal contact force exceeds the static friction resistance $R_{h,st}$ and the mass starts sliding. The dynamic response of the two systems (mass and structure) may then be evaluated separately under a constant contact force equal to the dynamic friction resistance $R_{h,dyn}$ (Fig. 3.7c). During this stage, the pallet moves on the mathematical deck and its position, velocity and acceleration ($=R_{h,dyn}/M$) can be evaluated at any time step. The sliding behaviour lasts until the relative velocity between the pallet and the structure becomes equal to zero. From that condition, it can then be evaluated whether and when the next "stick" phase is initiated.

In order to validate the sliding mass model, a series of very simple systems has been studied with FineLg and compared to equivalent MDOF systems solved with a semi-analytical approach (Denoël and Degée 2005). Some of the considered examples are presented in Fig. 3.8. The results obtained with FineLg and with the reference semi-analytical procedure are found in very good agreement. As an example, the results obtained with FineLg for the 3DOF model of Fig. 3.8a are plotted in Fig. 3.8b for $\mu/\alpha = 1.00$ (no sliding) and $\mu/\alpha = 0.5$ (μ is the friction

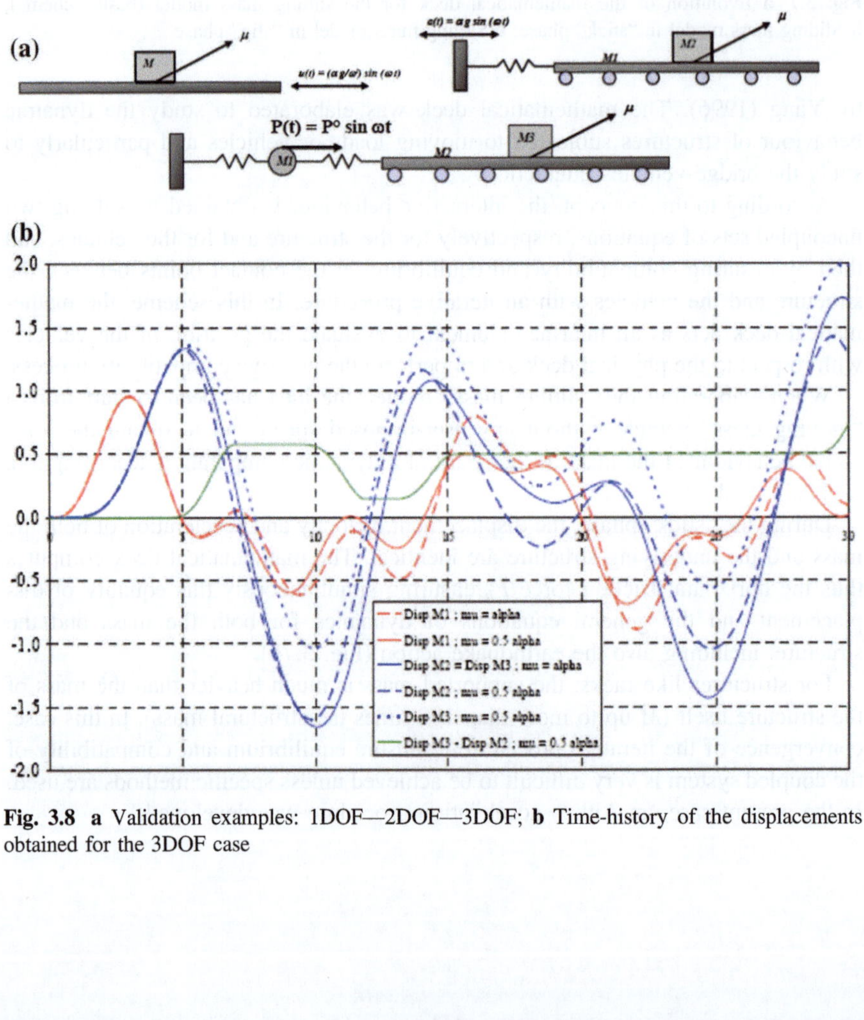

Fig. 3.8 **a** Validation examples: 1DOF—2DOF—3DOF; **b** Time-history of the displacements obtained for the 3DOF case

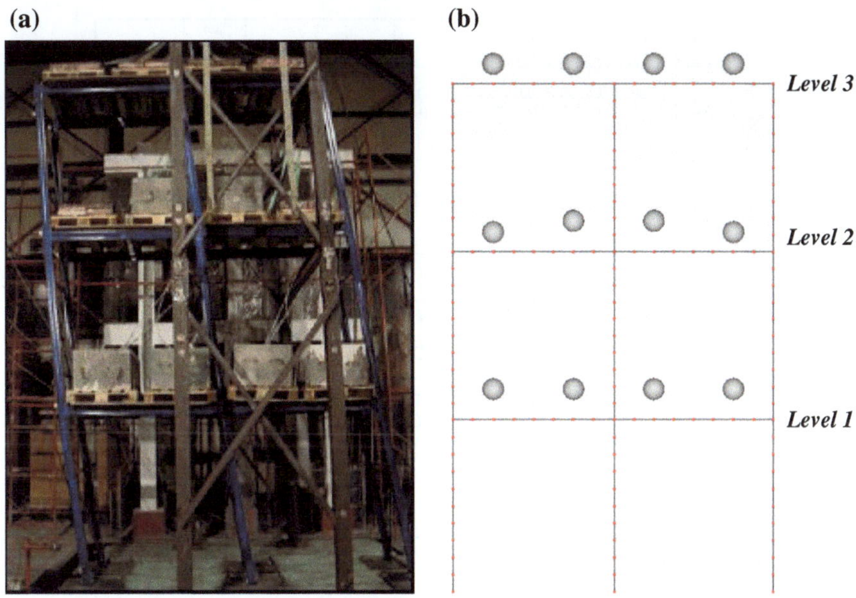

Fig. 3.9 **a** Test specimen, **b** corresponding numerical model

coefficient and α is the maximum imposed acceleration referred to gravity). In this second configuration, 4 sliding phases can be observed during which the relative displacement between M2 and M3 varies.

Linear elastic as well as non-linear simulation of the behaviour of a rack structure has also been performed (Degée and Denoël 2007). In particular, the numerical model has been used to reproduce test results obtained on the shaking table of the Laboratory of Earthquake Engineering of the NTU Athens within the SEISRACKS Project.

Figure 3.9 shows the tested specimen (2 bays—3 levels unbraced structure) and the corresponding numerical model.

In the model, the second-order geometrical non-linearities of the structure and the material non-linear behaviour of the joints are considered. The two main governing parameters have been identified as being (1) the plastic resistance of the base-anchorages and (2) the sliding properties of the masses.

Figure 3.10 shows the comparison, in terms of sliding, of the numerical results obtained for different assumptions on the friction factor (no sliding, $\mu = 0.40$ and $\mu = 0.30$) in the case of simulation of a test carried out with a PGA of 0.45g.

Figure 3.10a, b respectively refer to pallet sliding at the first and third level, while Fig. 3.10c shows the experimental measurements at the third level.

The numerical model of (Degée and Denoël 2007) is able to correctly predict both magnitude of the sliding and the timing of its occurrence. Amplitude of each individual sliding stage is overestimated, but the estimate could be improved by

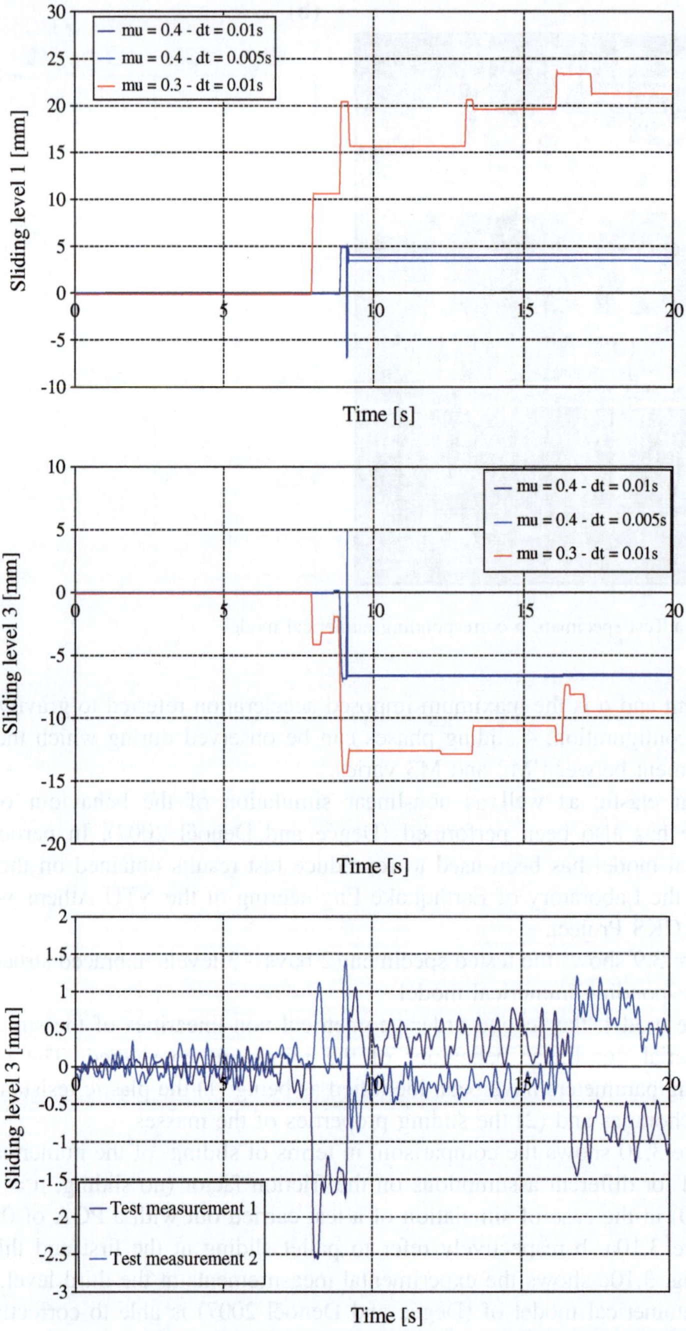

Fig. 3.10 Comparison between numerical simulation and experimental measurements of pallet sliding

using slightly higher values of the friction factor. This requirement for a higher friction factor is confirmed also by the experimental results.

3.1.3 Aims and Scopes of the Investigation

The research study presented hereafter has been developed within the SEISRACKS project. About 1260 static tests and 200 dynamic tests were carried out, aimed to an assessment of both the static and the dynamic friction factor developed between the pallet and the beam.

First, the influence of different parameters (such as the type of pallet and beam, the stored mass and the mass eccentricity) on the static friction factor is analysed.

Then, the dynamic tests carried out at the Laboratory for Earthquake Engineering of the National Technical University of Athens, on a small portion of pallet rack are presented. The structure is excited with a sinusoidal input signal of the shaking table, in cross and in down-aisle direction. The objective is to study the pallet behaviour under dynamic condition, and to identify the horizontal actions that are transferred to the rack structure by the live load (i.e. by the palletized goods) under dynamic conditions, as a function of the input motion characteristics (in terms of acceleration and frequency) as well as of the pallet-bam interface characteristics (i.e. pallet type and "condition", beam type and surface finish).

Some seismic tests were also carried out on the same structural subassemblage, in order to perform a comparison with the results of the sinusoidal tests and to validate the extrapolation of such results to actual design situations of racks under seismic loading.

3.2 Assessment of the Static Friction Factor

3.2.1 Overview

Quasi static sliding tests were performed at the Laboratory for Earthquake Engineering (LEE) of the National Technical University of Athens (NTUA), Greece. The aim of this group of tests was to obtain the static friction factor for different combinations of beams and pallets, and to study the influence of the mass and of its eccentricity.

The static, as well as the dynamic friction factor (that has been assessed by means of dynamic tests carried out in the second part of this research) are related to the behaviour at the interface between the pallet and the supporting beams.

The test set-up for quasi-static tests is shown in Fig. 3.11. Two horizontal beams were fixed on a rigid steel frame with pinned support. The frame was free to rotate about the pinned axis. The axial distance between the pinned axis and the free edge

Fig. 3.11 Experimental set-up for quasi-static sliding tests

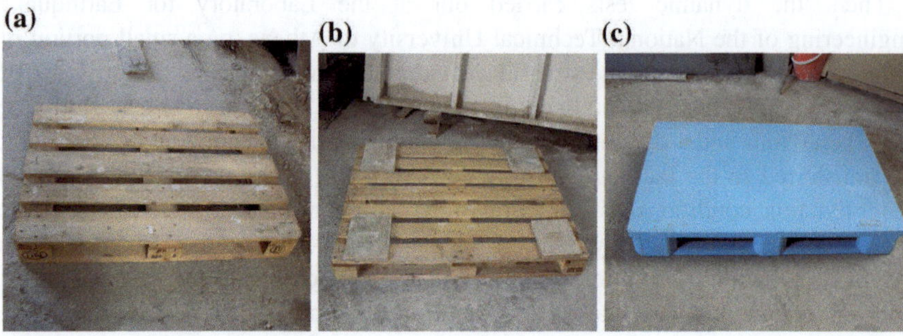

Fig. 3.12 Different types of pallet: **a** Wooden Euro pallet. **b** Wooden American. **c** Plastic Euro pallet

of the frame was 1575 mm. One pallet, with a rigidly fixed mass of 8 kN, was positioned on the beams. The system was gradually and slowly inclined with the use of a crane, that lifted the frame from its free edge, while the vertical displacement and the relative displacement between pallet and beam, were measured. Thirty repetitions of each test (combination of pallet and beam) were carried out.

These tests were performed in down and cross-aisle direction.

Different types of pallets and beams were used during the experimental tests as shown respectively in Fig. 3.12 and from Figs. 3.13, 3.14, 3.15, 3.16, 3.17, 3.18 and 3.19.

Three different values of the applied mass were considered (251, 785, 1036 kg) as well as the different position of the mass on the pallet (centered, eccentric downward, eccentric upward).

Seven types of pallets and six types of beam were used in the tests, with the following denomination (Table 3.1).

Pallet P1 is a wooden Euro pallet, new and dry. Usually, after being in use for a while, the lower faces of the pallet wear out. For this reason the normal situation is the one represented by pallet P2, a wooden Euro pallet old and dry. In order to

Fig. 3.13 Different type of beams used in the tests

Fig. 3.14 Section of beam type B1

investigate eventual environmental conditions, pallet P3 is a wooden Euro pallet, old, that was spread with water for a few minutes, before testing.

The same conditions were considered also for the American type of pallets, respectively P4, P5, and P6.

Pallet P7 is a plastic Euro pallet. This type of pallet is more and more adopted, as it is more resistant and can be more easily cleaned than the wooden one. It is widely adopted, for example, for storage of food, in particular in refrigeration units.

Fig. 3.15 Section of beam type B2

Fig. 3.16 Section of beam type B3

Six different types of beams, manufactured by three different companies were considered.

Beams B1, B2 and B3 were manufactured by the same company (A): the cross section was the same for all the three types, but surface treatment was different. B1 was powder coated beam, B2 was a hot dip coated beam and B3 a hot zinc coated beam.

Fig. 3.17 Section of beam type B4

Fig. 3.18 Section of beam type B5

Beam B4 was manufactured by a different company (B), with a different cross section than the previous ones, and a powder coating surface treatment.

Beam B5 and B6 were manufactured by another company (C) and differ for their geometry, although surface treatment was the same type of powder coating for both beams.

Fig. 3.19 Section of beam type B6

Table 3.1 Type of pallet and beam used in the different tests

Pallets	Beams
P1: Wooden Euro pallet 800 × 1200, new, dry	B1: Cold rolled, powder coated, new (Producer A)
P2: Wooden Euro pallet 800 × 1200, old, dry	B2: Cold rolled, hot dip coated, new (Producer A)
P3: Wooden Euro pallet 800 × 1200, old, wet	B3: Cold rolled, hot zinc coated, new (Producer A)
P4: Wooden American pallet, new, dry	B4: Cold rolled, hot dip coated, new (Producer B)
P5: Wooden American pallet, old, dry	B5: Cold rolled, hot dip coated, new (Producer C)
P6: Wooden American pallet, old, wet	B6: Cold rolled, hot dip coated, new (Producer C)
P7: Plastic Euro pallet	

Static tests are based on the principle of the inclined plane shown in Fig. 3.20, in which, when the pallet starts to slide, the component of the gravity force along the beam ($F_{\|}$) is equal to its perpendicular component ($F_{\perp}$) multiplied by a static friction factor, that can be obtained according to Eq. (3.15).

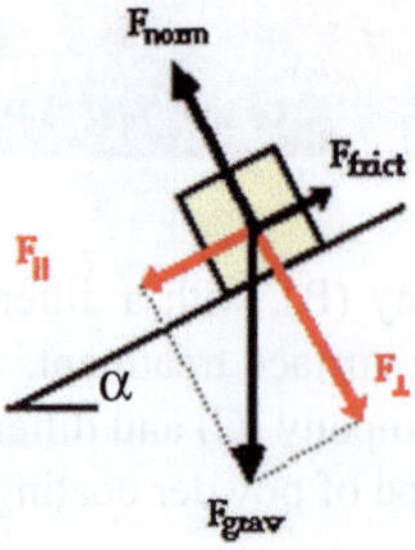

Φ

$$F_{\|} = F_{grav} \cdot \sin(\alpha) = F_{frict} = \mu \cdot F_{grav} \cdot \cos(\alpha)$$

hence :

$$\mu = \tan(\alpha) \tag{3.15}$$

Fig. 3.20 Principle of inclined plane

Within this study, the static friction factor was investigated for different combinations of beam, pallet, mass and mass eccentricity, both in cross-aisle and in down-aisle direction.

3.2.2 Friction in Cross-Aisle Direction

Table 3.21 shows the setup for tests in cross-aisle direction. Sliding in this direction is very dangerous because the pallet width is 1200 mm while the rack width is usually 1100 mm. Hence, a few mm of displacement, eventually correlated to a small eccentricity of positioning, can result in a loss of support of the pallet (Fig. 3.21).

In the next paragraphs all the figures show the mean values of the static friction factor, with the indication of the standard deviation, for every test type. All the tests are repeated 30 times in the same conditions, and the final table reports a statistical analysis of the experimental results. In the following, reference will be made to the statistical parameters described in Table 3.2.

3.2.2.1 Influence of the Pallet Type

Figure 3.22 shows the influence of the pallet type on the static friction factor in cross-aisle direction, while Table 3.3 presents the statistics of the results. All the tests are carried out with a mass of 785 kg centered on the pallet.

It can be observed that:

Fig. 3.21 Set up for cross-aisle tests

Table 3.2 Symbols used in the statistical analysis of the data

μ	Mean value	
σ	Standard deviation	
Cov %	Coefficient of variation %	$\mathrm{Cov}\,\% = \frac{\sigma}{\mu} \cdot 100$
Max	Maximum value	
Min	Minimum value	
Δ^+ %	Difference percentage between the maximum value and the mean one	$\Delta^+\,\% = \frac{\mathrm{Max}-\mu}{\mu} \cdot 100$
Δ^- %	Difference percentage between the minimum value and the mean one	$\Delta^-\,\% = \frac{\mu-\mathrm{Min}}{\mu} \cdot 100$
α^+	Difference between the maximum value and the mean one in terms of standard deviation	$\alpha^+ = \frac{\mathrm{Max}-\mu}{\sigma}$
α^-	Difference between the mean value and the minimum one in terms of standard deviation	$\alpha^- = \frac{\mu-\mathrm{Min}}{\sigma}$

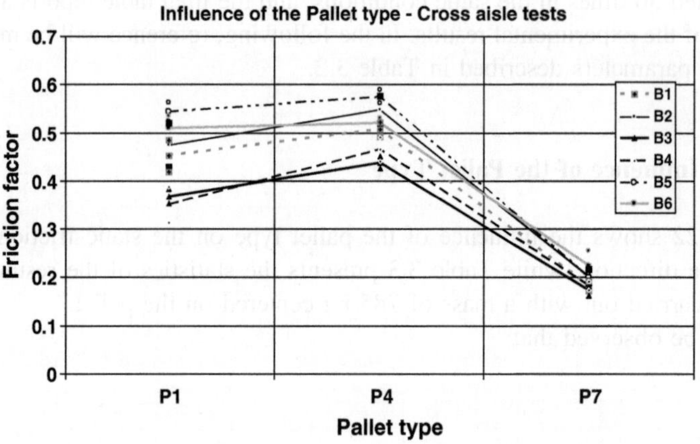

Fig. 3.22 Influence of the pallet type on the friction factor in cross-aisle direction, for different types of beam

Table 3.3 Statistics of test results for different types of pallet

	μ	σ	Cov %	Max	Min	Δ^+ %	Δ^- %	μ + σ	μ − σ	μ + 2σ	μ − 2σ	α^+	α^-
P1	0.45	0.08	16.9	0.58	0.27	28.7	39.7	0.53	0.37	0.60	0.30	1.70	2.35
P4	0.51	0.05	9.8	0.61	0.41	19.5	18.5	0.56	0.46	0.61	0.41	2.00	1.89
P7	0.19	0.02	11.8	0.34	0.16	72.9	19.6	0.22	0.17	0.24	0.15	6.18	1.66

- Pallet P4 (American Wooden Pallet) has the highest mean value of the friction factor (0.51), while pallet P7 (Plastic Euro Pallet) has the lowest. Pallet P1 (Wooden Euro Pallet) has an intermediate value, very close to that measured for Pallet P4.

- Scatter of the data for all pallets is limited, with c.o.v. ranging from 9.8 % (P4) to 16.9 % (P1), in particular P4 has the lowest scatter of the data, P1 has the highest one and P7 has an intermediate value.

3.2.2.2 Influence of the Beam Type

Similar considerations can be drawn with regards to Fig. 3.23 that shows the influence of the beam type. Tests are carried out with a centered mass of 785 kg. Test results and their statistical re-analysis are presented in tables from Tables 3.4, 3.5 and 3.6.

It can be observed that:

- Pallet P4 (American wooden Pallet) shows the highest value of the friction factor, pallet P1 (Wooden Euro Pallet) an intermediate one and pallet P7 (Plastic Euro Pallet) the lowest one.

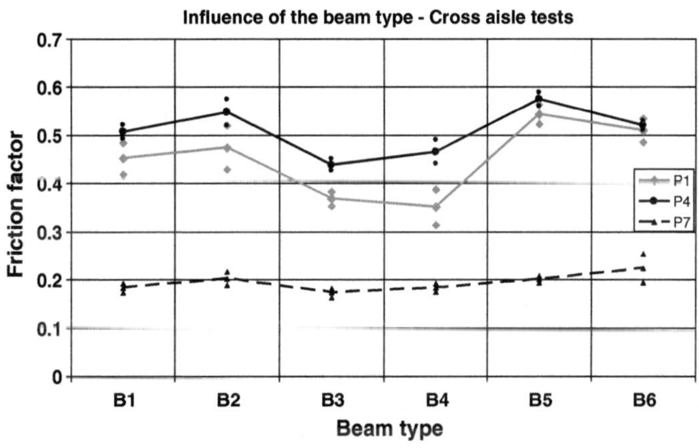

Fig. 3.23 Influence of the beam type on the friction factor in cross-aisle direction, for different types of pallet

Table 3.4 Statistics of test results for pallet P1 and different types of beam

	μ	σ	Cov %	Max	Min	Δ^+ %	Δ^- %	$\mu+\sigma$	$\mu-\sigma$	$\mu+2\sigma$	$\mu-2\sigma$	α^+	α^-
B1	0.45	0.03	7.1	0.49	0.38	9.3	16.1	0.48	0.42	0.52	0.39	1.30	2.25
B2	0.47	0.05	9.5	0.56	0.37	18.5	22.2	0.52	0.43	0.56	0.38	1.94	2.33
B3	0.37	0.02	4.1	0.39	0.33	6.5	10.1	0.38	0.35	0.40	0.34	1.57	2.45
B4	0.35	0.04	10.9	0.42	0.27	18.6	23.3	0.39	0.31	0.43	0.28	1.71	2.14
B5	0.54	0.02	3.7	0.58	0.51	6.5	5.8	0.56	0.52	0.58	0.50	1.77	1.58
B6	0.51	0.02	4.9	0.54	0.41	5.0	18.9	0.53	0.48	0.56	0.46	1.01	3.86

Table 3.5 Statistics of test results for pallet P4 and different types of beam

	μ	σ	Cov %	Max	Min	Δ^+ %	Δ^- %	μ + σ	μ − σ	μ + 2σ	μ − 2σ	α^+	α^-
B1	0.51	0.01	2.9	0.53	0.46	5.2	9.1	0.52	0.49	0.54	0.48	1.80	3.13
B2	0.55	0.03	5.1	0.59	0.47	8.3	13.3	0.58	0.52	0.60	0.49	1.63	2.62
B3	0.44	0.01	2.7	0.46	0.41	5.4	5.5	0.45	0.43	0.46	0.42	2.00	2.05
B4	0.47	0.02	5.2	0.51	0.42	9.2	9.1	0.49	0.44	0.51	0.42	1.77	1.73
B5	0.57	0.01	2.5	0.61	0.54	5.7	5.3	0.59	0.56	0.60	0.55	2.33	2.13
B6	0.52	0.01	2.0	0.54	0.49	3.3	5.0	0.53	0.51	0.54	0.50	1.65	2.45

Table 3.6 Statistics of test results for pallet P7 and different types of beam

	μ	σ	Cov %	Max	Min	Δ^+ %	Δ^- %	μ + σ	μ − σ	μ + 2σ	μ − 2σ	α^+	α^-
B1	0.18	0.01	5.2	0.20	0.16	11.7	11.0	0.19	0.17	0.20	0.16	2.24	2.10
B2	0.20	0.01	6.8	0.25	0.18	24.0	12.5	0.22	0.19	0.23	0.17	3.51	1.83
B3	0.17	0.01	5.7	0.19	0.16	10.5	9.3	0.18	0.16	0.19	0.15	1.85	1.63
B4	0.18	0.01	4.6	0.21	0.17	14.8	5.9	0.19	0.17	0.20	0.17	3.25	1.30
B5	0.20	0.01	3.6	0.22	0.19	11.7	4.6	0.21	0.19	0.22	0.19	3.22	1.27
B6	0.22	0.02	8.5	0.26	0.19	20.6	11.2	0.24	0.20	0.25	0.18	2.42	1.32

- Friction factor for pallet P7 is quite constant, while for the other two types of pallet the friction factor shows a strong dependence on the beam type. In particular the lowest values are obtained for beam types B3 and B4, while in the other cases the friction factor is similar.
- For pallets P1 and P4 the highest friction factor is obtained with beam B5.

The behaviour of the American wooden pallet and the wooden Euro pallet is very similar, and the following Table 3.7 shows the statistical analysis of their combined data, neglecting the Plastic Euro pallet. It can be noticed that scatter of the results is small with c.o.v. ranging from 3.8 % (beam B6) to 15.8 % (beam B4).

The following tables present the re-analysis of the results in case of pallet P1 and P4 (Wooden Euro pallet and American pallet) and of the groups of beams B1 + B2 + B5 + B6 (Table 3.8), B3 (Table 3.9) and B4 (Table 3.10).

It can be noticed that the response of the groups of beams (B1 + B2 + B5 + B6) is rather homogeneous. A mean value of the static friction factor of 0.52 was obtained, with a c.o.v. of 8.9 %.

Table 3.7 Statistics of test results for different types of beam considering pallet type P1 and P4

	μ	σ	Cov %	Max	Min	Δ^+ %	Δ^- %	μ + σ	μ − σ	μ + 2σ	μ − 2σ	α^+	α^-
B1	0.48	0.04	7.8	0.53	0.38	11.3	20.9	0.52	0.44	0.55	0.40	1.45	2.69
B2	0.51	0.05	10.2	0.59	0.37	16.0	27.8	0.56	0.46	0.62	0.41	1.56	2.71
B3	0.40	0.04	9.4	0.46	0.33	14.6	18.0	0.44	0.37	0.48	0.33	1.56	1.92
B4	0.41	0.06	15.8	0.51	0.27	24.3	33.8	0.47	0.34	0.54	0.28	1.53	2.14
B5	0.56	0.02	4.2	0.61	0.51	8.7	8.6	0.58	0.54	0.61	0.51	2.07	2.04
B6	0.51	0.02	3.8	0.54	0.41	4.3	19.7	0.53	0.49	0.55	0.48	1.11	5.13

Table 3.8 Statistics of test results considering pallet P1 + P4 and beams B1 + B2 + B5 + B6

	μ	σ	Cov %	Max	Min	Δ^+ %	Δ^- %	$\mu + \sigma$	$\mu - \sigma$	$\mu + 2\sigma$	$\mu - 2\sigma$	α^+	α^-
B1 + B2 + B5 + B6	0.52	0.05	8.9	0.61	0.37	17.8	28.5	0.56	0.47	0.61	0.42	2.01	3.21

Table 3.9 Statistics of test results considering pallet P1 + P4 and beam B3

	μ	σ	Cov %	Max	Min	Δ^+ %	Δ^- %	$\mu + \sigma$	$\mu - \sigma$	$\mu + 2\sigma$	$\mu - 2\sigma$	α^+	α^-
B3	0.40	0.04	9.4	0.46	0.33	14.6	18.0	0.44	0.37	0.48	0.33	1.56	1.92

Table 3.10 Statistics of test results considering pallet P1 + P4 and beam B4

	μ	σ	Cov %	Max	Min	Δ^+ %	Δ^- %	$\mu + \sigma$	$\mu - \sigma$	$\mu + 2\sigma$	$\mu - 2\sigma$	α^+	α^-
B4	0.41	0.06	15.8	0.51	0.27	24.3	33.8	0.47	0.34	0.54	0.28	1.53	2.14

In the case of the beam types B3 and B4, data seem to be less homogeneous than those of the other group of beams. Mean values of the static friction factor respectively 0.40 and 0.41, with a c.o.v. of 13.1 and of 15.8 %, were obtained.

It is evident that the static friction factor developed by the first group of beams is much higher than the one developed by the beam types B3 and B4.

Figure 3.24 shows the repetition of tests for pallet type P1 for every kind of considered beam.

It can be noticed that, in the first tests, the friction factor shows an increasing trend while, after 5–10 tests, the obtained value is practically constant. This is most probably due to the "wearing" of the surface of the beam. In the first tests, the beam is new, and the friction factor is low. Due to wearing, the surface roughness increases, together with the friction factor as well as the scatter of the results. Beyond a certain level, the phenomenon stabilizes.

3.2.2.3 Influence of the Applied Mass

Figure 3.25 shows two of the three masses used in the tests. The influence of the applied mass is measured with masses of 251, 785 and 1036 kg, for pallet type P1 and for different types of beam. The mass is fixed on the pallet so that there is no relative displacement.

Figure 3.26 shows the influence of the beam type on the friction factor in cross-aisle direction, for different values of the applied masses. The statistical re-analysis of these data is presented in Table 3.11.

It can be observed that:

- When a mass of 251 kg was used, it resulted in the highest value of the friction factor, for every beam type, with the exception of beam B4. This result is most probably influenced by the geometry of the masses adopted in the study. The centre of gravity of mass of 251 kg is, in fact, lower than the c.o.g. of the mass of 785 kg. This results in a lower "overturning moment". This factor influences the response of the system, as it will be shown in the case of down-aisle direction.
- Friction factor for mass of 785 kg and 1036 kg is similar, especially for beam B1, B2, B5 and B6. For beam B3 and B4 the value corresponding to the heavy mass is higher than the other ones.

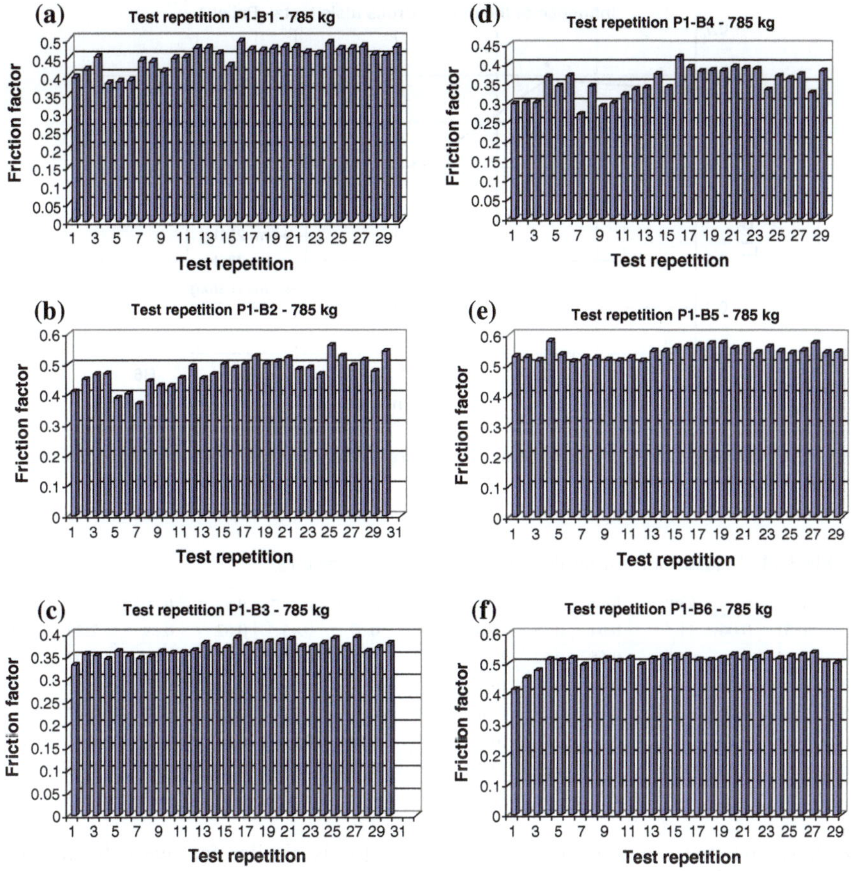

Fig. 3.24 Repetition of tests carried out with pallet type P1

Fig. 3.25 Different types of the applied mass: **a** 251 kg, **b** 785 kg

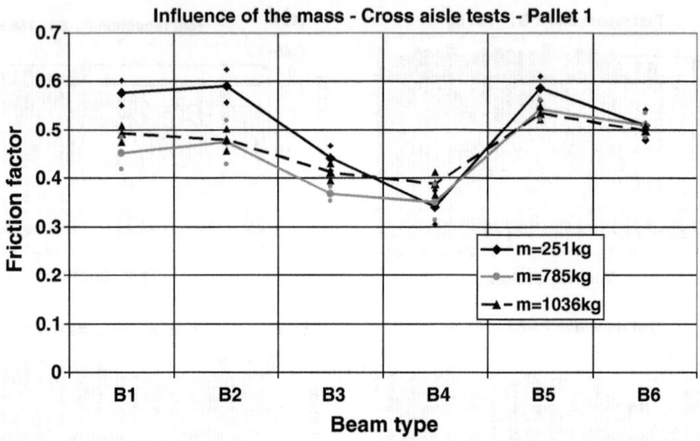

Fig. 3.26 Influence of the beam type on the friction factor in cross-aisle direction, for different values of the applied masses

Table 3.11 Statistics of test results for different types of beam

	μ	σ	Cov %	Max	Min	Δ^+ %	Δ^- %	μ + σ	μ − σ	μ + 2σ	μ − 2σ	α^+	α^-
B1	0.51	0.06	11.5	0.64	0.38	25.4	25.2	0.57	0.45	0.62	0.39	2.20	2.19
B2	0.52	0.06	12.5	0.65	0.37	25.9	28.4	0.58	0.45	0.64	0.39	2.07	2.27
B3	0.41	0.04	8.9	0.47	0.33	16.5	18.7	0.44	0.37	0.48	0.33	1.85	2.10
B4	0.36	0.04	10.7	0.43	0.27	18.8	24.6	0.40	0.32	0.44	0.28	1.77	2.31
B5	0.55	0.03	5.5	0.65	0.50	16.8	9.7	0.58	0.52	0.62	0.49	3.07	1.77
B6	0.49	0.03	5.7	0.54	0.39	9.0	21.5	0.52	0.47	0.55	0.44	1.56	3.75

- It seems that the trend of the friction factor is similar for every beam. If a constant mass is considered, beam B5 has the highest value while beam B4 the lowest one, as previously noticed.

Figure 3.27 shows the influence of the mass, for different types of beam, while the statistical re-analysis is presented in Table 3.12.

It can be observed that:

- Applied mass strongly influences the friction factor only for beams B1, B2 and B3, while in the other cases it is quite constant. In every case the standard deviation of all the data is rather limited, with a c.o.v. ranging from 11.5 to 19.4 %. It can be noticed that the scatter of the data decreases when the applied mass increases.
- Usually, the highest value of the friction factor is obtained with the mass of 251 kg independently of the beam type, while the lowest one with the intermediate mass. Exceptions are beams B4 and B6.

As expected, the value of the applied mass influences the response of the system less than the other analysed parameters.

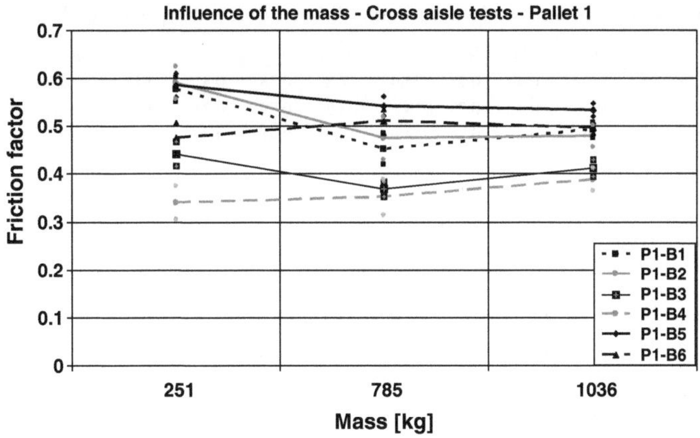

Fig. 3.27 Influence of the applied mass on the friction factor in cross-aisle direction, for different type of beam

Table 3.12 Statistics of test results for different types of the applied mass

Mass	μ	σ	Cov %	Max	Min	Δ^+ %	Δ^- %	μ + σ	μ − σ	μ + 2σ	μ − 2σ	α^+	α^-
251	0.50	0.10	19.4	0.65	0.28	29.3	44.3	0.60	0.40	0.70	0.31	1.51	2.28
785	0.45	0.08	16.9	0.58	0.27	28.7	39.7	0.53	0.37	0.60	0.30	1.70	2.35
1036	0.47	0.05	11.5	0.55	0.32	17.6	30.7	0.52	0.41	0.58	0.36	1.54	2.67

Tables 3.13, 3.14 and 3.15 show the statistics considering separately B1 + B2 + B5 + B6, B3 and B4.

The mean values of the static friction factor are practically non influenced by the value of the applied mass. They are very similar to those presented in Tables 3.8, 3.9 and 3.10 for the same groups of beams. In fact, if the results obtained with three different values of applied mass are re analysed as a single sample, the following results are obtained (Tables 3.16, 3.17 and 3.18).

Table 3.13 Statistics of test results considering B1 + B2 + B5 + B6 and different types of applied mass

Mass	μ	σ	Cov %	Max	Min	Δ^+ %	Δ^- %	μ + σ	μ − σ	μ + 2σ	μ − 2σ	α^+	α^-
251	0.56	0.06	10.0	0.65	0.39	16.3	30.6	0.61	0.50	0.67	0.45	1.63	3.05
785	0.49	0.05	9.6	0.58	0.37	17.0	25.3	0.54	0.45	0.59	0.40	1.78	2.65
1036	0.50	0.03	5.3	0.55	0.44	10.0	12.6	0.53	0.47	0.55	0.45	1.88	2.37

Table 3.14 Statistics of test results considering B3 and different types of applied mass

Mass	μ	σ	Cov %	Max	Min	Δ^+ %	Δ^- %	μ + σ	μ − σ	μ + 2σ	μ − 2σ	α^+	α^-
251	0.44	0.03	5.7	0.47	0.37	7.4	15.1	0.47	0.42	0.49	0.39	1.31	2.65
785	0.37	0.02	4.1	0.39	0.33	6.5	10.1	0.38	0.35	0.40	0.34	1.57	2.45
1036	0.41	0.02	4.4	0.43	0.35	4.9	15.0	0.43	0.39	0.45	0.38	1.11	3.44

Table 3.15 Statistics of test results considering B4 and different types of applied mass

Mass	μ	σ	Cov %	Max	Min	Δ^+ %	Δ^- %	μ + σ	μ − σ	μ + 2σ	μ − 2σ	α^+	α^-
251	0.34	0.04	10.3	0.43	0.28	25.6	17.8	0.38	0.31	0.41	0.27	2.49	1.73
785	0.35	0.04	10.9	0.42	0.27	18.6	23.3	0.39	0.31	0.43	0.28	1.71	2.14
1036	0.39	0.02	5.9	0.42	0.32	8.6	16.1	0.41	0.36	0.43	0.34	1.46	2.74

3.2.3 Friction in Down-Aisle Direction

In practice, sliding in this direction is less dangerous than in cross-aisle direction, because fall of the pallet can occur only if a rotation around the vertical axes is associated with the pallet displacement. In any case, with the test set up shown in Fig. 3.28, quasi-static sliding tests were carried out analysing the same parameters as in the tests in cross-aisle direction, in order to allow a comparison between static friction factor values in the two directions.

3.2.3.1 Influence of the Pallet Type

Figure 3.29 shows the influence of the pallet type on the friction factor in down-aisle direction, for different types of beam, while Table 3.19 shows the statistical re-analysis of the results.

It can be observed that:

- Trend of friction factor is similar for every considered beam with the exception of beam type B4. In all cases the highest value is obtained for pallet P1 (Wooden Euro Pallet). The lowest one is obtained for pallet P7 (Plastic Euro Pallet) in most cases, except for pallet P4 (Wooden American Pallet) that shows a minimum corresponding to beam type B4.
- In the case of pallet 7 (Plastic Euro Pallet) scatter of the results is very small (c.o.v. = 9.5 %). On the contrary, for pallet P1 c.o.v. = 20.5 % and for P4 c.o.v. = 32.1 %.

Table 3.19 shows the statistical analysis of the whole batch of data. Beam B4 and B3 show the lowest mean values of the friction factor. It is interesting to analyse the data neglecting those related to these types of beam.

Table 3.20 shows the analysis of the results only for beam types B1, B2, B3, B5 and B6, while Table 3.21 refers to beam types B1, B2, B5 and B6 and Table 3.22 to beam types B3 and B4.

It can be noticed that the mean values of the friction factor for pallet types P1 (Wooden Euro Pallet) and P4 (Wooden American Pallet) are very similar in the two cases of Tables 3.20 and 3.21, and the scatter of the results is limited.

In the case of the group of beams types B1 + B2 + B5 + B6, although manufactured by different producers, the static friction factor is very similar with a c.o.v. ranging from 8.1 % in the case of pallet type P1 to 8.8 % for the pallet type P7. On the contrary, beams B3 and B4 cannot be considered as homogeneous. In fact, in

Table 3.16 Statistics of test results considering B1 + B2 + B5 + B6 and all the applied mass

Mass	μ	σ	Cov %	Max	Min	Δ^+ %	Δ^- %	$\mu + \sigma$	$\mu - \sigma$	$\mu + 2\sigma$	$\mu - 2\sigma$	α^+	α^-
251 + 785 + 1036	0.52	0.05	10.3	0.65	0.37	25.4	28.7	0.57	0.46	0.62	0.41	2.46	2.78

Table 3.17 Statistics of test results considering B3 and all the applied mass

Mass	μ	σ	Cov %	Max	Min	Δ^+ %	Δ^- %	$\mu + \sigma$	$\mu - \sigma$	$\mu + 2\sigma$	$\mu - 2\sigma$	α^+	α^-
251 + 785 + 1036	0.41	0.04	8.9	0.47	0.33	16.5	18.7	0.44	0.37	0.48	0.33	1.85	2.10

Table 3.18 Statistics of test results considering B4 and all the applied mass

Mass	μ	σ	Cov %	Max	Min	Δ^+ %	Δ^- %	$\mu + \sigma$	$\mu - \sigma$	$\mu + 2\sigma$	$\mu - 2\sigma$	α^+	α^-
251 + 785 + 1036	0.36	0.04	10.7	0.43	0.27	18.8	24.6	0.40	0.32	0.44	0.28	1.77	2.31

Fig. 3.28 Set up for down-aisle tests

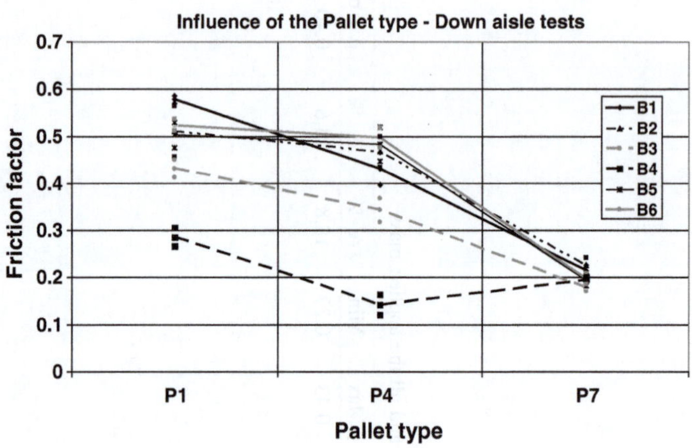

Fig. 3.29 Influence of the pallet type on the friction factor in down-aisle direction, for different types of beam

Table 3.19 Statistics of test results for different types of pallet and all types of beam

	μ	σ	Cov %	Max	Min	Δ^+ %	Δ^- %	$\mu + \sigma$	$\mu - \sigma$	$\mu + 2\sigma$	$\mu - 2\sigma$	α^+	α^-
P1	0.47	0.10	20.5	0.59	0.25	25.2	47.0	0.57	0.38	0.67	0.28	1.23	2.29
P4	0.39	0.13	32.1	0.55	0.11	39.0	71.1	0.52	0.27	0.65	0.14	1.21	2.21
P7	0.20	0.02	9.5	0.29	0.17	42.8	15.5	0.22	0.18	0.24	0.16	4.52	1.64

Table 3.22 c.o.v. as large as 42.5 % could be measured. For this reason, Tables 3.23 and 3.24 report the values of the statistical re-analysis of the test results respectively for beams B3 and B4, considered separately. It can be noticed that the c.o.v. strongly decreases.

Table 3.20 Statistics of test results for different types of pallet with data of beam types B1, B2, B3, B5 and B6

	μ	σ	Cov %	Max	Min	Δ^+ %	Δ^- %	$\mu+\sigma$	$\mu-\sigma$	$\mu+2\sigma$	$\mu-2\sigma$	α^+	α^-
P1	0.51	0.06	10.9	0.59	0.38	16.3	25.1	0.56	0.45	0.62	0.40	1.49	2.30
P4	0.45	0.06	14.1	0.55	0.29	23.2	34.0	0.51	0.38	0.57	0.32	1.65	2.42
P7	0.20	0.02	10.1	0.29	0.17	41.9	16.0	0.22	0.18	0.24	0.16	4.17	1.59

Table 3.21 Statistics of test results for different types of pallet with data of beam types B1, B2, B5 and B6

	μ	σ	Cov %	Max	Min	Δ^+ %	Δ^- %	$\mu+\sigma$	$\mu-\sigma$	$\mu+2\sigma$	$\mu-2\sigma$	α^+	α^-
P1	0.53	0.04	8.1	0.59	0.38	11.8	28.0	0.57	0.49	0.61	0.44	1.46	3.46
P4	0.47	0.04	8.5	0.55	0.35	16.7	25.4	0.51	0.43	0.55	0.39	1.97	3.00
P7	0.21	0.02	8.8	0.29	0.17	38.1	17.6	0.23	0.19	0.24	0.17	4.31	1.99

Table 3.22 Statistics of test results for different types of pallet with data of beam types B3 and B4

	μ	σ	Cov %	Max	Min	Δ^+ %	Δ^- %	$\mu+\sigma$	$\mu-\sigma$	$\mu+2\sigma$	$\mu-2\sigma$	α^+	α^-
P1	0.36	0.07	20.7	0.46	0.25	27.3	30.6	0.44	0.29	0.51	0.21	1.32	1.48
P4	0.24	0.10	42.5	0.40	0.11	65.0	52.8	0.34	0.14	0.45	0.04	1.53	1.24
P7	0.19	0.01	5.8	0.21	0.17	10.9	9.1	0.20	0.18	0.21	0.17	1.88	1.57

Table 3.23 Statistics of test results for different types of pallet with data of beam type B3

	μ	σ	Cov %	Max	Min	Δ^+ %	Δ^- %	$\mu+\sigma$	$\mu-\sigma$	$\mu+2\sigma$	$\mu-2\sigma$	α^+	α^-
P1	0.43	0.02	4.4	0.46	0.39	6.7	9.6	0.45	0.41	0.47	0.39	1.51	2.19
P4	0.34	0.03	7.3	0.40	0.29	16.1	14.4	0.37	0.32	0.39	0.29	2.19	1.96
P7	0.18	0.01	3.6	0.19	0.17	8.9	5.0	0.19	0.17	0.19	0.17	2.45	1.36

Table 3.24 Statistics of test results for different types of pallet with data of beam type B4

	μ	σ	Cov %	Max	Min	Δ^+ %	Δ^- %	$\mu+\sigma$	$\mu-\sigma$	$\mu+2\sigma$	$\mu-2\sigma$	α^+	α^-
P1	0.29	0.02	6.6	0.32	0.25	11.1	12.6	0.31	0.27	0.32	0.25	1.67	1.90
P4	0.14	0.02	15.2	0.20	0.11	38.7	20.4	0.17	0.12	0.19	0.10	2.55	1.34
P7	0.19	0.01	4.0	0.21	0.18	6.4	7.8	0.20	0.19	0.21	0.18	1.58	1.93

3.2.3.2 Influence of the Beam Type

Similar results can be obtained showing the influence of the beam type on the friction factor, for a mass of 785 kg centered on the pallet, as evidenced in Fig. 3.30 and Table 3.25.

- Friction factor is generally higher for pallet P1 (wooden Euro pallet). Pallet P7 (plastic Euro pallet) shows the lowest friction factor. Pallet P4 (wooden American pallet) has an intermediate behaviour.

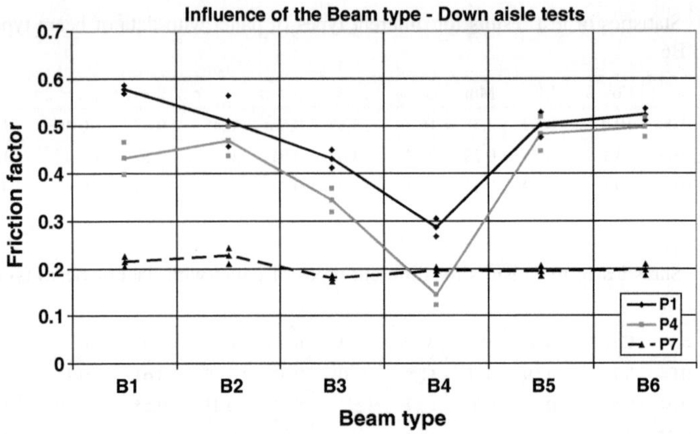

Fig. 3.30 Influence of the beam type on the friction factor in down-aisle direction, for different types of pallet

Table 3.25 Statistics of test results for different types of beam

	μ	σ	Cov %	Max	Min	Δ^+ %	Δ^- %	μ + σ	μ − σ	μ + 2σ	μ − 2σ	α^+	α^-
B1	0.41	0.15	37.0	0.59	0.19	44.8	53.4	0.56	0.26	0.71	0.11	1.21	1.44
B2	0.40	0.13	32.4	0.59	0.21	46.7	47.7	0.53	0.27	0.66	0.14	1.44	1.47
B3	0.32	0.11	33.1	0.46	0.17	43.0	47.2	0.43	0.22	0.53	0.11	1.30	1.42
B4	0.21	0.06	29.6	0.32	0.11	52.8	45.2	0.27	0.15	0.33	0.08	1.78	1.53
B5	0.39	0.14	36.4	0.55	0.18	39.0	55.2	0.54	0.25	0.68	0.11	1.07	1.52
B6	0.41	0.15	36.7	0.54	0.17	33.9	57.8	0.56	0.26	0.70	0.11	0.92	1.57

- Plastic pallet (P7) shows practically the same friction factor independently on the beam type.
- Behaviour of the friction factor for pallet P1 and P4 is quite similar: the lowest value is obtained for beam type B4; beam types B1, B2, B5 and B6 have more or less the same value.
- Scatter of the results is very large with a c.o.v. ranging from 29.6 to 37.0 %.

The next tables show the results for all the beam types and for the two different types of wooden pallet, P1 and P4 (Table 3.26), and for the plastic pallet P7

Table 3.26 Statistics of test results for different types of beam for pallet types P1 and P4

	μ	σ	Cov %	Max	Min	Δ^+ %	Δ^- %	μ + σ	μ − σ	μ + 2σ	μ − 2σ	α^+	α^-
B1	0.51	0.08	15.3	0.59	0.35	17.0	30.5	0.58	0.43	0.66	0.35	1.11	1.99
B2	0.49	0.05	10.0	0.59	0.38	20.7	22.3	0.54	0.44	0.59	0.39	2.07	2.22
B3	0.39	0.05	12.6	0.46	0.29	18.1	24.5	0.44	0.34	0.49	0.29	1.43	1.95
B4	0.21	0.07	34.6	0.32	0.11	48.1	46.9	0.29	0.14	0.36	0.07	1.39	1.35
B5	0.49	0.03	6.8	0.55	0.42	11.2	14.0	0.53	0.46	0.56	0.43	1.65	2.07
B6	0.51	0.02	4.3	0.54	0.45	6.5	11.3	0.53	0.49	0.55	0.47	1.53	2.66

(Table 3.27). It can be noticed that these batches are more homogeneous, and the scatter is heavily reduced. With exception of beam B4, c.o.v. ranges between 4.3 and 15.3 % in the case of the wooden pallets P1 and P4, while for the plastic pallet P7 is always very low ranging from 3.6 to 7.2 %.

Tables 3.28 and 3.29 consider the statistics of data of test carried out with pallets P1 and P4 and respectively beams B1 + B2 + B5 + B6 and beams B3 + B4.

When the group of beams B1, B2, B5 and B6 is considered (Table 3.28), the scatter of the results reduced to 10.1 %, and a mean value of the static friction factor of 0.5 is obtained.

As already observed, beams B3 and B4 (Table 3.29) should be considered separately due to the large c.o.v. of 35.5 %. The results are shown respectively in Tables 3.30 and 3.31. A large scatter is always present for the data of beam B4.

3.2.3.3 Influence of the Applied Mass

The influence of the mass is measured on the same pallet type P1 (Wooden Euro pallet) positioning the mass on the pallet without eccentricity and considering different types of beams.

Figure 3.31 shows the influence of the beam type on the friction factor in down-aisle direction for different values of the applied mass. Table 3.32 shows the statistics of the same results for all beam types, while the statistics presented in Table 3.33 consider only data of beam types B1 + B2 + B5 + B6.

- Considering each single beam type separately, the static friction factor assumes more or less the same value independently of the mass. This behaviour can be observed in Fig. 3.32.
- Independently of the applied mass, the lowest value of the static friction factor is obtained for beam type B4. For a same type of beam, the standard deviation is very small. With the exception of beam type B4 (c.o.v. = 16.6 %), all the other types show c.o.v. ranging from 5.8 to 8.6 %.
- When beam types B3 and B4 are removed from the batch of data (Table 2.33), it can be noticed that the scatter of results is very limited, with c.o.v. = 7.8 %.

The following tables show the re-analysis of the results in terms of applied mass, for all beam types (Table 3.34) and for beam types B1 + B2 + B5 + B6 (Table 3.35).

Table 3.27 Statistics of test results for different types of beam for pallet type P7

	μ	σ	Cov %	Max	Min	Δ^+ %	Δ^- %	$\mu+\sigma$	$\mu-\sigma$	$\mu+2\sigma$	$\mu-2\sigma$	α^+	α^-
B1	0.21	0.01	5.4	0.24	0.19	12.8	11.3	0.23	0.20	0.24	0.19	2.38	2.09
B2	0.23	0.02	7.2	0.29	0.21	26.9	6.9	0.24	0.21	0.26	0.19	3.73	0.96
B3	0.18	0.01	3.6	0.19	0.17	8.9	5.0	0.19	0.17	0.19	0.17	2.45	1.36
B4	0.19	0.01	4.0	0.21	0.18	6.4	7.8	0.20	0.19	0.21	0.18	1.58	1.93
B5	0.19	0.01	5.7	0.21	0.18	10.6	8.9	0.21	0.18	0.22	0.17	1.85	1.55
B6	0.20	0.01	5.9	0.23	0.17	14.2	13.0	0.21	0.19	0.22	0.17	2.40	2.20

Table 3.28 Statistics of test results considering pallet P1 + P4 and beams B1 + B2 + B5 + B6

	μ	σ	Cov %	Max	Min	Δ^+ %	Δ^- %	$\mu + \sigma$	$\mu - \sigma$	$\mu + 2\sigma$	$\mu - 2\sigma$	α^+	α^-
B1 + B2 + B5 + B6	0.50	0.05	10.1	0.59	0.35	18.3	29.8	0.55	0.45	0.60	0.40	1.81	2.94

Table 3.29 Statistics of test results considering pallet P1 + P4 and beams B3 + B4

	μ	σ	Cov %	Max	Min	Δ^+ %	Δ^- %	$\mu + \sigma$	$\mu - \sigma$	$\mu + 2\sigma$	$\mu - 2\sigma$	α^+	α^-
B3 + B4	0.30	0.11	35.5	0.46	0.11	51.8	62.3	0.41	0.20	0.52	0.09	1.46	1.75

Table 3.30 Statistics of test results considering pallet P1 + P4 and beam B3

	μ	σ	Cov %	Max	Min	Δ^+ %	Δ^- %	μ + σ	μ − σ	μ + 2σ	μ − 2σ	α^+	α^-
B3	0.39	0.05	12.6	0.46	0.29	18.1	24.5	0.44	0.34	0.49	0.29	1.43	1.95

Table 3.31 Statistics of test results considering pallet P1 + P4 and beam B4

	μ	σ	Cov %	Max	Min	Δ^+ %	Δ^- %	μ + σ	μ − σ	μ + 2σ	μ − 2σ	α^+	α^-
B4	0.21	0.07	34.6	0.32	0.11	48.1	46.9	0.29	0.14	0.36	0.07	1.39	1.35

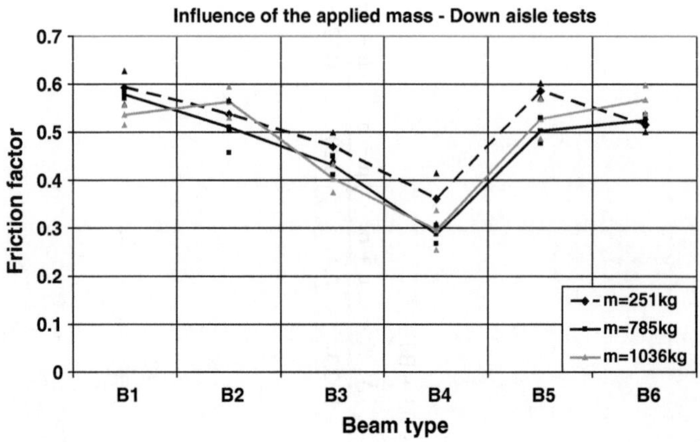

Fig. 3.31 Influence of the beam type on the friction factor in down-aisle direction, for different values of the applied masses

Table 3.32 Statistics of test results for different types of beam

	μ	σ	Cov %	Max	Min	Δ^+ %	Δ^- %	μ + σ	μ − σ	μ + 2σ	μ − 2σ	α^+	α^-
B1	0.57	0.03	6.0	0.64	0.50	12.6	12.5	0.60	0.53	0.64	0.50	2.09	2.09
B2	0.54	0.05	8.5	0.61	0.38	13.8	29.0	0.58	0.49	0.63	0.44	1.62	3.39
B3	0.44	0.04	8.6	0.51	0.31	16.5	28.4	0.47	0.40	0.51	0.36	1.92	3.31
B4	0.31	0.05	16.6	0.46	0.20	45.2	35.5	0.37	0.26	0.42	0.21	2.73	2.14
B5	0.54	0.05	8.5	0.61	0.45	14.2	16.7	0.58	0.49	0.63	0.45	1.66	1.97
B6	0.53	0.03	5.9	0.61	0.46	14.4	14.6	0.57	0.50	0.60	0.47	2.42	2.46

The same consideration drawn in case of cross-aisle direction holds: the mass doesn't influence the static friction factor in down-aisle direction. As shown in Table 3.36, the mean value of the friction factor for the sample made by results of tests carried out with beams B1 + B2 + B5 + B6 and all the three values of applied mass is 0.54 with a c.o.v. of 7.8 %. Same consideration (the mass doesn't influence much the static friction factor) holds also for beams B3 and B4, that however

Table 3.33 Statistics of test results considering the three masses and beams B1 + B2 + B5 + B6

	μ	σ	Cov %	Max	Min	Δ^{+} %	Δ^{-} %	$\mu + \sigma$	$\mu - \sigma$	$\mu + 2\sigma$	$\mu - 2\sigma$	α^{+}	α^{-}
B1 + B2 + B5 + B6	0.54	0.04	7.8	0.64	0.38	17.7	30.0	0.59	0.50	0.63	0.46	2.27	3.86

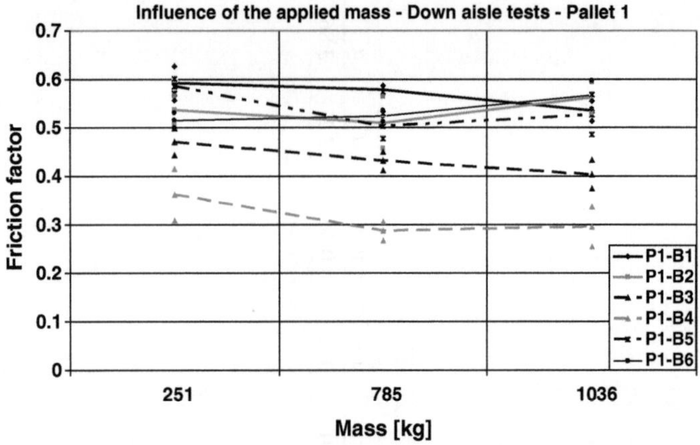

Fig. 3.32 Influence of the applied mass on the friction factor in down-aisle direction, for different types of beam

Table 3.34 Statistics of test results for different types of applied mass and all types of beam

Mass	μ	σ	Cov %	Max	Min	Δ^+ %	Δ^- %	$\mu+\sigma$	$\mu-\sigma$	$\mu+2\sigma$	$\mu-2\sigma$	α^+	α^-
251	0.51	0.08	16.6	0.64	0.29	25.7	43.7	0.59	0.43	0.68	0.34	0.13	0.22
785	0.47	0.10	20.5	0.59	0.25	25.2	47.0	0.57	0.38	0.67	0.28	0.12	0.22
1036	0.48	0.11	21.9	0.61	0.20	27.5	57.9	0.59	0.38	0.69	0.27	0.13	0.28

Table 3.35 Statistics of test results for different types of applied mass, beam types B1 + B2 + B5 + B6

Mass	μ	σ	Cov %	Max	Min	Δ^+ %	Δ^- %	$\mu+\sigma$	$\mu-\sigma$	$\mu+2\sigma$	$\mu-2\sigma$	α^+	α^-
251	0.56	0.04	7.7	0.64	0.44	15.0	20.7	0.60	0.51	0.64	0.47	1.96	2.70
785	0.53	0.04	8.0	0.59	0.38	11.0	28.5	0.58	0.49	0.62	0.45	1.37	3.57
1036	0.55	0.04	6.6	0.61	0.46	12.0	16.2	0.58	0.51	0.62	0.48	1.81	2.45

should be considered separately due to influence of their surface treatment on the results (Tables 3.37 and 3.38).

Finally, from the exam of the previous figures and tables, it can be concluded that the variation of the mass has a limited influence on the value of the friction factor. In any case, such an influence is lower than the one of other parameters like pallet and beam types.

Table 3.36 Statistics of test results for all types of applied mass, beam types B1 + B2 + B5 + B6

Mass	μ	σ	Cov %	Max	Min	Δ^+ %	Δ^- %	μ + σ	μ − σ	μ + 2σ	μ − 2σ	α^+	α^-
251 + 785 + 1036	0.54	0.04	7.8	0 64	0.38	17.7	30.0	0.59	0.50	0.63	0.46	2.27	3.86

Table 3.37 Statistics of test results for all types of applied mass, beam type B3

Mass	μ	σ	Cov %	Max	Min	Δ^+ %	Δ^- %	μ + σ	μ − σ	μ + 2σ	μ − 2σ	α^+	α^-
251 + 785 + 1036	0.44	0.04	8.6	0.51	0.31	16.5	28.4	0.47	0.40	0.51	0.36	1.92	3.31

Table 3.38 Statistics of test results for all types of applied mass, beam type B4

Mass	μ	σ	Cov %	Max	Min	Δ^+ %	Δ^- %	$\mu + \sigma$	$\mu - \sigma$	$\mu + 2\sigma$	$\mu - 2\sigma$	α^+	α^-
251 + 785 + 1036	0.31	0.05	16.6	0.46	0.20	45.2	35.5	0.37	0.26	0.42	0.21	2.73	2.14

3.2.3.4 Influence of the Mass Eccentricity

The influence of the mass eccentricity was investigated only in the down-aisle direction. The position of the mass on the pallet determines a different distribution of the weight force on the beam, that can influences the value of the friction factor.

Figure 3.33 shows the position of the mass in the three analysed cases.

The weight of the mass can be divided in two components, $F_\perp$ and $F_\parallel$ due to the inclined plane. The former component decreases during the test performed increasing the inclination θ of the plane on the horizontal, the latter increases, being:

$$\begin{aligned} F_\perp &= F \cdot \cos(\vartheta) \\ F_\parallel &= F \cdot \sin(\vartheta) \end{aligned} \tag{3.16}$$

The orthogonal component $F_\perp$ can be considered distributed on the three series of blocks of the pallet with the three components $F'_\perp$, $F''_\perp$ and $F'''_\perp$ (as shown in Fig. 3.34c). The parallel component $F_\parallel$ (applied in the c.o.g. of the mass) is resisted by the "friction", on the beam-to-pallet interface. As a consequence $F_\parallel$ has a lever arm with respect to the sliding plane, where the friction reaction develops. Hence, the effect of $F_\parallel$ can be represented as shown in Fig. 3.34b, where M is the transport moment of $F_\parallel$ that has been "moved" from the c.o.g. to the beam-to-pallet interface. Effect of such a moment (an overturning moment) is to increase the reaction

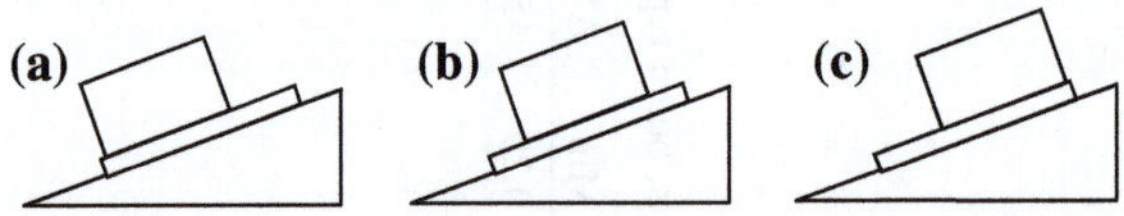

Fig. 3.33 Mass eccentricity: **a** eccentric downward, **b** centred, **c** eccentric upward

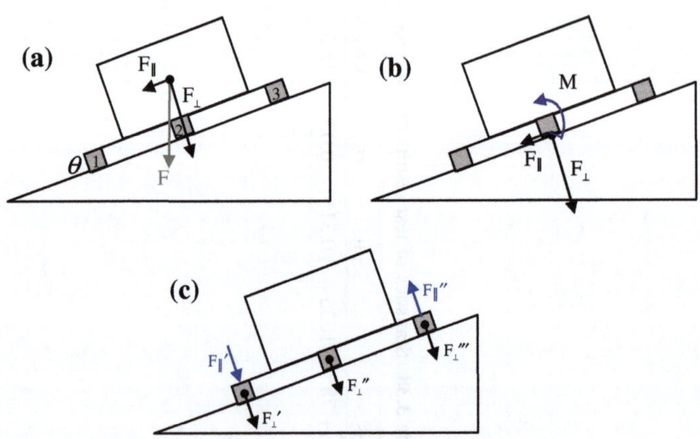

Fig. 3.34 Different component of the forces on the mass during the test

on the wooden block no. 1 with the force $F'_{\parallel}$, and to decrease the one on the third block with the $F''_{\parallel}$ component, as shown in Fig. 3.34c.

When the mass is positioned downward the orthogonal reaction on the block no. 3 is lower than that on blocks no. 1 and 2. Furthermore, the effect of the overturning moment is to increase the reaction on block no. 1 and reduce that on block no. 3. The result is an "uplift" of the block no. 3, i.e. a reduction of the contact surface (Fig. 3.35a).

If the mass is centered on the pallet this effect is reduced. On the contrary, when the mass is positioned with an "upward" eccentricity, the effect due to the overturning moment somehow compensates the non-uniform distribution of the reactions perpendicular to the sliding surface resulting in a more uniform distribution of the weight of the pallet, "maximizing" the contact surface (Fig. 3.35b).

Figure 3.36 shows the results of the tests, carried out with the combination of Pallet type P4 (Wooden American Pallet) with a beam type B6.

Experimental results confirm the previous considerations. When the mass is positioned with an upward eccentricity, the measured friction factor is larger than in the case with the downward eccentricity. The centered mass develops a friction factor larger than in the case of downward eccentricity (Table 3.39).

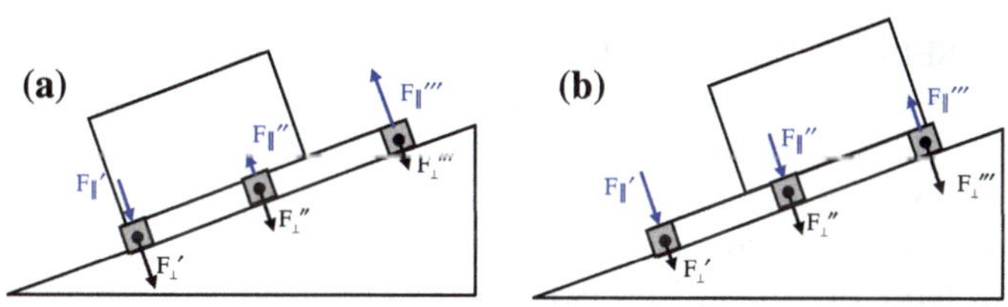

Fig. 3.35 Components of the forces in a downward (**a**) and upward (**b**) position

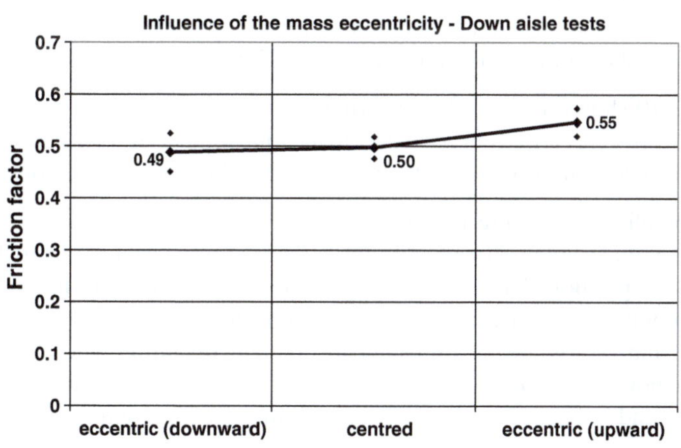

Fig. 3.36 Influence of the mass eccentricity on the friction factor in down-aisle direction

Table 3.39 Statistics of test results for different positions of applied mass

	μ	σ	Cov %	Max	Min	Δ^+ %	Δ^- %	μ + σ	μ − σ	μ + 2σ	μ − 2σ	α^+	α^-
Down	0.49	0.04	7.5	0.55	0.43	13.0	11.9	0.53	0.45	0.56	0.42	1.73	1.59
Centre	0.50	0.02	4.2	0.54	0.45	9.4	8.9	0.52	0.48	0.54	0.46	2.22	2.10
Up	0.55	0.03	4.9	0.60	0.51	9.0	7.4	0.57	0.52	0.60	0.49	1.83	1.50

The variations of the friction coefficient due to differences in the eccentricity of the mass are in any case very small. The slightly increasing trend of the parameter can be explained with the force distribution as shown before. Passing from the downward to the upward position of the mass, the reaction forces on the beam-pallet interface are more balanced, the contact area is maximized and the friction factor increases.

Although this trend has been investigated only for one combination of beam and pallet, the feeling is that this conclusion can be generalized.

3.3 Assessment of the Sliding Conditions of the Pallets Under Dynamic Loading

Within the “SEISRACKS” research Project 182 tests were carried out in order to investigate the dynamic sliding conditions of pallets on racks. Dynamic tests were performed at the Laboratory for Earthquake Engineering (LEE) of the National Technical University of Athens (NTUA), for the assessment of the dynamic “pallet-beam” friction factor and sliding conditions.

Different types of “sliding” test were performed, considering different combinations of beams and pallets, both in cross-aisle and down-aisle direction.

The following types of excitations were considered:

- Sinusoidal with constant frequency and increasing acceleration (133 tests)
- Sinusoidal with constant acceleration and increasing frequency (27 tests)
- Seismic tests with recorded input motions (22 tests)

Table 3.40 shows a list of the performed tests.

All the tests were carried out according to the following instructions/specifications, in order to try to minimize bias of the results due to “operating conditions”:

- Horizontality of structure is checked at every test.
- Masses are absolutely “fixed” on pallet (no relative movement or rocking).
- Pallets are positioned at equal spacing on the beams and “re-positioned” in their initial configuration at every test (a time consuming procedure).
- Pallets and beams are instrumented in order to obtain as much the pertinent information as possible (not only translation in excitation direction, but eventual rotation and acceleration of pallets and beams in both x and y directions).

Table 3.40 Different types of sliding test carried out: (a) constant frequency, (b) constant acceleration, (c) seismic

(a) Tests with constant frequency [*f*]

Cross-aisle				Down-aisle			
f [Hz]	Pallet	Beam	No. of test	*f* [Hz]	Pallet	Beam	No. of test
1.0	P2	B1	10	1.0	P2	B2	4
2.0	P2	B1	20	2.0	P2	B2	11
3.0	P2	B1	10	3.0	P2	B2	8
4.0	P2	B1	6	4.0	P2	B2	5
1.0	P2	B3	10	1.0	P2	B3	6
1.5	P2	B3	10	2.0	P2	B3	1
2.0	P2	B3	10	3.0	P2	B3	1
2.5	P2	B3	11				
3.0	P2	B3	10				
Total			97	Total			36

(b) Tests with constant acceleration

Down-aisle				
Acceleration [g]	Pallet	Beam	Frequency interval [Hz]	No. of tests
0.10	P2	B2	1–16	5
0.20	P2	B2	1–16	5
0.25	P2	B2	1–8	5
0.30	P2	B2	1–8	5
0.35	P2	B2	2–7	4
0.40	P2	B2	2–7	3
Total				27

(c) Seismic tests

Cross-aisle				Down-aisle			
Accelerogram	Pallet	Beam	No. of tests	Accelerogram	Pallet	Beam	No. of tests
Argostoli	P2	B1	4	Argostoli	P2	B2	2
Edessa	P2	B1	9	Edessa	P2	B2	3
Kalamata	P2	B1	4				
Total			17	Total			5

- Every 10 tests, both the beam-to-upright connection and the base connections are checked for possible damage (looseness or permanent deformation) and eventually replaced.
- Minimum 5 repetitions are carried out for each test.

3.3.1 Tests in Cross-Aisle Direction

3.3.1.1 Test Set up

In cross-aisle direction tests were carried out only with a sinusoidal excitation of the table, with constant frequency. The test set up is shown in Figs. 3.37 and 3.38. Three pallets are supported by two beams and instrumented by means of:

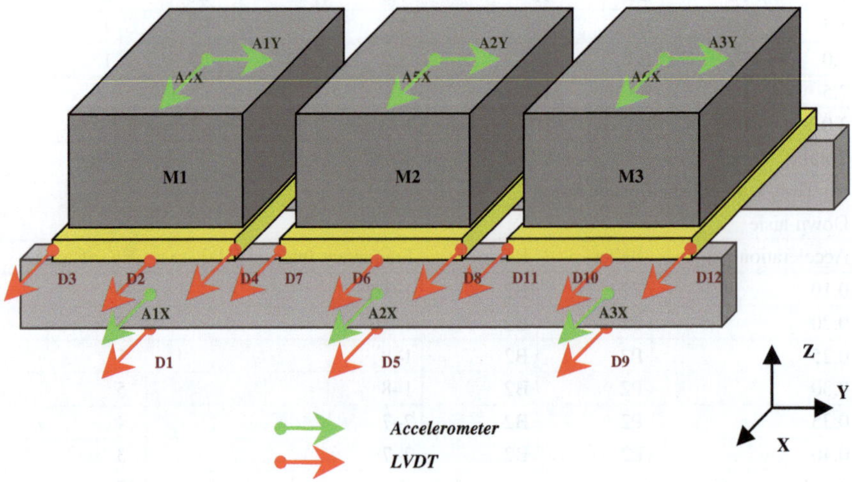

Fig. 3.37 Instrumentation set up for cross-aisle tests

Fig. 3.38 Instrumentation set up for cross-aisle test

- 6 accelerometers in X direction (1 on each mass and 3 on a beam, under each mass);
- 3 accelerometers in Y direction (one for each mass);
- 12 LVDTs (two on every pallet and two on a beam below each mass).

In this way the possible displacements and accelerations of the pallets are monitored.

3.3.1.2 Procedure for Re-analysis of the Tests Results

In the cross-aisle tests only one type of pallet was adopted: the Wooden Euro Pallet 800 × 1200 mm old and dry (type P2). Two types of beams were considered: a cold rolled, powder coated, new beam (type B1), and a cold rolled, hot zinc coated, new beam (type B3).

Due to the nature of the phenomenon and to the large number of variables involved in the problem, a large scatter is to be expected in the results of sliding tests.

Hence, in order to reduce to a minimum, the possibility of introducing further bias of the results, a standardized re-analysis procedure has been identified, and systematically applied. Such a procedure is explained hereafter, with reference to a typical set of sliding test results. Figure 3.39 represents the typical response of a test carried out under constant frequency conditions. Figure 3.39a, c refer to the two lateral masses, while Fig. 3.39b refers to the central mass. In each figure, the acceleration of the shaking table is compared with the acceleration of the mass, and to the one of the beam, measured by the accelerometer positioned on the beam, under the mass. It is possible to notice how the accelerations of the mass and of the beam are higher than the table one. The acceleration of the central mass, however, remains equal to that of the beam for a longer time than the two lateral ones.

Difference between acceleration of the beam and of the mass means that the mass is sliding on the beam. If the sliding exceeds a limit value, the pallet can fall off the support. In Fig. 3.39 it can be noticed that the acceleration of the mass initially follows the acceleration measured on the beam, then increases up to reach a sort of "limit value of acceleration". In this condition it is possible an assessment of the upper value of the sliding acceleration of the pallet on the beam.

Figure 3.40 presents the maximum and the minimum values of the acceleration of the masses, measured at every cycle of excitation. The phenomenon in which the acceleration of the mass reaches a maximum value is not evident in the central pallet (pallet 2), because sliding of the two lateral pallets (pallet 1 and 3) usually starts earlier, and tests were stopped once sliding of pallets was detected. For this reason a clear assessment of the upper bound of the sliding acceleration of the central mass was not always possible. Only in some tests, in which the acceleration of the table is higher or the duration of the test is longer, this phenomenon is more evident.

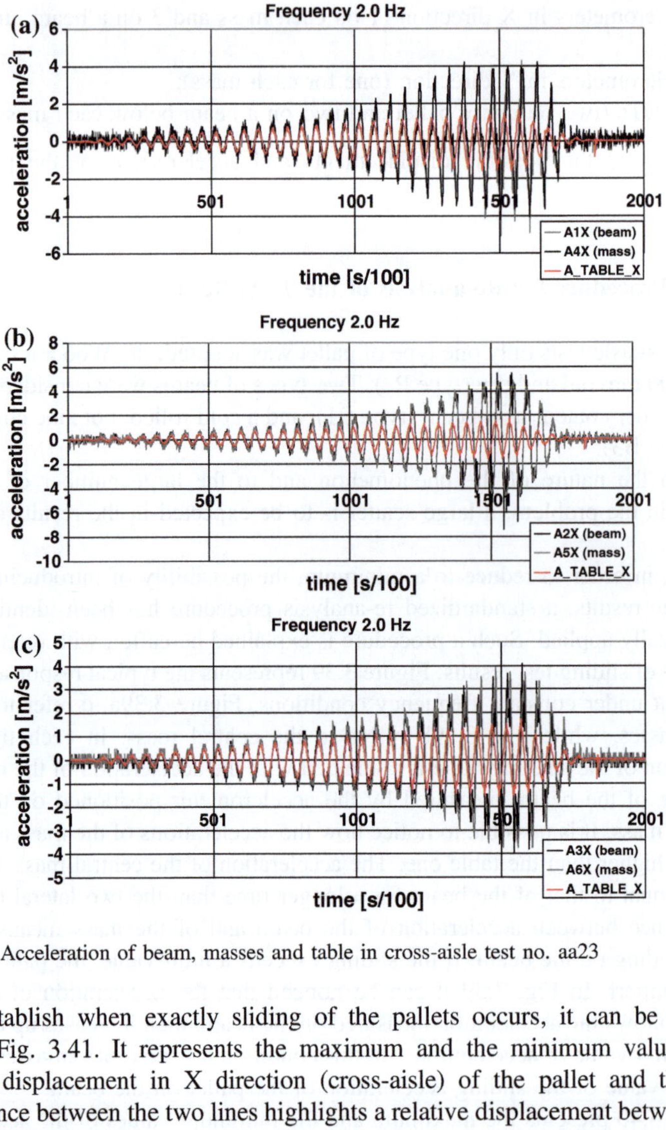

Fig. 3.39 Acceleration of beam, masses and table in cross-aisle test no. aa23

To establish when exactly sliding of the pallets occurs, it can be useful to observe Fig. 3.41. It represents the maximum and the minimum values of the absolute displacement in X direction (cross-aisle) of the pallet and the beam. A difference between the two lines highlights a relative displacement between pallet and beam, i.e. "sliding".

The sliding (i.e. the relative displacement between the pallet and the beam) of the three pallets is shown in Fig. 3.42. For the two lateral pallets, it has a sinusoidal trend and can reach a considerable final value. In the case of the central pallet this phenomenon is not so evident. Probably, as shown in Fig. 3.42b, sliding starts approximately 15 s after test initiation; however, the shaking table was stopped at approximately 17 s, too early to allow a large sliding of pallet 2. For pallet one and three a high frequency response superimposed to a low frequency one can be observed in Fig. 3.42a, b respectively.

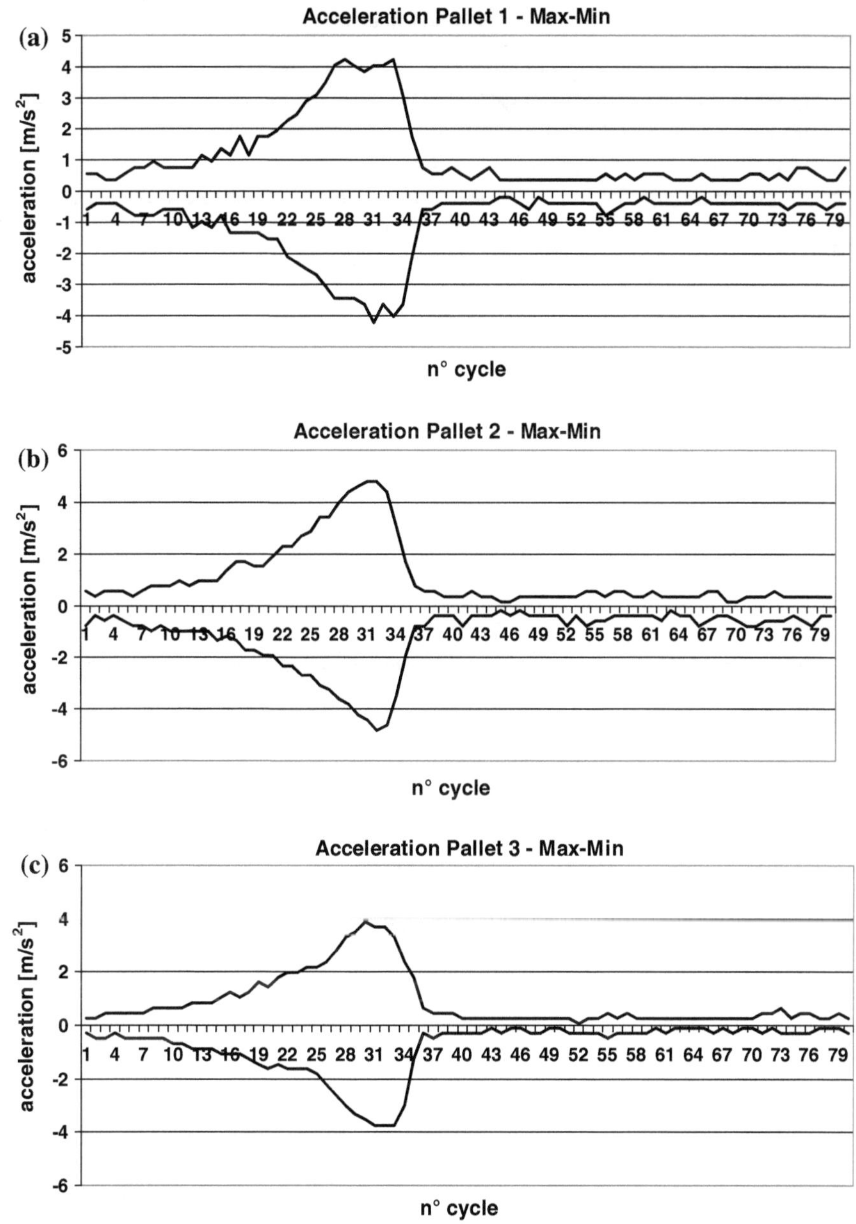

Fig. 3.40 Maximum and minimum values of acceleration of pallets (test no. aa23): **a** pallet 1, **b** pallet 2, **c** pallet 3

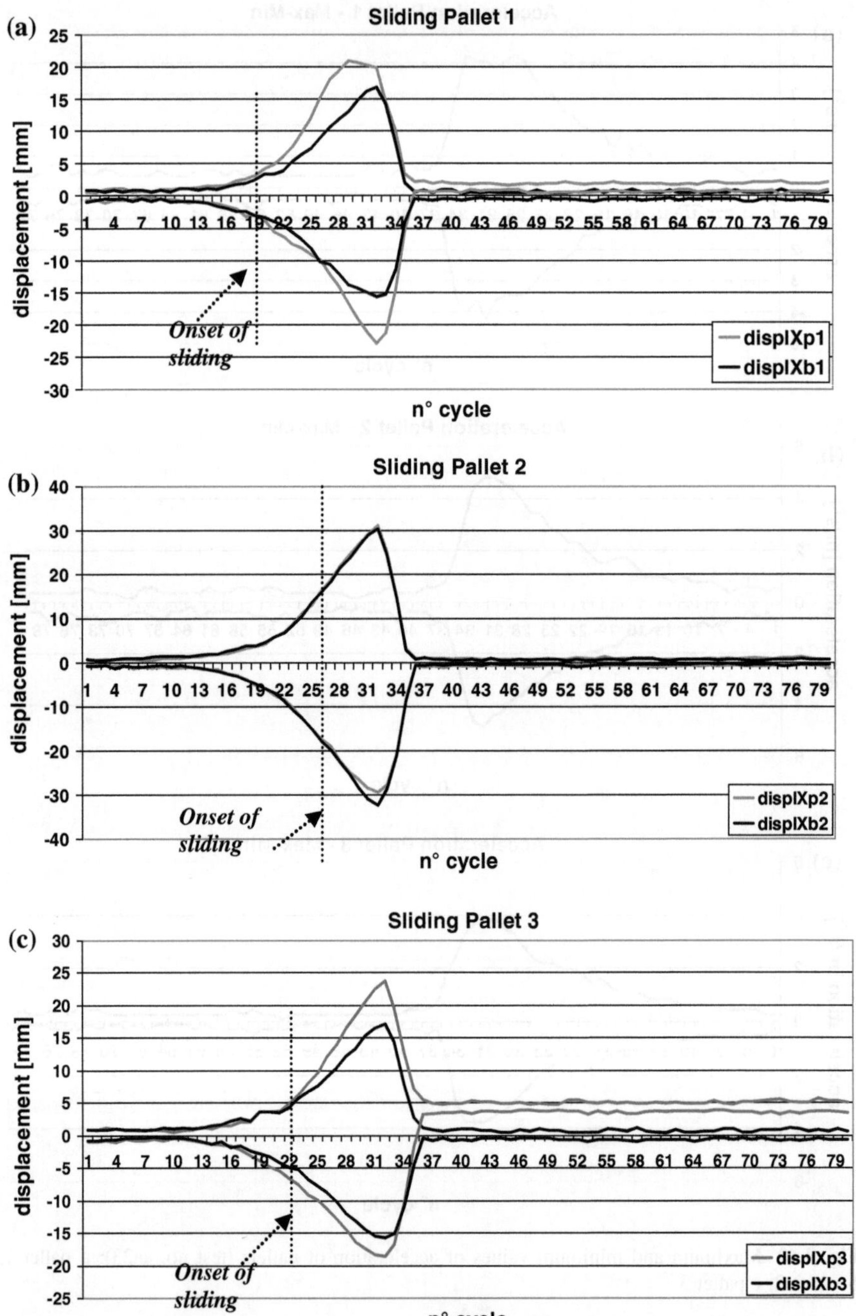

Fig. 3.41 Displacement of pallets and beam (test no. aa23): **a** pallet 1, **b** pallet 2, **c** pallet 3

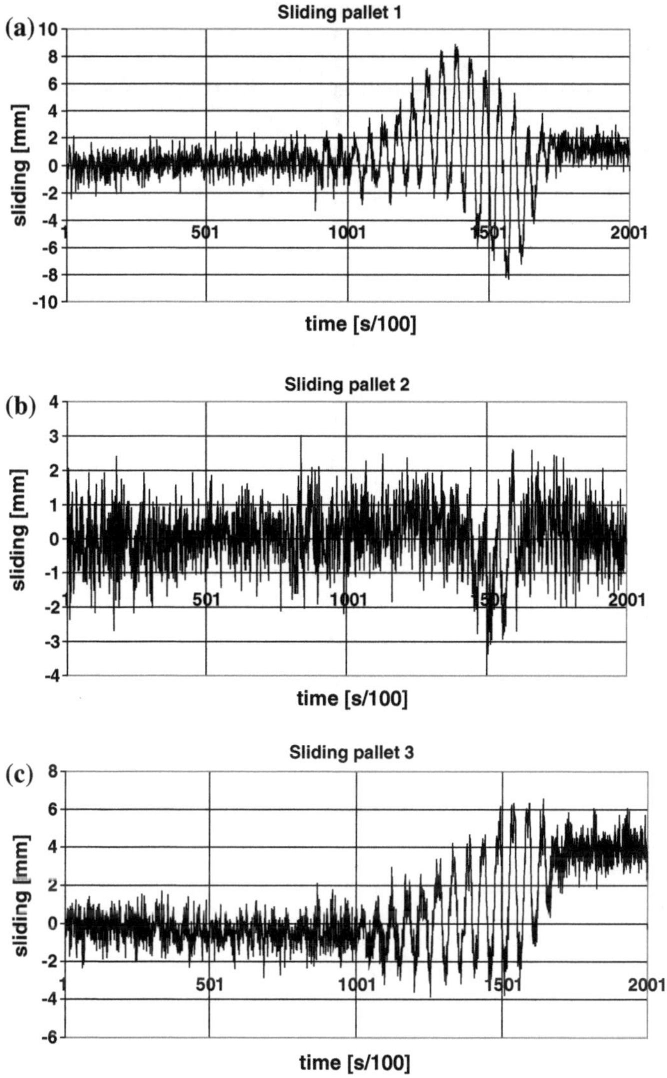

Fig. 3.42 Sliding of pallets (test no. aa23): **a** pallet 1, **b** pallet 2, **c** pallet 3

However, while in the first case at the end of the test a relative displacement lower than 2 mm occurs (Fig. 3.42a), in the second case a sort of "accumulation trend" takes place with a final value of sliding of 4 mm (Fig. 3.42c). If the excitation was longer or the frequency was higher, probably the sliding would increase.

Figure 3.43 shows the hysteresis loop in terms of pallet sliding versus acceleration; such hysteresis loops show that the sliding phenomenon is associated with energy dissipation.

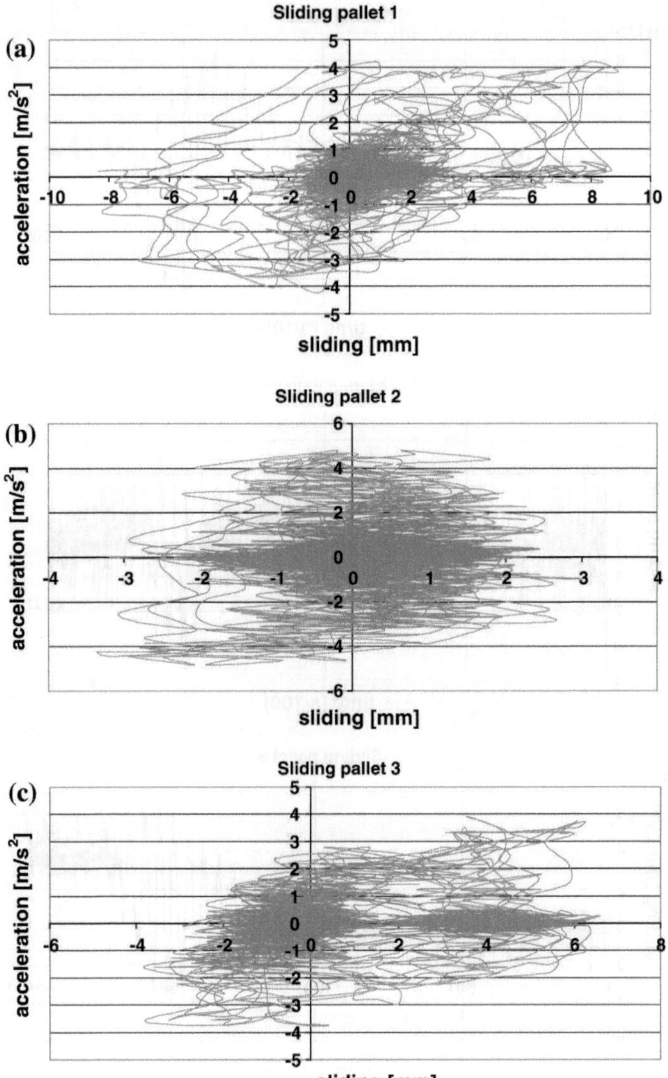

Fig. 3.43 Hysteresis loops for the three pallets on the structure (test aa23): **a** pallet 1, **b** pallet 2, **c** pallet 3

By examining Fig. 3.43 a limit value of the acceleration of the mass can be observed. In particular, in this specific case, the upper value is 4.0 m/s^2 for pallet 1 and 3, and 5.0 m/s^2 for pallet 2.

Figure 3.44 summarizes the maximum and the minimum values of the acceleration of table, masses and beams, as well as the sliding of the pallets. Sliding starts exactly when the acceleration of the mass differs from the beam one.

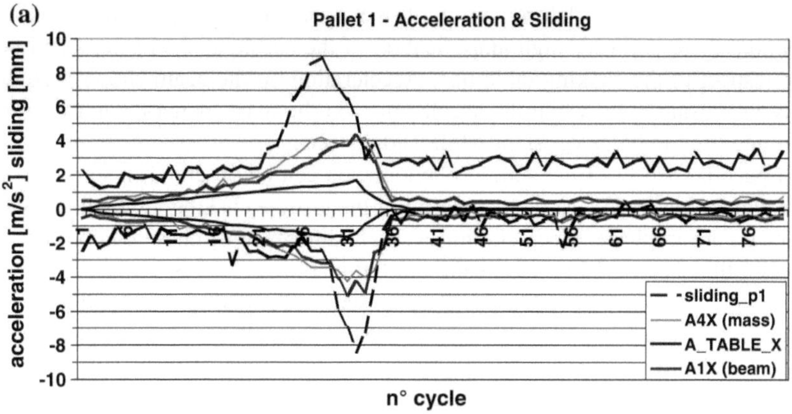

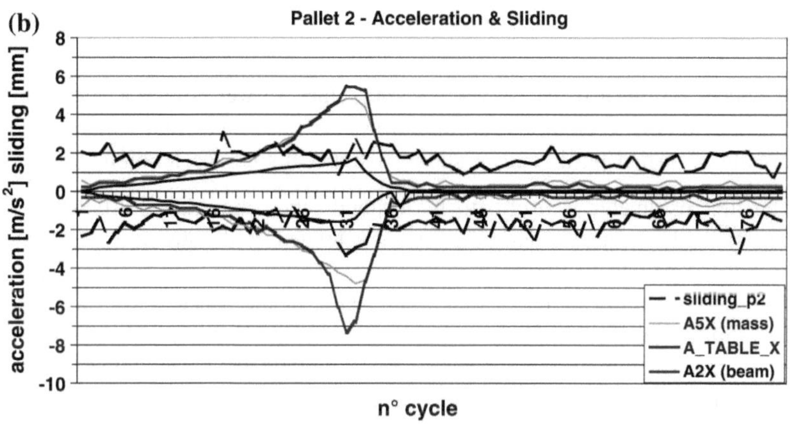

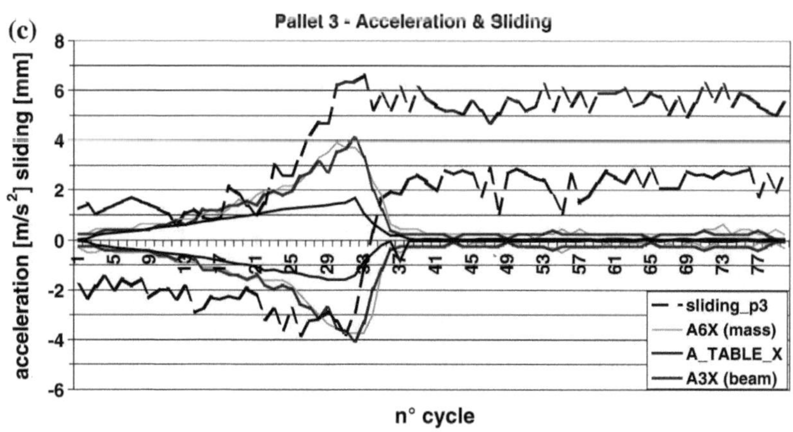

Fig. 3.44 Acceleration of table, mass M3 and beam, and sliding (test no. aa23): **a** pallet 1, **b** pallet 2, **c** pallet 3

Due to the lateral deformability of the beam, especially in the cross-aisle direction, there is a very high amplification of the accelerations of mass and beam with respect to the shaking table one. Acceleration on the beam shows a continuously increasing trend while the acceleration on the mass increases up to a certain level and then remains constant while sliding occurs.

A peculiar aspect of the sliding phenomenon can be observed in some test (see for example Fig. 3.45). At the beginning of the test the shaking table, the mass and the beam are in phase; when sliding occurs a difference of phase can be observed between the shaking table, the beam and the masses.

Figure 3.46 shows the situation of the pallets at the end of a test. Sliding of the pallet on the beam is evident; in some cases, the pallet was nearly loosing the support. It is also evident the "deterioration" of the surface of the beams, where the paint film is scratched and removed due to the movement of the wooden pallets.

The cross-aisle tests have been carried out with constant frequency and increasing acceleration for two different types of beam and for the wooden Euro Pallet. When re-analysing sliding test results, two parameters have been assessed:

- the acceleration of the mass when sliding starts (lower bound acceleration);
- the maximum acceleration of the mass, i.e. the value that, once achieved, remains constant until "end of sliding" (upper bound acceleration).

The first parameter is useful to understand when there is possibility to have sliding: is the lower acceleration that, when achieved on the structure, can generate sliding of the pallets on the beam.

The second parameter is the maximum value of the acceleration of the pallet, during sliding. Hence, this value is associated to the maximum force eventually acting on the structure during a dynamic event: beyond this value, pallets are sliding and their mass can be considered independent of the structure. Higher acceleration values affect only the structural masses that are just a small percentage of the masses stored on the pallets.

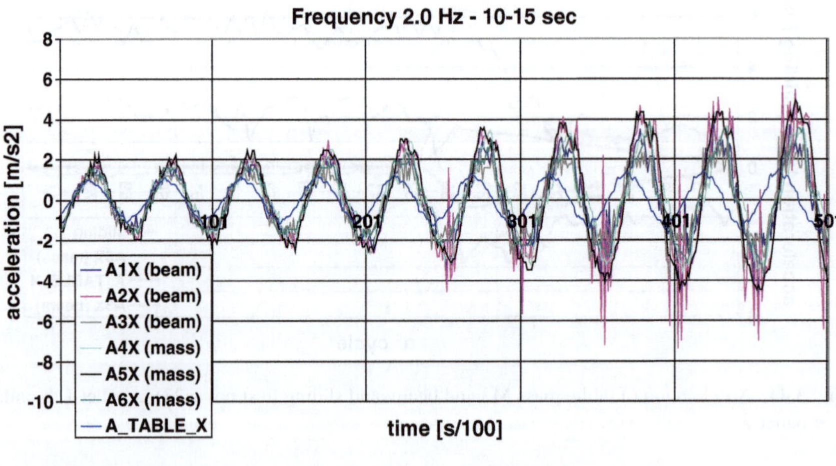

Fig. 3.45 Acceleration signals in the time interval 10–15 s (test no. aa21)

Fig. 3.46 Images of beam and pallets at the end of the tests

Figures 3.47 and 3.48 show the procedure for assessment of the lower bound sliding acceleration. This is obtained as the value of the mass acceleration corresponding to the time instant when a relative displacement occurs between the pallet and the beam.

As first thing, the cycle in which a relative displacement between the pallet and the beam occurs should be identified, on a plot like the one in Fig. 3.47. Then, by re-analysing the record of the pallet acceleration, is possible to identify the amplitude of the signal in the cycle previously defined, as shown in Fig. 3.48.

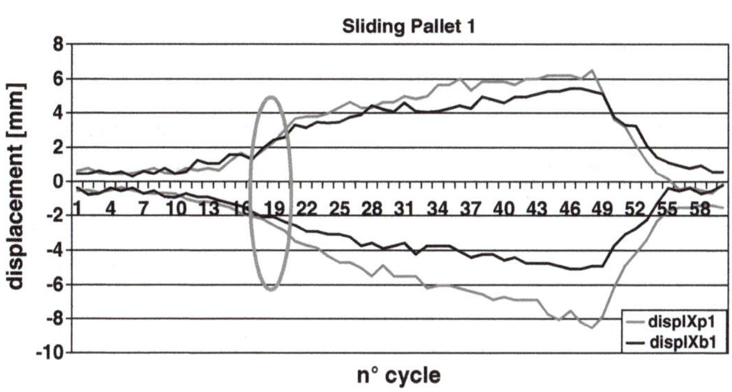

Fig. 3.47 Displacement of the pallet 1 and the beam under the pallet (test no. aa78)

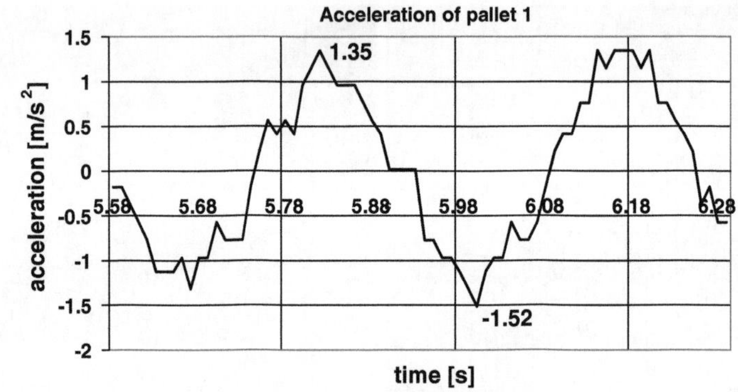

Fig. 3.48 Acceleration of pallet 1 at the considered instant—tests aa78

In the example shown in Fig. 3.48, the amplitude is the mean value between 1.35 and 1.52 m/s^2. Hence, the lower bound of the sliding acceleration is 1.44 m/s^2.

The upper bound of the sliding acceleration can be obtained only by re-analysis of those tests in which its trend becomes constant, as in the example shown in Fig. 3.49, where acceleration reaches an upper bound value of about 2.3 m/s^2.

In some tests, especially those carried out at low frequency, sliding did not occur or was limited and in most cases acceleration didn't reach an upper bound value.

The tests, in fact, had to be stopped when the maximum horizontal excursion of the shaking table reached ±100 mm. For low frequency (e.g. 1.0 Hz) this condition occurred for acceleration of the shaking table of approximately 0.2g, and this value is lower than the upper bound of the sliding acceleration. Hence, in many cases, no sliding of the central pallet was observed in low frequency tests.

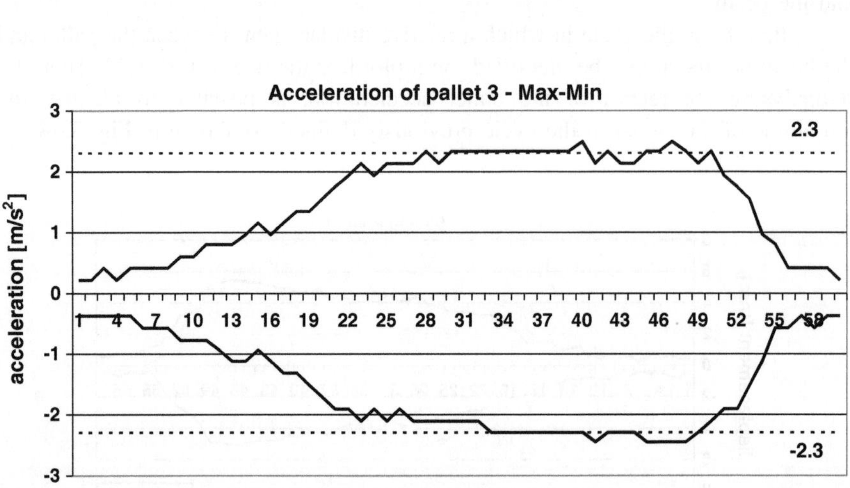

Fig. 3.49 Maximum values of the acceleration of pallet 3 in test aa78

3.3.1.3 Results of the Cross-Aisle Tests

3.3.1.3.1 Test with Pallet Type P2—Beam Type B1

In this paragraph all the results of the cross-aisle tests carried out with a wooden Euro Pallet old (type P2) and a cold rolled, powder coated, new beam (type B1) are shown. For every frequency of the shaking table the following data are presented:

1. The characteristics of the input signal (maximum acceleration, frequency and duration);
2. A table that shows, for each of the three pallets, the acceleration of sliding initiation (Lower bound acceleration), the upper bound of the sliding acceleration (if applicable) as well as the time when sliding starts (evaluated from the beginning of the test), in order to give an indication about the sequence of sliding of the different pallets. Mean value, standard deviation and c.o.v. of these data are also presented.
3. A graphical representation of the numerical data, for every group of tests.

3.3.1.3.2 Frequency = 1.0 Hz

The increment rate of the shaking acceleration, shown in Fig. 3.50 is $0.07\frac{m/s^2}{s}$ (Fig. 3.51).

3.3.1.3.2.1 Frequency = 2.0 Hz

Tests carried out with this value of frequency were excited with the three different types of signal shown in Figs. 3.52, 3.53 and 3.54, characterised by a different increment rates of the table acceleration, respectively (Tables 3.41 and 3.42)

$$0.1\frac{m/s^2}{s}, 0.075\frac{m/s^2}{s}, 0.05\frac{m/s^2}{s}.$$

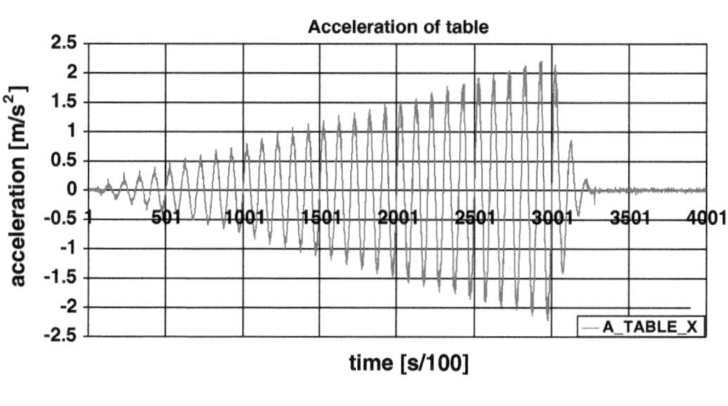

Fig. 3.50 Table acceleration in tests from aa52 to aa61

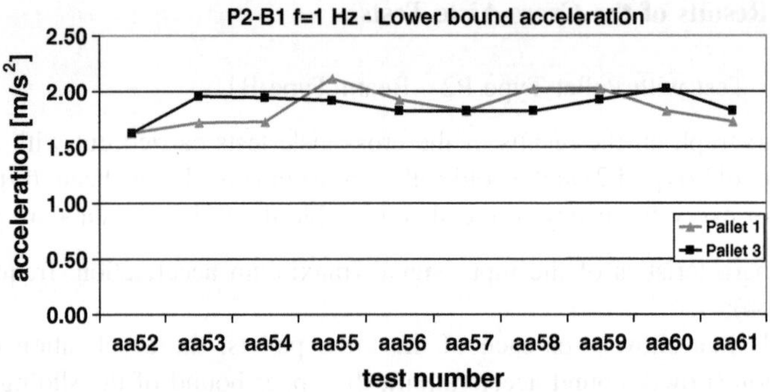

Fig. 3.51 Acceleration of sliding initiation in different tests—P2B1 f = 1.0 Hz

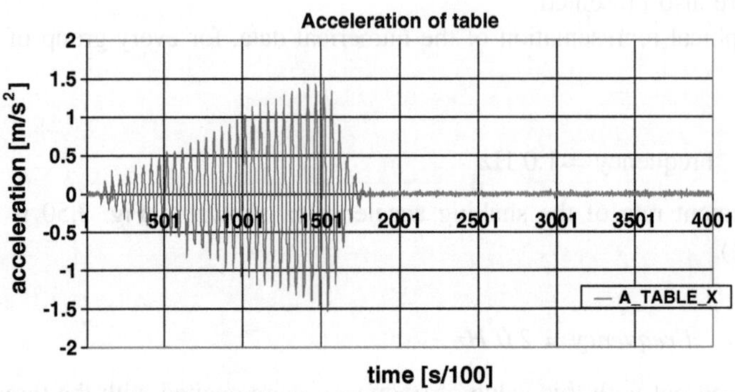

Fig. 3.52 Table acceleration in tests from aa62 to aa71, with increment rate 0.1 (m/s^2)/s

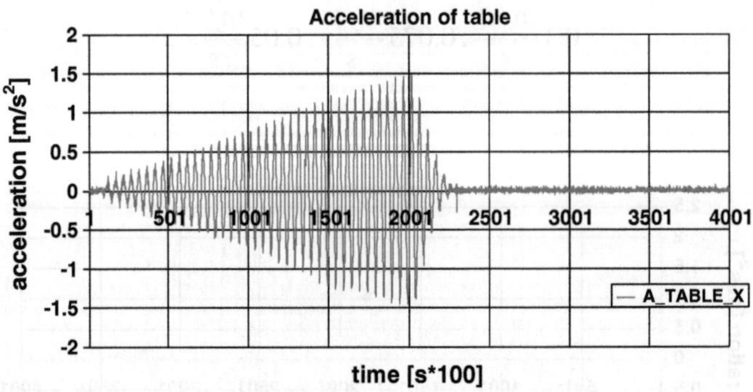

Fig. 3.53 Table acceleration in tests aa88 and from aa106 to aa108, with increment rate 0.075 (m/s^2)/s

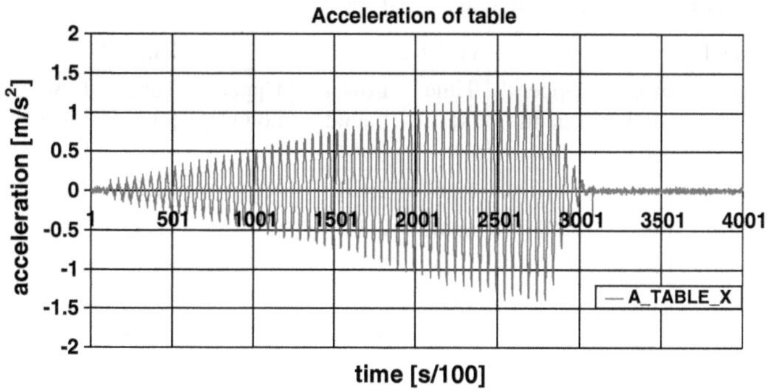

Fig. 3.54 Table acceleration in tests from aa109 to aa114, with increment rate 0.05 $(m/s^2)/s$

Table 3.41 Data of P2-B1 test at frequency 1.0 Hz

Test	Pallet 1			Pallet 2			Pallet 3		
	Time [s]	Lower bound [m/s^2]	Upper bound [m/s^2]	Time [s]	Lower bound [m/s^2]	Upper bound [m/s^2]	Time [s]	Lower bound [m/s^2]	Upper bound [m/s^2]
aa52	18.48	1.63	2.80				19.46	1.63	2.50
aa53	19.17	1.72	2.80				21.32	1.95	2.60
aa54	18.51	1.73	3.00				21.24	1.94	2.90
aa55	19.24	2.11					20.22	1.91	
aa56	19.54	1.92					20.49	1.83	2.50
aa57	19.50	1.82					19.50	1.83	
aa58	19.53	2.02					20.49	1.83	
aa59	20.52	2.02	2.90				21.52	1.92	2.70
aa60	18.51	1.83					21.54	2.02	2.70
aa61	17.51	1.73					19.47	1.83	2.60
μ	19.05	1.85	2.88				20.53	1.87	2.64
σ		0.16	0.10					0.11	0.14
Cov %		8.63	3.33					5.81	5.29

Tables 3.43, 3.44 and 3.45 show the statistical re-analysis of the test data obtained with a same input signal, respectively characterized by the shaking table acceleration shown in Figs. 3.52, 3.53 and 3.54, while Fig. 3.55 shows a summary of all the results obtaines with a frequency input f = 2.0 Hz.

3.3.1.3.2.2 Frequency = 3.0 Hz

In these tests the increment rate of the shaking table acceleration, shown in Fig. 3.56, is $0.1 \frac{m/s^2}{s}$. The test results are shown in Figs. 3.57 and 3.58.

Table 3.42 Data of P2-B1 test at frequency 2.0 Hz

Test	Pallet 1			Pallet 2			Pallet 3		
	Time [s]	Lower bound [m/s^2]	Upper bound [m/s^2]	Time [s]	Lower bound [m/s^2]	Upper bound [m/s^2]	Time [s]	Lower bound [m/s^2]	Upper bound [m/s^2]
aa62	9.94	1.52		13.62	2.69		11.54	1.63	2.60
aa63	11.56	1.82	2.40	13.85	2.39		12.55	1.92	
aa64	12.35	1.92		14.36	2.79		13.15	1.94	
aa65	11.82	1.83		14.49	2.72		13.06	1.92	2.70
aa66	9.80	1.43		13.86	2.59		13.57	2.12	
aa67	8.80	1.33		13.19	2.30		11.82	1.92	
aa68	10.06	1.43		13.60	2.49		12.68	1.92	
aa69	10.29	1.53		14.12	2.69		11.56	1.82	
aa70	11.33	1.63		14.64	2.96		12.70	1.93	
aa71	11.33	1.73		14.38	2.79		12.29	1.82	
aa88	16.34	1.63		19.87	3.06		15.82	1.73	
aa106	17.07	2.02					17.07	2.00	
aa107	16.32	1.73					17.33	1.92	
aa108	16.34	1.82					15.32	1.63	
aa109	24.08	1.92					22.19	1.74	
aa110	22.34	1.82					19.69	1.50	
aa111	21.68	1.50		28.61	2.86		22.68	1.71	
aa112	22.57	1.63					20.46	1.53	
aa113	24.35	1.83					21.56	1.53	
aa114	24.34	1.73		28.13	3.06		21.07	1.63	
μ	15.63	1.69	2.40	16.67	2.72		15.91	1.79	2.65
σ		0.19			0.24			0.18	0.07
Cov %		11.13			8.74			9.89	2.67

Table 3.43 Statistics of data for tests from aa62 to aa71

	Pallet 1			Pallet 2			Pallet 3		
	Time [s]	Lower bound [m/s^2]	Upper bound [m/s^2]	Time [s]	Lower bound [m/s^2]	Upper bound [m/s^2]	Time [s]	Lower bound [m/s^2]	Upper bound [m/s^2]
μ	10.73	1.62	2.40	14.01	2.64		12.49	1.90	2.65
σ		0.20			0.20			0.12	0.07
Cov %		12.31			7.59			6.52	2.67

Table 3.44 Statistics of data for tests aa88 and from aa106 to aa108

	Pallet 1			Pallet 2			Pallet 3		
	Time [s]	Lower bound [m/s²]	Upper bound [m/s²]	Time [s]	Lower bound [m/s²]	Upper bound [m/s²]	Time [s]	Lower bound [m/s²]	Upper bound [m/s²]
μ	16.52	1.80		19.87	3.06		16.38	1.82	
σ		0.17						0.17	
Cov %		9.31						9.45	

Table 3.45 Statistics of data for tests from aa109 to aa114

	Pallet 1			Pallet 2			Pallet 3		
	Time [s]	Lower bound [m/s²]	Upper bound [m/s²]	Time [s]	Lower bound [m/s²]	Upper bound [m/s²]	Time [s]	Lower bound [m/s²]	Upper bound [m/s²]
μ	23.23	1.74		28.37	2.96		21.27	1.61	
σ		0.15			0.14			0.10	
Cov %		8.78			4.68			6.24	

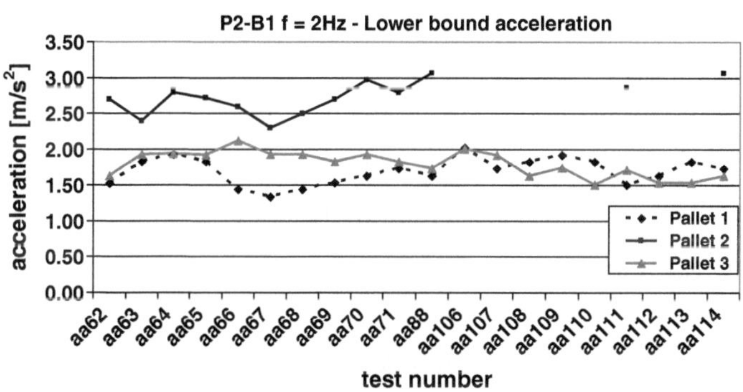

Fig. 3.55 Acceleration of sliding initiation in different tests—P2B1 f = 2.0 Hz

3.3.1.3.2.3 Frequency = 4.0 Hz

In these tests the increment rate of the table acceleration is $0.085\frac{m/s^2}{s}$ (Fig. 3.59).

3.3.1.3.2.4 Re-analysis of P2-B1 Test Results

All the values obtained from the analysis of test of pallet type P2 and beam type B1 are summarised hereafter. Table 3.48 shows the statistics of the mass acceleration, at sliding initiation (lower bound) and its upper bound value, independently of the

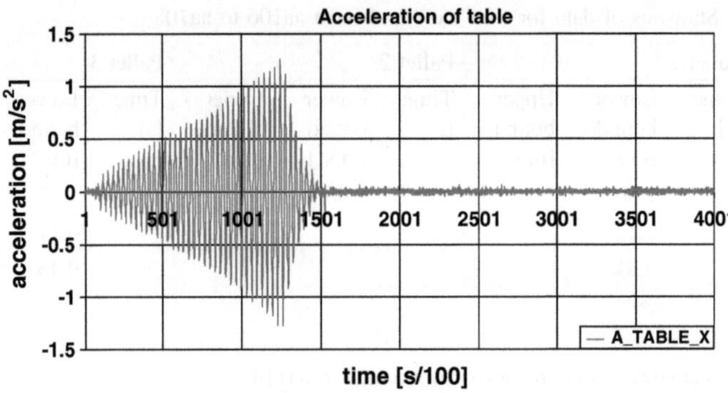

Fig. 3.56 Acceleration of table in tests from aa72 to aa81

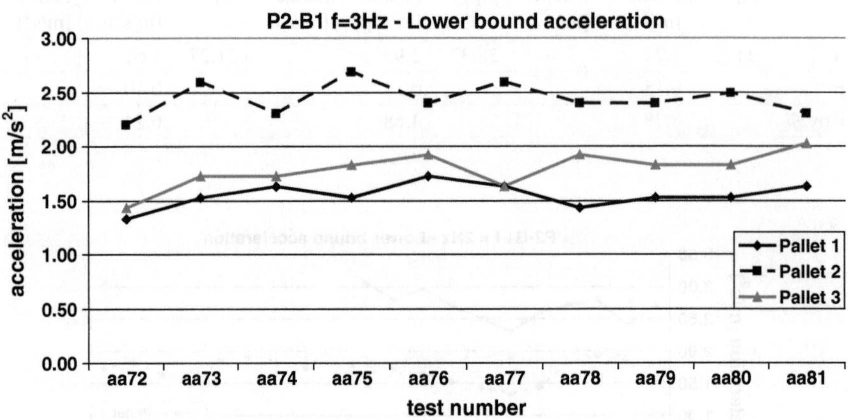

Fig. 3.57 Acceleration of sliding initiation in different tests—P2B1 f = 3.0 Hz

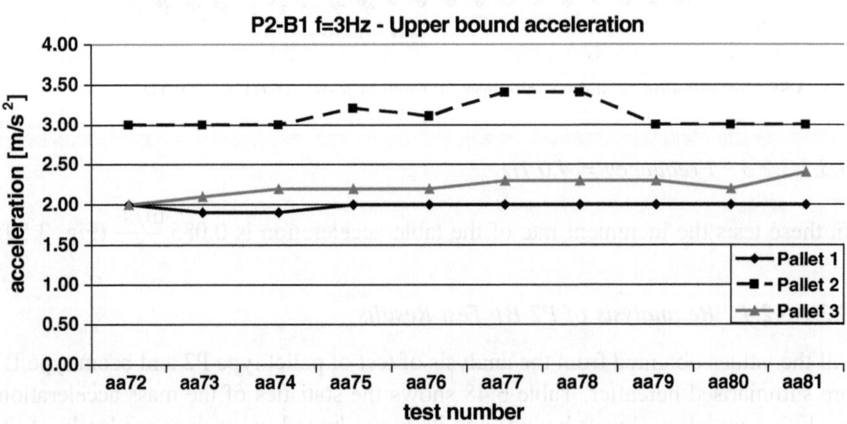

Fig. 3.58 Upper bound of the acceleration in different tests—P2B1 f = 3.0 Hz

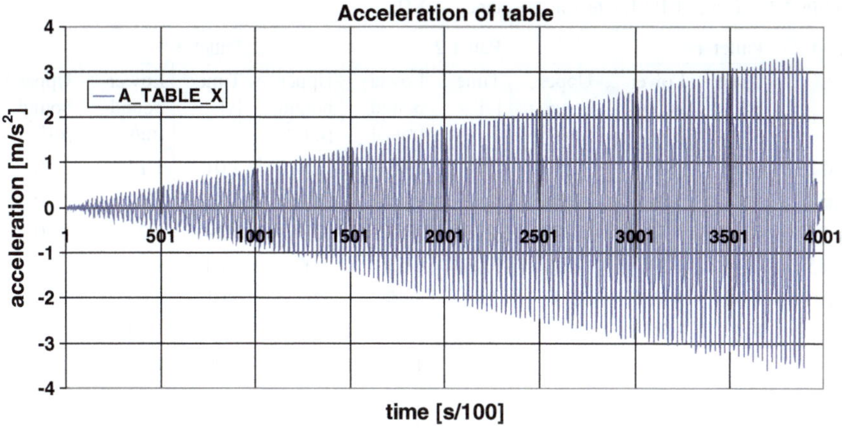

Fig. 3.59 Acceleration of table in tests from aa82 to aa87

Table 3.46 Data of P2-B1 test at frequency 3.0 Hz

Test	Pallet 1			Pallet 2			Pallet 3		
	Time [s]	Lower bound [m/s²]	Upper bound [m/s²]	Time [s]	Lower bound [m/s²]	Upper bound [m/s²]	Time [s]	Lower bound [m/s²]	Upper bound [m/s²]
aa72	5.77	1.33	2.00	6.78	2.20	3.00	5.77	1.43	2.00
aa73	6.25	1.53	1.90	7.98	2.59	3.00	6.61	1.73	2.10
aa74	6.42	1.63	1.90	7.64	2.30	3.00	6.62	1.73	2.20
aa75	7.42	1.53	2.00	9.15	2.69	3.20	6.95	1.83	2.20
aa76	6.93	1.73	2.00	8.15	2.39	3.10	6.96	1.92	2.20
aa77	6.41	1.63	2.00	7.81	2.59	3.40	6.44	1.63	2.30
aa78	5.91	1.43	2.00	7.80	2.39	3.40	7.13	1.92	2.30
aa79	6.10	1.53	2.00	7.47	2.39	3.00	6.63	1.82	2.30
aa80	6.07	1.53	2.00	7.80	2.49	3.00	7.79	1.82	2.20
aa81	6.42	1.63	2.00	7.12	2.30	3.00	7.64	2.02	2.40
μ	6.37	1.55	1.98	7.77	2.43	3.11	6.85	1.79	2.22
σ		0.11	0.04		0.15	0.17		0.17	0.11
Cov %		7.19	2.13		6.36	5.35		9.41	5.11

frequency of the table acceleration. In Table 3.49 the same parameters are presented for each frequency considered in this group of tests (Figs. 3.60 and 3.61; Tables 3.46 and 3.47).

From the analysis of the results obtained with the use of beams type B1 (cold rolled, powder coated, new beam) graphically represented in Figs. 3.62 and 3.63 some observations can be drawn.

Table 3.47 Data of P2-B1 test at frequency 4.0 Hz

Test	Pallet 1			Pallet 2			Pallet 3		
	Time [s]	Lower bound [m/s²]	Upper bound [m/s²]	Time [s]	Lower bound [m/s²]	Upper bound [m/s²]	Time [s]	Lower bound [m/s²]	Upper bound [m/s²]
aa82	8.64	1.43	1.50	3.88	1.63	3.10	4.01	1.14	2.20
aa83	5.12	1.24	1.60	4.91	1.63	3.00	4.15	1.14	2.20
aa84	6.14	1.33	1.90	3.39	1.55	3.00	4.89	1.24	2.20
aa85	4.39	1.24	1.50	3.26	1.55	3.00	3.64	1.14	2.30
aa86	6.06	1.30	2.00	3.90	1.45	3.30	3.51	1.04	2.10
aa87	8.75	1.35	1.60	3.69	1.51	3.00	3.78	1.14	2.00
μ	6.52	1.32	1.68	3.84	1.55	3.07	4.00	1.14	2.17
σ		0.08	0.21		0.07	0.12		0.06	0.10
Cov %		5.71	12.69		4.45	3.95		5.45	4.77

Table 3.48 Statistics of the cross-aisle test, P2-B1

	Pallet 1		Pallet 2		Pallet 3	
	Lower bound	Upper bound	Lower bound	Upper bound	Lower bound	Upper bound
Statistics P2B1 cross-aisle						
μ	1.65	2.09	2.38	3.09	1.72	2.36
σ	0.22	0.44	0.48	0.15	0.27	0.24
Cov %	13.6	21.3	20.4	4.8	15.9	10.4

When the frequency of the table acceleration is very low (e.g. 1.0 Hz) sliding occurs only on the lateral pallets, while no relative displacement occurs between the central pallet and the beam. Table 3.41, for example, shows that there is no sliding for the central pallet. In the same way also the upper bound of the sliding acceleration of the masses is reached earlier for the lateral pallets, and later by the central one. This situation can be observed in Table 3.42, in which only in some tests the central pallets reach the upper bound value. This behaviour is due to the fact that the lateral pallets reach a higher value of the sliding acceleration earlier than the central one.

Figure 3.64 shows the structure when it is excited in cross-aisle direction. The relative displacement of the two beams in a lateral position, near the joints, is larger than in a central position. In fact the joint, shown in Fig. 3.64, behaves in two different ways depending on the sign of the excitation signal.

The rotation α is larger than β due to the shape of the joint and to its non symmetrical behaviour. This effect, however, doesn't influence much the behaviour of the beams far from the supports. Hence, the lateral pallets are not only subjected to a component of the acceleration in cross-aisle direction, but, due to the rotation of the beam near the supports, they are also subjected to a torsional component, in

Table 3.49 Statistics of the cross-aisle test, P2-B1 for every frequency

	Pallet 1		Pallet 2		Pallet 3	
	Lower bound	Upper bound	Lower bound	Upper bound	Lower bound	Upper bound
Statistics f = 1.0 Hz						
μ	1.85	2.88			1.87	2.64
σ	0.16	0.10			0.11	0.14
Cov %	8.6	3.3			5.8	5.3
Characteristics f = 2.0 Hz						
μ	1.69	2.40	2.72		1.79	2.65
σ	0.19		0.24		0.18	0.07
Cov %	11.1		8.7		9.9	2.7
Characteristics f = 3.0 Hz						
μ	1.55	1.98	2.43	3.11	1.79	2.22
σ	0.11	0.04	0.15	0.17	0.17	0.11
Cov %	7.2	2.1	6.4	5.3	9.4	5.1
Characteristics f = 4.0 Hz						
μ	1.32	1.68	1.55	3.07	1.14	2.17
σ	0.08	0.21	0.07	0.12	0.06	0.10
Cov %	5.7	12.7	4.5	3.9	5.4	4.8

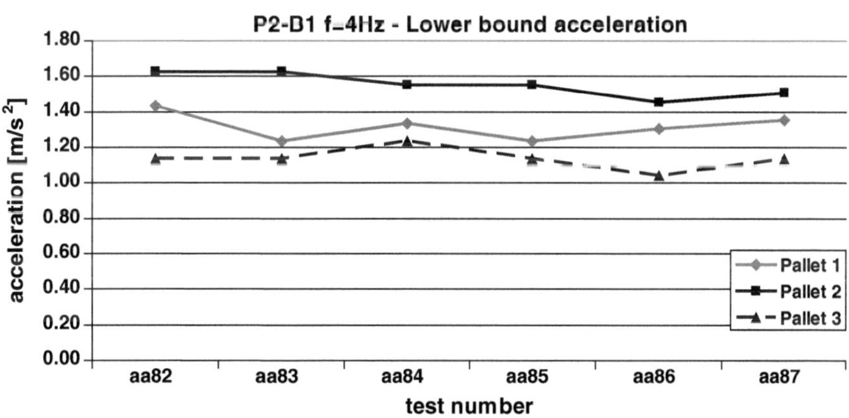

Fig. 3.60 Acceleration of sliding initiation in different tests—P2B1 f = 4.0 Hz

order to follow the beam without any sliding. For this reason, sliding of the lateral pallets occurs already under rather low values of acceleration. Experimental observation confirms presence of a "torsional" sliding of the lateral pallets.

On the contrary, in the central position the rotation of the beam is smaller, as is the possibility of sliding on the pallet. So for the central pallet sliding starts with

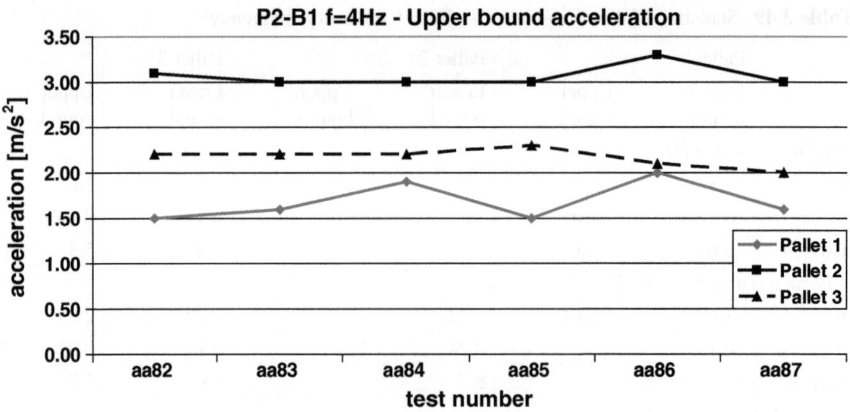

Fig. 3.61 Upper bound of the acceleration in different tests—P2B1 f = 4.0 Hz

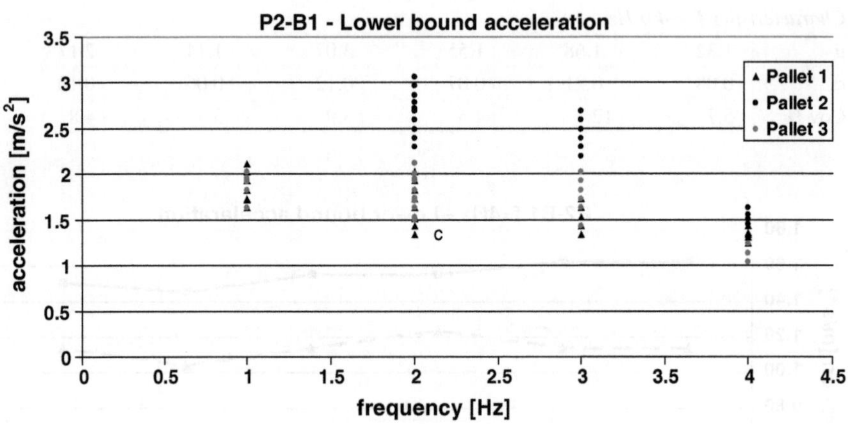

Fig. 3.62 Acceleration of sliding initiation versus frequency of the P2-B1 test

higher values of the table acceleration and possibility of torsional effects is small. For the low frequencies, such values of acceleration couldn't be reached due to the limited excursion of the shaking table.

A similar response is expected for both the lateral pallets. In some cases, however, they show slightly different values of sliding initiation acceleration and upper bound acceleration (e.g. Figs. 3.57 and 3.58). Observing Figs. 3.65 and 3.67 it can be noticed that the displacement of the beam, recorded in the same test under the two lateral pallets, is quite different. It can be estimated that the stiffness's of the joints are lightly different from one another, due to damage accumulated during testing. Figures 3.65, 3.66 and 3.67, related to test aa79, compared with Table 3.46, show that if the displacement of the beam follows for a longer time the displacement of the mass, sliding starts at a higher value of acceleration. In fact, in the

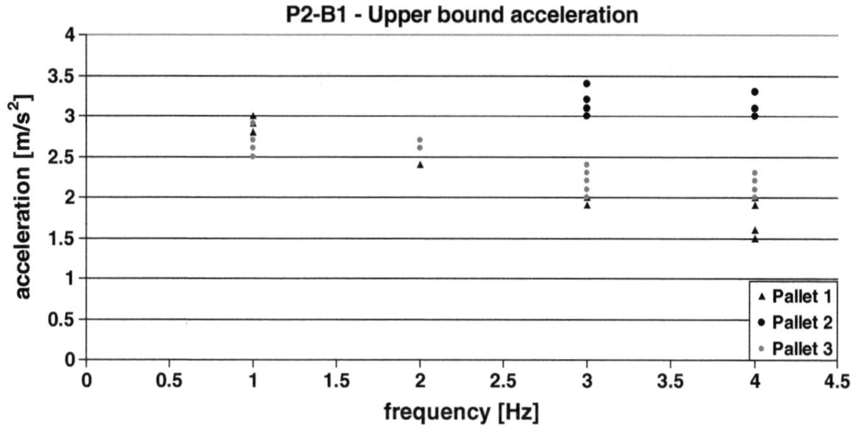

Fig. 3.63 Upper bound of sliding acceleration versus frequency of the P2B1 test

analysed case, pallet 1 slides after 6.1 s, before pallet 3 (6.6 s). The central pallet (pallet 2) slides at 7.5 s, after the two lateral ones, due to the higher deformation of the beam.

Many tests show a very low scatter of the results, with a c.o.v. lower than 13 %. The highest c.o.v. (12.7 %) is associated for the upper bound of the acceleration for pallet 1 at 4.0 Hz, but in most cases the c.o.v.% value is lower than 10 %. This is probably the result of the frequent replacement of the beams, as soon as deterio ration of the surface paint was detected. Unfortunately, this was not possible for the whole set of tests, as the number of the available beams of the same type was limited.

Figure 3.55 ($f = 2.0$ Hz) shows that the test results are independent of the loading history. In fact, analysing the statistics of the results obtained with different input motions (Figs. 3.52, 3.53 and 3.54 as well as Tables 3.43, 3.44 and 3.45) it can be noticed that the values of the lower and the upper bounds of the sliding acceleration are very similar, and with a low scatter.

Analysing the global results for this group of tests (Figs. 3.62 and 3.63) it is possible to notice that the acceleration of sliding initiation shows a decreasing trend when the frequency increases. This behaviour is evident for the two lateral pallets; the central pallet starts to slide at higher values of acceleration. This behaviour is highlighted in Figs. 3.68, 3.69 and 3.70 that show the response in terms of sliding initiation and upper bound acceleration for each pallet. The linear decreasing trend approximates all the analysed cases, with the exception of the upper bound for the central pallet, in which the values are more or less constant.

Figures 3.71 and 3.72 show, for each pallet, the mean values of the lower and upper bound of the sliding acceleration plotted versus the frequency of the excitation.

It can be noticed that this combination of beam and pallet shows a decreasing trend of the sliding acceleration when the frequency increases.

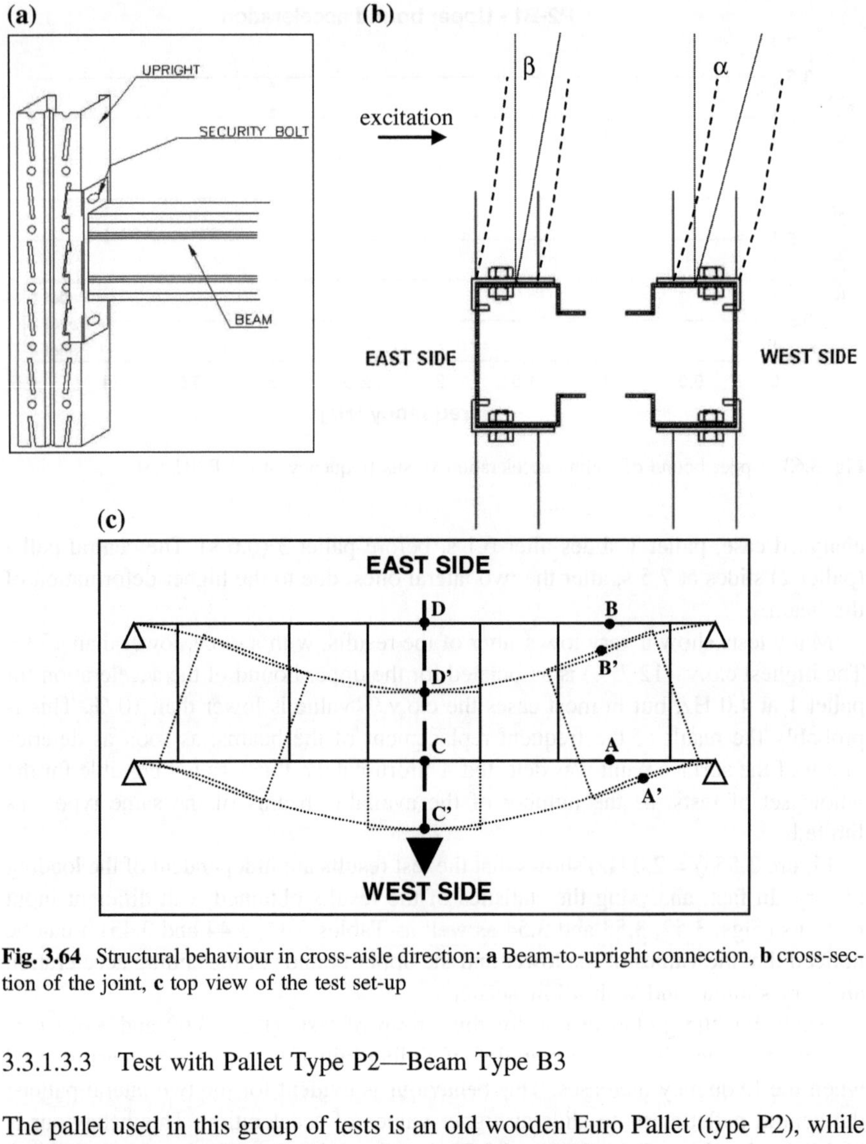

Fig. 3.64 Structural behaviour in cross-aisle direction: **a** Beam-to-upright connection, **b** cross-section of the joint, **c** top view of the test set-up

3.3.1.3.3 Test with Pallet Type P2—Beam Type B3

The pallet used in this group of tests is an old wooden Euro Pallet (type P2), while the beam is a new cold rolled, hot zinc coated (type B3).

3.3.1.3.3.1 Frequency = 1.0 Hz

In these tests the increment rate of the shaking table acceleration is 0.075 $\frac{m/s^2}{s}$, as shown in Fig. 3.73 (Fig. 3.74; Table 3.50).

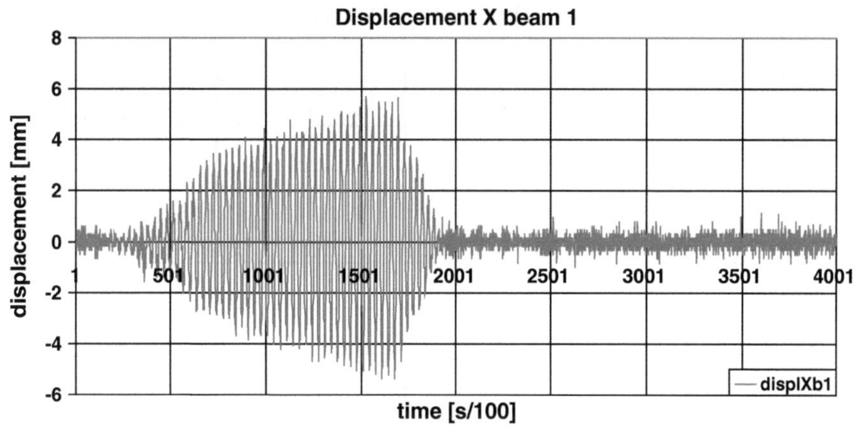

Fig. 3.65 Displacement of beam under pallet 1 in cross-aisle test no. aa79

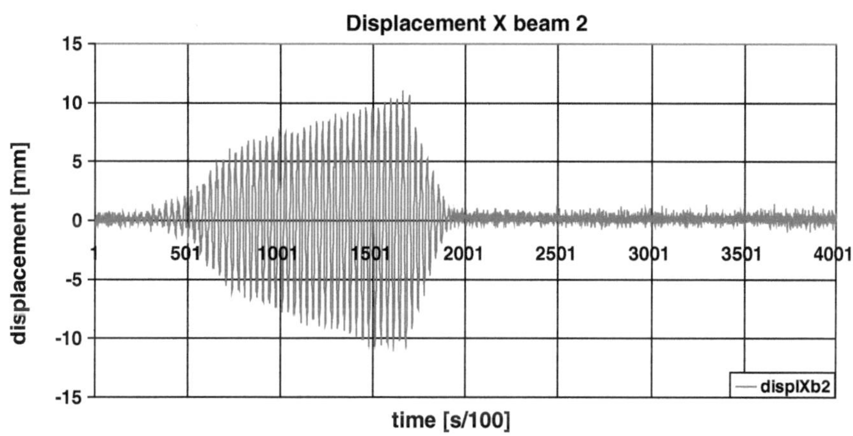

Fig. 3.66 Displacement of beam under pallet 2 in cross-aisle test no. aa79

3.3.1.3.3.2 *Frequency = 1.5 Hz*

In these tests the increment rate of the shaking table acceleration (Fig. 3.75) is $0.1\,\frac{\mathrm{m/s^2}}{\mathrm{s}}$ (Figs. 3.76 and 3.77; Table 3.51).

3.3.1.3.3.3 *Frequency = 2.0 Hz*

The loading history for this group of tests is included into the two "limit cases" shown in Fig. 3.78a, b, with the same increment rate of the table acceleration, equal

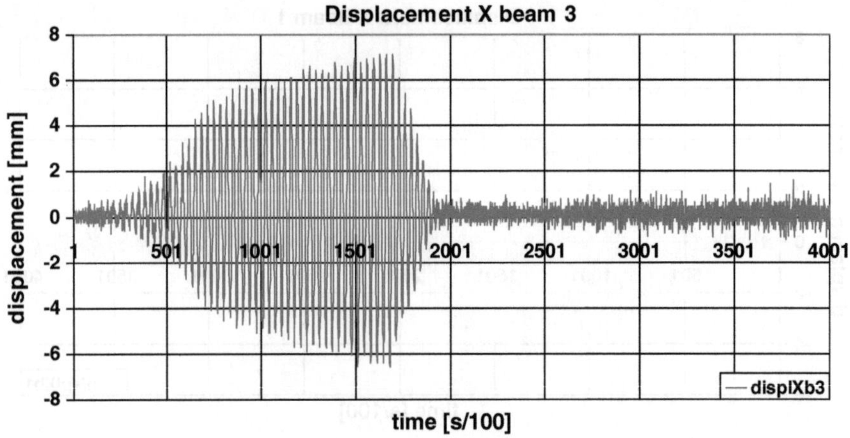

Fig. 3.67 Displacement of beam under pallet 3 in cross-aisle test no. aa79

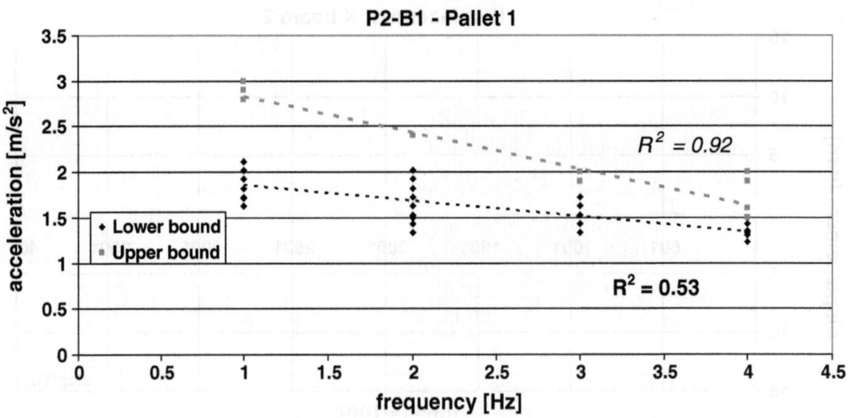

Fig. 3.68 Acceleration of sliding initiation and upper bound for pallet 1 in cross-aisle P2-B1 tests

to 0.1 $\frac{m/s^2}{s}$, but different durations. Tests were stopped when sliding of at least one pallet could be visually detected. This operative decision was taken for the following reasons:

- to minimize displacement of the pallets, in order to reduce to a minimum the time for their "repositioning" between one test and the other;
- to minimize damage to the beams.

The batch of hot zinc coated beams was limited, because some beams were damaged, probably during shipping. For this reason, run short of new beams, it was impossible to replace the beams after 10 tests, according to the standard procedure,

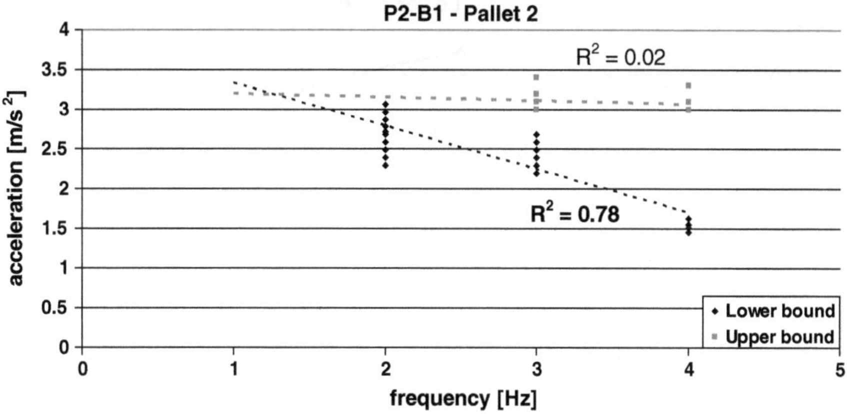

Fig. 3.69 Acceleration of sliding initiation and upper bound for pallet 2 in cross-aisle P2-B1 tests

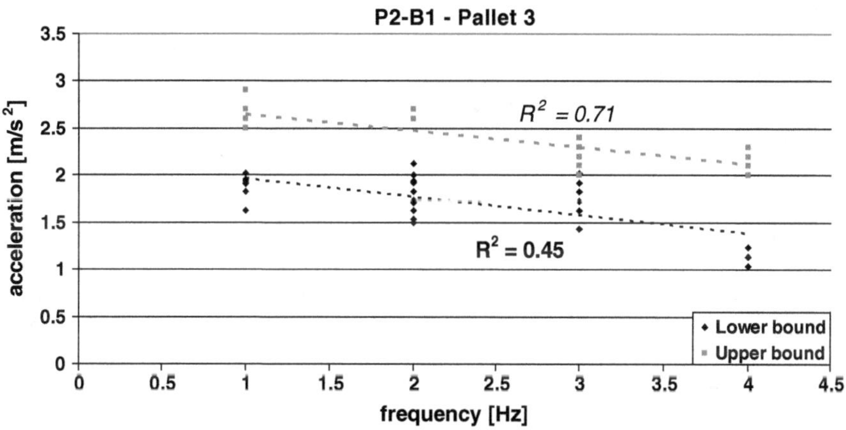

Fig. 3.70 Acceleration of sliding initiation and upper bound for pallet 3 in cross-aisle P2-B1 tests

and the same (last) beam was used for the last 30 tests. However, it was decided to run first 5 tests for each frequency (2.0, 2.5 and 3.0 Hz respectively) and then repeat other 5 tests for the same frequencies (starting again from 2.0 Hz, then 2.5 and 3.0 Hz).

Hence, for each frequency, the first test of the first group of 5, and the first test of the second group are spaced of 15 tests, carried out on the same beam. Possible scatter in the results of the two groups of data might be due to deterioration either of the beam surface or of the lateral end-plate connections (or of both these effects) (Fig. 3.79).

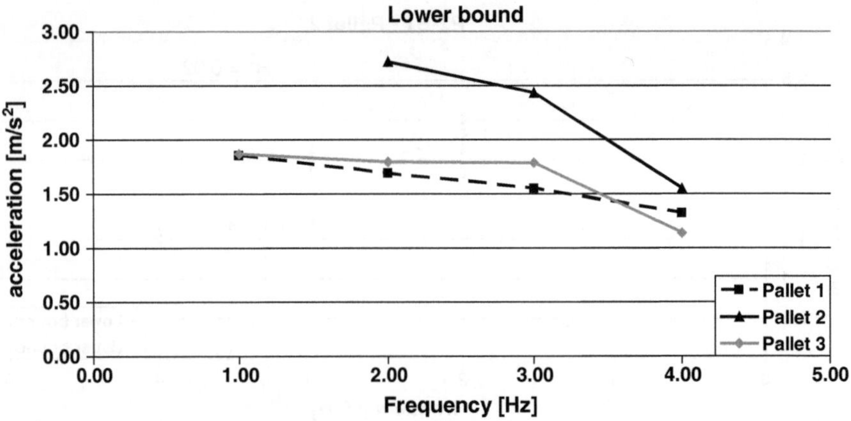

Fig. 3.71 Lower bound of sliding acceleration in cross-aisle P2-B1 tests

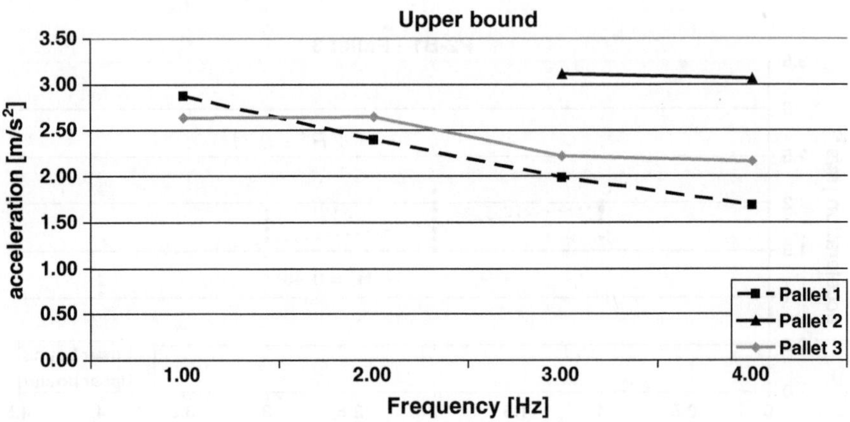

Fig. 3.72 Upper bound of sliding acceleration in cross-aisle P2-B1 tests

Possible influence of such a damage was investigated, by considering not only the full batch of the results (Table 3.52), but also by evaluating separately the statistics for the two groups of 5 tests, as shown in Tables 3.53 and 3.54.

In general, it can be noticed that the mean values increase and the scatter of the results decreases when passing from the first series of 5 tests to the second one (Fig. 3.80).

3.3.1.3.3.4 Frequency = 2.5 Hz

Duration of these tests was 13 s with an acceleration increment rate of 0.1 $\frac{\mathrm{m/s^2}}{\mathrm{s}}$ (Fig. 3.82). Only for test no. aa26 the duration was 20 s with a table acceleration increment rate of 0.075 $\frac{\mathrm{m/s^2}}{\mathrm{s}}$ (Fig. 3.81).

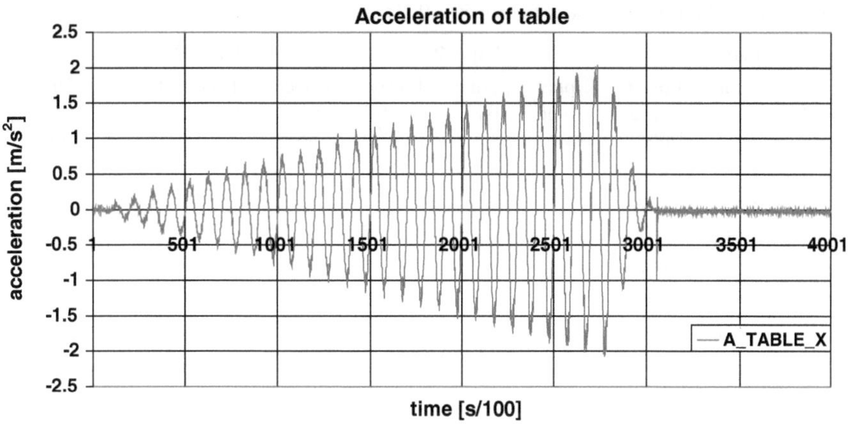

Fig. 3.73 Table acceleration in tests from aa1 to aa10

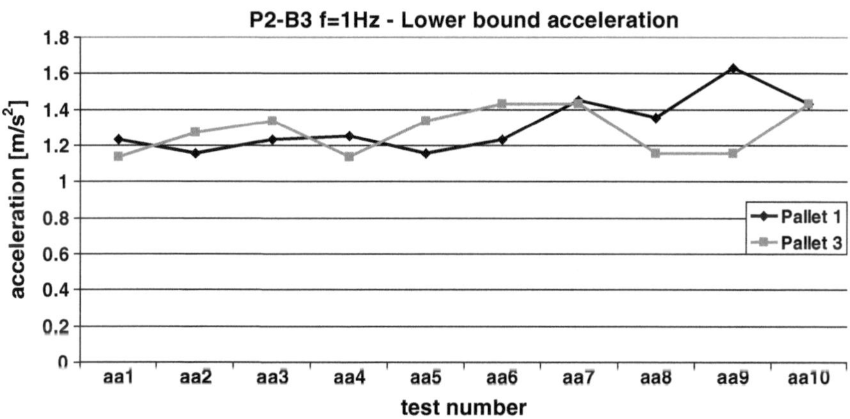

Fig. 3.74 Acceleration of sliding initiation in different tests—P2B3 f = 1.0 Hz

Results of this group of tests are summarized in Table 3.55, while Tables 3.56 and 3.57 show the statistical analysis of the data for the tests respectively from aa26 to aa30 and from aa41 to aa46 (Figs. 3.83 and 3.84).

3.3.1.3.3.5 Frequency = 3.0 Hz

Duration was 22 s for tests from no. aa31 to aa35 (Fig. 3.85) with a table acceleration increment rate of $0.075 \frac{m/s^2}{s}$, while duration was 28 s with an acceleration increment rate of $0.1 \frac{m/s^2}{s}$ for tests from no. aa47 to aa51 (Figs. 3.86, 3.87 and 3.88).

Table 3.50 Data of P2-B3 test at frequency 1.0 Hz

Test	Pallet 1			Pallet 2			Pallet 3		
	Time [s]	Lower bound [m/s²]	Upper bound [m/s²]	Time [s]	Lower bound [m/s²]	Upper bound [m/s²]	Time [s]	Lower bound [m/s²]	Upper bound [m/s²]
aa1	12.53	1.24					12.52	1.14	
aa2	12.26	1.16					12.78	1.28	
aa3	12.54	1.24					14.51	1.33	
aa4	11.01	1.26					12.52	1.14	
aa5	10.51	1.16					12.51	1.33	2.50
aa6	11.49	1.24	3.10				13.53	1.43	3.20
aa7	14.52	1.45	3.00				14.52	1.43	
aa8	12.50	1.35	2.90				12.56	1.16	
aa9	14.55	1.63					12.54	1.16	
aa10	14.56	1.43					14.53	1.43	
μ	12.64	1.31	3.00				13.25	1.28	2.85
σ		0.15	0.10					0.13	0.49
Cov %		11.46	3.33					9.91	17.37

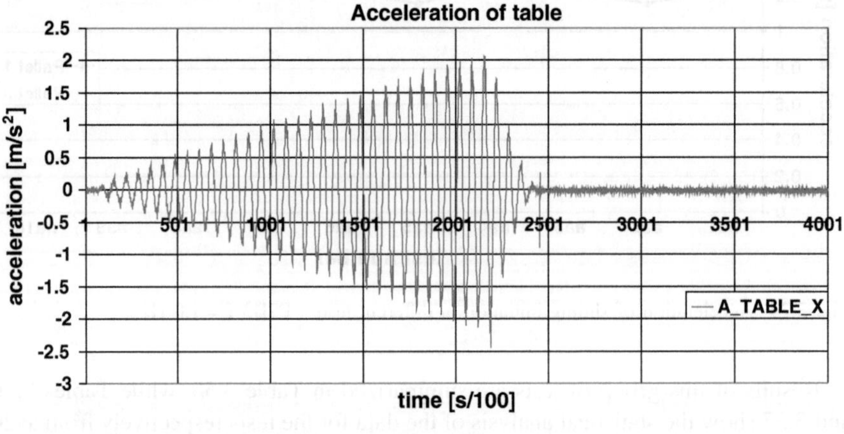

Fig. 3.75 Table acceleration in tests from aa11 to aa20

Results of this group of tests are summarized in Table 3.58, while Tables 3.59 and 3.60 show the statistical analysis of the data for the tests respectively from aa31 to aa35 and from aa47 to aa51.

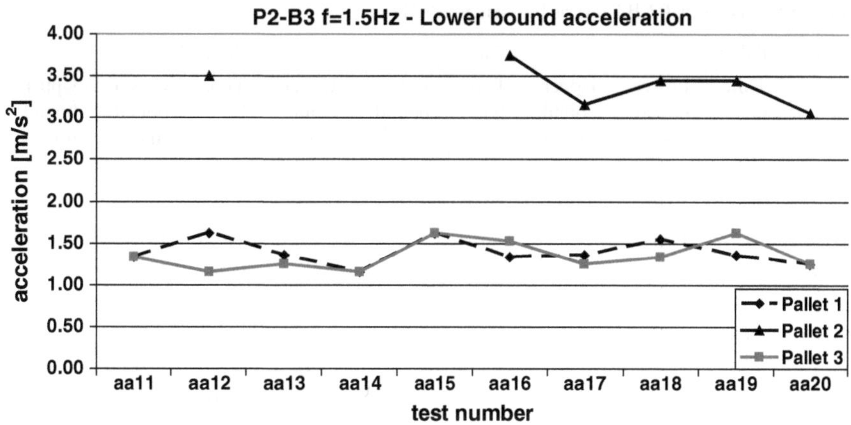

Fig. 3.76 Acceleration of sliding initiation in different tests—P2B3 f = 1.5 Hz

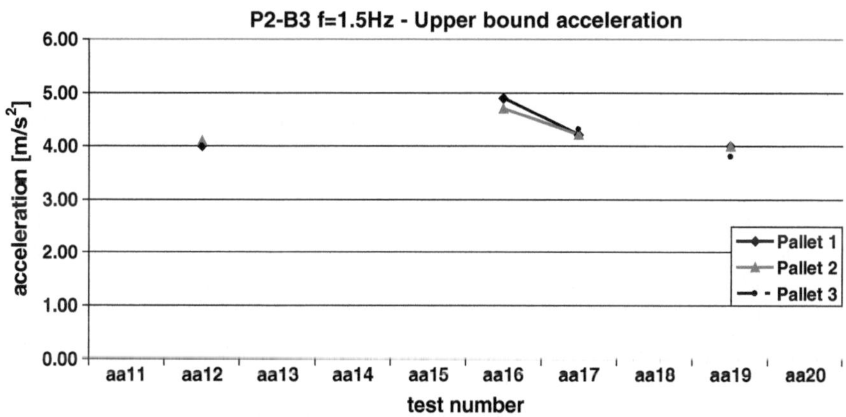

Fig. 3.77 Upper bound of the acceleration in different tests—P2B3 f = 1.5 Hz

3.3.1.3.3.6 Re-analysis of P2-B3 Test Results

Tables 3.61 and 3.62 show the statistical re-analysis of all the values obtained from tests with pallet P2 and beam type B3. Table 3.61 shows the mean values of the mass acceleration, at sliding initiation (lower bound) and its upper bound value, independently of the frequency of the table acceleration. Table 3.62 presents the same parameters for every frequency considered for this group of tests. Figures 3.89 and 3.90 are a graphical representation of these results.

The acceleration of sliding initiation (lower bound) has a more or less constant value independent of the frequency for the two lateral pallets, while for the central one it shows a decreasing trend when the frequency increases. At low frequency,

Table 3.51 Data of P2-B3 test at frequency 1.5 Hz

Test	Pallet 1			Pallet 2			Pallet 3		
	Time [s]	Lower bound [m/s²]	Upper bound [m/s²]	Time [s]	Lower bound [m/s²]	Upper bound [m/s²]	Time [s]	Lower bound [m/s²]	Upper bound [m/s²]
aa11	10.41	1.33					10.40	1.33	
aa12	9.43	1.63	4.00	18.96	3.50	4.10	9.50	1.15	4.00
aa13	10.74	1.35					10.39	1.26	
aa14	7.74	1.16					9.94	1.16	
aa15	11.08	1.63					12.42	1.63	
aa16	10.87	1.33	4.90	19.13	3.75	4.70	11.75	1.53	
aa17	9.74	1.35	4.20	17.79	3.16	4.20	9.71	1.26	4.30
aa18	11.02	1.55		18.46	3.45		11.07	1.33	
aa19	9.70	1.35	4.00	18.78	3.45	4.00	12.42	1.63	3.80
aa20	9.75	1.26		17.12	3.06		10.39	1.26	
μ	10.04	1.39	4.28	18.37	3.40	4.25	10.80	1.35	4.03
σ		0.16	0.43		0.25	0.31		0.18	0.25
Cov %		11.25	9.99		7.33	7.32		13.29	6.24

e.g. 1.0 Hz, only the two lateral pallets slide on the beam, while the central one doesn't show a relative displacement.

For the same reasons discussed with reference to tests carried out with beam type B1, also in this case sliding starts first for the lateral pallets. Sliding of the central mass occurs under much higher acceleration. In some cases, sliding of the central mass couldn't be developed for because the tests were stopped before the attainment of the limit conditions.

Table 3.62 shows that the two lateral pallets start sliding more or less for the same acceleration, for every considered frequency, with the exception of the case of 2.5 Hz, in which the sliding acceleration of pallet 3 (1.30 m/s^2) is lower than that of pallet 1 (1.58 m/s^2), and 3.0 Hz in which pallet 1 begins to slide at 1.40 m/s^2 while pallet 3 at 1.52 m/s^2.

The upper bound value of the acceleration of the two lateral pallets shows a strongly decreasing trend when increasing the frequency, although values at 1.0 Hz are in countertendency. The upper bound of the sliding acceleration is similar for the three pallets in the case of f = 1.5 Hz.

Figures 3.91, 3.92 and 3.93 show the values of the lower bound and of the upper bound of the acceleration, respectively for each pallet.

The behaviour of pallet 1 is similar to the pallet 3. Sliding starts with an acceleration of about 1.5 m/s^2, for both pallets, and this value remains constant independently of the frequency. Table 3.63 shows the statistics considering the data of both the lateral pallets. The standard deviation is 0.20 m/s^2 and the c.o.v. is 14 %. Figure 3.94 plots the sliding initiation acceleration versus the frequency of the excitation, for both pallet1 and pallet 3 considered together as a single sample.

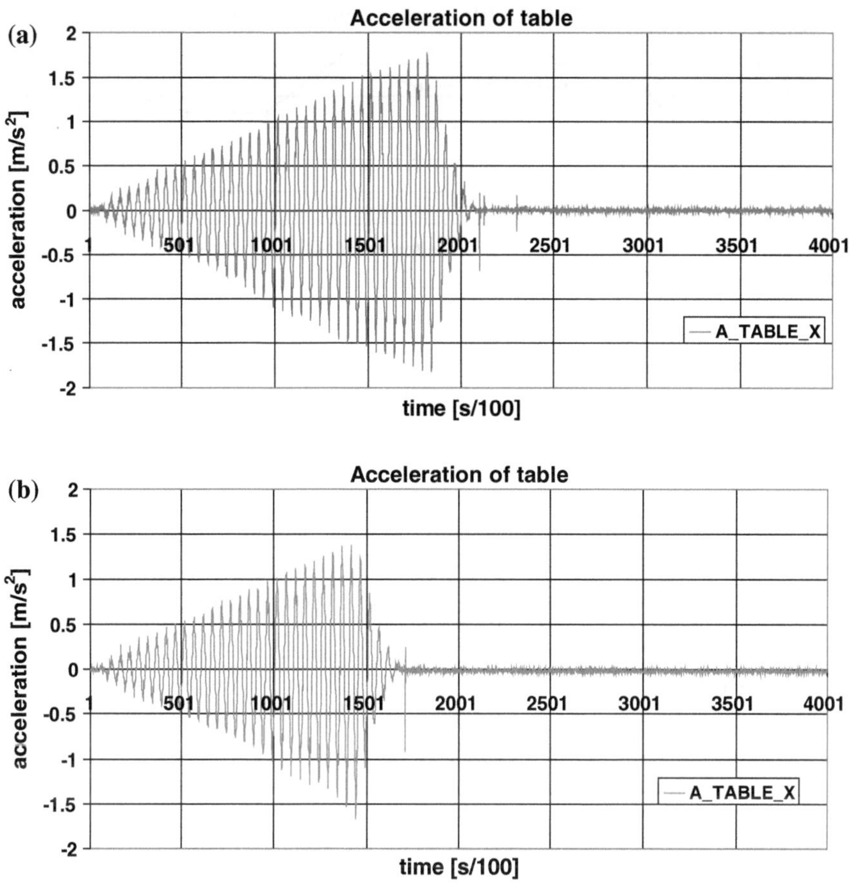

Fig. 3.78 Table acceleration for tests from aa21 to aa25 and from aa36 to aa40: **a** longest duration, **b** shortest duration

Homogeneity of the data can be noticed. In Figs. 3.95 and 3.96 the mean values (of the test results obtained under the same frequency) of lower and the upper bound of the sliding acceleration are plotted for each pallet versus the frequency of the excitation. Also in this case similarity between the behaviours of the two lateral pallets can be observed, while the central pallet stands alone.

3.3.2 *Tests in Down-Aisle Direction*

3.3.2.1 Test Set up

In down-aisle direction tests were carried out with two different types of excitation of the shaking table:

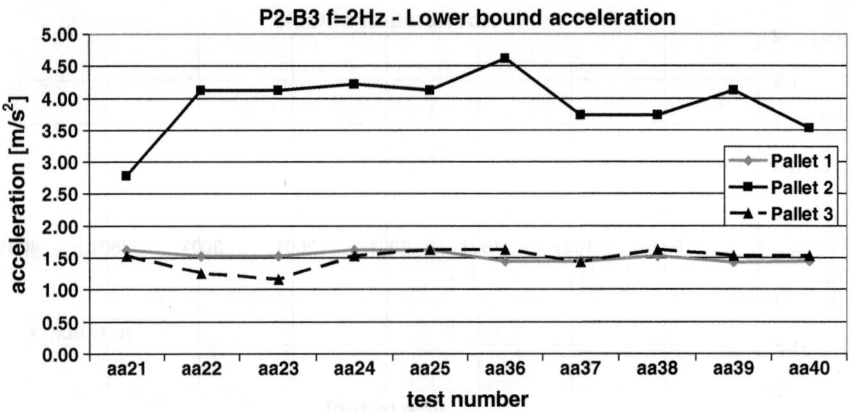

Fig. 3.79 Acceleration of sliding initiation in different tests—P2B3 f = 2.0 Hz

Table 3.52 Data of P2-B3 test at frequency 2.0 Hz

Test	Pallet 1			Pallet 2			Pallet 3		
	Time [s]	Lower bound [m/s^2]	Upper bound [m/s^2]	Time [s]	Lower bound [m/s^2]	Upper bound [m/s^2]	Time [s]	Lower bound [m/s^2]	Upper bound [m/s^2]
aa21	10.08	1.63	3.80	12.12	2.79	4.80	9.84	1.53	3.80
aa22	9.34	1.53	4.00	13.65	4.12	5.00	8.59	1.26	4.00
aa23	9.61	1.53	3.90	14.14	4.12	4.70	8.59	1.16	3.80
aa24	9.86	1.63		14.65	4.22		10.34	1.53	
aa25	10.84	1.63	4.00	14.65	4.12	5.00	10.32	1.63	3.90
aa36	10.36	1.45	4.60	15.63	4.61	5.00	10.80	1.63	
aa37	10.30	1.45	4.30	14.63	3.73	4.80	10.31	1.43	3.90
aa38	10.32	1.53	4.00	14.88	3.73		10.83	1.62	
aa39	11.33	1.43		15.14	4.12		10.32	1.53	
aa40	10.33	1.45		14.12	3.53		9.83	1.53	
μ	10.24	1.53	4.09	14.36	3.91	4.88	9.97	1.48	3.88
σ		0.08	0.27		0.50	0.13		0.16	0.08
Cov %		5.21	6.69		12.75	2.72		10.80	2.16

- a sinusoidal excitation with constant frequency and increasing acceleration;
- a sinusoidal excitation with constant acceleration and increasing frequency.

The test set up, shown in Figs. 3.97 and 3.98 is similar to that adopted for the previous series of tests in cross-aisle direction, but with a different position of the instrumentation, that it is composed by:

Table 3.53 Statistical analysis of data for tests from aa21 to aa25

	Pallet 1			Pallet 2			Pallet 3		
	Time [s]	Lower bound [m/s²]	Upper bound [m/s²]	Time [s]	Lower bound [m/s²]	Upper bound [m/s²]	Time [s]	Lower bound [m/s²]	Upper bound [m/s²]
μ	9.95	1.59	3.93	13.84	3.87	4.88	9.53	1.42	3.88
σ		0.05	0.10		0.61	0.15		0.20	0.10
Cov %		3.40	2.44		15.73	3.08		14.25	2.47

Table 3.54 Statistical analysis of data for tests from aa36 to aa40

	Pallet 1			Pallet 2			Pallet 3		
	Time [s]	Lower bound [m/s²]	Upper bound [m/s²]	Time [s]	Lower bound [m/s²]	Upper bound [m/s²]	Time [s]	Lower bound [m/s²]	Upper bound [m/s²]
μ	10.58	1.47	4.30	15.07	4.05	4.90	10.56	1.55	3.90
σ		0.04	0.30		0.42	0.14		0.09	
Cov %		2.96	6.98		10.36	2.89		5.99	

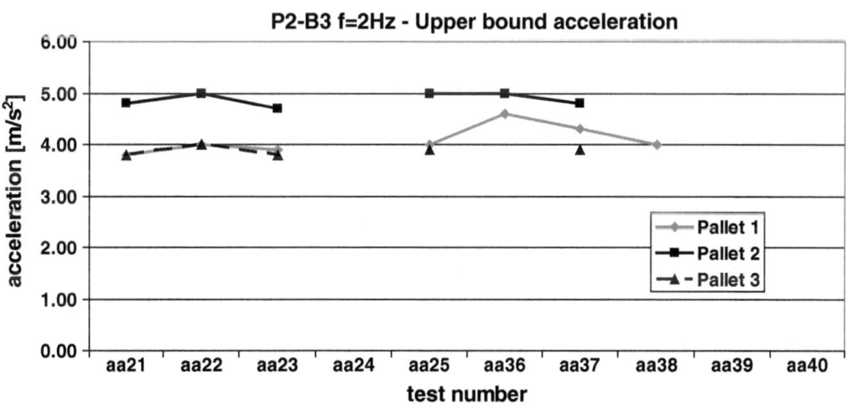

Fig. 3.80 Upper bound of the acceleration in different tests—P2B3 f = 2.0 Hz

- 5 accelerometers in Y direction, one for each mass and one for each beam;
- 4 LVDTs in X direction, one on each lateral pallet and two on the central one;
- 8 LVDTs in Y direction, two on each pallet and one on each beam.

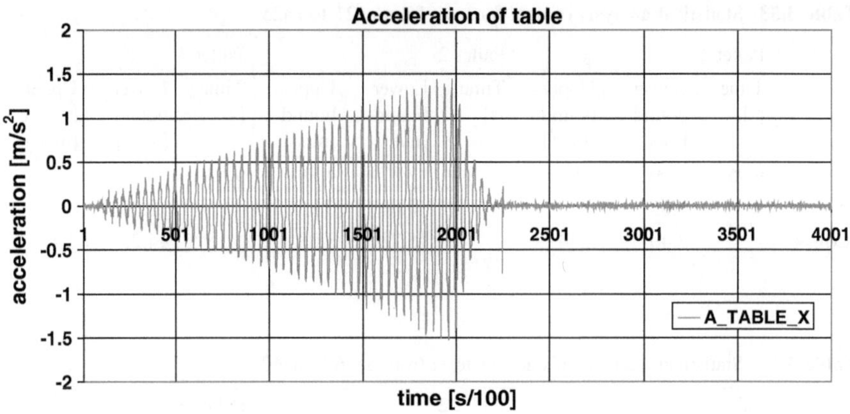

Fig. 3.81 Table acceleration in tests aa26

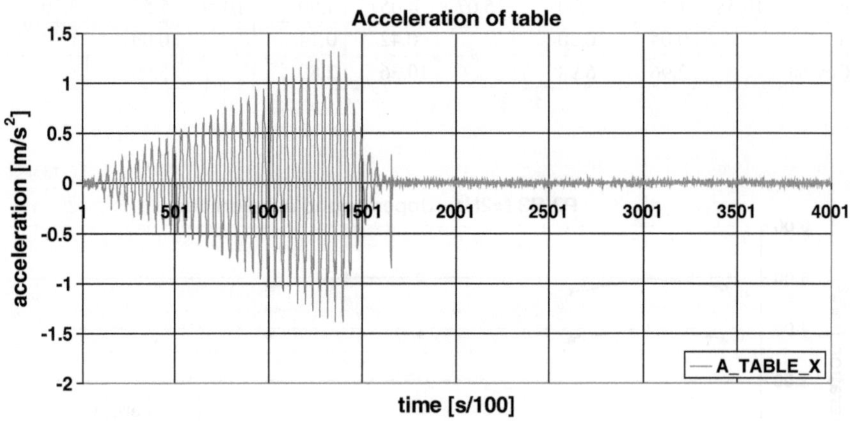

Fig. 3.82 Table acceleration in tests aa27 to aa30 and aa41 to aa46

3.3.2.2 Procedure for Re-Analysis of the Test Results

The procedure to re-analyse the test results is practically the same adopted for the cross-aisle tests. In this case however, sliding initiation is rather evident, while the upper bound of the sliding acceleration is more difficult to be detected. The following Figs. 3.99 and 3.100 show a typical case of structural response under constant frequency and increasing acceleration. The re-analysis is presented and discussed with reference to this case. Figure 3.99 shows the acceleration measured on the central pallet, while Fig. 3.100 those measured on the two beams, left (BL) and right (BR) respectively.

Table 3.55 Data of P2-B3 test at frequency 2.5 Hz

Test	Pallet 1			Pallet 2			Pallet 3		
	Time [s]	Lower bound [m/s²]	Upper bound [m/s²]	Time [s]	Lower bound [m/s²]	Upper bound [m/s²]	Time [s]	Lower bound [m/s²]	Upper bound [m/s²]
aa26	10.12	1.53	2.90	16.16	3.16	3.60	8.67	1.26	2.20
aa27	7.10	1.35	2.80	11.78	3.16	3.90	6.28	1.26	2.40
aa28	7.51	1.43	2.80	10.15	2.88	3.90	6.27	1.16	2.40
aa29	6.70	1.33	2.90	9.74	2.69	3.80	7.48	1.35	2.30
aa30	7.11	1.43	2.90	11.76	3.26	4.00	8.10	1.71	2.40
aa41	7.08	1.26	3.00	12.17	2.98	3.80	6.65	1.16	2.60
aa42	8.12	1.55	2.80	12.57	3.08	3.50	7.70	1.43	2.50
aa43	8.74	1.90	2.70	12.15	3.06		6.67	1.16	2.40
aa44	8.33	1.63	2.90	12.98	3.26	3.80	7.08	1.35	2.50
aa45	8.72	2.00		12.95	3.26		6.28	0.96	
aa46	8.95	2.00	3.00	12.96	3.26	4.20	7.87	1.55	2.60
μ	8.04	1.58	2.87	12.30	3.09	3.83	7.18	1.30	2.43
σ		0.27	0.09		0.18	0.21		0.21	0.13
Cov %		17.00	3.31		5.93	5.38		15.92	5.15

Table 3.56 Statistical analysis of data for tests from aa26 to aa30

Test	Pallet 1			Pallet 2			Pallet 3		
	Time [s]	Lower bound [m/s²]	Upper bound [m/s²]	Time [s]	Lower bound [m/s²]	Upper bound [m/s²]	Time [s]	Lower bound [m/s²]	Upper bound [m/s²]
μ	7.71	1.42	2.86	11.92	3.03	3.84	7.36	1.35	2.34
σ		0.08	0.05		0.24	0.15		0.21	0.09
Cov %		5.49	1.92		7.79	3.95		15.86	3.82

Table 3.57 Statistical analysis of data for tests from aa41 to aa46

Test	Pallet 1			Pallet 2			Pallet 3		
	Time [s]	Lower bound [m/s²]	Upper bound [m/s²]	Time [s]	Lower bound [m/s²]	Upper bound [m/s²]	Time [s]	Lower bound [m/s²]	Upper bound [m/s²]
μ	8.32	1.72	2.88	12.63	3.15	3.83	7.04	1.27	2.52
σ		0.30	0.13		0.12	0.29		0.22	0.08
Cov %		17.30	4.53		3.90	7.51		16.99	3.32

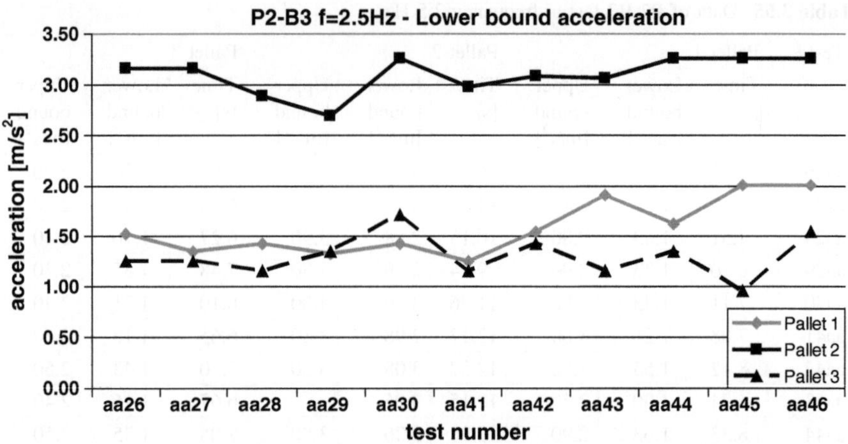

Fig. 3.83 Acceleration of sliding initiation in different tests—P2B3 f = 2.5 Hz

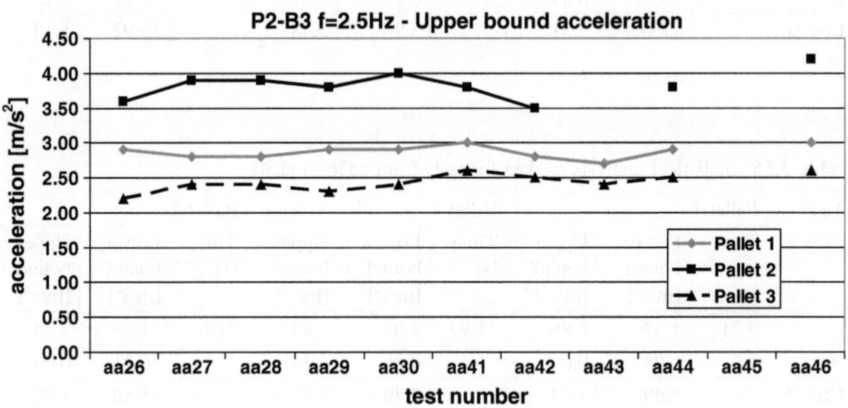

Fig. 3.84 Upper bound of the acceleration in different tests—P2B3 f = 2.5 Hz

With the employed instrumentation it is possible to measure the rotation of the different elements and consequently the relative displacement between pallet and beam (Figs. 3.101, 3.102, 3.103 and 3.104).

As the applied excitation is in down-aisle direction, no transversal (cross-aisle) component of deformation of the beams has to be expected. The two beams, under the weight of the pallets, undergo a deformation in the vertical plane (Fig. 3.105). Due to the symmetrical disposition of the three pallets on the beams, rotation is larger at the support and lower at the centre. This type of deformed shape is such that the two lateral pallets may slide "down hill" or "up hill", depending on the sign of the applied acceleration. On the contrary, the central pallet can only slide "uphill", independently on the direction of the applied acceleration. Hence, the two

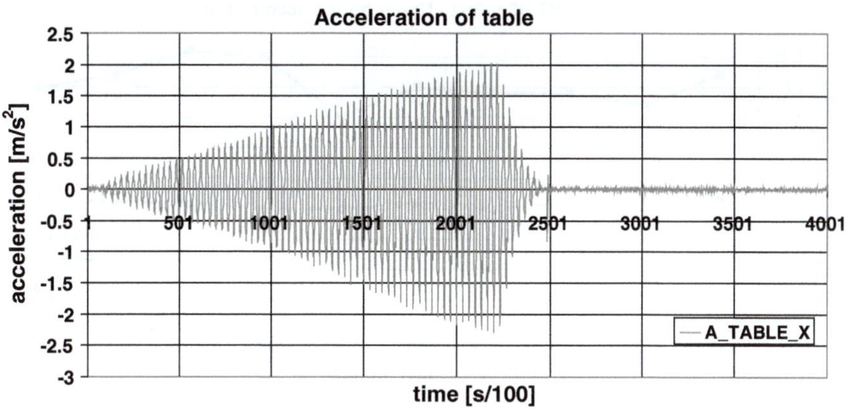

Fig. 3.85 Table acceleration in tests from aa31 to aa35

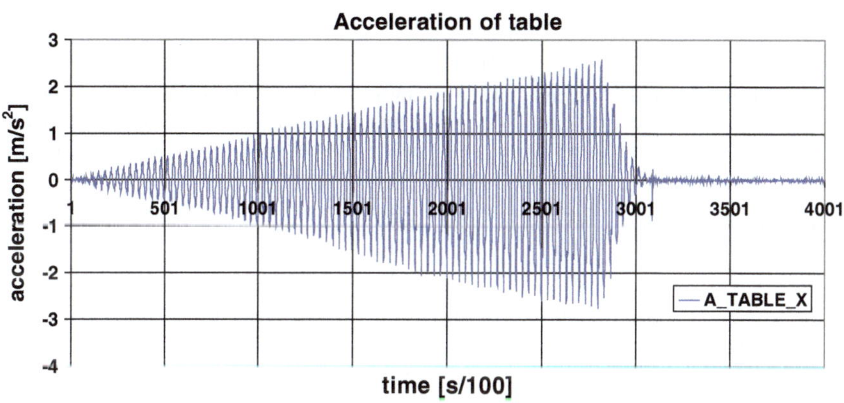

Fig. 3.86 Table acceleration in tests from aa47 to aa51

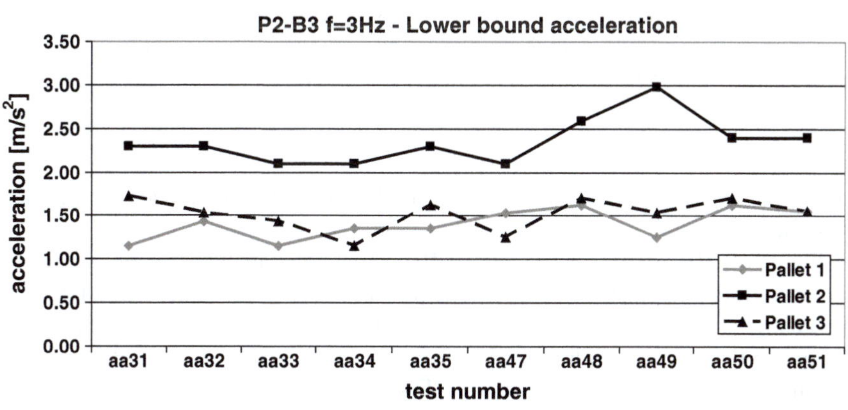

Fig. 3.87 Acceleration of sliding initiation in different tests—P2B3 f = 3 Hz

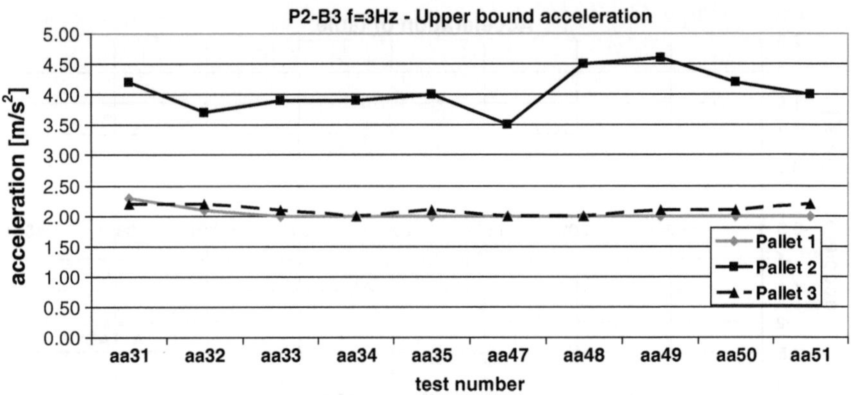

Fig. 3.88 Upper bound of the acceleration in different tests—P2B3 f = 3 Hz

Table 3.58 Data of P2-B3 test at frequency 3 Hz

Test	Pallet 1			Pallet 2			Pallet 3		
	Time [s]	Lower bound [m/s^2]	Upper bound [m/s^2]	Time [s]	Lower bound [m/s^2]	Upper bound [m/s^2]	Time [s]	Lower bound [m/s^2]	Upper bound [m/s^2]
aa31	5.91	1.16	2.30	9.48	2.30	4.20	7.31	1.73	2.20
aa32	7.29	1.43	2.10	8.80	2.30	3.70	5.94	1.53	2.20
aa33	4.90	1.16	2.00	7.12	2.10	3.90	5.95	1.43	2.10
aa34	6.58	1.35	2.00	6.77	2.10	3.90	5.09	1.16	2.00
aa35	6.24	1.35	2.00	8.15	2.30	4.00	6.62	1.63	2.10
aa47	6.56	1.53	2.00	6.96	2.10	3.50	5.27	1.26	2.00
aa48	6.41	1.63	2.00	10.16	2.59	4.50	6.45	1.71	2.00
aa49	6.61	1.26	2.00	15.17	2.98	4.60	6.29	1.53	2.10
aa50	7.24	1.63	2.00	7.73	2.40	4.20	6.62	1.71	2.10
aa51	6.74	1.55	2.00	8.82	2.39	4.00	6.29	1.55	2.20
μ	6.44	1.40	2.04	8.91	2.35	4.05	6.18	1.52	2.10
σ		0.18	0.10		0.27	0.34		0.19	0.08
Cov %		12.72	4.74		11.49	8.33		12.66	3.89

lateral pallets are expected to slide "earlier" than the central one, for physical ("gravity") reason.

Figures 3.101, 3.102, 3.103 and 3.104 show that, in the specific test under consideration, sliding starts earlier in the pallet 3, then in the pallet 1 and finally in the central one (pallet 2). It can be noticed that there is one order of magnitude difference between the sliding of the two lateral pallets (Pallet 1 and Pallet 3) and the sliding of the central one (Pallet 2). Figures 3.106, 3.107 and 3.108 show that

Table 3.59 Statistical analysis of data for tests from aa31 to aa35

	Pallet 1			Pallet 2			Pallet 3		
	Time [s]	Lower bound [m/s²]	Upper bound [m/s²]	Time [s]	Lower bound [m/s²]	Upper bound [m/s²]	Time [s]	Lower bound [m/s²]	Upper bound [m/s²]
μ	6.18	1.29	2.08	8.06	2.22	3.94	6.18	1.50	2.12
σ		0.13	0.13		0.11	0.18		0.22	0.08
Cov %		9.76	6.27		4.84	4.61		14.60	3.95

Table 3.60 Statistics of sliding acceleration for tests from aa47 to aa51

	Pallet 1			Pallet 2			Pallet 3		
	Time [s]	Lower bound [m/s²]	Upper bound [m/s²]	Time [s]	Lower bound [m/s²]	Upper bound [m/s²]	Time [s]	Lower bound [m/s²]	Upper bound [m/s²]
μ	6.71	1.52	2.00	9.76	2.49	4.16	6.18	1.55	2.08
σ		0.15			0.33	0.44		0.18	0.08
Cov %		10.12			13.05	10.56		11.92	4.02

Table 3.61 Statistics of the cross-aisle test, P2-B3

	Pallet 1		Pallet 2		Pallet 3	
	Lower bound [m/s²]	Upper bound [m/s²]	Lower bound [m/s²]	Upper bound [m/s²]	Lower bound [m/s²]	Upper bound [m/s²]
Sliding acceleration—P2B3 Cross-aisle—all frequencies						
μ	1.45	3.05	3.16	4.18	1.39	2.75
σ	0.20	0.87	0.67	0.46	0.20	0.77
Cov %	13.8	28.6	21.1	11.0	14.1	28.0

only in case of pallet P3 an upper value of the sliding acceleration could be detected. In fact, while the table acceleration increases, that of pallet P3, after cycle 100 remains constant at about 5 m/s^2 (Fig. 3.108). A similar behaviour is not evident for pallets P1 and P2 that, although showing a difference between their acceleration and the one of the beam (indicating sliding), don't show any upper bound value of their acceleration, that increases together with the one of the shaking table (Figs. 3.106 and 3.107).

Table 3.62 Statistics of the cross-aisle test, P2-B3 for every frequency

	Pallet 1		Pallet 2		Pallet 3	
	Lower bound [m/s^2]	Upper bound [m/s^2]	Lower bound [m/s^2]	Upper bound [m/s^2]	Lower bound [m/s^2]	Upper bound [m/s^2]
Sliding acceleration—P2B3 Cross-aisle—f = 1.0 Hz						
μ	1.31	3.00			1.28	2.85
σ	0.15	0.10			0.13	0.49
Cov %	11.5	3.3			9.9	17.4
P2B3 Cross-aisle—f = 1,5 Hz						
μ	1.39	4.28	3.40	4.25	1.35	4.03
σ	0.16	0.43	0.25	0.31	0.18	0.25
Cov %	11.2	10.0	7.3	7.3	13.3	6.2
P2B3 Cross-aisle—f = 2.0 Hz						
μ	1.53	4.09	3.91	4.88	1.48	3.88
σ	0.08	0.27	0.50	0.13	0.16	0.08
Cov %	5.2	6.7	12.8	2.7	10.8	2.2
P2B3 Cross-aisle—f = 2,5 Hz						
μ	1.58	2.87	3.09	3.83	1.30	2.43
σ	0.27	0.09	0.18	0.21	0.21	0.13
Cov %	17.0	3.3	5.9	5.4	15.9	5.2
P2B3 Cross-aisle—f = 3.0 Hz						
μ	1.40	2.04	2.35	4.05	1.52	2.10
σ	0.18	0.10	0.27	0.34	0.19	0.08
Cov %	12.7	4.7	11.5	8.3	12.7	3.9

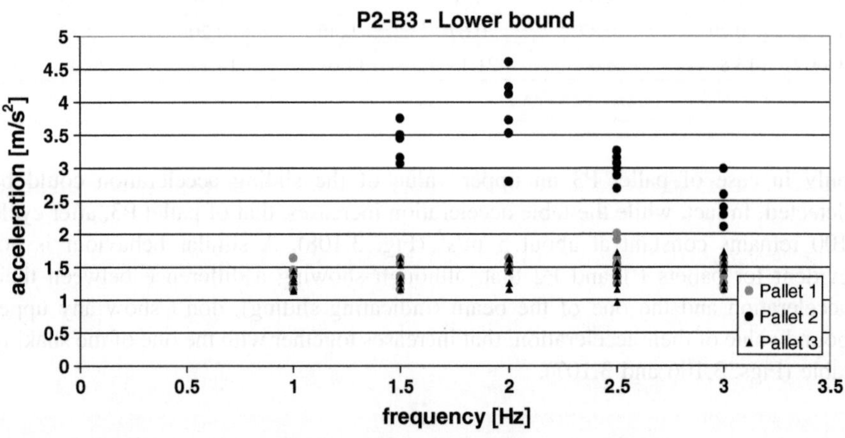

Fig. 3.89 Acceleration of sliding initiation versus frequency of P2-B3 tests

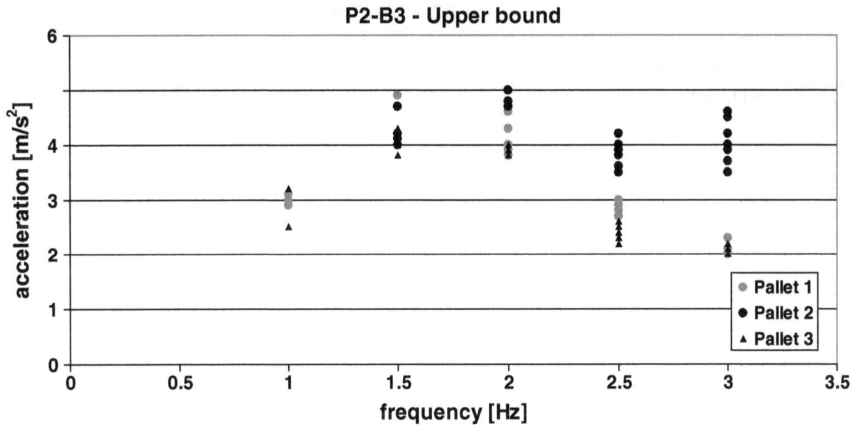

Fig. 3.90 Upper bound of sliding acceleration versus frequency of P2-B3 tests

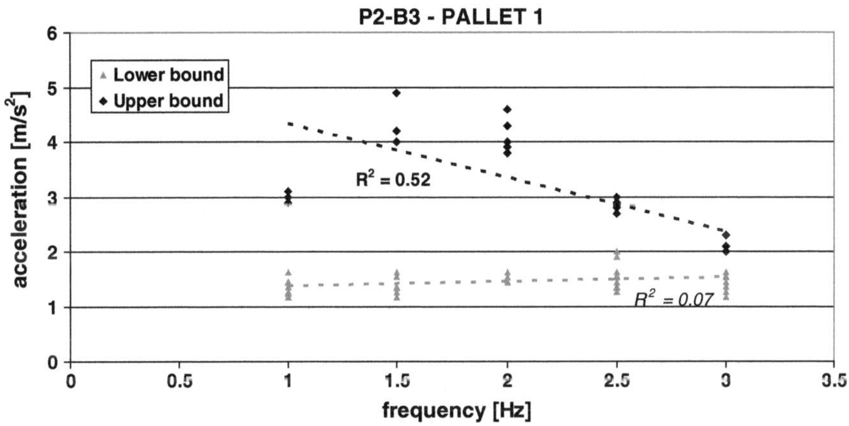

Fig. 3.91 Lower bound and upper bound of sliding acceleration for pallet 1 in cross-aisle P2B3 tests

3.3.2.3 Results of the Down-Aisle Tests

3.3.2.3.1 Constant Frequency and Increasing Acceleration Tests

The following paragraphs show the results of the down-aisle tests with constant frequency. All these tests were performed with the use of an old, dry, wooden Euro pallet (type P2), combined in most cases with a new hot dip coated cold rolled beam (type B2). Only few tests were carried out with a new hot zinc coated cold rolled beam (type B3), but are neglected in the following analysis because considered statistically non significant, because of the small numerousity of the batch, due to the low number of beams available beams for testing.

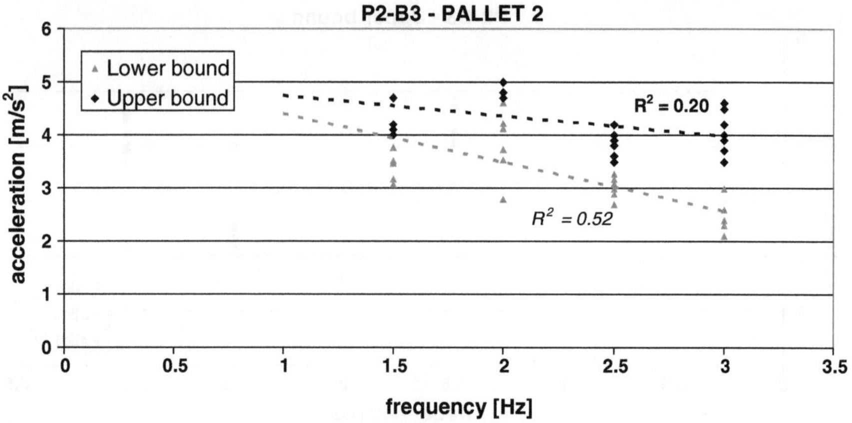

Fig. 3.92 Lower bound and upper bound of sliding acceleration for pallet 2 in cross-aisle P2B3 tests

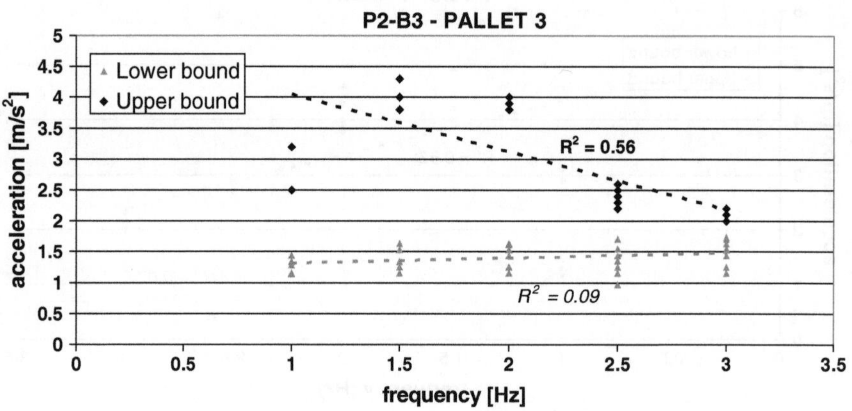

Fig. 3.93 Lower bound and upper bound of sliding acceleration for pallet 3 in cross-aisle P2B3 tests

Table 3.63 Mean value of lower bound acceleration in cross-aisle P2-B3 tests for the lateral pallets

Mean	1.42
St. dev.	0.20
Cov %	14.0

3.3.2.3.1.1 Frequency = 1.0 Hz

The table acceleration for these tests is shown in Fig. 3.109, while the test results are summarized in Table 3.64 as well as in Figs. 3.110 and 3.111.

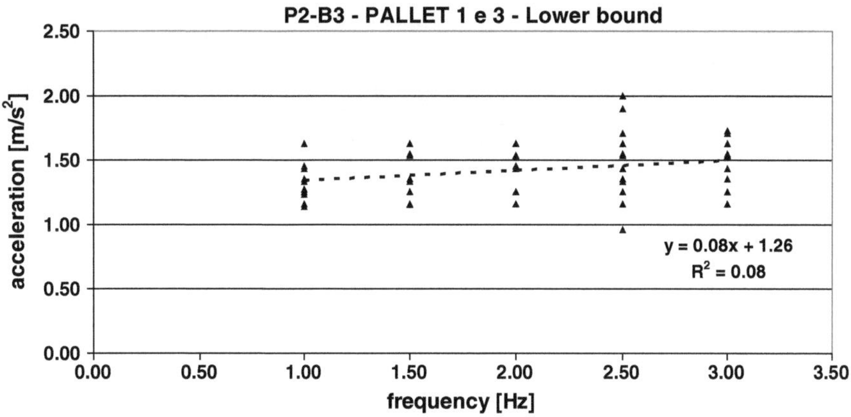

Fig. 3.94 Acceleration of sliding initiation versus frequency of excitation for pallet 1 and 3 in cross-aisle P2-B3 tests

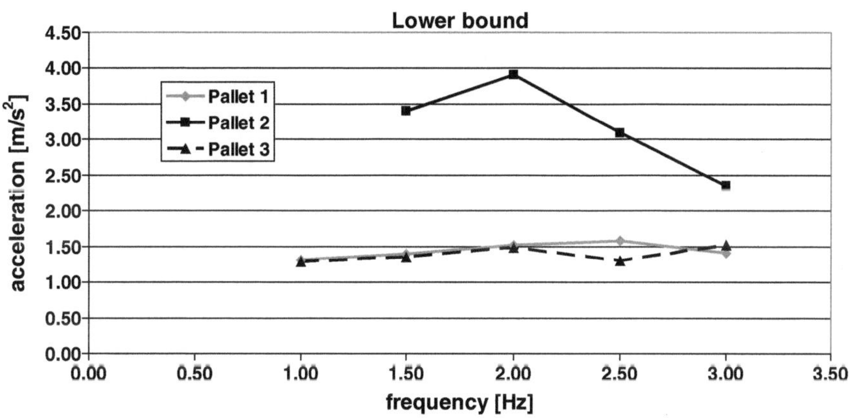

Fig. 3.95 Lower bound of sliding acceleration in the cross P2-B3 test

3.3.2.3.1.2 Frequency = 2.0 Hz

Figures 3.112 and 3.113 show the two different time histories of the table acceleration adopted respectively in tests from aa127 to aa135 and from aa150 to aa152. The second time history is characterized by a trend of increment of the shaking table acceleration ($0.2\ \frac{m/s^2}{s}$) double than the first one ($0.1\ \frac{m/s^2}{s}$). Test results are summarized in Table 3.65 and in Figs. 3.114 and 3.115.

Table 3.66 and 3.67 show the statistical analysis of the data for the tests respectively from aa127 to aa135 and from aa150 to aa152.

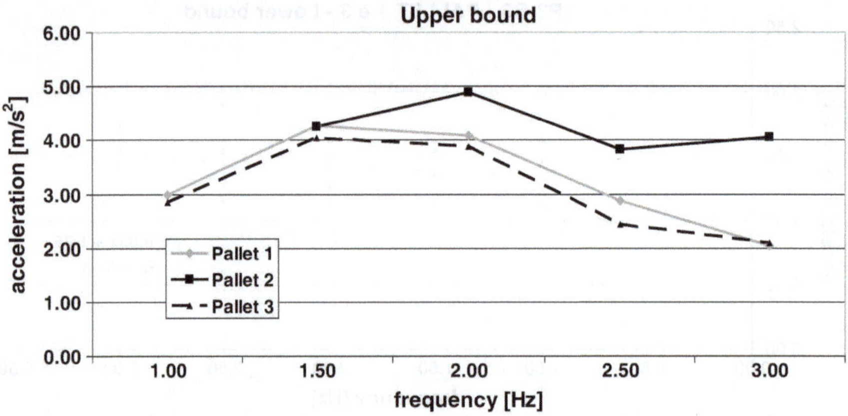

Fig. 3.96 Upper bound of sliding acceleration in the cross P2-B3 test

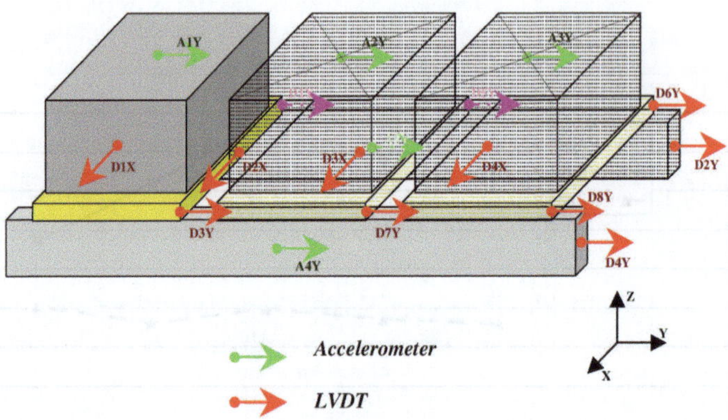

Fig. 3.97 Instrumentation set up for down-aisle tests

Fig. 3.98 Experimental set-up and instrumentation for the down-aisle tests

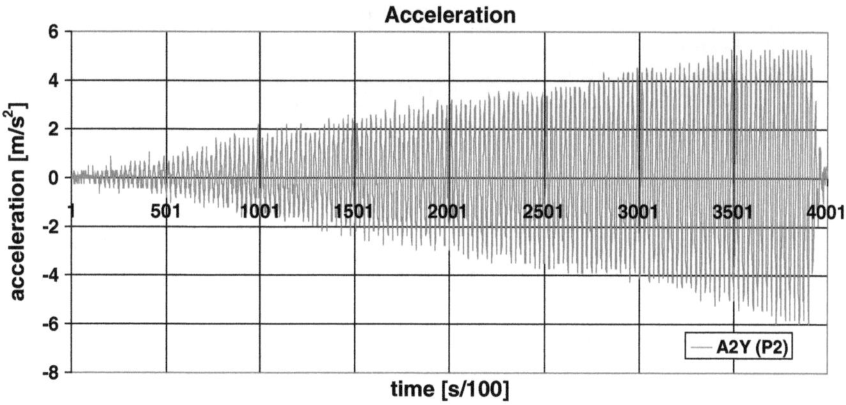

Fig. 3.99 Acceleration of Pallet 2 in the test no. aa144

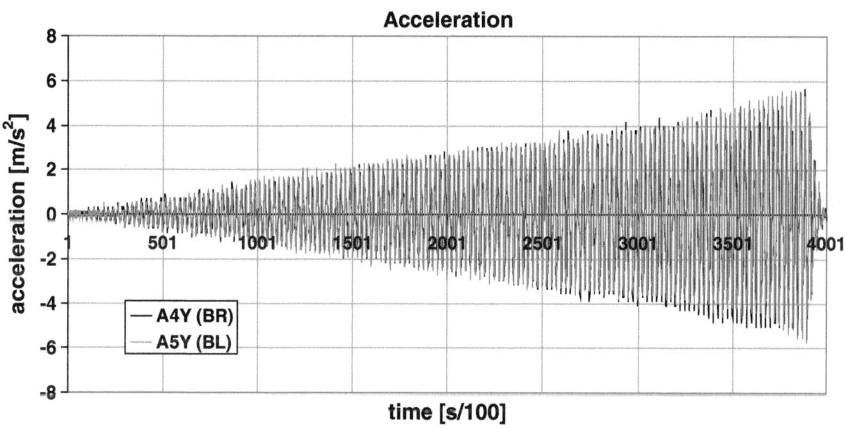

Fig. 3.100 Acceleration of the two beams in test no. aa144

3.3.2.3.1.3 Frequency = 3.0 Hz

Figures 3.116 and 3.117 show the two different time histories for the table acceleration adopted respectively in tests from aa136 to aa139 and from aa153 to aa156. The second time history is characterized by a trend of increment of the shaking table acceleration of 0.18 $\frac{m/s^2}{s}$, while that of the first one is 0.12 $\frac{m/s^2}{s}$.

Test results are summarized in Table 3.68 as well as in Figs. 3.118 and 3.119.

Tables 3.69 and 3.70 show the statistical analysis of the data for the tests respectively from aa136 to aa139 and from aa153 to aa155.

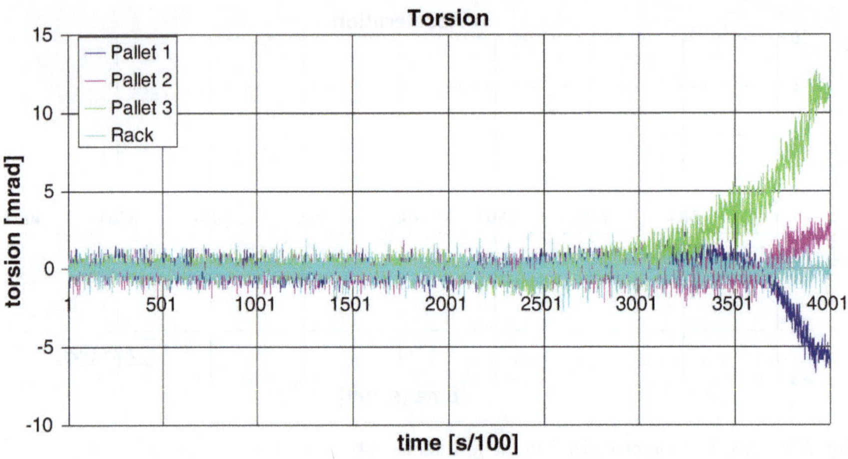

Fig. 3.101 Torsion of the pallets in the test no. aa144

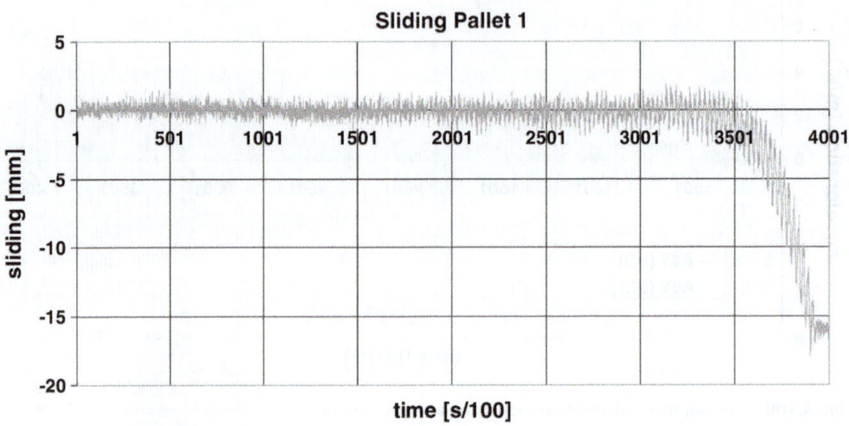

Fig. 3.102 Sliding of pallet 1 in test no. aa144

3.3.2.3.1.4 Frequency = 4.0 Hz

Figure 3.120 shows the table acceleration of these tests, characterized by a trend of increment of the shaking table acceleration of 0.11 $\frac{m/s^2}{s}$. Test results are summarized in Table 3.71 as well as in Figs. 3.121 and 3.122.

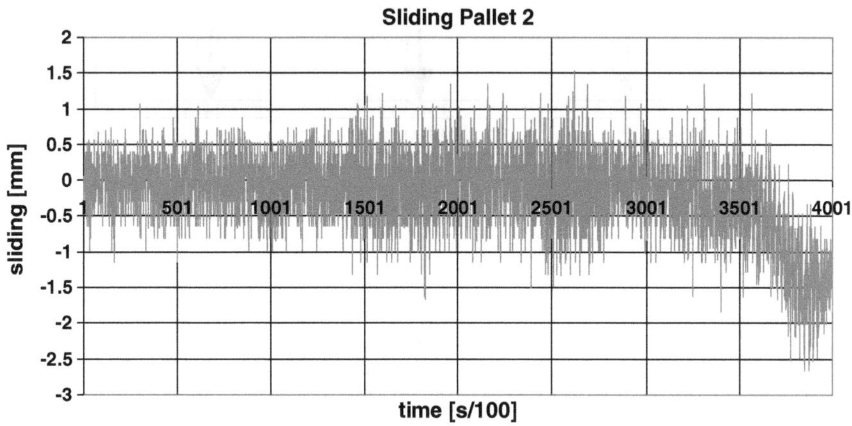

Fig. 3.103 Sliding of pallet 2 in test no. aa144

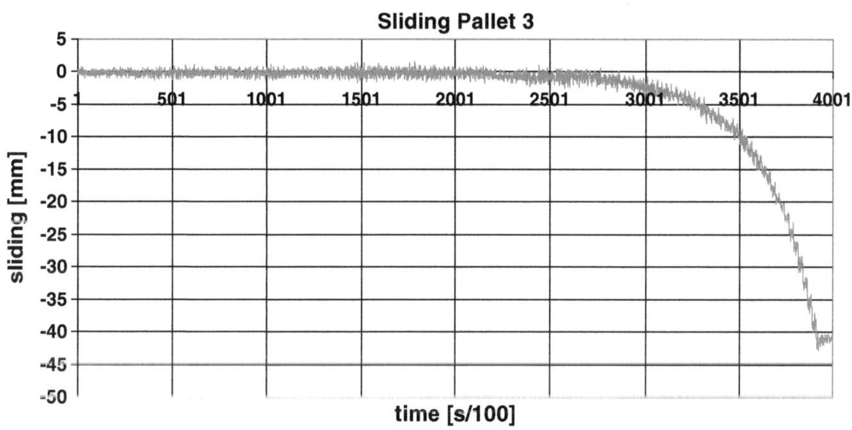

Fig. 3.104 Sliding of pallet 3 in test no. aa144

3.3.2.3.1.5 Re-analysis of P2-B2 Down-Aisle Test Results

Tables 3.72 and 3.73 as well as Figs. 3.123 and 3.124 summarize the results of tests performed in down-aisle direction, with constant frequency and increasing acceleration, on P2-B2 specimens.

Table 3.72 presents the mean values, standard deviations and c.o.v. of the lower and upper bound of the sliding acceleration, considering all the data, independently on the frequency of the excitation. Table 3.73 reports the same parameters, for each frequency considered in this group of tests. Figures 3.123 and 3.124 are a graphical representation of these data.

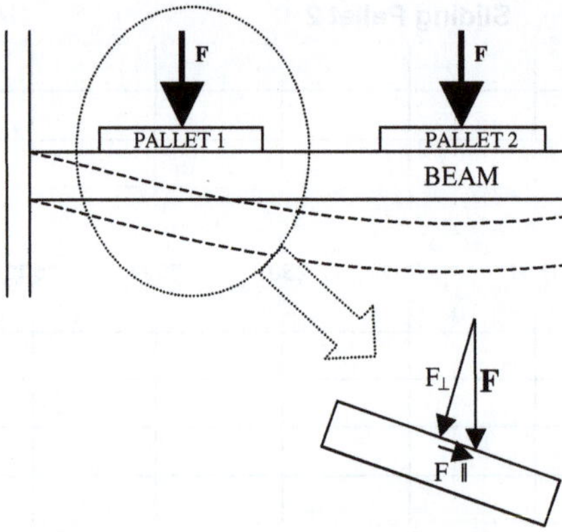

Fig. 3.105 Forces on the structure during a down-aisle test

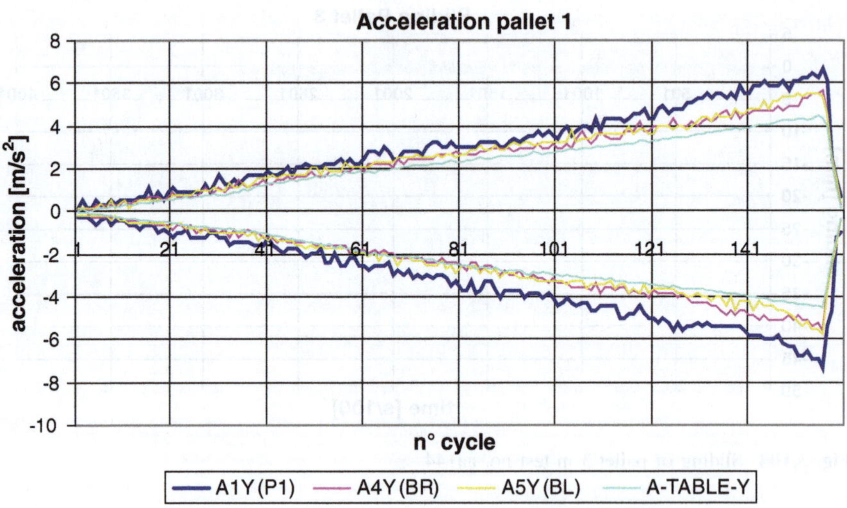

Fig. 3.106 Maximum and minimum values of acceleration of table, beam and pallet P1 (test no. aa144)

Tests in down-aisle direction show some characteristics similar to those of cross-aisle tests. The data confirm that the first pallet to slide is one of the two lateral ones, as can be observed in Tables 3.64, 3.65, 3.68 and 3.71.

In the tests at frequency 1.0 Hz, sliding of the central pallet never occurred, as the tests had to be terminated before the acceleration reached a value large enough to cause sliding.

The values of sliding initiation obtained for the three pallets in tests performed at a frequency not lower than 2.0 Hz are very close to each other. The difference that

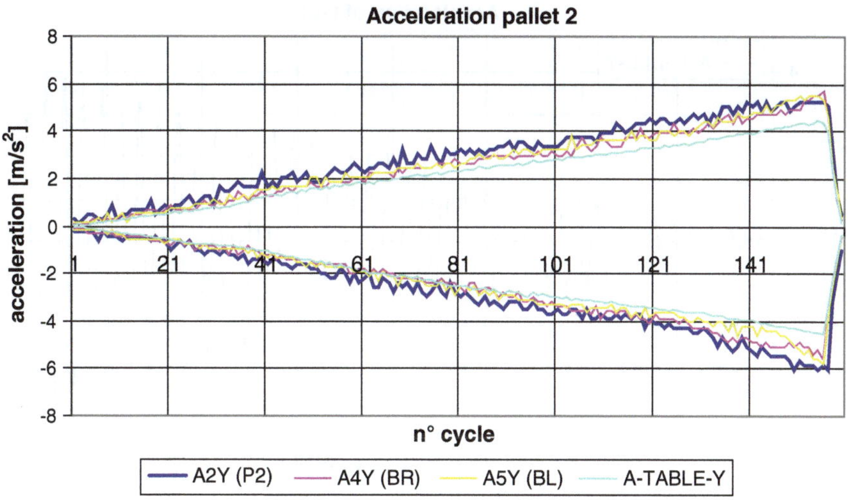

Fig. 3.107 Maximum and minimum values of acceleration of table, beam and pallet P2 (test no. aa144)

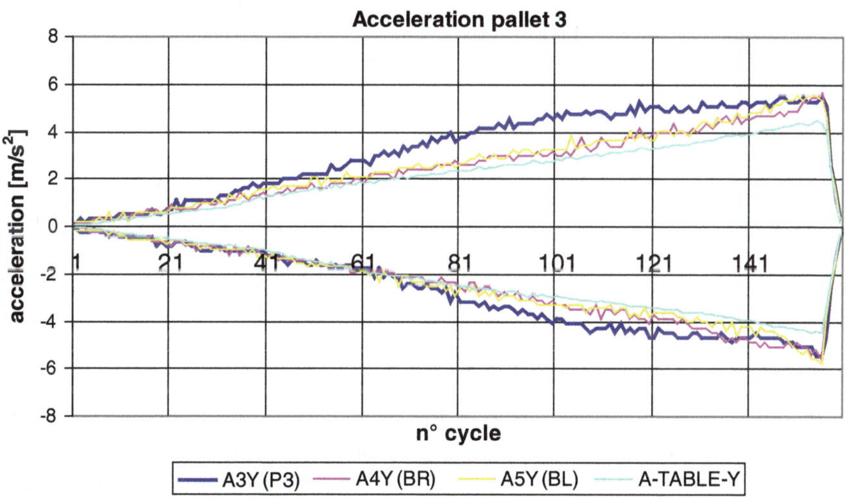

Fig. 3.108 Maximum and minimum values of acceleration of table, beam and pallet P3 (test no. aa144)

was evident in the cross-aisle tests between the sliding acceleration of the central pallet and the two lateral ones is not evident in this case.

In down-aisle tests, the effect of the deformation in the X-Y plane is limited, and doesn't influence much the pallet behaviour.

Also the upper bound of the acceleration shows similar values for all pallets.

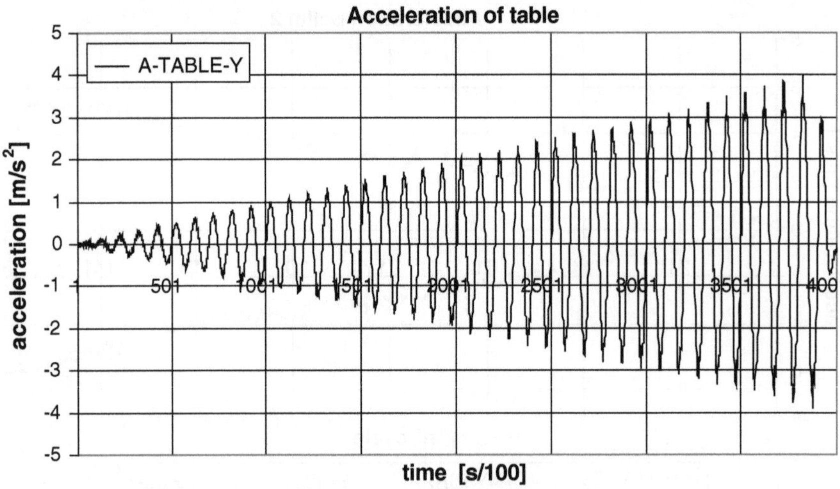

Fig. 3.109 Table acceleration in tests from aa123 to aa126

Table 3.64 Data of tests with beam type B2 at frequency 1.0 Hz

Test	Pallet 1			Pallet 2			Pallet 3		
	Time [s]	Lower bound [m/s^2]	Upper bound [m/s^2]	Time [s]	Lower bound [m/s^2]	Upper bound [m/s^2]	Time [s]	Lower bound [m/s^2]	Upper bound [m/s^2]
aa123	29.24	3.11	3.20				28.75	2.94	3.00
aa124	31.24	3.45	3.50				32.19	3.33	3.50
aa125	34.66	3.40	3.80				33.22	3.35	3.40
aa126	33.26	3.49	3.50				32.7	3.18	3.60
μ	32.10	3.37	3.50				31.72	3.20	3.38
σ		0.17	0.24					0.19	0.26
Cov %		5.17	7.00					5.80	7.79

In order to evaluate the effect of the beam-surface deterioration, tests were carried out for the same frequency, on the same beam, in two different groups, spaced of a number of tests from each other. This means that, after performing a first group of tests at 2.0 Hz, the frequencies of 3.0 and 4.0 Hz were investigated (running the first group of tests for each frequency). Later, a second group of tests was carried out for the frequencies of 2.0 and 3.0 Hz.

A statistical re-analysis of the results of the sample made of the two batches gives information related to the pallet sliding behaviour in presence of an average "deterioration" of the beam surface (Tables 3.65 and 3.68).

On the contrary, re-analysis of the two separated batches (Tables 3.66 and 3.67 at 2.0 Hz, Tables 3.69 and 3.70 at 3.0 Hz) give information about the effect of the beam surface deterioration on the sliding of the pallets in down-aisle direction.

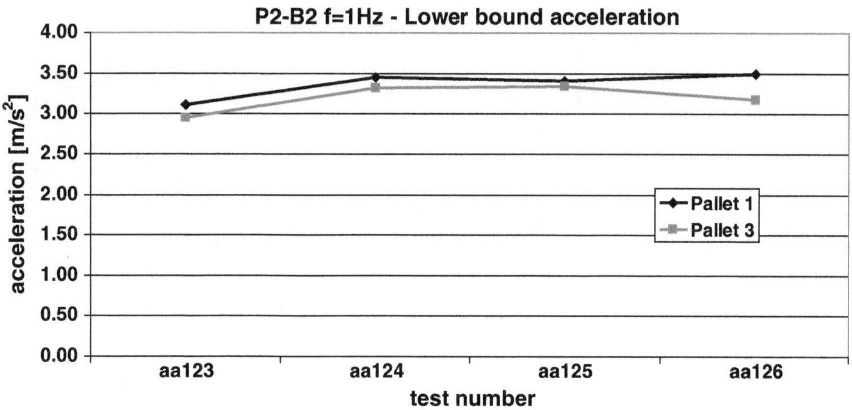

Fig. 3.110 Acceleration of sliding initiation in different tests

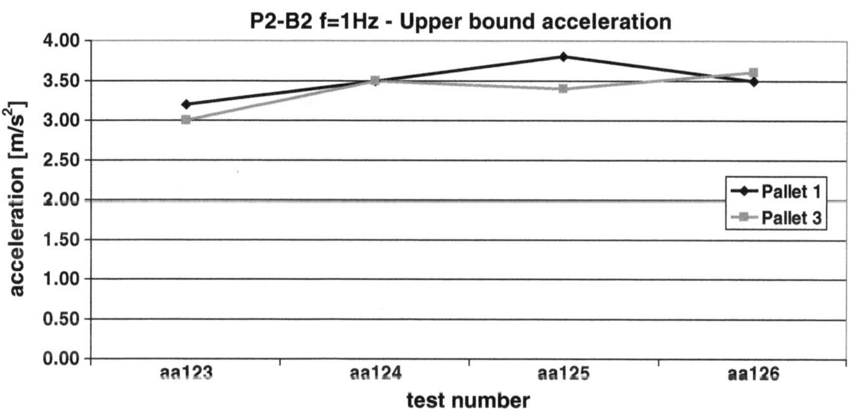

Fig. 3.111 Upper bound of sliding acceleration in different tests

Deterioration of the beam surface influences very much the sliding behaviour of the pallets; the mean values of both the lower and upper bound of the acceleration of all pallets increase when passing from the first to the second group of tests.

Figures 3.125, 3.126 and 3.127 show the values of the lower bound and of the upper bound of the sliding acceleration separately for each pallet. The linear approximation suggests that both the lower and the upper bound of the sliding acceleration of the lateral pallets increase with the frequency of the excitation, while remain more or less constant in the case of the central pallet.

Figures 3.128 and 3.129 show the mean values of the results obtained in terms of lower bound (Fig. 3.128) and upper bound (Fig. 3.129) of the sliding acceleration plotted versus the frequency of the excitation.

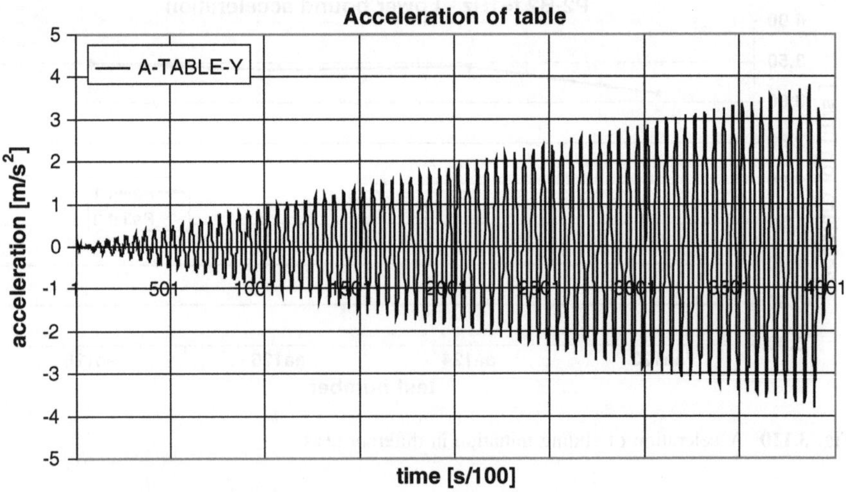

Fig. 3.112 Table acceleration in tests from aa127 to aa135

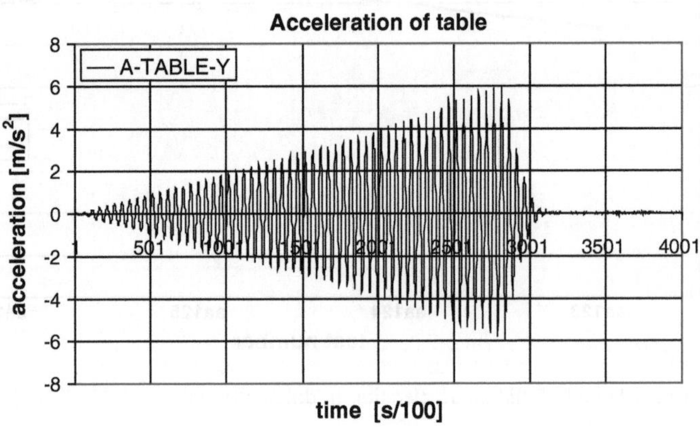

Fig. 3.113 Table acceleration in tests from aa150 to aa152

3.3.2.3.2 Constant Acceleration and Increasing Frequency Tests

Tests with a constant value of the acceleration and variable frequency were carried out on the combination pallet type P2 (wooden Euro pallet) and beam type B2 (cold rolled, hot dip coated, new beam). The instrumentation set up is the same used in the tests with constant frequency (Fig. 3.97). The purpose of this kind of tests is to analyse the eventual influence of the type of excitation (constant frequency and increasing acceleration or constant acceleration and increasing frequency) on the sliding acceleration. In this group of tests useful data could be obtained only with regard to the lower bound of the sliding acceleration. It was impossible to obtain any information about the upper bound of the sliding acceleration because the table acceleration is kept constant while increasing the frequency. Table 3.74 presents a list of the tests carried out according to this procedure.

During the test the frequency was increased according to the following law:

$$f_1 = f_0 \cdot 2^{\frac{t_1}{60}} \tag{3.15}$$

where:

f_0 is the frequency of the initial excitation of the shaking table;
t_1 is a generic acquisition time.

The frequency of the shaking table doubles every minute. A typical plot of this law is shown in Fig. 3.130, obtained with f_0 = 1.0 Hz.

Frequencies as high as 16 Hz were reached in the first 12 tests. Actually, during an earthquake, the typical frequencies of the event are generally lower.

The risk with this type of testing procedure is that, increasing the frequency of the excitation, the first eigen-frequency of the structure can be reached during the test, causing resonance and consequent serious damage. In order to avoid this problem most tests were terminated before reaching resonance.

Hence, the results shown hereafter, obtained with the same re-analysis procedure used in the tests in down-aisle direction with constant frequency, take into account only the response before the resonance peak.

Table 3.75 presents for each pallet the frequency of the excitation and the mass acceleration when sliding occurs. Due to dynamic amplification, such acceleration can be larger than the one of the shaking table. Acceleration values reported in Table 3.75 can be considered lower bound values of the sliding acceleration. Upper bound of the sliding acceleration could not be assessed with this testing procedure, because testing was terminated before attaining resonance. Usually this occurred before the sliding acceleration could reach a constant "upper bound" value.

Figures 3.131, 3.132 and 3.133 show the values of acceleration and frequency at sliding initiation, for each pallet. In the case of pallet 1 the measured values of the sliding acceleration are concentrated at the frequency of about 2.5 and 5 Hz. No relevant value is present between these two frequencies. For pallet 2 and 3, values could be measured also for intermediate frequencies. Sliding initiation acceleration of the central pallet is slightly higher than that of the two lateral ones.

It seems that sliding initiation occurs at the same value of acceleration independently of the frequency. No pallet shows sliding for input motions with a low frequency.

3.4 Comparison and Discussion of the Tests Results

Hereafter, all the sliding tests performed within the "SEISRACKS" project are briefly summarized and the results are compared and discussed aiming to an assessment of the limit conditions (namely lower and upper bound of the acceleration) for pallet sliding on the beams of a storage rack.

3.4.1 Cross-Aisle Tests

Tests in cross-aisle direction have been carried out with a sinusoidal excitation of the shaking table with constant frequency and an increasing acceleration. Only one type of pallet was considered, the wooden Euro pallet, old (type P2). Two different types of beams were adopted, beam B1, hot rolled, powder coated, and beam B3, hot rolled, hot zinc coated, both manufactured by the same producer.

In Figs. 3.135 and 3.136 the pallet is referenced with regard to his position on the beam, as it is shown in Fig. 3.134.

Figures 3.135 and 3.136 summarize the results of all tests carried out in cross-aisle direction, respectively on beam type B1 and type B3. It can be observed that, in the case of beam B1 (powder coated), the minimum measured value of the acceleration causing sliding of at least one of the pallets decreases when increasing the frequency of the excitation, ranging from approximately 0.17*g* for a frequency of 1.0 Hz, to 0.1*g* for a frequency of 4.0 Hz. On the contrary, in the case of beam B3 (hot zinc coated) the minimum measured value of the acceleration causing sliding of at least one of the pallets is approximately 0.1*g* and this value is practically independent of the frequency.

Figures 3.137, 3.138 and 3.139 show a comparison of the mean values of the acceleration causing sliding initiation of the pallets. Each figure presents for a single pallet (respectively pallet 1, pallet 2 and pallet 3) and for the two types of beam considered (B1 and B3), the mean value of the test results obtained with a same frequency of excitation.

Figure 3.140 summarizes the mean values of the sliding accelerations (obtained with a same frequency) for the set of both pallet 1 and 3, for the two types of beams B1 and B3. As already evidenced, the set of data is rather homogeneous, and seems to be not much influenced by the frequency of the excitation, at least up to 3.0 Hz. Examining Fig. 3.140, it can be stated that no sliding of the lateral pallets occurs if the acceleration is lower than 0.12*g*.

By examining Figs. 3.138 and 3.140, as well as Table 3.76, that summarizes the same data, the following considerations can be drawn.

The behaviour of the lateral pallets (pallet 1 and 3) is similar for all tests; for both pallets, sliding starts more or less at the same value of acceleration, and this trend is independent of the frequency of the excitation, at last for frequencies up to 3.0 Hz.

Sliding of the central pallet 2, starts under values of acceleration higher than for the two lateral pallets.

When a structure is subjected to a seismic event (or to a dynamic excitation), the first pallets to slide are the lateral ones.

The minimum value of the acceleration causing sliding of the pallet is important because gives an idea about initiation of a phenomenon that can lead to a loss of support of the pallet, with consequent fall, spillage, loss of the goods, as well as structural damage due to impact of the falling masses.

However, as previously discussed, aim of this study was the assessment of the design values of the horizontal actions to be considered in the seismic design of pallet racks. Hence, focus should be on the maximum values of acceleration recorded on the pallets. These values, in the previous paragraphs named "upper bound" of the sliding acceleration, are associated to the maximum forces that a "sliding" pallet can transfer to the beam by which it is supported.

Figures 3.141 and 3.142 show, respectively for the two lateral pallets 1 and 3, and for the central pallet 2, the mean values of the "upper bound" of the sliding acceleration, obtained for different frequencies of the excitation, and for the two types of beam B1 and B3. The same data are summarized in Table 3.77.

Upper bound of sliding acceleration for the central pallet is larger than the one of the two lateral ones. There is a slightly decreasing trend of the upper bound of the sliding acceleration with the frequency. It can be noticed that the lower bound of the sliding acceleration for beam type B3 is lower than the one of beam type B1. On the contrary, the upper bound of the sliding acceleration for beam B3 is higher than the one of beam B1.

The maximum of the mean values of the acceleration that could be measured on a mass in cross-aisle direction is $0.45g$ for the lateral pallets, and $0.5g$ for the central one.

3.4.2 Down-Aisle Tests

Tests in down-aisle direction are carried out with two different types of beam: type B2 (hot dip coated) and type B3 (hot zinc coated) both manufactured by the same producer. In the first case it is possible to compare a large batch of data of tests performed both at constant frequency and at constant acceleration. In the second case the low number of tests doesn't allow a significant re-analysis.

Figure 3.143 shows the test results for the case of beam B2. Both sets of data obtained in tests performed with constant frequency and with constant acceleration are considered. It seems that the two sets of data are fully compatible. It can be seen that no sliding occurs for accelerations lower than $0.3g$.

Figures 3.144, 3.145 and 3.146 show a comparison of the mean values of acceleration causing sliding initiation of the pallets.

Each figure presents the mean values of the test results obtained with a same frequency of excitation, for a single pallet (respectively pallet 1, 2 and 3) and with beam type B2. It can be noticed that, for frequencies larger than 1.0 Hz, the response seems to be independent of the frequency.

At a frequency of 1.0 Hz, the sliding acceleration is lower than for higher frequencies. Most probably, tests results for 1.0 Hz frequency are biased because during the test execution, the maximum excursion of the shaking table was reached in most tests before sliding of pallets occurred. For this reason, the set of data at 1.0 Hz is incomplete and it's reliably poor. If data related to 1.0 Hz frequency are disregarded, the results are consistent with each other.

Figure 3.147 summarizes the mean values of the sliding acceleration for the batch of data of both the two lateral pallets 1 and 3, for type of beam B2.

By examining Figs. 3.145 and 3.147, as well as Table 3.78 that summarizes the same data, the following considerations can be drawn:

- There is no pallet sliding for acceleration lower than 0.4*g*, for the lateral pallets.
- The central pallet (pallet 2) slides for accelerations larger than 0.5g, higher than those of the two lateral ones.

Figures 3.148 and 3.149 show, respectively for the two lateral pallets 1 and 3 and for the central pallet 2, the mean values of the "upper bound" of the sliding acceleration, obtained for different frequencies of the excitation. The same data are summarized in Table 3.79.

The upper bound of the sliding acceleration for the central pallet is slightly higher than the value obtained from the two lateral ones.

It seems that there is a sort of increasing trend of the upper bound of the sliding acceleration with the frequency. This behaviour is more evident for the lateral pallets. The highest mean value of the upper bound of the sliding acceleration measured in down-aisle direction is 0.55*g* for the lateral pallets, and 0.6*g* for the central one. These values are higher than those obtained for the cross-aisle direction.

3.5 Seismic Tests

Seismic tests were carried out both in down-aisle and cross-aisle direction adopting different types of input motions.

Tests set-up was the same used in the tests with sinusoidal excitation, in cross and down-aisle direction as shown in Figs. 3.37 and 3.97, using respectively beams type B1 and B2, always with a wooden Euro pallets (type P2).

The structure was excited with three different types of seismic signal, samples of Greek earthquakes, appropriately scaled. The seismic input motions were selected to have quite different characteristics and were registered during destructive, shallow earthquake of magnitude around 6.0. Namely:

EDESSA signal: it was registered in Edessa during the Griva earthquake of December 21st, 1990. It had a magnitude 5.9 and was recorded on a soft soil at distance of 31 km from the epicenter; the effect of local geology is evident on the

Table 3.65 Data of tests with beam type B2 at frequency 2.0 Hz

Test	Pallet 1			Pallet 2			Pallet 3		
	Time [s]	Lower bound [m/s^2]	Upper bound [m/s^2]	Time [s]	Lower bound [m/s^2]	Upper bound [m/s^2]	Time [s]	Lower bound [m/s^2]	Upper bound [m/s^2]
aa127	32.65	3.45	3.80				34.12	3.44	
aa128	31.15	3.54		35.67	3.99		34.91	3.61	
aa129	32.13	3.67		35.17	4.11		32.18	3.40	3.90
aa130	32.66	3.71		35.43	4.12		34.67	3.56	
aa131	30.67	4.43	4.50	31.54	4.41		30.63	3.89	4.00
aa132	28.55	4.12	4.50	33.93	4.89		32.91	4.37	4.60
aa133	29.92	4.19	4.50	34.81	4.98	5.00	31.9	4.42	5.00
aa134	33.54	4.79	4.90	36.80	5.08		34.01	4.59	
aa135	34.16	4.41	4.80	38.54	5.08		34.91	4.66	
aa150	24.45	5.17		25.43	5.98		25.14	5.42	
aa151	25.44	5.59		24.93	5.59		25.63	5.51	
aa152	20.93	4.98	6.00	25.31	5.75	6.00	25.44	5.42	6.00
μ	29.69	4.34	4.71	32.51	4.91	5.50	31.37	4.36	4.70
σ		0.69	0.67		0.69	0.71		0.79	0.85
Cov %		15.89	14.15		14.01	12.86		18.17	18.18

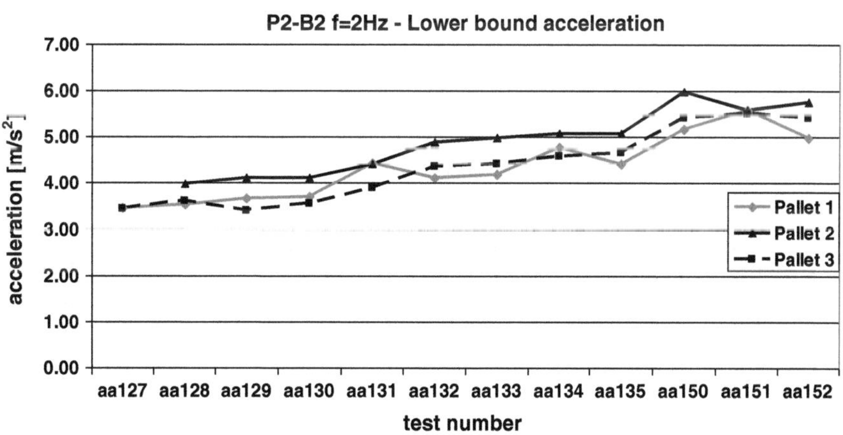

Fig. 3.114 Acceleration of sliding initiation in different tests

sustained strong motion part of the record (Fig. 3.150). It is a typical far field earthquake. The Fourier analysis (Fig. 3.151) highlights presence in the input signal of a predominant frequency at 1.53 Hz.

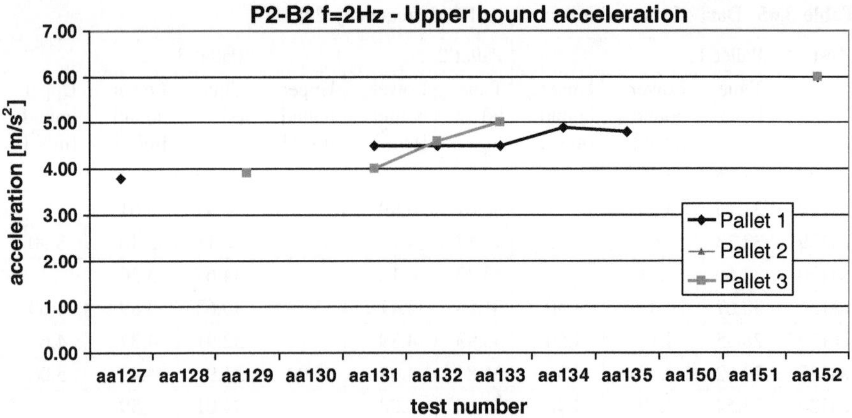

Fig. 3.115 Upper bound of sliding acceleration in different tests

Table 3.66 Data of tests from aa127 to aa135

Test	Pallet 1			Pallet 2			Pallet 3		
	Time [s]	Lower bound [m/s²]	Upper bound [m/s²]	Time [s]	Lower bound [m/s²]	Upper bound [m/s²]	Time [s]	Lower bound [m/s²]	Upper bound [m/s²]
μ	31.71	4.04	4.50	35.24	4.59	5.00	33.36	3.99	4.38
σ		0.46	0.38		0.47			0.52	0.52
Cov %		11.50	8.55		10.32			12.91	11.86

Table 3.67 Data of tests from aa150 to aa152

Test	Pallet 1			Pallet 2			Pallet 3		
	Time [s]	Lower bound [m/s²]	Upper bound [m/s²]	Time [s]	Lower bound [m/s²]	Upper bound [m/s²]	Time [s]	Lower bound [m/s²]	Upper bound [m/s²]
μ	23.61	5.25	6.00	25.22	5.77	6.00	25.40	5.45	6.00
σ		0.31			0.20			0.05	
Cov %		5.95			3.41			0.99	

KALAMATA signal: it was registered on September 13th, 1986 at a distance of 9 km from the epicenter. It had a magnitude of 6.2. The signal (Fig. 3.152) is characterized by the frequency content shown is Fig. 3.153.

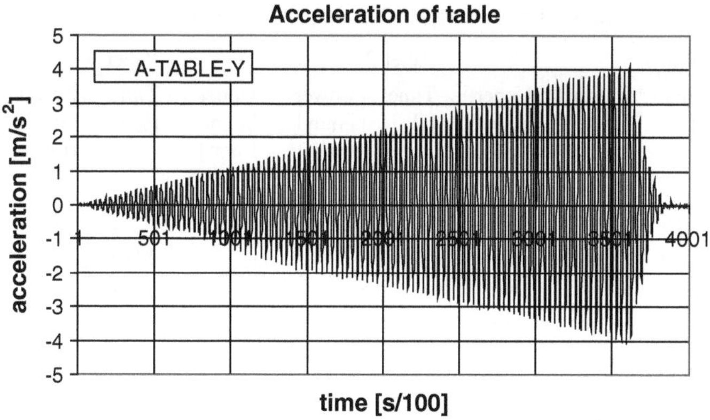

Fig. 3.116 Table acceleration in tests from aa136 to aa139

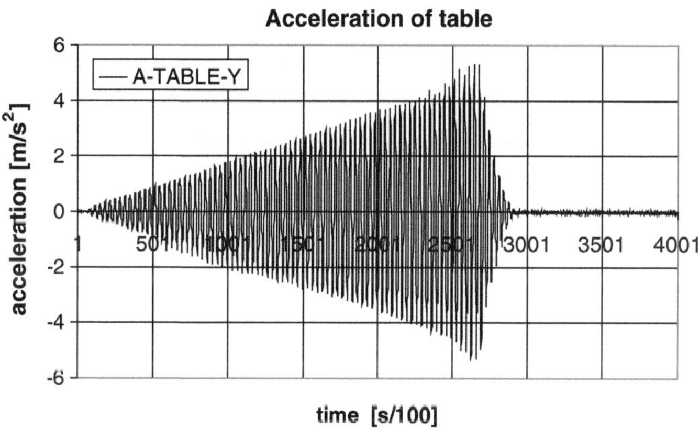

Fig. 3.117 Table acceleration in tests from aa153 to aa155

ARGOSTOLI signal: it was registered during the Cephalonia earthquake of 24 March 1983 at a distance of 10 km from the epicenter. It had a magnitude 5.5 and it's characterized by an impulse of acceleration with a time length of about half second, while the signal is null just before and after, up to the end (Fig. 3.154). It's a typical near field earthquake, with the frequency content shown in Fig. 3.155.

Table 3.80 summarizes the data of the Greek earthquake considered in the tests.

Seismic tests have been carried out in cross-aisle direction for beam type B1, and in down-aisle direction for beam type B2. In some cases, a bi-directional excitation of the shaking table was adopted, with acceleration components in both in cross (X) and down-aisle (Y) directions.

Table 3.68 Data of tests with beam type B2 at frequency 3.0 Hz

Test	Pallet 1			Pallet 2			Pallet 3		
	Time [s]	Lower bound [m/s²]	Upper bound [m/s²]	Time [s]	Lower bound [m/s²]	Upper bound [m/s²]	Time [s]	Lower bound [m/s²]	Upper bound [m/s²]
aa136	22.05	3.83	4.50	35.05	4.71		34.37	4.69	
aa137	22.15	3.97	5.00				23.13	3.68	4.80
aa138	20.39	3.63	4.80				27.05	3.92	4.90
aa139	21.15	3.76	4.80				27.05	4.02	5.10
aa153	17.06	4.89	5.30	25.20	5.26		23.04	5.08	
aa154	17.40	4.59	5.50						
aa155	16.72	4.39	5.80	19.05	4.40		25.37	5.26	6.00
μ	19.56	4.15	5.10	26.43	4.79		26.67	4.44	5.20
σ		0.48	0.45		0.44			0.66	0.55
Cov %		11.45	8.91		9.13			14.80	10.53

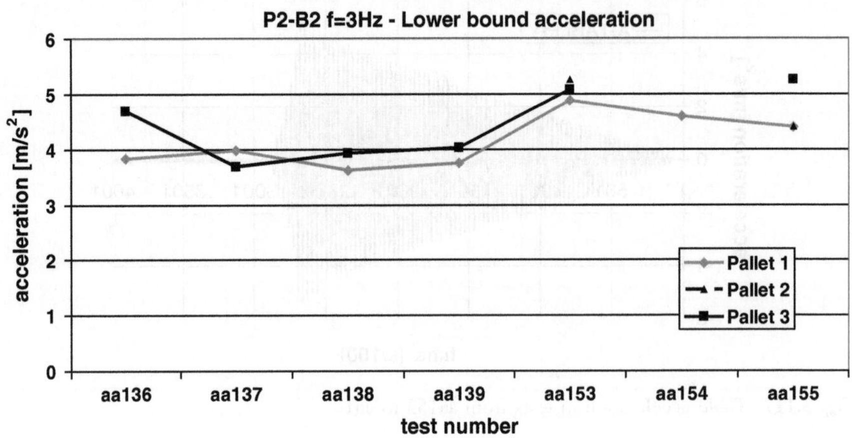

Fig. 3.118 Acceleration of sliding initiation in different tests

The amplitudes of the earthquake motions were scaled in order to apply peak accelerations (PGA) of increasing values until sliding of the pallets was recorded.

Table 3.81 shows the list of the test and the input motions with the different components of the acceleration.

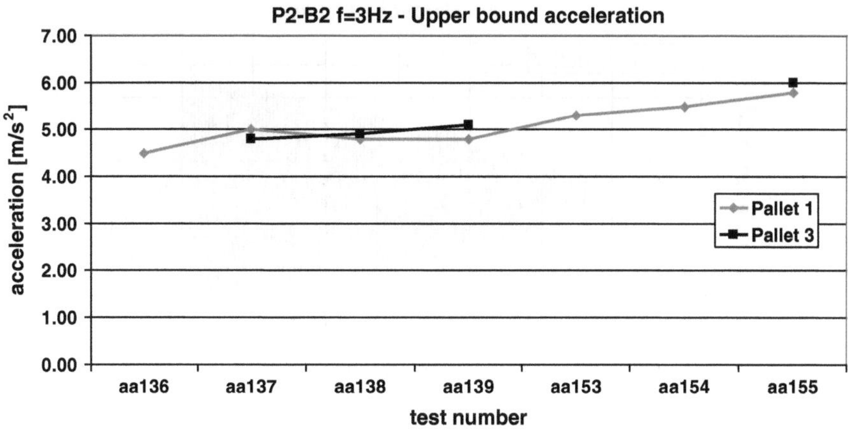

Fig. 3.119 Upper bound of sliding acceleration in different tests

Table 3.69 Data of tests from aa136 to aa139

Test	Pallet 1			Pallet 2			Pallet 3		
	Time [s]	Lowe bound [m/s^2]	Upper bound [m/s^2]	Time [s]	Lower bound [m/s^2]	Upper bound [m/s^2]	Time [s]	Lower bound [m/s^2]	Upper bound [m/s^2]
μ	21.44	3.80	4.78	35.05	4.71		27.90	4.08	4.93
σ		0.14	0.21					0.43	0.15
Cov %		3.77	4.32					10.58	3.10

Table 3.70 Data of tests from aa153 to aa155

Test	Pallet 1			Pallet 2			Pallet 3		
	Time [s]	Lower bound [m/s^2]	Upper bound [m/s^2]	Time [s]	Lower bound [m/s^2]	Upper bound [m/s^2]	Time [s]	Lower bound [m/s^2]	Upper bound [m/s^2]
μ	17.06	4.62	5.53	22.13	4.83		24.21	5.17	6.00
σ		0.25	0.25		0.61			0.12	
Cov %		5.34	4.55		12.65			2.41	

3.5.1 Seismic Test—Cross-Aisle Direction

Occurrence of sliding in a seismic test can be detected in analogy with the procedure adopted for the cross-aisle tests with constant frequency.

The sliding acceleration is that corresponding to occurrence of relative displacement between pallet and beam. Figures 3.156 and 3.157 show, for example,

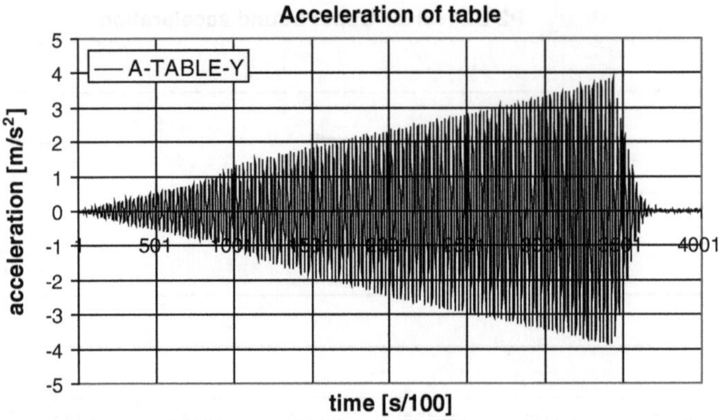

Fig. 3.120 Table acceleration in tests from aa140 to aa144

Table 3.71 Data of tests with beam type B2 at frequency 4.0 Hz

Test	Pallet 1			Pallet 2			Pallet 3		
	Time [s]	Lower bound [m/s²]	Upper bound [m/s²]	Time [s]	Lower bound [m/s²]	Upper bound [m/s²]	Time [s]	Lower bound [m/s²]	Upper bound [m/s²]
aa140	32.51	5.17	5.70	34.94	5.06	5.80	26.06	4.22	4.80
aa141	29.05	4.89					25.06	4.22	5.00
aa142	31.81	5.08		36.13	5.00		25.56	4.51	4.80
aa143	31.12	4.92	6.00	35.56	5.06		26.56	4.41	4.80
aa144	33.02	5.22		35.13	5.09		25.31	4.22	5.00
μ	31.50	5.06	5.85	35.44	5.05	5.80	25.71	4.32	4.88
σ		0.15	0.21		0.04			0.14	0.11
Cov %		2.91	3.63		0.72			3.20	2.24

the sliding of pallet 1, measured during test aa96, under an Edessa type of excitation.

In an EDESSA test it is possible to analyse the sliding phenomenon both when the acceleration is in the increasing phase and in the decreasing phase. In this latter case, the acceleration of interest is that of the table when sliding ends. In this way, two relevant data can be obtained from the same test. This type of analysis is possible only when the peak acceleration is much higher than the sliding one, for example in test no. aa96 and aa97. In Fig. 3.158 there is a graphical description of how this data can be obtained.

Also in a KALAMATA signal it is possible to obtain two values of the pallet sliding acceleration, from corresponding to peaks in the same test, as sketched in Figs. 3.159. Figure 3.160 shows the presence of a relative displacement between

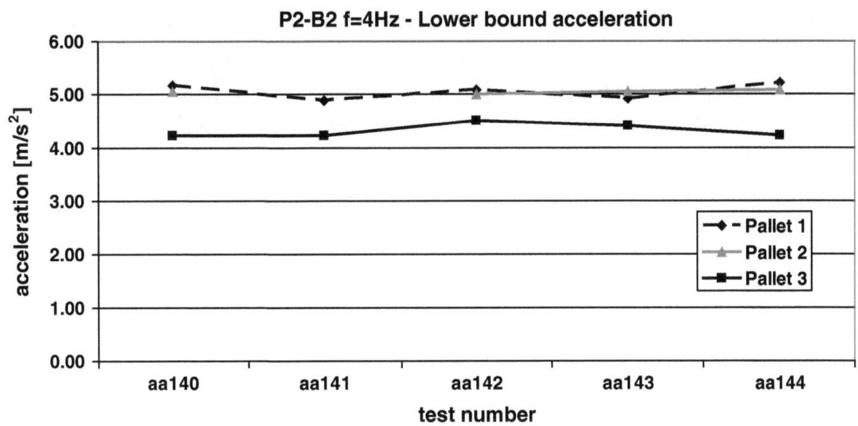

Fig. 3.121 Acceleration of sliding initiation in different tests

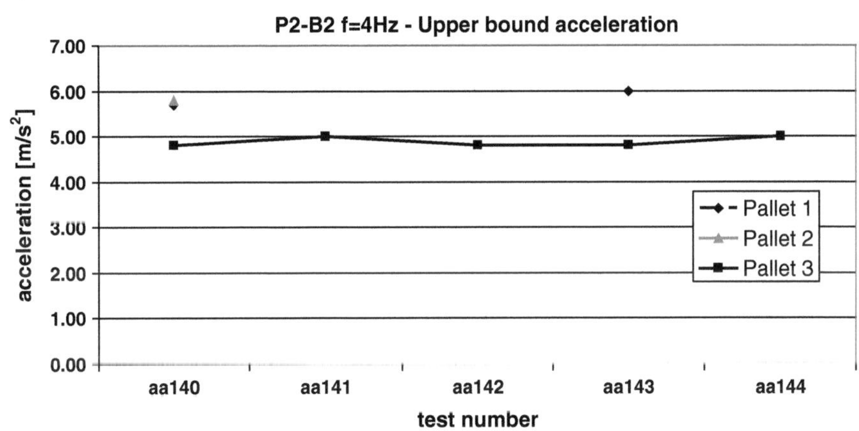

Fig. 3.122 Upper bound of sliding acceleration in different tests

Table 3.72 Statistics of the down-aisle tests

	Pallet 1		Pallet 2		Pallet 3	
	Lower bound [m/s^2]	Upper bound [m/s^2]	Lower bound [m/s^2]	Upper bound [m/s^2]	Lower bound [m/s^2]	Upper bound [m/s^2]
Sliding acceleration—P2B2 Down-aisle—all frequencies						
μ	4.28	4.72	4.92	5.60	4.20	4.57
σ	0.70	0.85	0.56	0.53	0.73	0.84
Cov %	16.4	18.0	11.3	9.4	17.5	18.4

Table 3.73 Mean parameters of the down-aisle test for each frequency of the excitation

	Pallet 1		Pallet 2		Pallet 3	
	Lower bound [m/s^2]	Upper bound [m/s^2]	Lower bound [m/s^2]	Upper bound [m/s^2]	Lower bound [m/s^2]	Upper bound [m/s^2]
Sliding acceleration—P2B2 Down-aisle—f = 1.0 Hz						
μ	3.37	3.50			3.20	3.38
σ	0.17	0.24			0.19	0.26
Cov %	5.2	7.0			5.8	7.8
Sliding acceleration—P2B2 Down-aisle—f = 2.0 Hz						
μ	4.34	4.71	4.91	5.50	4.36	4.70
σ	0.69	0.67	0.69	0.71	0.79	0.85
Cov %	15.9	14.1	14.0	12.9	18.2	18.2
Sliding acceleration—P2B2 Down-aisle—f = 3.0 Hz						
μ	4.15	5.10	4.79		4.44	5.20
σ	0.48	0.45	0.44		0.66	0.55
Cov %	11.4	8.9	9.1		14.8	10.5
Sliding acceleration—P2B2 Down-aisle—f = 4.0 Hz						
μ	5.06	5.85	5.05	5.80	4.32	4.88
σ	0.15	0.21	0.04		0.14	0.11
Cov %	2.9	3.6	0.7		3.2	2.2

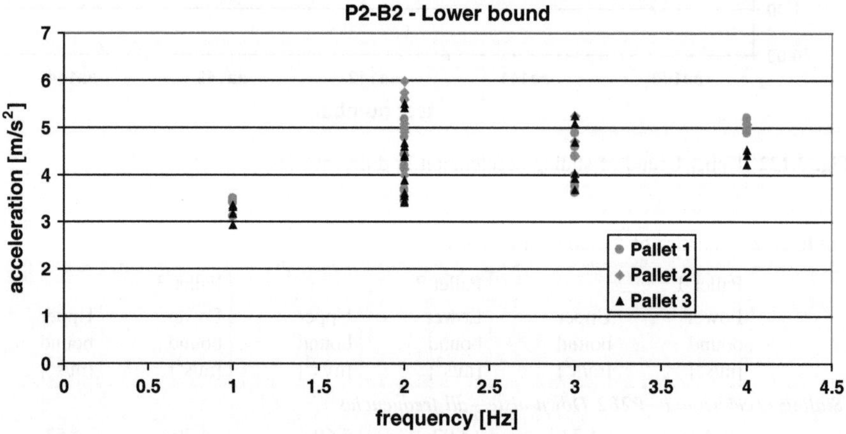

Fig. 3.123 Acceleration of sliding initiation versus frequency

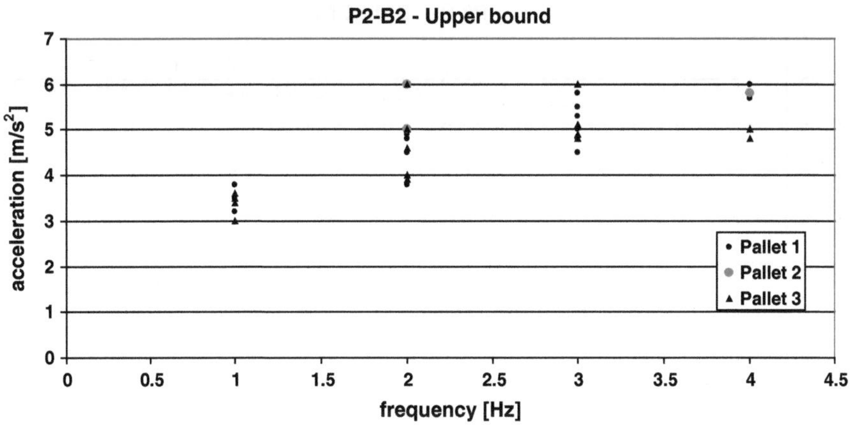

Fig. 3.124 Upper bound of sliding acceleration versus frequency

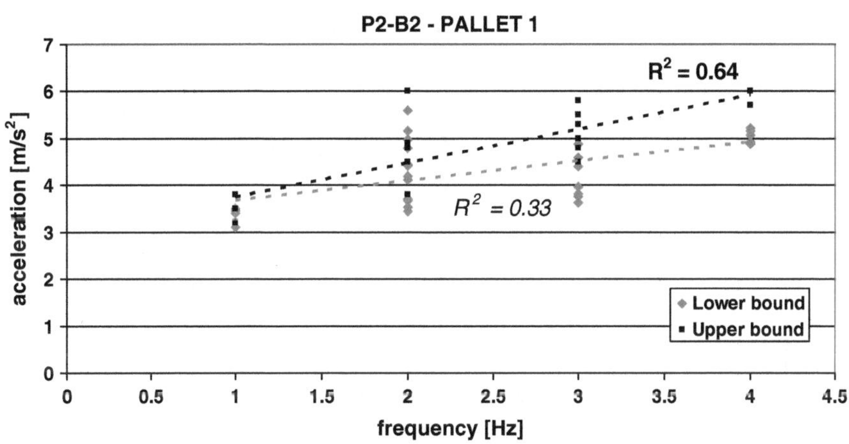

Fig. 3.125 Lower and upper bound of sliding acceleration for pallet 1 in down-aisle P2B2 tests

pallet and beam in two different moments. Sliding doesn't occur between cycles 7 and 8.

An important consideration can be done observing the general behaviour of the pallet in the different types of test. In an EDESSA or a KALAMATA test (Fig. 3.162), if sliding occurs the final amplitude of the pallet displacement depends on the time length of the input signal. If it is sufficiently large, the pallet can lose support and fall from the beam. In an ARGOSTOLI test (Fig. 3.161), on the contrary, when the pallet slides its displacement is instantaneous and limited to the short duration of the excitation.

All the data of the cross-aisle seismic tests are summarized in Table 3.82. The acceleration of the pallets reported in the table refers to the acceleration of sliding

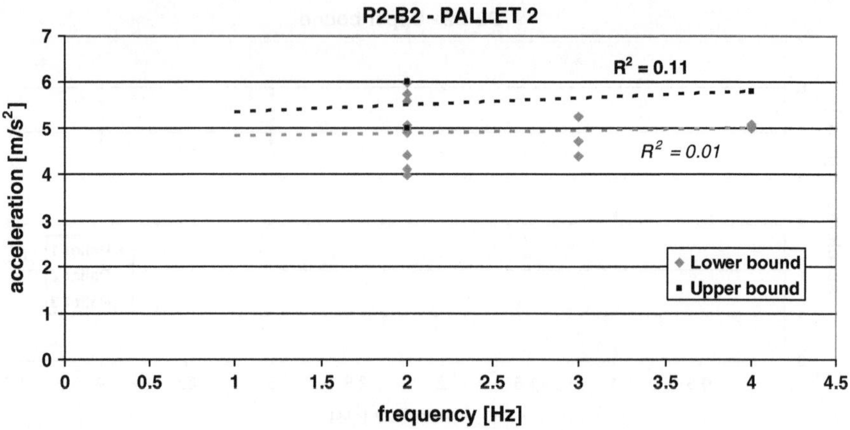

Fig. 3.126 Lower and upper bound of sliding acceleration for pallet 2 in down-aisle P2B2 tests

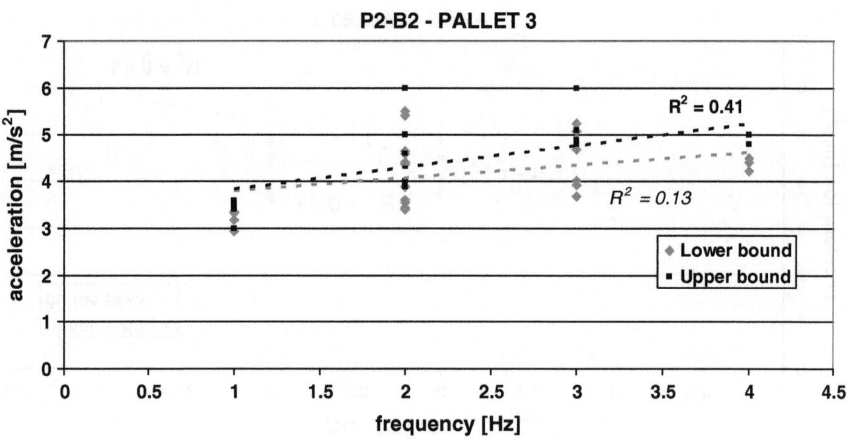

Fig. 3.127 Lower and upper bound of sliding acceleration for pallet 3 in down-aisle P2B2 tests

initiation if sliding occurs, or to the maximum acceleration reached during the test if no sliding could be detected during the test.

Figures 3.163 and 3.164 show the lower bound of the sliding acceleration for all tests. Figure 3.163 considers the data for the two lateral pallets, while Fig. 3.164 the data of the central pallet.

These results can be compared with the sliding domain obtained by means of sinusoidal tests in cross-aisle direction, presented in Fig. 3.135. It can be noticed that in the seismic tests sliding occurs first for ARGOSTOLI signal at around 0.15*g*. For the other types of seismic test the acceleration of sliding initiation is higher. The central pallet starts sliding with acceleration higher than the two lateral ones, for the same reason previously described with regard to sinusoidal cross-aisle

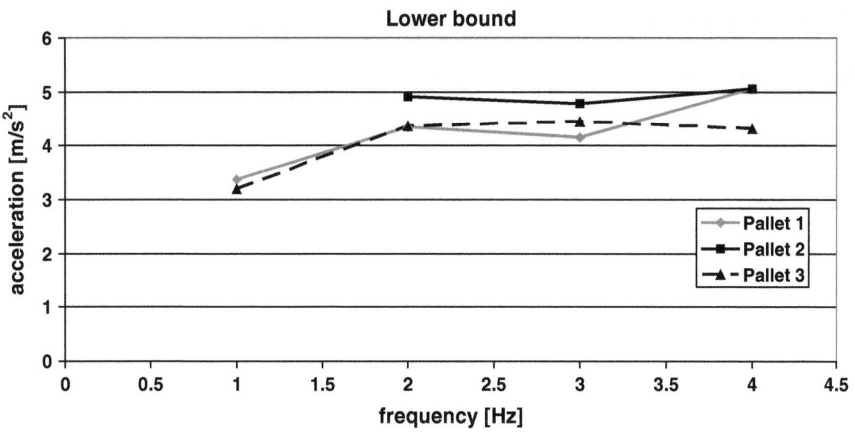

Fig. 3.128 Acceleration of sliding initiation in the down-aisle P2-B2 tests

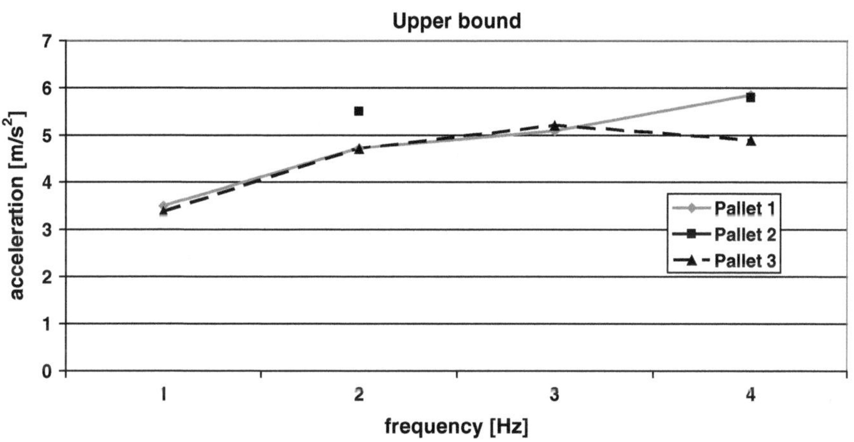

Fig. 3.129 Upper bound of sliding acceleration in the down-aisle P2-B2 tests

tests. If sliding occurs on the central pallet, its final displacement can be as large as 20 mm.

3.5.2 Seismic Test—Down-Aisle Direction

Seismic tests in down-aisle direction were carried out with beams type B2 (hot dip coated beam). Test set-up is the same shown in Figs. 3.97and 3.98.

Table 3.74 Tests with constant acceleration and increasing frequency

Test	Direction	Acceleration	Frequency [Hz]
1	Y	0.10g	1–16
2	Y	0.10g	1–16
3	Y	0.10g	1–16
4	Y	0.10g	1–16
5	Y	0.10g	1–16
6	Y	0.20g	1–16
7	Y	0.20g	1–16
8	Y	0.20g	1–16
9	Y	0.20g	1–16
10	Y	0.20g	1–16
11	Y	0.25g	1–16
12	Y	0.25g	1–16
13	Y	0.25g	1–8
14	Y	0.25g	1–8
15	Y	0.25g	1–8
16	Y	0.30g	1–8
17	Y	0.30g	1–8
18	Y	0.30g	1–8
19	Y	0.30g	1–8
20	Y	0.30g	1–8
21	Y	0.35g	2–7
22	Y	0.35g	2–7
23	Y	0.35g	2–7
24	Y	0.35g	2–7
25	Y	0.40g	2–7
26	Y	0.40g	2–7
27	Y	0.40g	2–7

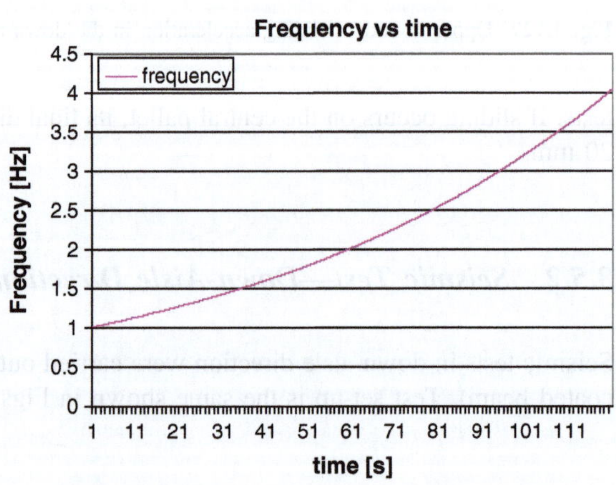

Fig. 3.130 Trend of excitation frequency in constant acceleration tests

Table 3.75 Data of the tests with constant acceleration

Test	Pallet 1			Pallet 2			Pallet 3		
	Time [s]	*f* [Hz]	Lower bound [m/s^2]	Time [s]	*f* [Hz]	Lower bound [m/s^2]	Time [s]	*f* [Hz]	Lower bound [m/s^2]
1									
2									
3									
4									
5									
6	166	6.81	5.00				168	6.96	4.75
7	150	5.66	4.40				164	6.65	4.50
8	146	5.40	4.00	169.5	7.09	5.40	166	6.81	4.85
9	150	5.66	4.50	168	6.96	5.25	161	6.42	4.25
10	150	5.66	4.10						
11	140	5.04	4.50						
12	142	5.16	4.75	158	6.20	5.25	154	5.92	5.75
13	140	5.04	4.90	160	6.35	5.50	154	5.92	5.75
14	139	4.98	4.75	159	6.28	5.50	153	5.86	5.00
15									
16	85	2.67	4.75	152	5.79	5.75	130	4.49	3.90
17	82.5	2.59	5.00						
18	83	2.61	4.75				115	3.78	4.15
19	83	2.61	4.75						
20	90	2.83	4.25	100	3.17	4.75	100	3.17	3.75
21	25	2.67	4.75				52	3.65	4.65
22	20	2.52	5.00				50	3.56	4.50
23	25	2.67	4.75	73	4.65	5.00	50	3.56	4.40
24	23	2.61	4.75				62	4.09	4.25
25	17.5	2.45	5.00	21.5	2.56	5.25	36	3.03	5.00
26	10	2.24	5.50	10	2.24	5.00	38	3.10	5.00
27	32	2.89	5.00	12	2.30	5.00	38	3.10	5.00

Only two tests were carried out, as shown in Table 3.83, test no. aa145, with Edessa time history, and test no. aa148, with Argostoli time history. Figure 3.165 and Table 3.83 show the results in terms of sliding acceleration of the various pallets. It can be noticed that sliding starts for acceleration of the pallet higher than 0.4*g*. This is in good agreement with the results of down-aisle tests carried out under sinusoidal excitation, previously presented.

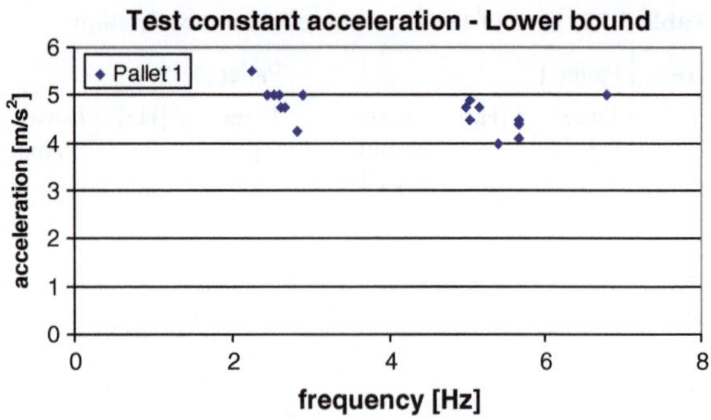

Fig. 3.131 Lower bound of sliding acceleration for pallet 1 in the tests with constant acceleration and variable frequency

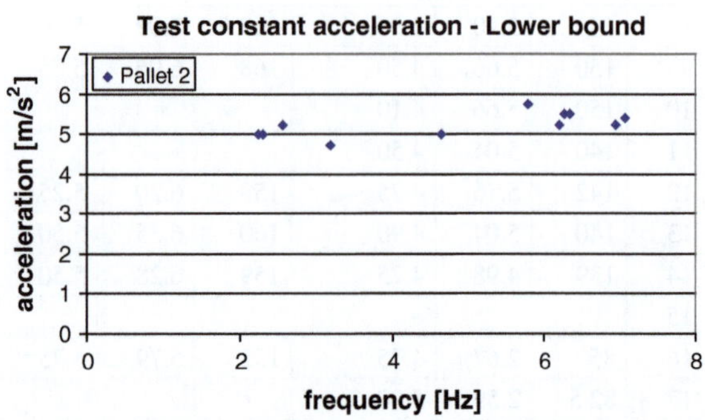

Fig. 3.132 Lower bound of sliding acceleration for pallet 2 in the tests with constant acceleration and variable frequency

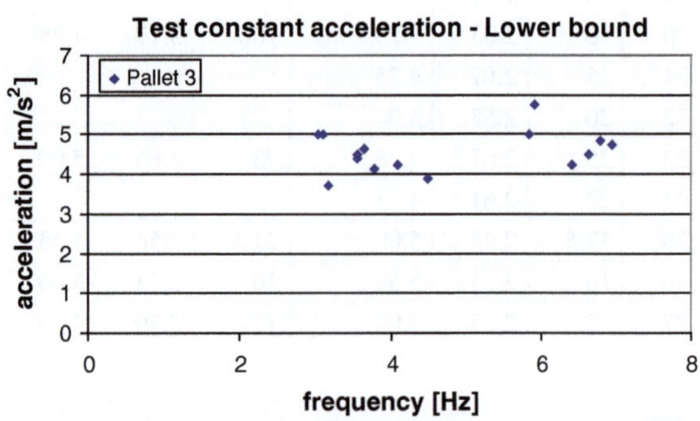

Fig. 3.133 Lower bound of sliding acceleration for pallet 3 in the tests with constant acceleration and variable frequency

Figures 3.166 and 3.167 show the sliding of the pallets in the two tests. In both tests the three pallets start sliding nearly contemporary. The final displacement of the pallets is larger in the case of Argostoli earthquake (nearly 9 mm for pallet 2)

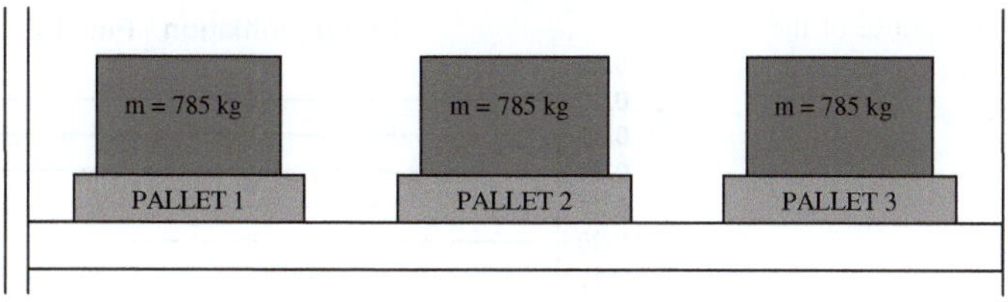

Fig. 3.134 Sketch of the analysed structure

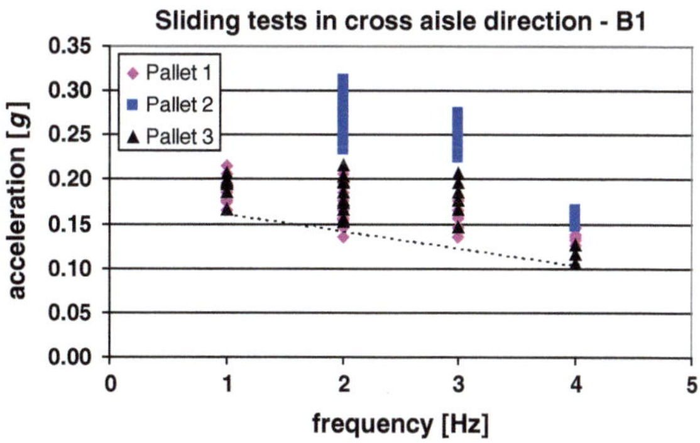

Fig. 3.135 Sliding tests in cross-aisle direction for beam type B1

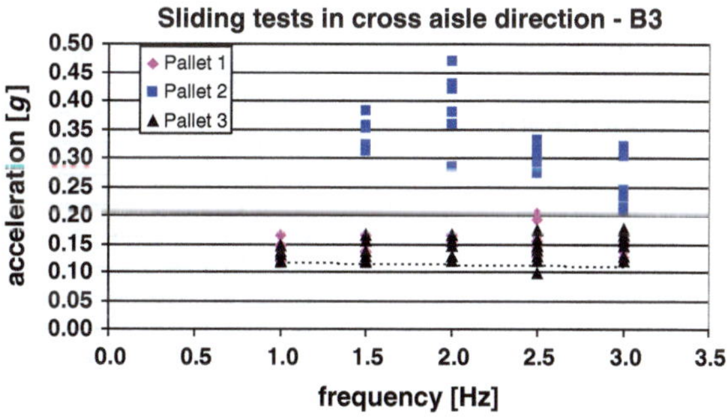

Fig. 3.136 Sliding tests in cross-aisle direction for beam type B3

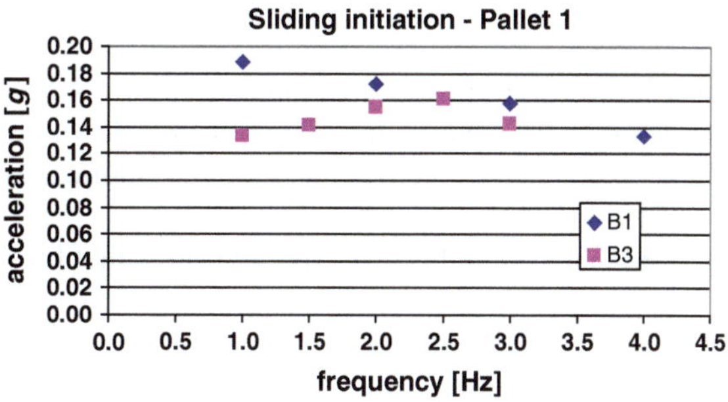

Fig. 3.137 Response of the cross-aisle tests for pallet 1 (mean values)

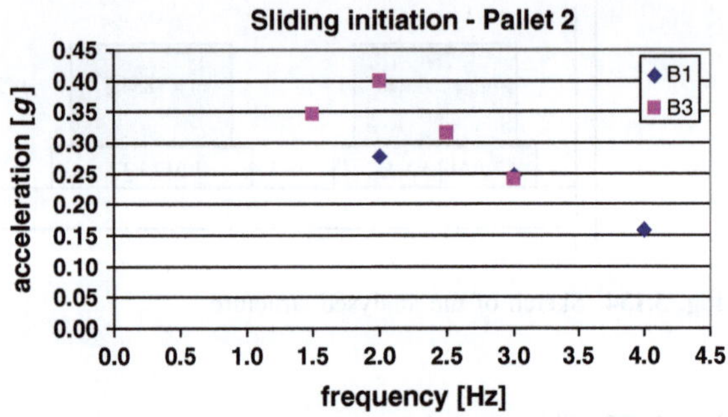

Fig. 3.138 Response of the cross-aisle tests for pallet 2 (mean values)

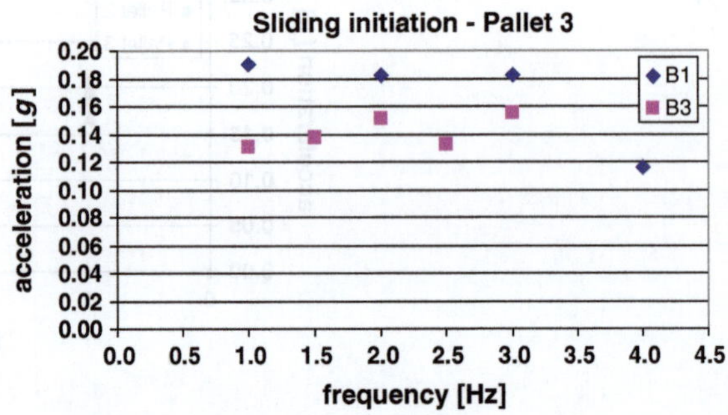

Fig. 3.139 Response of the cross-aisle tests for pallet 3 (mean values)

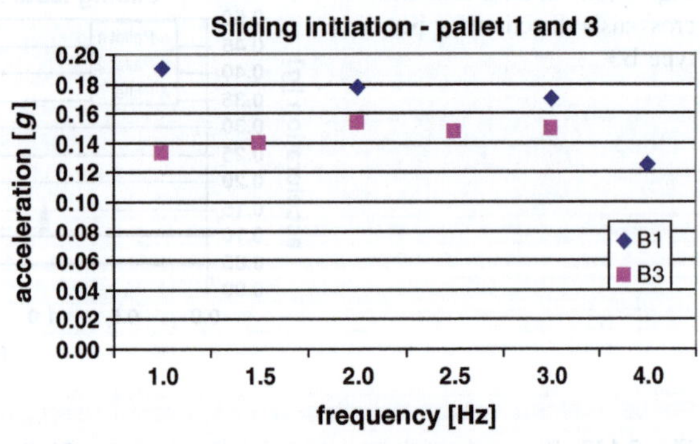

Fig. 3.140 Sliding of the lateral pallets in cross-aisle tests (mean values)

than in the case of Edessa earthquake (nearly 4.5 mm for pallet 2). The values of acceleration of sliding initiation for the down-aisle seismic tests presented in Table 3.83 can be compared with the sliding domain in down-aisle direction obtained by means of sinusoidal excitation, presented in Fig. 3.143. It can be noticed that also in this case, sliding occurs first for ARGOSTOLI signal, at around

Table 3.76 Acceleration of sliding initiation [*g*] in cross-aisle tests (mean values)

Test type		Pallet position	Frequency [Hz]		
			1.00	1.50	2.00
Cross-aisle	B1	Lateral	0.19 ± 0.01	–	0.18 ± 0.02
		Central	–	–	0.28 ± 0.02
	B3	Lateral	0.13 ± 0.01	0.14 ± 0.02	0.15 ± 0.01
		Central	–	0.35 ± 0.03	0.40 ± 0.05
Test type		Pallet position	Frequency [Hz]		
			2.50	3.00	4.00
Cross-aisle	B1	Lateral	–	0.17 ± 0.02	0.12 ± 0.02
		Central	–	0.25 ± 0.02	0.16 ± 0.01
	B3	Lateral	0.15 ± 0.03	0.15 ± 0.02	–
		Central	0.32 ± 0.02	0.24 ± 0.03	–

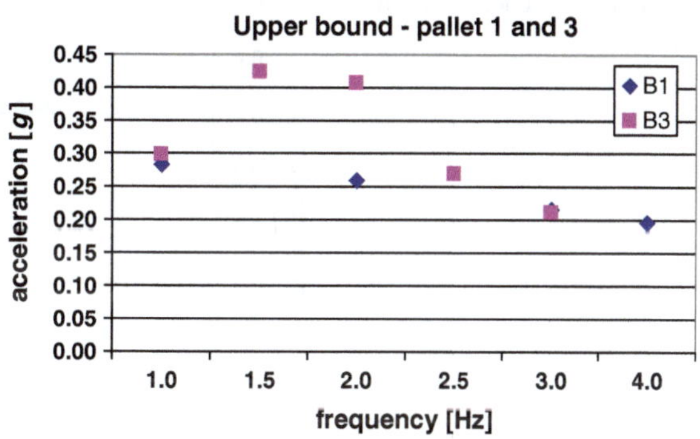

Fig. 3.141 Upper bound of sliding acceleration versus frequency in cross-aisle direction for the lateral pallets (mean values)

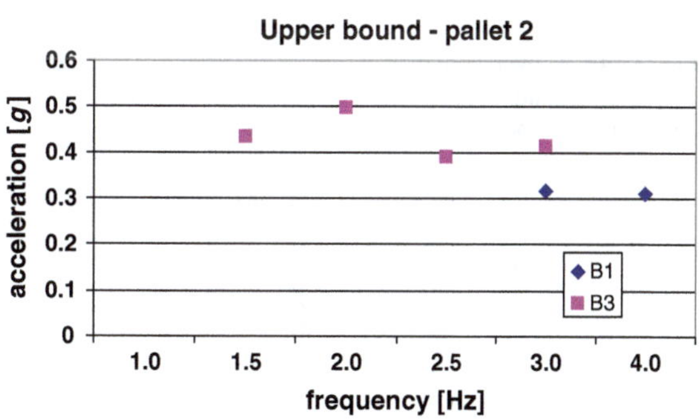

Fig. 3.142 Upper bound of sliding acceleration versus frequency in cross-aisle direction for the central pallet (mean values)

Table 3.77 Upper bound of sliding acceleration [*g*] in cross-aisle tests (mean values)

Test type		Pallet position	Frequency [Hz]		
			1.00	1.50	2.00
Cross-aisle	B1	Lateral	0.28 ± 0.02	–	0.26 ± 0.02
		Central	–	–	–
	B3	Lateral	0.30 ± 0.03	0.43 ± 0.04	0.41 ± 0.02
		Central	–	0.43 ± 0.03	0.50 ± 0.01
Test type		Pallet position	Frequency [Hz]		
			2.50	3.00	4.00
Cross-aisle	B1	Lateral	–	0.21 ± 0.02	0.20 ± 0.03
		Central	–	0.32 ± 0.02	0.31 ± 0.01
	B3	Lateral	0.27 ± 0.03	0.21 ± 0.01	–
		Central	0.39 ± 0.02	0.41 ± 0.03	–

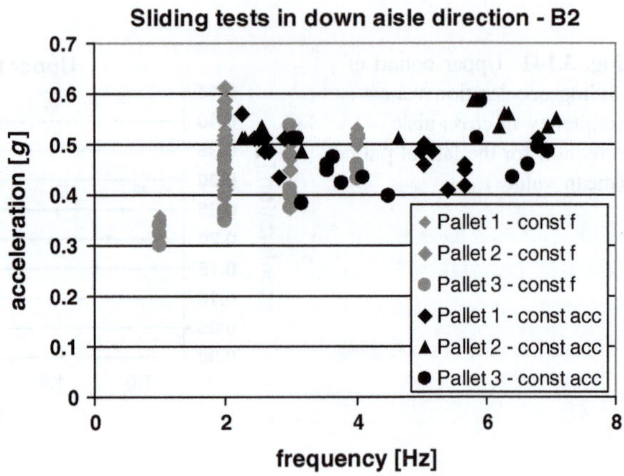

Fig. 3.143 Sliding tests in down-aisle direction for beam type B2

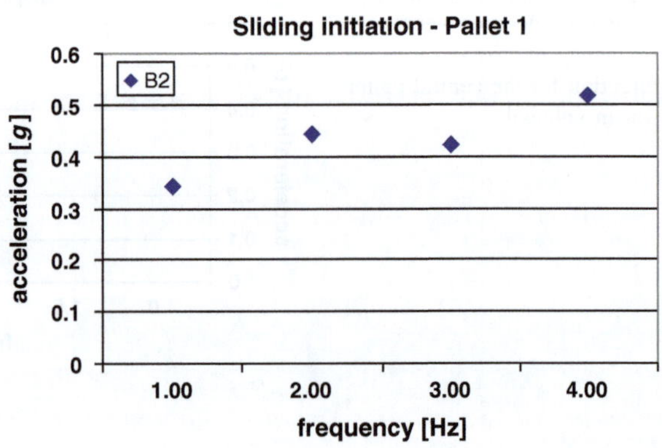

Fig. 3.144 Lower bound of sliding acceleration in down-aisle for pallet 1 (mean values)

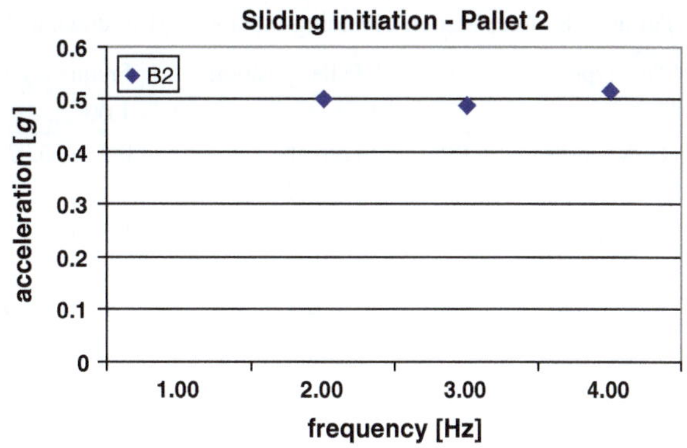

Fig. 3.145 Lower bound of sliding acceleration in down-aisle for pallet 2 (mean values)

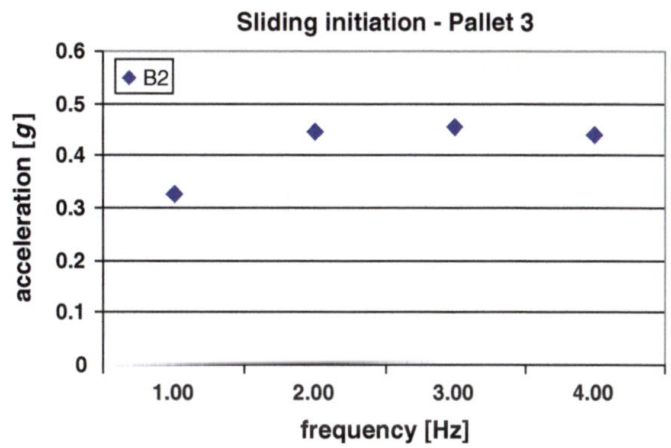

Fig. 3.146 Lower bound of sliding acceleration in down-aisle for pallet 3 (mean values)

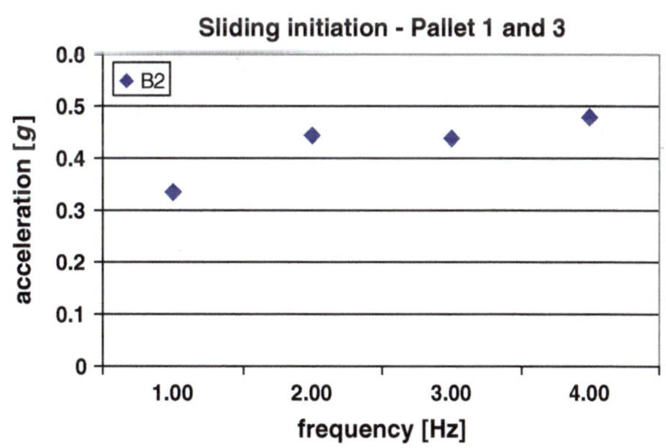

Fig. 3.147 Lower bound of sliding acceleration in down-aisle for lateral pallets (mean values)

Table 3.78 Acceleration of sliding initiation [g] in down-aisle tests (mean values)

Test type		Pallet position	Frequency [Hz]		
			1.00	1.50	2.00
Down-aisle	B2	Lateral	0.33 ± 0.02	–	0.44 ± 0.07
		Central	–	–	0.50 ± 0.07
Test type		Pallet position	Frequency [Hz]		
			2.50	3.00	4.00
Down-aisle	B2	Lateral	–	0.44 ± 0.06	0.48 ± 0.04
		Central	–	0.49 ± 0.04	0.52 ± 0.00

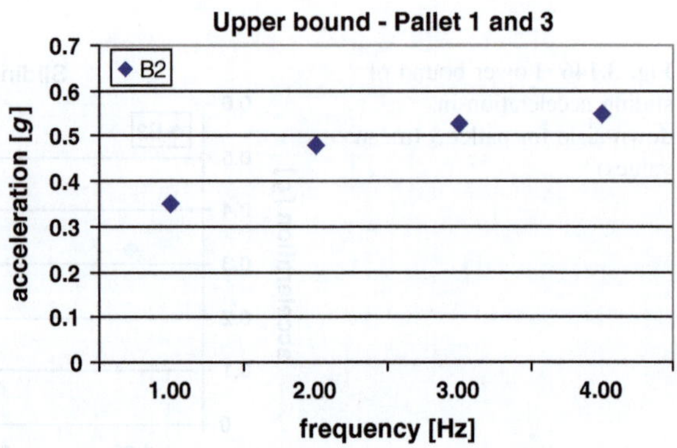

Fig. 3.148 Upper bound of sliding acceleration versus frequency in down-aisle direction for the lateral pallets (mean values)

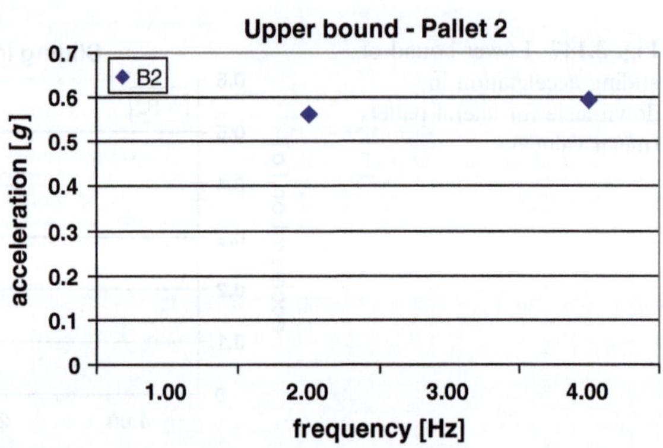

Fig. 3.149 Upper bound of sliding acceleration versus frequency in down-aisle direction for the central pallet (mean values)

Table 3.79 Upper bound of sliding acceleration [*g*] in down-aisle tests (mean values)

Test type		Pallet position	Frequency [Hz]					
			1.00	1.50	2.00	2.50	3.00	4.00
Down-aisle	B2	Lateral	0.35 ± 0.02	–	0.48 ± 0.07	–	0.52 ± 0.05	0.53 ± 0.05
		Central	–	–	0.56 ± 0.07	–	–	–

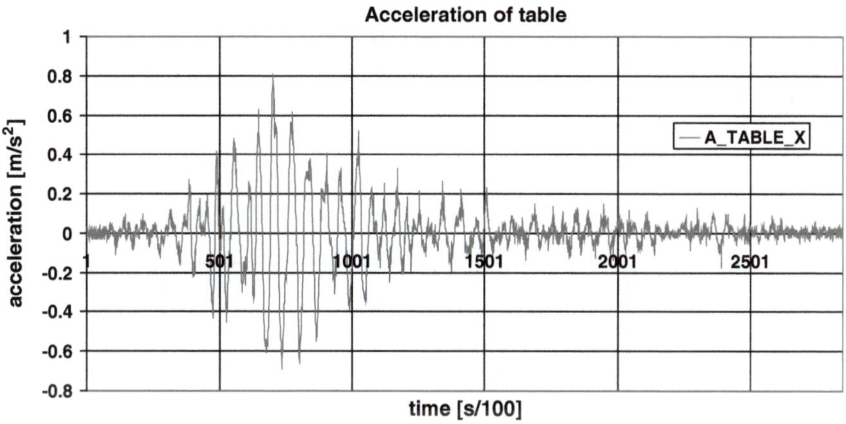

Fig. 3.150 Acceleration of the shaking table in EDESSA test no. aa89

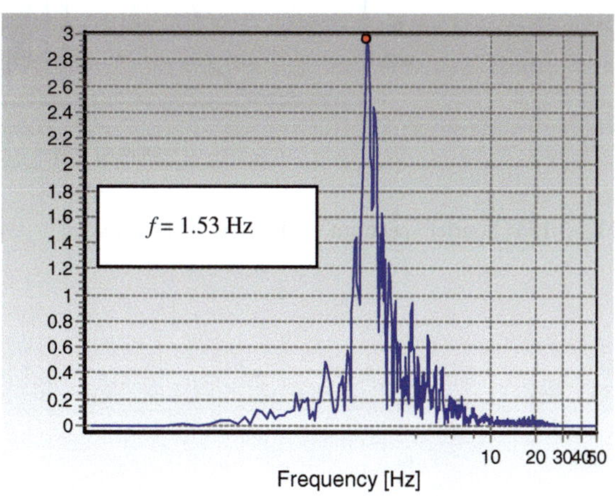

Fig. 3.151 Fourier spectrum of the EDESSA signal—test no. aa96

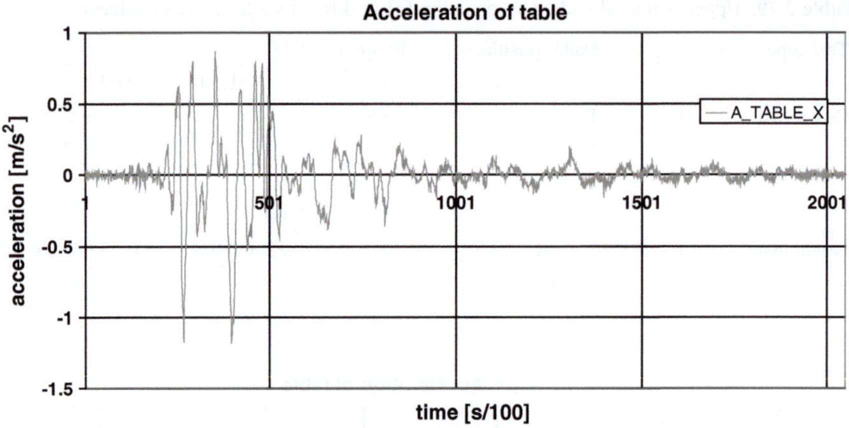

Fig. 3.152 Acceleration of the shaking table in KALAMATA test no. aa98

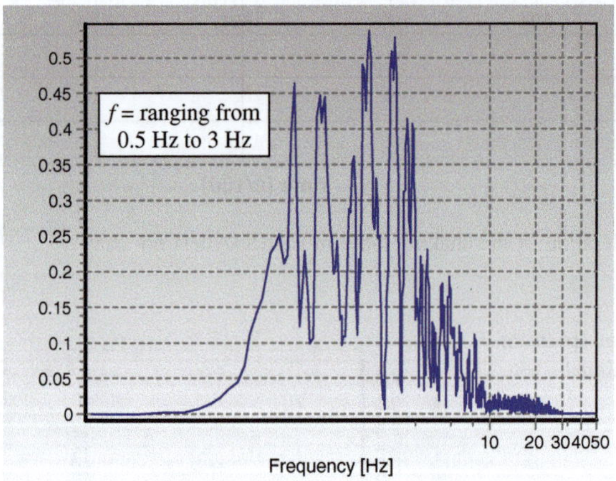

Fig. 3.153 Fourier spectrum of the KALAMATA signal—test no. aa98

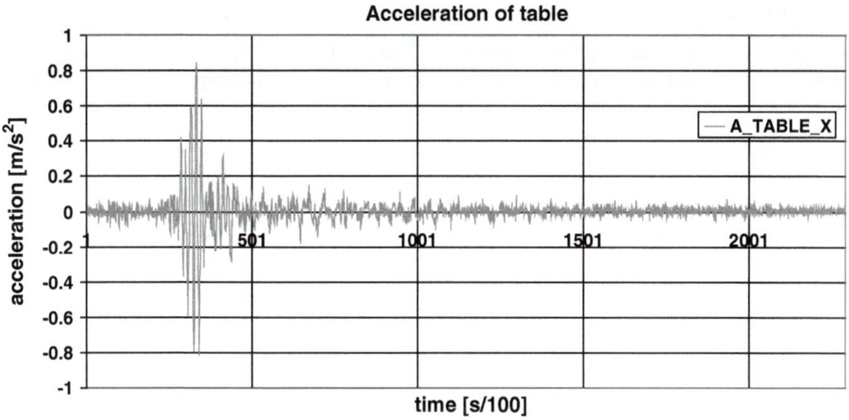

Fig. 3.154 Acceleration of the shaking table in ARGOSTOLI test no. aa102

Fig. 3.155 Fourier spectrum of the ARGOSTOLI signal—test no. aa105

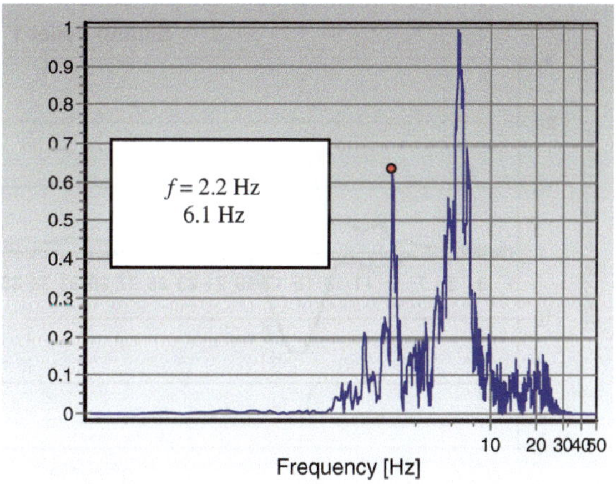

Table 3.80 Earthquake data

Event	Date	Ms	Δ [km]
Edessa	21 December 1990	5.9	31
Kalamata	13 September 1986	6.2	9
Argostoli	24 Marhc 1983	5.5	10

Table 3.81 Input motion data

Beam type	Test	Type of motion	PGA cross aisle [m/s^2]	PGA down aisle [m/s^2]	Beam type	Test	Type of motion	PGA Cross aisle [m/s^2]	PGA Down aisle [m/s^2]
B1	aa89	edessa	0.8	–	B1	aa100	kalamata	1.3	–
	aa90	edessa	1.0	–		aa101	kalamata	1.5	–
	aa91	edessa	1.2	–		aa102	argostoli	0.8	–
	aa92	edessa	1.4	–		aa103	argostoli	1.9	–
	aa93	edessa	1.7	–		aa104	argostoli	2.9	–
	aa94	edessa	1.9	–		aa105	argostoli	4.0	–
	aa95	edessa	2.1	–	B2	aa145	edessa	–	5.7
	aa96	edessa	2.3	–		aa146	edessa	2.0	2.6
	aa97	edessa	2.5	–		aa147	edessa	2.5	3.2
	aa98	kalamata	0.8	–		aa148	argostoli	–	7.8
	aa99	kalamata	1.0	–		aa149	argostoli	7.1	6.0

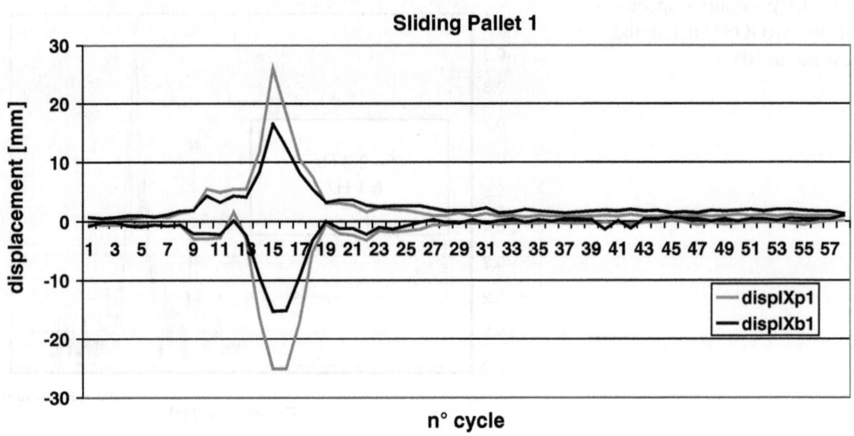

Fig. 3.156 Displacement of beam and pallet 1 in EDESSA seismic test no. aa96

0.45*g*. For the other types of seismic input motions the acceleration of sliding initiation is higher. The central pallet starts sliding more or less at the same value of acceleration of the two lateral ones, as previously described with regard to sinusoidal down-aisle tests.

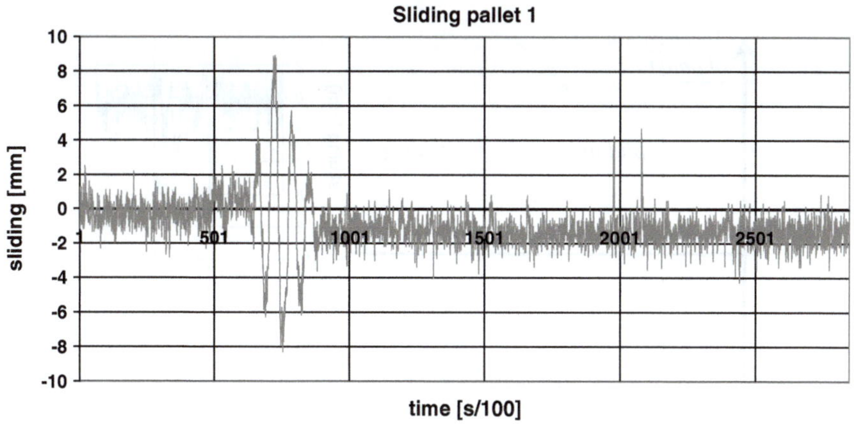

Fig. 3.157 Sliding of pallet 1 in EDESSA seismic test no. aa96

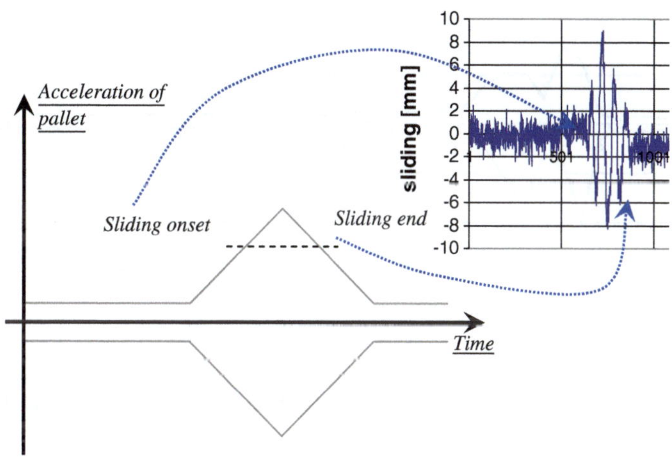

Fig. 3.158 Sketch of the analysis of the data in the EDESSA test

3.5.3 Bidirectional Seismic Test

Bidirectional seismic tests were carried out with Edessa and Argostoli time history, as shown in Table 3.84.

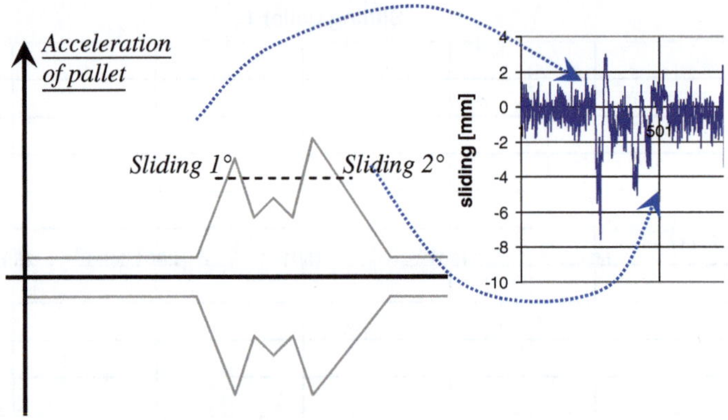

Fig. 3.159 Sketch of the analysis of the data in the KALAMATA test

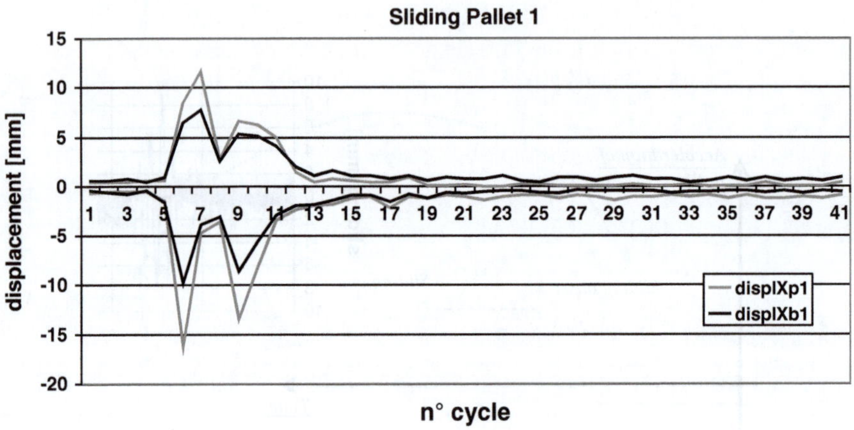

Fig. 3.160 Displacement of beam and pallet 1 in KALAMATA seismic test no. aa101

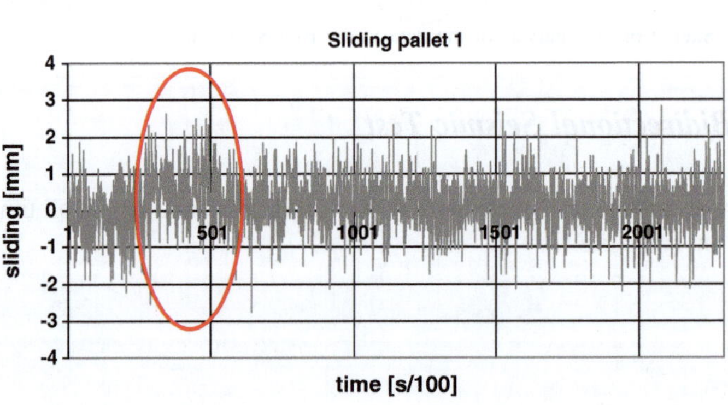

Fig. 3.161 Sliding of pallet 1 in ARGOSTOLI test no. aa104

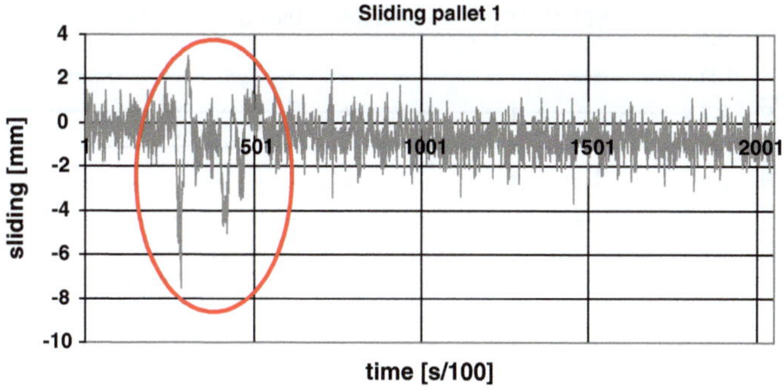

Fig. 3.162 Sliding of pallet 1 in KALAMATA test no. aa101

Table 3.82 Sliding acceleration for cross-aisle seismic tests

Test		Pallet 1		Pallet 2		Pallet 3	
		a [*g*]	Sliding	a [*g*]	Sliding	a [*g*]	Sliding
EDESSA	aa89	0.16	No	0.15	No	0.11	No
	aa90	0.16	No	0.20	No	0.15	No
	aa91	0.17	No	0.20	No	0.18	No
	aa92	0.21	No	0.25	No	0.19	No
	aa93	0.20	Yes	0.24	Yes	0.20	Yes
	aa94	0.25	Yes	0.27	Yes	0.21	Yes
	aa95	0.22	Yes	0.26	Yes	0.27	Yes
	aa96 start	0.22	Yes	0.22	Yes	0.20	Yes
	aa96 end	0.20	Yes	0.23	Yes	0.19	Yes
	aa97 start	0.27	Yes	0.30	Yes	0.25	Yes
	aa97 end	0.20	Yes	0.31	Yes	0.25	Yes
KALAMATA	aa98	0.17	No	0.20	No	0.19	No
	aa99	0.22	Yes	0.26	Yes	0.26	Yes
	aa100/1°	0.31	Yes	0.32	Yes	0.30	Yes
	aa100/2°	0.28	Yes	0.30	Yes	0.27	Yes
	aa101/1°	0.35	Yes	0.36	Yes	0.30	Yes
	aa101/2°	0.28	Yes	0.31	Yes	0.29	Yes
ARGOSTOLI	aa102	0.07	No	0.13	No	0.10	No
	aa103	0.14	No	0.20	No	0.11	No
	aa104	0.15	Yes	0.25	No	0.14	No
	aa105	0.19	Yes	0.30	Yes	0.17	Yes

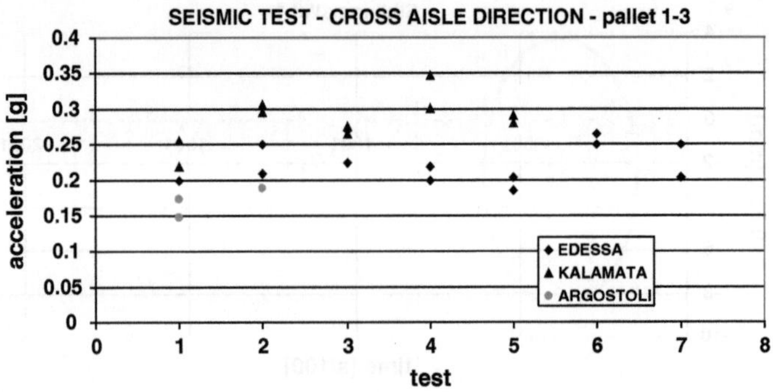

Fig. 3.163 Acceleration of the lateral pallets when sliding occurs in cross-aisle seismic tests with the beam type B1

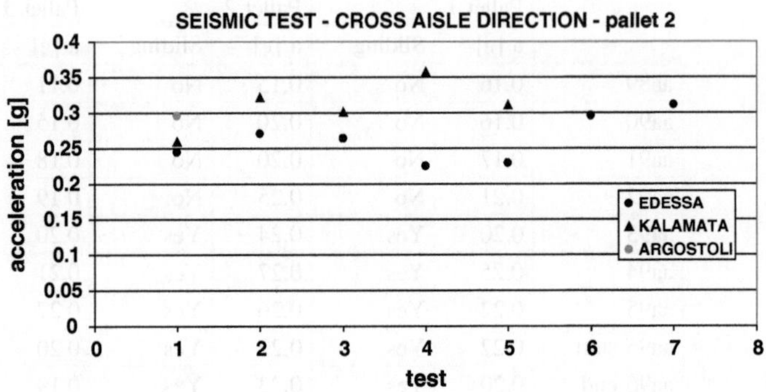

Fig. 3.164 Acceleration of the central pallet when sliding occurs in cross-aisle seismic tests with the beam type B1

Table 3.83 Sliding acceleration for down-aisle seismic tests

Test		Pallet 1		Pallet 2		Pallet 3	
		a [*g*]	Sliding	a [*g*]	Sliding	a [*g*]	Sliding
EDESSA	aa145	0.51	Yes	0.52	Yes	0.53	Yes
ARGOSTOLI	aa148	0.61	Yes	0.45	Yes	0.59	Yes

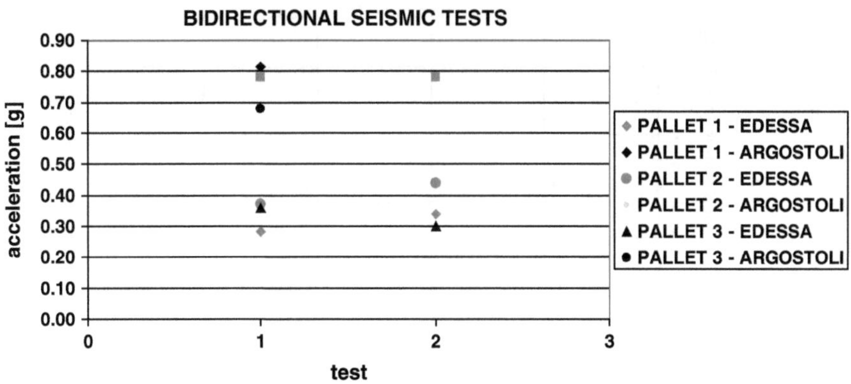

Fig. 3.165 Lower bound of sliding acceleration in down-aisle seismic tests with beam type B2

Table 3.85 shows the cross-aisle (a_x) and down-aisle (a_y) components of the sliding acceleration of the three pallets. In all tests sliding occurs with acceleration in cross-aisle direction higher than that obtained as "lower bound" in the cross-aisle sinusoidal tests with constant frequency excitation, considering the beam type B1 (that develops a static friction factor very similar to that of beam type B2, as shown while discussing the static tests). In down-aisle direction, on the contrary, only in test no. aa149 the acceleration of sliding initiation exceeds the value of the lower bound acceleration in down-aisle direction for the beam type B2.

Figure 3.168 shows the values of the sliding initiation acceleration of the three pallets obtained by vectorial composition of a_x and a_y. These values should be compared with a sliding domain obtained by sinusoidal tests with a bidirectional excitation of the table. Such a domain was not available, because only mono-directional sinusoidal tests were carried out. Hence it was derived by vectorial composition of the two lower bounds and of the two upper bounds of the sliding acceleration (respectively in cross and down-aisle direction), obtained by sinusoidal excitation. The resulting lower and upper bound values are plotted in Fig. 3.168 with dotted lines. It can be noticed that the results of the bi-directional seismic tests are fully compatible with the derived "bi-directional sliding domain".

As can be noticed in Fig. 3.169 that shows the maximum and minimum peaks of acceleration of pallet 1 and of the two beams in test no. aa149, the sliding initiation acceleration of the pallet remains lower than those of the beams, confirming the validity of the previously defined upper bound value of the sliding acceleration.

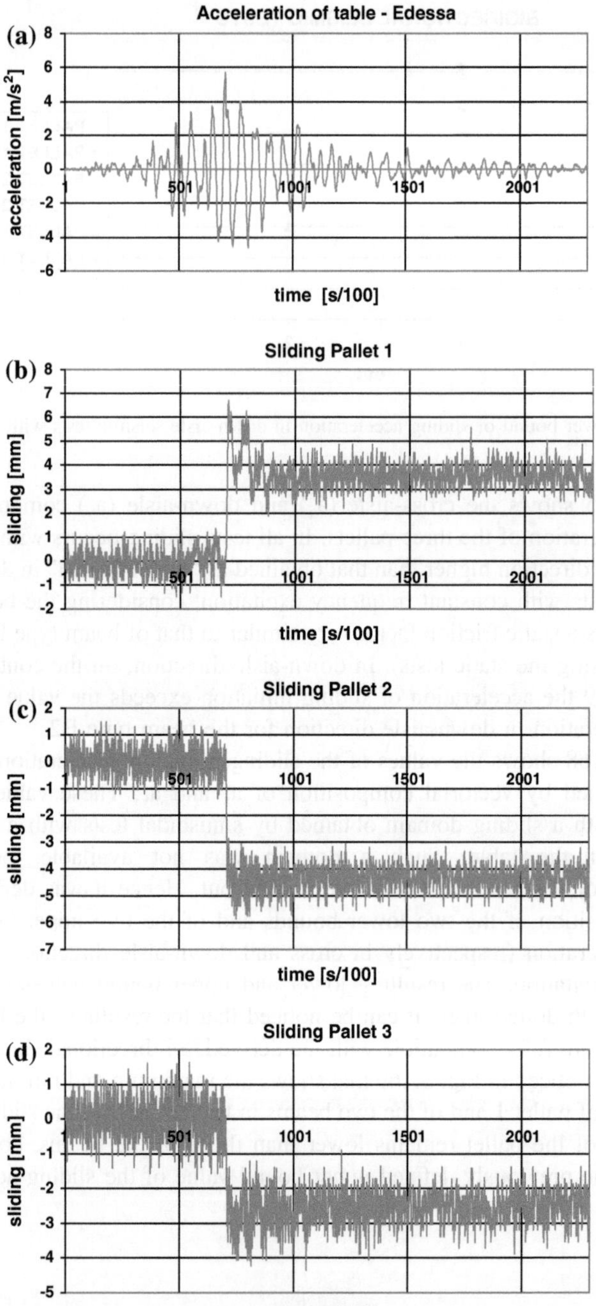

Fig. 3.166 Sliding of the pallets in the EDESSA seismic tests no. aa145. **a** Input motion, **b**, **c**, **d** sliding of pallets 1, 2 and 3 respectively

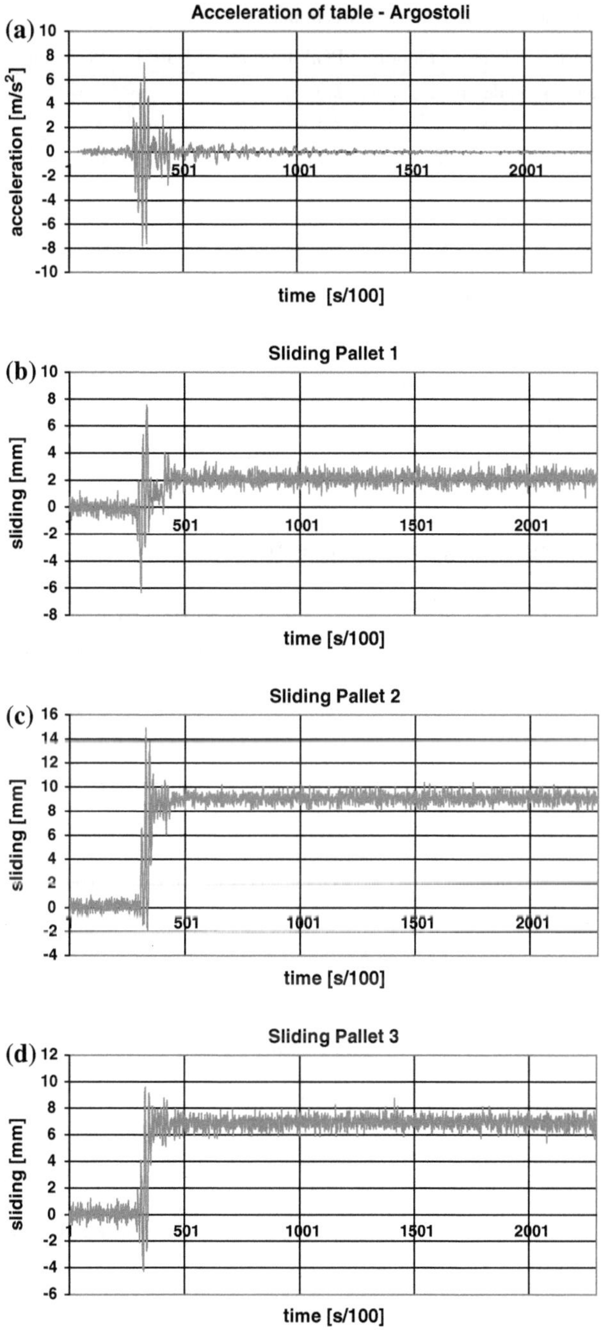

Fig. 3.167 Sliding of the pallets in the ARGOSTOLI seismic tests no. aa148. **a** Input motion, **b**, **c**, **d** sliding of pallets 1,2 and 3 respectively

Table 3.84 Input motion data for the bidirectional tests

Beam type	Test	Type of motion	PGA transversal [m/s²]	PGA longitudinal [m/s²]
B2	aa146	edessa	2.0	2.6
	aa147	edessa	2.5	3.2
	aa149	argostoli	7.1	6.0

Table 3.85 Sliding acceleration for bidirectional seismic tests—P2-B2 type

Test		Pallet 1			Pallet 2			Pallet 3		
		a_X [*g*]	a_Y [*g*]	Sliding	a_X [*g*]	a_Y [*g*]	Sliding	a_X [*g*]	a_Y [*g*]	Sliding
EDESSA	aa146	0.19	0.21	Yes	0.24	0.28	No	0.24	0.27	No
	aa147	0.21	0.27	Yes	0.27	0.34	Yes	0.19	0.23	Yes
ARGOSTOLI	aa149	0.63	0.52	Yes	0.60	0.50	Yes	0.52	0.43	Yes

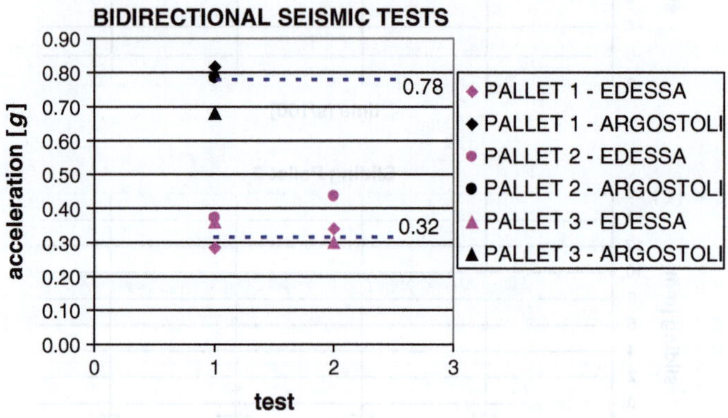

Fig. 3.168 Sliding initiation acceleration in bidirectional seismic tests—type P2-B2

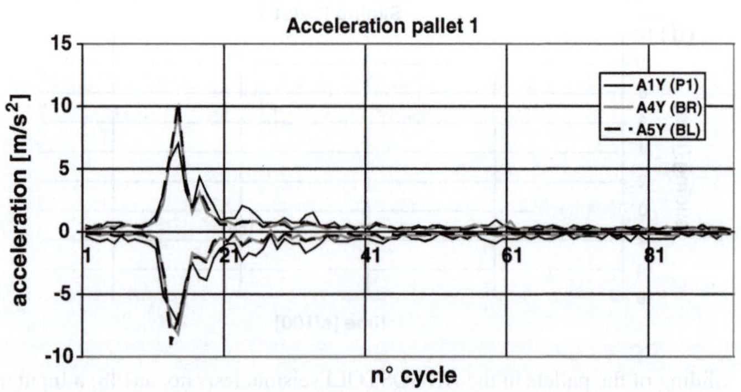

Fig. 3.169 Acceleration of the two beams and of pallet 1 in test no. aa149

3.6 Conclusions

Assessment of both the static and the dynamic sliding conditions of pallets stored on steel racking systems was carried out within the SEISRACKS research project, by means of static as well as dynamic tests performed at the Earthquake Engineering Laboratory of the National Technical University of Athens.

Static tests were carried out in both down and cross-aisle direction, by means of an "inclined plane" device, by slowly increasing the inclination of the plane, and measuring the sliding of the pallet on the rack steel beams.

Influence of the following parameters was investigated:

- Type of beam (namely type of surface finish of the beam)
- Type of pallet (namely geometry and wear conditions)
- Geometry and weight of mass resting on the pallet

Influence of the type of beam was investigated by adopting six different types of beam specimens, manufactured by three different producers, with different types of surface finish. In particular, hot zinc, hot dip and powder coated steel beams were considered.

In both cross and down-aisle direction, the surface finish influenced very much the static friction factor, with differences as large as 20–30 % from one type to the other, in the case of wooden pallets.

Influence of the type of pallet was investigated by adopting three different types of pallets, namely: wooden Euro pallets, wooden American-pallet and plastic Euro pallet. In both cross and down-aisle direction the plastic Euro pallet showed a very low friction factor (of the order of 0.2), practically being non-influenced by the type of beam surface finish. The wooden pallets show a very similar friction factor (of the order of 0.5), and are similarly influenced by the beam surface finish.

In both cross and down-aisle direction, the mass weight didn't affect much the results. However, its geometry (height of the c.o.g.) and its "placement" on the pallet (centered or eccentric) resulted in small variations of the measured friction factor.

Dynamic tests were carried out on the shaking table facility of the National Technical University of Athens, on a simplified set-up, made of two uprights, connected by two horizontal beams, at approximately 0.30 m from the shaking table. Three wooden Euro pallets were positioned on the beams, with concrete blocks rigidly fixed on top. Most tests were carried out with a sinusoidal excitation, and with constant frequency and increasing acceleration. Some tests, in down-aisle direction only, were carried out with sinusoidal excitation, with constant acceleration and increasing frequency, in order to verify independence of the obtained results on the type of adopted excitation. A lower bound of the acceleration exists, beyond which pallets start sliding on the steel beams. When acceleration of the mass is lower than such "lower bound", the pallet "sticks" on the beams, and no sliding occurs. When the "lower bound" of acceleration is exceeded, increasing the acceleration of the input motion results in a lower increment in the mass acceleration, until an "upper bound" is reached of the mass acceleration. Any further increase in the acceleration of the input motion doesn't affect the acceleration of the mass that is "free" to slide on the beams. "Stiction" between pallet and beam is not resumed until a reduction of the acceleration occurs. The "upper bound" of the sliding acceleration is, in general, lower than the static friction factor.

In both cross and down-aisle direction lateral pallets slide systematically earlier than the central one.

Dynamic behaviour in cross-aisle direction is completely different to the one in down-aisle direction.

In cross-aisle direction, the flexural stiffness of the beams in the horizontal plane as well as their torsional stiffness influence very much the results. In particular, such stiffnesses are affected by the out-of-plane and torsional behaviour of the beam-to-upright connections, whose stiffness rapidly deteriorates under cycling. Test results show, in general, a dependence of the sliding acceleration on the frequency of the input motion. Both the lower and the upper bound of the sliding acceleration seem to decrease when increasing the frequency of the excitation.

Lower bound sliding acceleration as low as 0.1g was measured, for wooden pallets on hot dip coated steel beams. Upper bound values of the acceleration ranging from 0.3g to 0.5g were measured depending on the type of beam surface finish as well as on the position of the pallet (laterals or central one).

In down-aisle direction, the sliding acceleration is in general higher than the one measured in cross-aisle direction, under the same testing conditions, with a lower bound of the measured sliding acceleration of nearly 0.3g, and an upper bound of nearly 0.6g. Also in down-aisle direction, test results show, in general, a dependence of the sliding acceleration on the frequency of the input motion. However, in this case, both the lower and the upper bound of the sliding acceleration seem to increase when increasing the frequency of the excitation.

Results of tests carried out with constant acceleration and increasing frequency are fully compatible with those obtained in tests with constant frequency and increasing acceleration.

A few *seismic tests* were carried out, adopting three different input motions recorder in Greece during recent earthquakes, and characterized by different durations and frequency contents. Both mono-directional and bi-directional tests were carried out. The obtained results were compared with those of tests carried out with a sinusoidal excitation, showing full compatibility. Measured sliding accelerations range from 0.15g to 0.35g in cross-aisle direction and from 0.45g to 0.6g in the down-aisle direction. Similar compatibility was also obtained for bi-directional tests, when comparing the resultants of the vector-compositions of the components of the sliding accelerations in the two orthogonal directions.

References

Armstrong, B., Dupont, P., & Canudas de Wit, C. (1994). A survey of models, analysis tools and compensation methods for the control of machines with friction. *Automatica, 30*(7), 1083–1138.

Bliman P. A., & Sorine M. (1991). Friction modelling by hysteresis operators. Application to Dahl, sticktion and Stribeck effects. In *Proceedings of the Conference "Models of Hysteresis", Trento, Italy.*

Bliman P. A., & Sorine M. (1993). A system-theoretic approach of systems with hysteresis. Application to friction modelling and compensation. In *Proceedings of the second European Control Conference, Groningen, TheNetherlands* (pp. 1844–49).

Bliman P. A., & Sorine M. (1995). Easy-to-use realistic dry friction models for automatic control. In *Proceedings of the 3rd European. Control Conference* (pp. 3788–3794). ECC'95, Rome, Italy.

Canudas de Wit, C., Olsson, H., Åström, K. J., & Lischinsky, P. (1995). A new model for control of systems with friction. *IEEE Transaction on Automatic Control, 40*(3), 419–425.

Coulomb C. A. (1776). Essai sur une application des regles de maximis et minimis à quelques problemes relatifs à l'architecture, Memoires de Mathèmatique et de Physique (Vol. 7, pp. 343–382). Academie Royale des Sciences, Paris.

Dahl P. R. (1968). *A Solid Friction Model*, The Aerospace Corporation, El Segundo, CA, Technical Report, TOR-0158(3107-18).

Dahl P. R. (1975). Solid friction damping of spacecraft oscillations. In *AIAA Guidance and Control Conference*. AIAA Paper No. 75-1104, Boston, USA.

Dahl, P. R. (1976). Solid friction damping of mechanical vibrations. *AIAA Journal, 14*(12), 1675–1682.

Degée, H., & Denoël, V. (2007). Numerical modelling of storage racks subjected to earthquake. In M. Papadrakakis, D.C. Charmpis, N.D. Lagaros & Y. Tsompanakis (Eds.), *Proceedings of ECCOMAS Thematic Conference on Computational Methods in Structural Dynamics and Earthquake Engineering (COMPDYN)*. Crete, paper 1089, 11 pp.

Denoël, V., & Degée, H. (2005). Cas particulier d'étude analytique de l'élément à frottement, Internal report 2005-1, Department M&S, University of Liege.

Karnopp, D. (1985). Computer simulation of stick-slip friction in mechanical dynamic systems. *Journal of Dynamic systems, Measurement and Control, 107*(1), 100–103.

Morin A. J. (1833). New friction experiments carried out at Metz in 1831–1833. In Proceedings of the French Royal Academy of Sciences (Vol. 4, pp. 1–128).

Olsson, H., Aström, K. J., Canudas de Wit, C., Gäfvert, M., & Lischinsky, P. (1998). Friction models and friction compensation. *European Journal of Control, 4*(12), 176–195.

Rabinowicz, E. (1951). The nature of the static and kinetic coefficients of friction. *Journal of Applied Physics, 22*(11), 1373–1379.

Stribeck R. (1902). The key qualities of sliding and roller bearings, Zeitschrifts des Vereines Deutscher Ingenieure, *46*(38, 39), 1342–1348, 1432–1437.

Yang F. (1996). Vibrations of cable-stayed bridges under moving vehicles (Ph.D. thesis, University of Liege).

Further Readings

Armstrong B. (1988). Friction: Experimental determination, modelling and compensation, *IEEE International Conference on Robotics and Automation*, *3*(1), 422–1427.

Barahanov, N., & Ortega, R. (2000). Necessary and sufficient conditions for passivity of the LuGre friction model. *IEEE Transactions on Automatic Control, 45*(4), 830–832.

Baril C. B. (1993). Control of Mechanical Systems Affected by Friction and Other Non-differentiable Nonlinearities (Ph.D. thesis, Technion, Israel Institute of Technology, Haifa, Israel).

Bliman, P. A. (1992). Mathematical study of the Dahl's friction model. *European Journal of Mechanics. A/Solids, 11*(6), 835–848.

Bowden, F., & Tabor, D. (1956). *Friction and lubrication*. New York: Wiley.

Canudas de Wit C., Olsson H., Åström K. J., & Lischinsky P. (1993). Dynamic friction models and control design. In *Proceedings of the 1993 American Control Conference* (pp. 1920–1926). San Francisco, California.

Caundas de Wit C., & Lischinsky P. (1996). Adaptive friction compensation with dynamic friction model. In *Proceedings of 13'th IFAC World Congress* (pp. 197–202). San Francico.

Canudas de Wit C., & Ge S. S. (1997). Adaptive friction compensation for systems with generalized velocity/position friction dependency. In *Conference on Decision and Control*. San Diego, CA, USA.

Canudas de Wit, C., & Lischinsky, P. (1997). Adaptive friction compensation with partially known dynamic friction model. *International Journal of Adaptive Control and Signal Processing, Special Issue on Adaptive Systems with Nonsmooth Nonlinearities, 11*(1), 65–80.

Caundas de Wit, C. (1998). Comments on "a new model for control of systems with friction". *IEEE Transaction on Automatic Control, 43*(8), 1189–1190.

Courtney-Pratt J., & Eisner E. (1957). The effect of a tangential force on the contact of metallic bodies. In *Proceedings of the Royal Society* (Vol. A238, pp. 529–550).

Dohring, M. E., Eunjeong, L., & Newman, W. S. (1993). Load-dependent transmission friction model: Theory and experiments. In *IEEE International Conference on Robotics and Automation* (pp. 430–436).

Dupont, P., Hayward, V., Armstrong, B., & Altpeter, F. (2002). Single state elastoplastic friction models. *IEEE Transactions Automatic Control, 47*, 683–687.

Ferretti G., Magnani G., Martucci G., Rocco P., & Stampacchia V. (2003). Friction model validation in sliding and presliding regimes with high resolution encoders. In B. Siciliano & P. Dario (Eds.), *Experimental Robotics VIII* (pp. 328–337). New York: Springer.

Ferretti, G., Magnani, G., & Rocco, P. (2004). Single and multistate integral friction models. *IEEE Transactions Automatic Control, 49*, 2292–2297.

FineLg User's manual, V9.2. (2003). Greisch Info, department ArGEnCo, University of Liege.

Friedland B., & Park Y. J. (1991). On adaptive friction compensation. In *Proceedings of the IEEE Conference on Decision and Control* (pp. 2899–2902).

Friedland, B. (1992). On adaptive friction compensation. *IEEE Transaction on Automatic Control, 37*(10), 1609–1612.

Gäfvert M. (1996). Comparison of two friction models. Master's thesis, Lund Institute of Technology.

Hägglund T. (1997). Stiction compensation in control valves. In *European Control Conference*. Brussels, Belgium.

Haessig, D. A., & Friedland, B. (1991). On the modelling and simulation of friction. *ASME Transactions, Journal of Dynamic Systems, Measurement, and Control, 113*(3), 354–362.

Hess, D. P., & Soom, A. (1990). Friction at a lubricated line contact operating at oscillating sliding velocities. *Journal of Tribology, 112*, 147–152.

Johannes, V. I., Green, M. A., & Brockley, C. A. (1973). The role of the rate of application of the tangential force in determining the static friction coefficient. *Wear, 24*, 381–385.

Krasnoselskij, M. A., & Pokrovskij, A. V. (1980). *Systems with hysteresis*. New York: Springer.

Lampaert, V., Swevers, J., & Al-Bender, F. (2002). Modification of the Leuven integrated friction model structure. *IEEE Transactions on Automatic Control, 47*(4), 683–687.

Lazan, B. J. (1968). *Damping of materials and members in structural mechanics*. London: Pergamom Press.

Lee, S. W., & Kim, Y. H. (1995). Robust adaptive stick slip compensation. *IEEE Transactions on Industrial Electronics, 42*(5), 474–479.

Leonard N. E., & Krishnaprasad P. (1992a). Adaptive friction compensation for bidirectional low-velocity position tracking. In *Proceedings of the 31st Conference on Decision and Control* (pp. 267–273).

Leonard, N. E., & Krishnaprasad, P. S. (1992b) Adaptive friction compensation for bidirectional low-velocity position tracking. In *Proceedings of 30th IEEE Conference on Decision and Control* (pp. 267–273). New York.

Liao, T. L., & Chien, T. I. (2000). An exponentially stable adaptive friction compensator. *IEEE Transactions on Automatic Control, 45*(5), 977–980.

Popovic, M. R., & Goldenberg, A. A. (1998). Modelling of friction using spectral analysis. *IEEE Transactions on Robotics Automation, 14*(1), 114–122.

Rabinowicz, E. (1995). *Friction and wear of materials* (2nd ed.). New York: Wiley.

Ramberg W., & Osgood W. R. (1943). Description of stress-strain curves by three parameters. Technical Note 902, National Advisory Committee for Aeronautics, Washington.

Suh, N. P., & Sin, H. C. (1981). The genesis of friction. *Wear, 69*(1), 91–114.

Swevers, J., Al-Bender, F., Ganseman, C., & Prajogo, T. (2000). An integrated friction model structure with improved presliding behaviour for accurate friction compensation. *IEEE Transactions on Automatic Control, 45*(4), 675–686.

Walrath, C. (1984). Adaptive bearing friction compensation based on recent knowledge of dynamic friction. *Automatica, 20*(6), 717–727.

Chapter 4
Pushover Tests

4.1 Overview

Two push-over tests (one in down-aisle, the other in cross-aisle direction) were carried out, in order to evaluate the possibility to propose in a Standard Design Code for Racks in Seismic Areas, static push-over analyses (currently available in many commercially available software packages for structural analysis) as alternative to dynamic (linear or non-linear) analyses, as it seems to be the current trend in many codes for seismic design of building structures.

Tests were carried out at the European Laboratory for Structural Assessment of the Joint Research Center of Ispra, on three levels, two bays full scale specimens, 6.0 m high and 3.6 m large, with 130 × 45 × 1.5 mm beams and 100/20b uprights. No bracing system is present in the down-aisle direction (Fig. 4.1). In the cross-aisle directions frames are made of the uprights braced with a system of diagonals and struts. Diagonals are all positioned in the same direction (Fig. 4.2; Table 4.1).

Details of the beam-to-upright connections as well as of column bases were the same tested at Instituto Superior Tecnico of Lisbon (see Chap. 2).

Four pallets were placed at each level, with an 8.5 kN load on each pallet, for a total vertical load on the structure of 102 kN (34 kN at each floor).

In order to avoid possible fall of the pallets/masses in case of collapse of the specimen, these were hanged to a safety frame by means of steel bars connected to the (steel) masses by means of eye-plates (Fig. 4.3). In order to avoid friction at the eye-plates, in case of large deformation of the specimen, some Teflon was placed around the bars.

C.A. Castiglioni, *Seismic Behavior of Steel Storage Pallet Racking Systems*, Research for Development, DOI 10.1007/978-3-319-28466-8_4

Fig. 4.1 Test set up in down-aisle direction

Fig. 4.2 Test set up in down-aisle direction

Table 4.1 Tests on beam-to-upright connections

Id.	Beam	N. lev.	Direction
I1	130/15	3	Down-aisle
I2	130/15	3	Cross-aisle

Fig. 4.3 "Safety" mass support system

4.2 Push-Over Test in Down-Aisle Direction

4.2.1 Test Set-up

An inverted triangular force distribution was adopted for the pushover test in down-aisle direction, as shown in Fig. 4.4.

Displacements in the down-aisle direction were imposed to the various levels of the specimen by means of three actuators (one per level).

By means of load distributors, the forces of the actuators were transferred to the uprights directly at the beam-to-upright joints.

In order to avoid local buckling of the uprights due to direct application of the forces, the load distributors were connected to the uprights by means of 200 mm long portions of beams, with the usual endplates and safety bolts (Fig. 4.5).

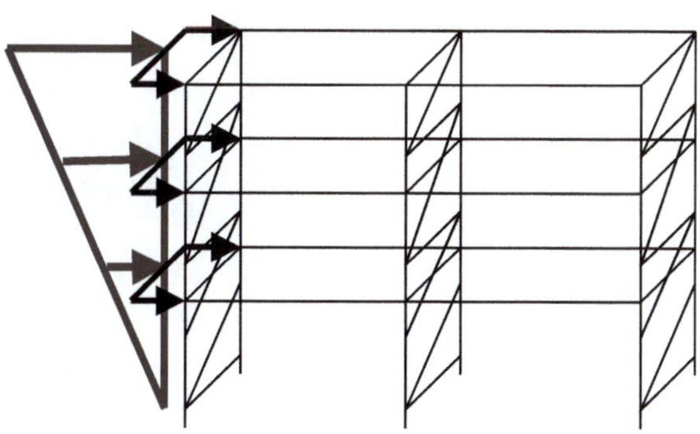

Fig. 4.4 Force distribution for the down-aisle pushover test

Fig. 4.5 Connection detail of the actuator to the uprights

The actuators were connected to the reaction wall by means of spherical hinges and were hanged to the safety frame, in order to avoid influence of their load on the specimen (Fig. 4.6).

It was decided to perform the test with steps of loading up to a given top displacement, followed by unloading and by a subsequent reloading up to a larger top displacement value, as shown in Fig. 4.7.

In this test configuration the base connection is subjected to bending in the direction of the line of the bolts, as shown in Fig. 4.8.

The specimen was instrumented in order to measure the displacements at all levels (Fig. 4.9), the forces at the actuators (Fig. 4.5) and the beam-to-upright rotations (Fig. 4.10).

Figure 4.11 summarizes the instrumentation applied to the specimen, while Fig. 4.12 shows the reference system.

Fig. 4.6 Actuators hanging system

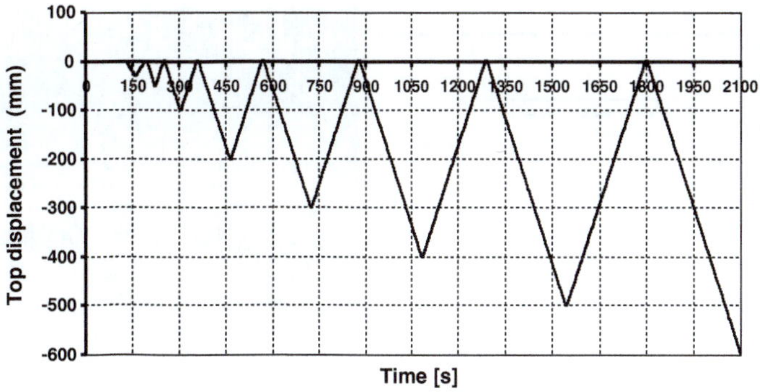

Fig. 4.7 Displacement history for the down-aisle pushover test

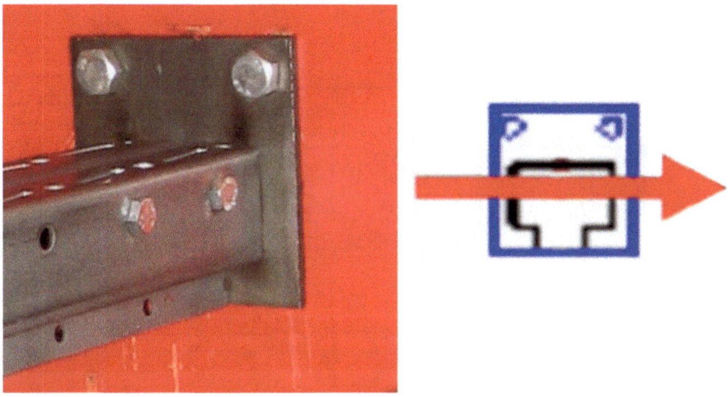

Fig. 4.8 Bending direction of the base connections

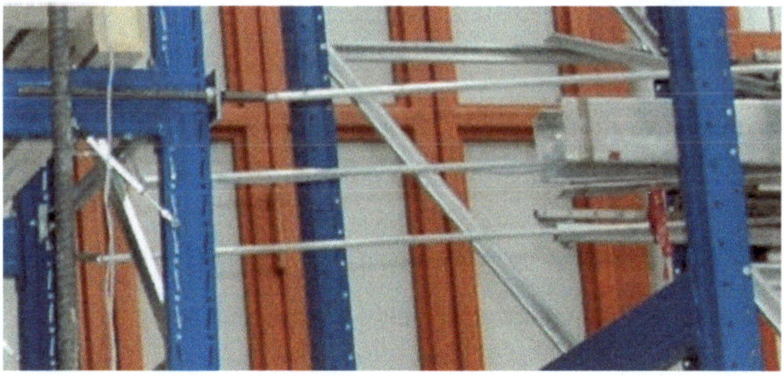

Fig. 4.9 LVDT's for global displacement measurement, at each level

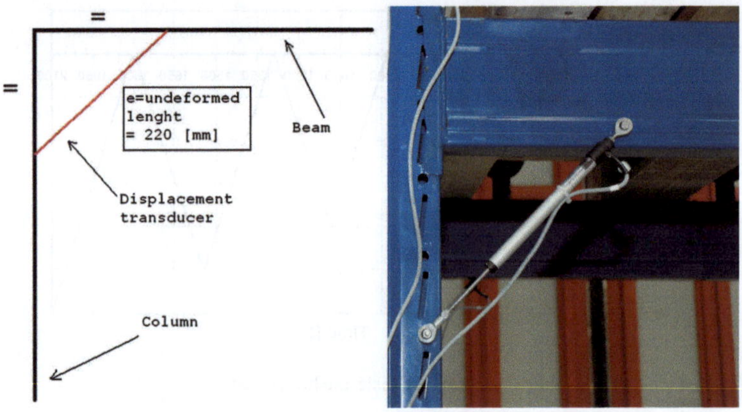

Fig. 4.10 LVDT's for beam-to-upright rotation measurement

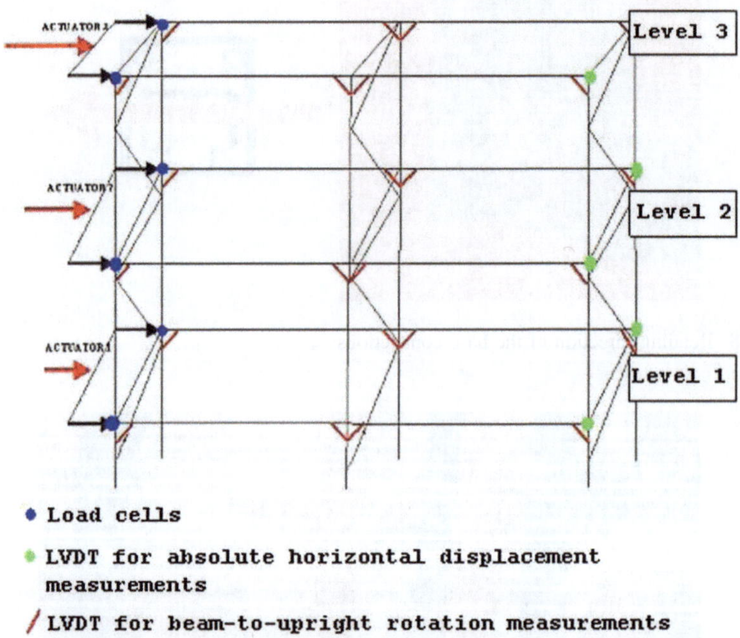

Fig. 4.11 Summary of the instrumentation applied to the specimen

4.2.2 Structural Behaviour

After the cycles shown in Fig. 4.7, the specimen was brought to collapse under monotonic loading. Specimen collapse was achieved corresponding to a top level transversal displacement of approximately 650 mm.

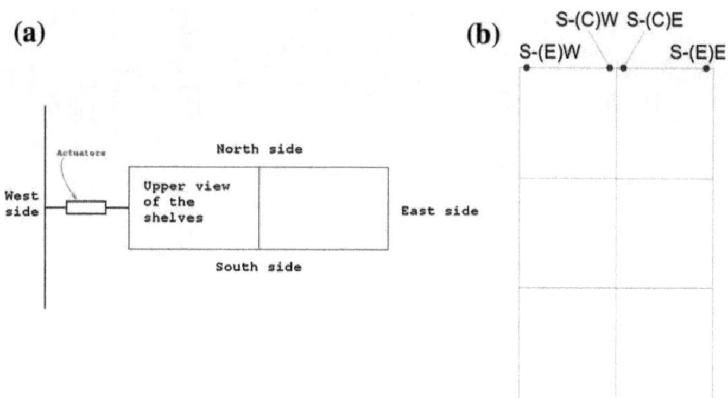

Fig. 4.12 Reference system for the down-aisle pushover test

Figure 4.13 shows the deformed shapes at various stages during the test, while Fig. 4.14 shows the collapse mechanism, due to subsequent failure of various components.

Initially, failure occurs in the base plate connections, that lose stiffness, and behave like hinges (Fig. 4.15). Immediately afterwards, plastic hinges are formed in the uprights, at the top of the lower level, just below the beam-to-upright connection at first level (Fig. 4.16). The structural collapse is hence due to a "soft-floor" mechanism type.

Finally, local buckling occurs in the uprights of the W(est)-side (to which the actuators are connected), between the first and the second level. This is evident in Fig. 4.14, noticing the abrupt change in the concavity of the deformed shape. This mechanism is caused by some local effect connected with the concentrated-load transfer mechanism between actuators and structure. In the beam-to-upright connection zones, in fact, short portions of SHS were introduced inside the upright section, in order to avoid distortion of the cross section, due to the concentrated load. Local buckling occurred in the "unstiffened" portion of the upright.

When further increasing the transversal displacement, collapse of the beam-to-upright connection is also achieved (Fig. 4.17).

4.2.3 Analysis of Test Results

According to the reference system presented in Figs. 4.12 and 4.22 the following symbols and abbreviations will be used in the presentation of the test results:

South = S; North = N; East = E; West = W
I Level = I; II Level = II; III Level = III
External = (E); Central = (C).

Fig. 4.13 Deformed shape at various stages during the test

Figure 4.18 shows the absolute displacements of the various levels of the structure, while Fig. 4.19 shows the inter-storey drifts. As the structural response is symmetric for the N and S sides, without torsional effects, in the following, only the results for the S-side are presented.

In agreement with the deformed shape, the inter-storey drifts are lager for the lower level, and reduce with the height. The large inter-storey drift of the first level is the cause of the "soft floor" collapse mechanism of the structure.

Figure 4.20 shows the relative rotations of the beam-to-upright connections at the various levels. A difference can be noticed, for all three levels, between the external joint of the west side (S-(E)W) and the central one of the east side (S-(C) E). Both joints develop positive rotations (opening of the angle), but with different absolute values. This fact can be explained noticing the eccentricity (Fig. 4.21) of

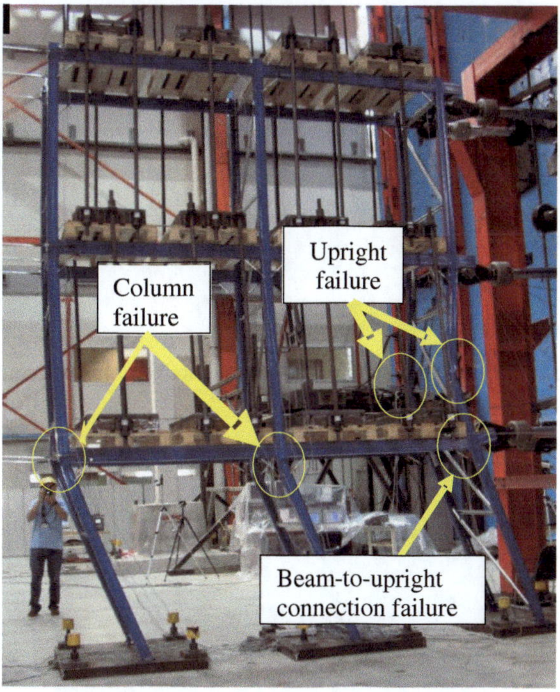

Fig. 4.14 Collapse mechanism

Fig. 4.15 Plastic hinge at the base plate connection

the load applied by the actuator connected to the (S-(E)W) joint. Such eccentricity generates an additional "local" bending moment, which affects the rotation of the joint.

On the contrary, practically no difference can be observed between the two joints showing negative rotations (closure of the angle): (S-(C)W) and (S-(E)E).

Fig. 4.16 Plastic hinge in the upright, near the beam to upright connection

Fig. 4.17 Failure of the beam-to-upright connection

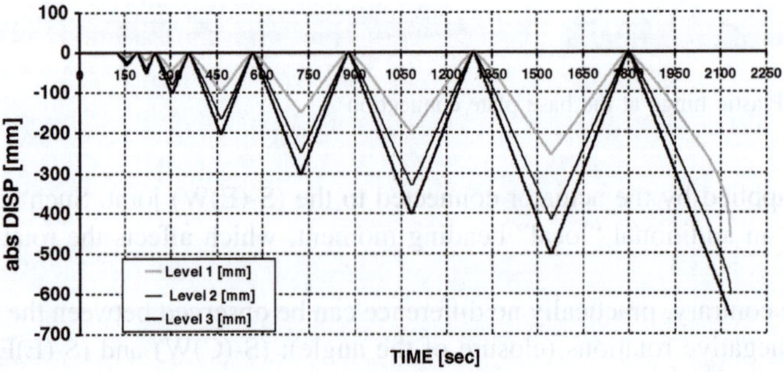

Fig. 4.18 Absolute transversal displacements of the various levels

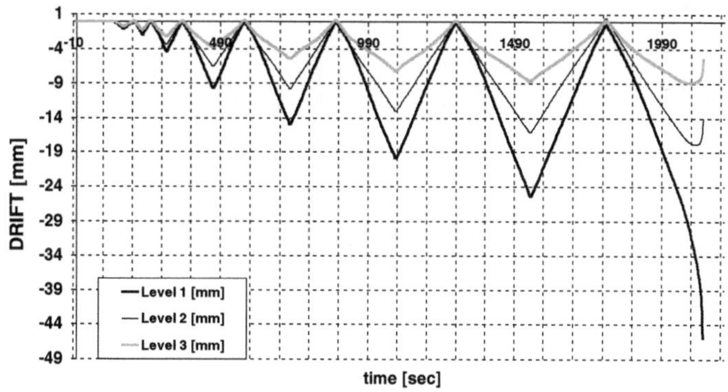

Fig. 4.19 Inter-storey drifts of the various levels

Furthermore, by observing the rotations of the joints on the central upright (S-(C)W) and (S-(C)E) it can be noticed that the joints on the W-side show smaller absolute values of the rotations than those on the east side. As the former undergo negative rotations, and the latter positive ones, this difference can be explainable with reference to the different stiffness of the connections in the two directions of bending (evidenced also by the component tests described in Chap. 2): under sagging bending, the connections develop a larger stiffness than in the case of hogging bending.

Similar considerations can be drawn by examining Figs. 4.22, 4.23 and 4.24, showing the hysteresis loops of the beam-to-upright connections expressed in terms of shear force applied at each level and beam-to-upright relative rotation.

The four joints at the same level show practically the same behaviour, with similar initial stiffness. The two connections (S-(C)W) and (S-(E)E), under hogging bending show permanent deformations also when the deformed shape of the structure is brought back to the initial (undeformed) state. These permanent deformations are larger at the first level, and decrease at the upper levels.

As already evidenced, the interstorey drifts increase from the third (upper) level, to the first (lower) one.

The behaviour of the connections of the first level is scattered, and less uniform than those of the upper levels. It can be noticed, for example, that the (S-(E)W) connection shows lower strength deterioration than the other connections. This is clearly evident when observing Fig. 4.24 that shows the hysteresis loops of the connections at each level plotted versus the absolute value of the beam-to-column rotation. The larger plastic stiffness of the (S-(E)W) connection is probably due to the fact that, as already described, the upright has been locally reinforced in order to prevent local buckling due to the applied concentrated load.

It can also be noticed that the two connection on the central upright (S-(C)W) and (S-(C)E), forming a four way joint, show a larger deterioration trend than the two external connections (S-(E)W) and (S-(E)E). This fact is not only due to the larger bending moments challenging the upright in the case of a four way joint, but

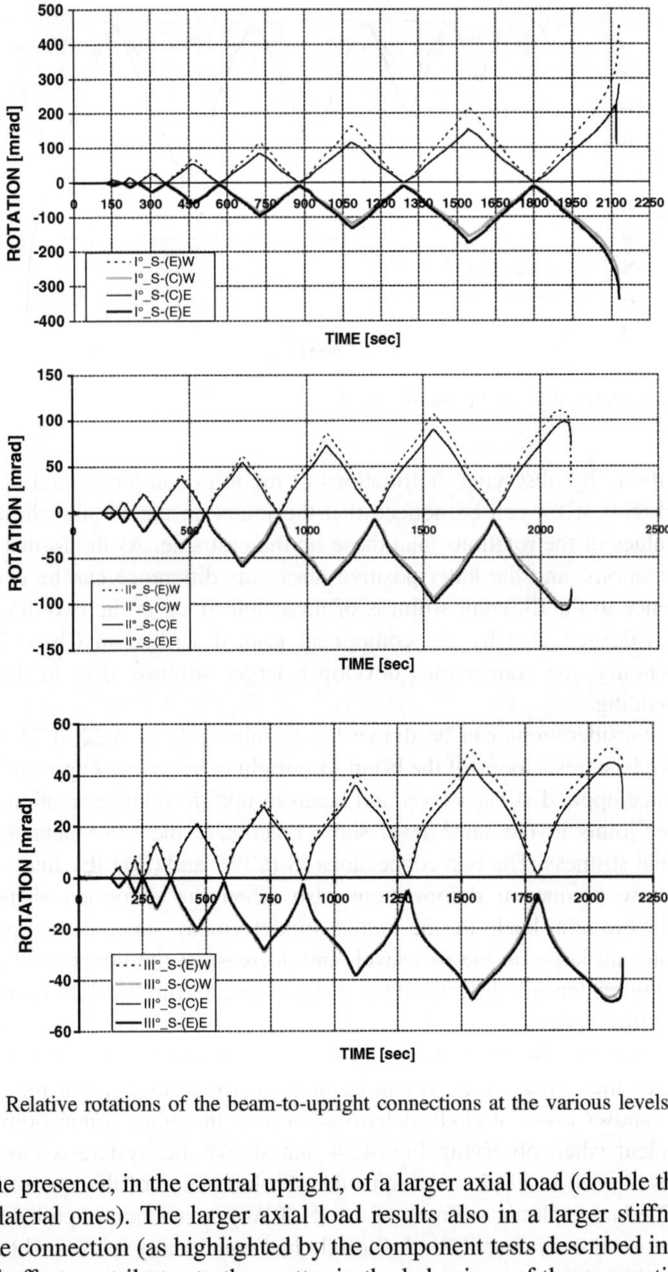

Fig. 4.20 Relative rotations of the beam-to-upright connections at the various levels

also to the presence, in the central upright, of a larger axial load (double than that in the two lateral ones). The larger axial load results also in a larger stiffness of the base plate connection (as highlighted by the component tests described in Chap. 2); this local effect contributes to the scatter in the behaviour of the connections of the first level.

Figure 4.25 shows the energy absorbed at the various levels plotted versus the inter-storey drift. The amount of energy absorbed at the first level is the largest, and is explainable with reference to the plastic deformations occurring in the structural

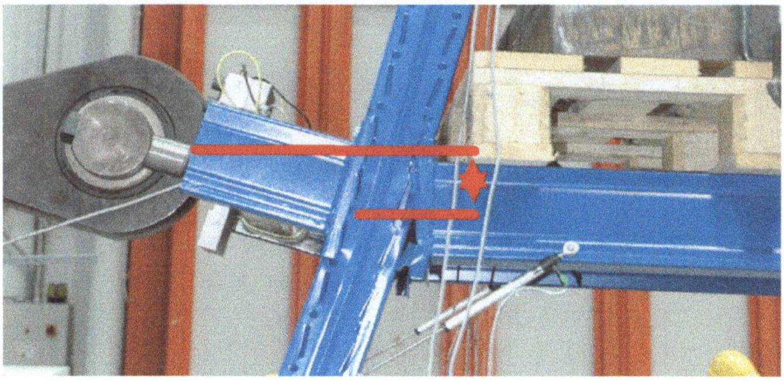

Fig. 4.21 Failure of the beam-to-upright connection

components of this level: not only in the beam-to-upright connections, but also in the base plate connections and the uprights.

The structural behaviour is non-linear since the first cycle; this is probably due to the small movements allowed by the beam-to-upright connections.

At the beginning of the test, the structure is subjected to small displacements applied in a quasi-static way and behaves locally in different ways. Some of the joints take advantage of the looseness in the hooked connections that allow small displacements and "slippage"; in other joints, internal friction develops a "grip" such that the seismic action is dissipated through plastic deformation of the connection components which are "in bearing". As these phenomena occur at different timing among the various joints, the input energy is dissipated only by the components "in bearing" of those joints where slippage does not occur.

Hence, distribution of the external actions is not uniform among the connections, as well as the consequent plastic deformations.

Figure 4.26 shows the trend of the total absorbed energy during the test, compared to the energy absorbed at the different levels. It can be noticed that the first level absorbs nearly twice the energy absorbed at the second level, which is nearly three times that absorbed by the third level.

Figure 4.27 shows the trend of variation of the stiffness as a function of the imposed displacement amplitude and of the plastic deformation. A stiffness index has been defined as the ratio of the shear at the various levels to the inter-storey rotation, defined as the ratio of the inter-storey drift to the inter-storey height:

$$Stiff-index = \frac{Shear}{\left(\frac{I\text{-}S\,Drift}{Storey\text{-}height}\right)}.$$

It can be noticed that increasing the imposed displacement amplitude (and consequently the plastic deformations) the stiffness of the various levels diminishes, and that the three levels show a similar behaviour and loss of stiffness during the test.

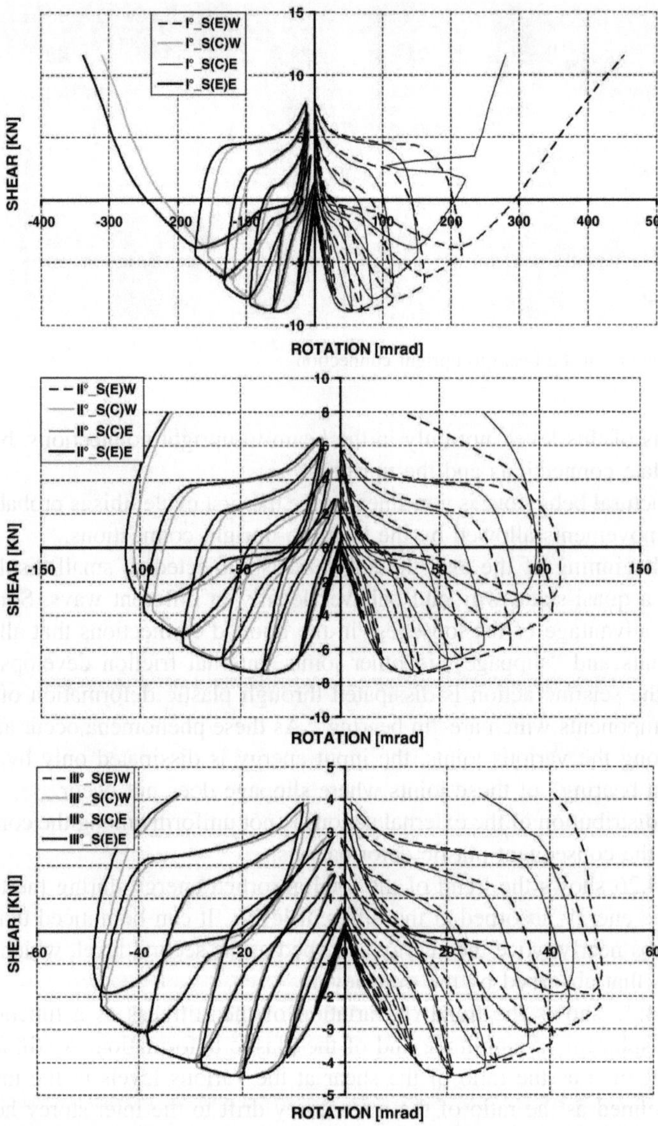

Fig. 4.22 Hysteresis loops of the beam-to-upright connections at the various levels in terms of shear and relative rotation

4.2.4 *Assessment of the q-Factor*

Assessment of the q-factor was carried out by analysing the structural response in terms of base shear plotted versus the horizontal top displacement, as shown in Fig. 4.28.

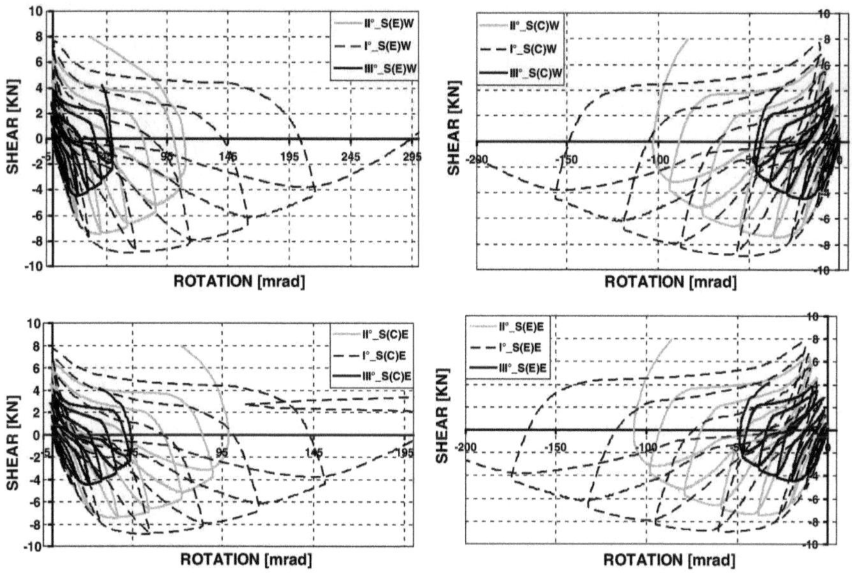

Fig. 4.23 Comparison of the hysteresis loops of the beam-to-upright connections at the various levels in terms of shear and relative rotation

The yield force F_y and the yield displacement v_y are conventionally defined according to the ECCS Recommendations.

Two values can be identified for the behaviour factor, based on ductility considerations: a value

$$q_{\mu_{\max}} = \frac{v_{\max}}{v_y} = \frac{193.9}{52.1} = 3.7$$

defined as the ratio of the displacement v_{max} corresponding to the maximum load carrying capacity of the structure to the yield displacement (v_y), and a value

$$q_{\mu_u} = \frac{v_u}{v_y} = \frac{363.8}{52.1} = 6.98$$

defined as the ratio of the displacement v_u (the maximum displacement bearable by the structure with a load carrying capacity larger than the yield force F_y) to the yield displacement (v_y).

With reference to the ductility factor theory, it is also possible to define a value of the q-factor based on strength, as the ratio of the ideal strength $F_{max,el}$

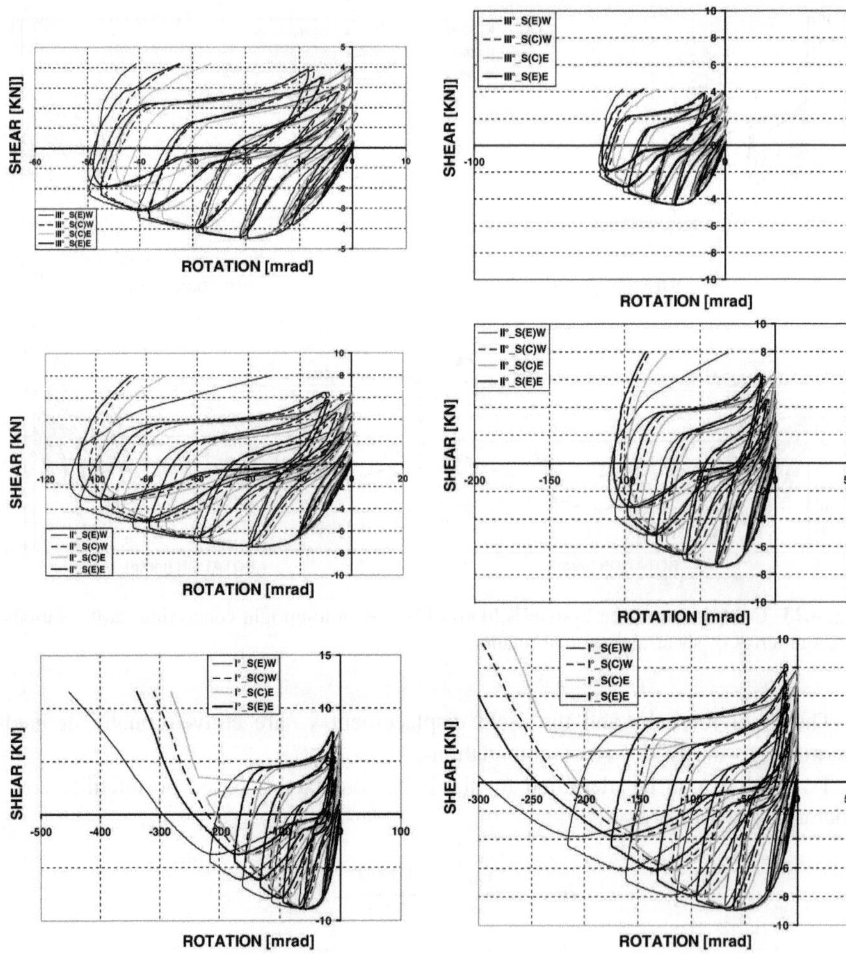

Fig. 4.24 Comparison of the hysteresis loops of the beam-to-upright connections at the various levels in terms of shear and absolute value of the relative rotation

(corresponding to v_{max} and evaluated on the basis of the initial elastic stiffness) to the actual maximum load carrying capacity F_{max}:

$$q_{f_{max}} = \frac{F_{max,el}}{F_{max}} = \frac{27.6}{8.9} = 3.1$$

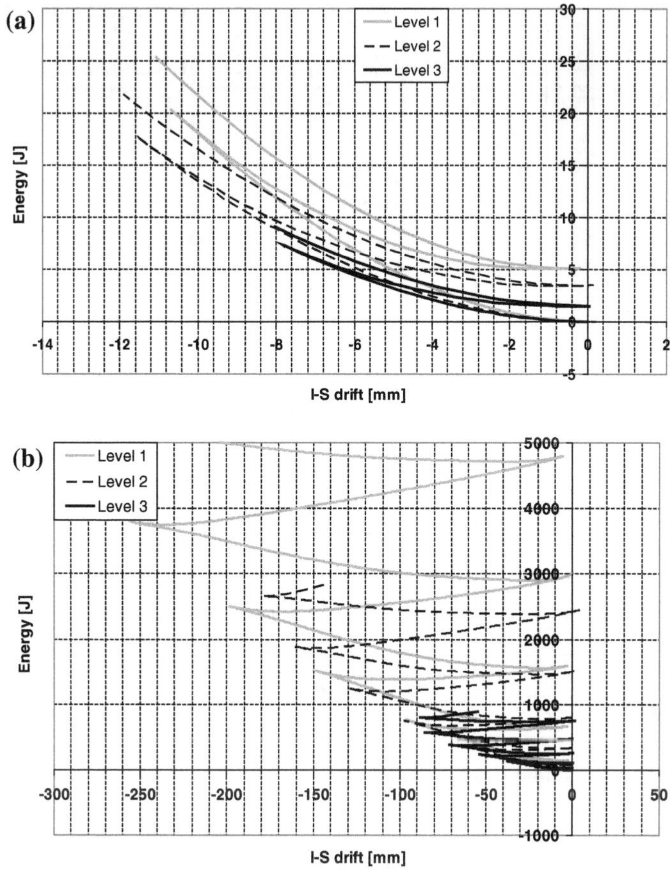

Fig. 4.25 Absorbed energy: in the first three cycles **a** in the whole test, **b** plotted versus interstorey-drift

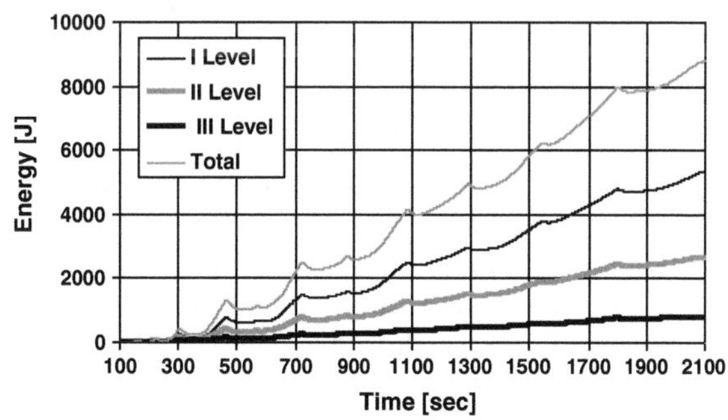

Fig. 4.26 Absorbed energy (total and at different levels) during the test

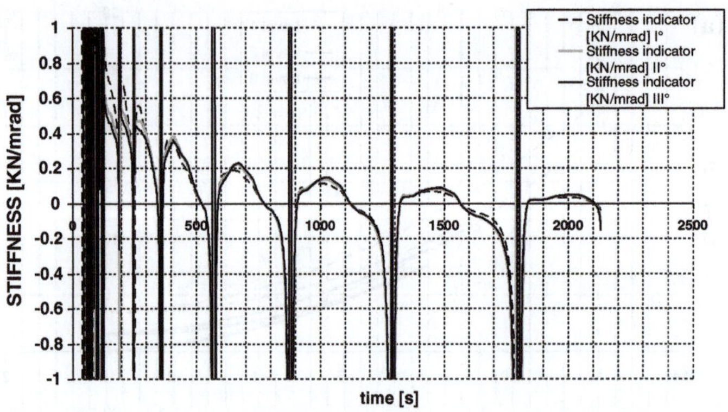

Fig. 4.27 Trend of the stiffness index during the test

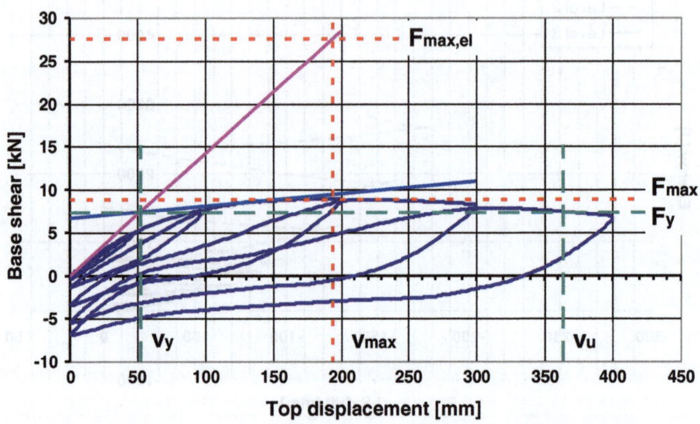

Fig. 4.28 Base shear versus the top displacement. Assessment of the q-factor

4.3 Push-Over Test in Cross-Aisle Direction

4.3.1 Test Set-up

In the test in cross-aisle direction the base plates are subjected to bending in a direction perpendicular to the line of the bolts. Due to the eccentric position of the upright on the base plate, as well as with respect to the bolt line, under transverse loading (in cross-aisle direction) one base plate has bolts in tension (preventing bending of the plate) while the other base plate has bolts in compression (hence not preventing bending of the plate), as shown in Fig. 4.29. Furthermore, it has to be noticed that the diagonal members of the upright were positioned (according to erection common practice) all in the same direction. The structure was hence

Fig. 4.29 Test set-up for the push-over test in cross-aisle direction

positioned in such a way to resist to the applied loads with diagonals in compressions, this being the most severe loading condition.

Figure 4.30 shows the instrumentation as well as the test set up, characterised by two jacks on the top level, and by one jack at each of the two lower levels. This choice is due to the need of avoiding possible torsion effects of the structure, at least by controlling the rotation of the top level. Loads are applied directly to the beams; this allows simulation of the "real" loading condition, in which loads are transferred from the pallets to the beams and from the beams to the uprights. Furthermore, as shown in Fig. 4.30, the jacks were positioned in such a ways to obtain a uniform redistribution of the applied horizontal forces among the transversal upright frames.

In addition to LVDT's measuring the transversal displacements at each level, inclinometers were positioned on the lower part of the uprights and on the column bases, in order to measure their rotations.

The horizontal load distribution among the different levels was assumed according to the first eigenvector. The loading history is shown in Fig. 4.31. As in the case of the down-aisle direction, the test was performed with a sequence of loadings and unloadings. In particular, the first two loading-unloading sequences were performed symmetrical to the origin, and with a small absolute top displacement (approximately ±35 mm).

This allowed an assessment of the elastic stiffness and of the structural response in the elastic range, when diagonal of the upright are respectively in tension or in compression. From the third loading-unloading sequence, however, the test was carried out under increasing values of the maximum displacement, always "pushing" in such a way to have diagonals in compression.

In each loading-unloading sequence, once reached the maximum imposed displacement, the structure was unloaded until zero value of the applied force, and then

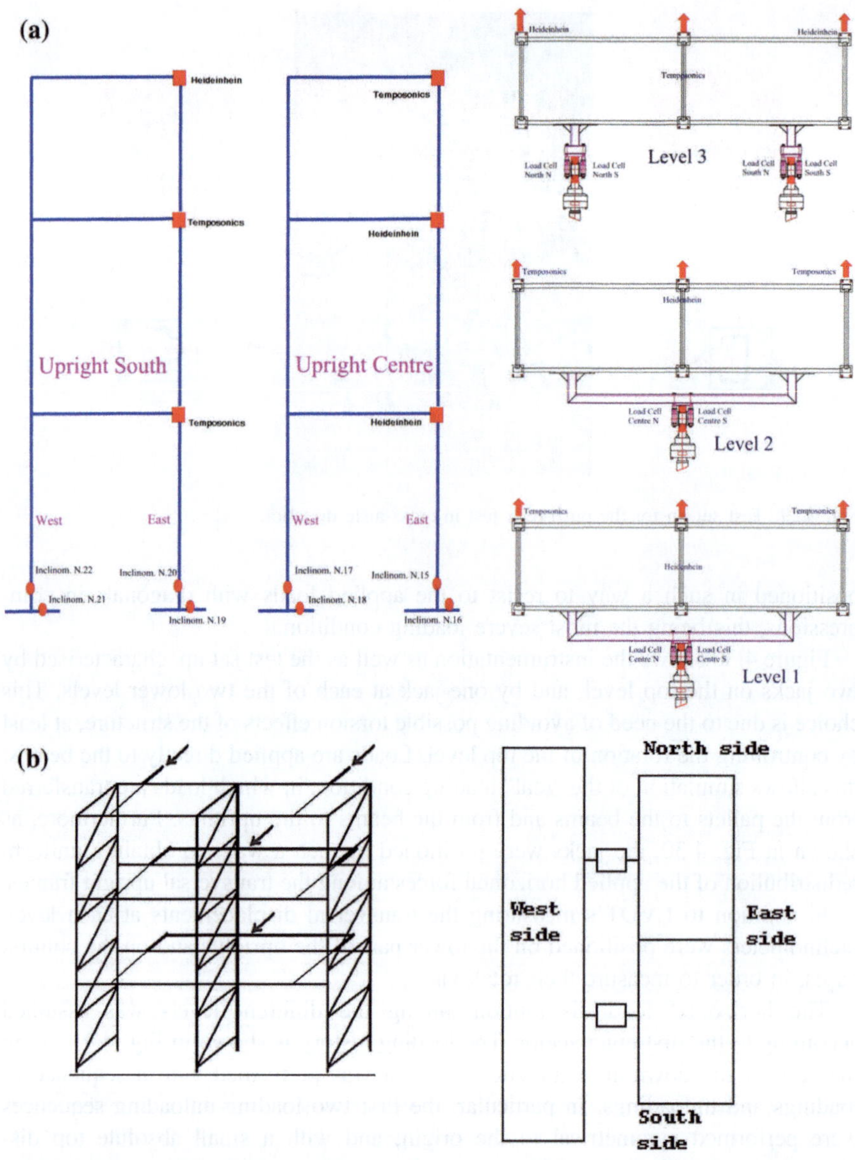

Fig. 4.30 Instrumentation and positions of the jacks (**a**); set-up and reference system for the push-over test in cross-aisle direction (**b**)

reloaded up to a larger imposed displacement. Being the structural behaviour in the plastic range, this resulted in permanent deformations, as evident in Fig. 4.31.

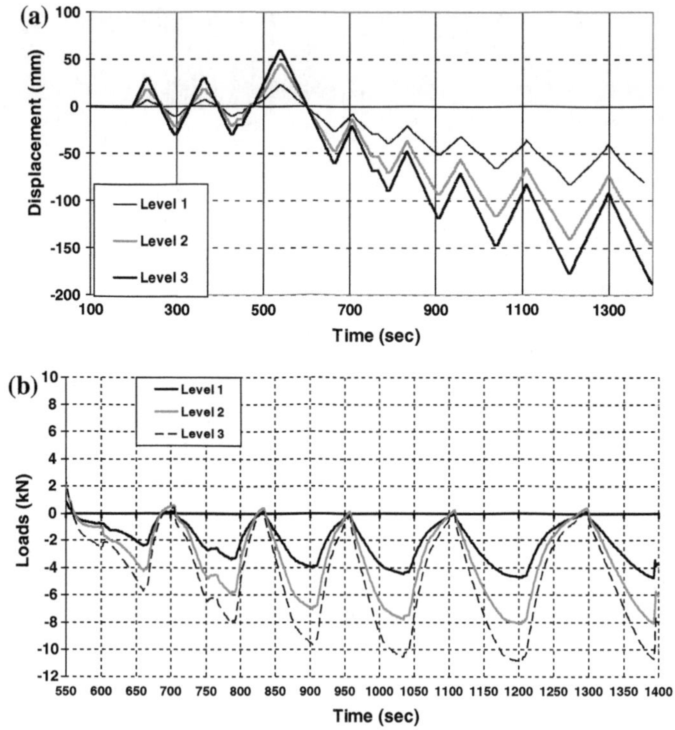

Fig. 4.31 Applied loading history: **a** displacements; **b** forces

4.3.2 *Structural Behaviour*

The pushover test in the cross-aisle direction was carried out up to collapse of the specimen.

Failure occurred because of buckling of the diagonal members of the transversal frames. These members were all positioned in the same direction (as from common erection practice), and loads were applied in order to stress the diagonals in compression.

Buckling started in the diagonals of the lower level, where the inter-storey drift is larger. The test was however continued until buckling of the diagonal members of all levels occurred.

Figure 4.32 shows the deformed shape of the structure during the test. Buckling of the diagonal members of the uprights is clearly evident. Buckled diagonals can be observed also in the following Fig. 4.33, together with base-plate bending (Fig. 4.34) and upright buckling near to the diagonal-to-upright connection as well as failure of the bolts connecting the diagonal to the column (Fig. 4.35).

Buckling of the diagonal members occurred around mid-test, arising not simultaneously in the three transversal frames; afterwards, the structure is subjected

Fig. 4.32 Deformed shapes of the specimen in the pushover test

to torsional effects, due to difference in the stiffness of the transversal frames with and without buckled diagonals. In any case, no relevant difference can be noticed among the three frames, as the horizontal displacements of the central one are similar to those of the lateral ones.

Buckling of the diagonals started in a lateral frame, most probably because the member geometric imperfections were larger than in the members of the other frames. When the diagonal buckled, the whole structure was subjected to torsion, most evident at the third level, where the presence of two actuators allows the structure to continue to respond according to the first modal shape.

The structural response in the cross-aisle direction is strongly influenced by the orientation of the diagonal members of the transversal frames, by the behaviour of the base-plate connections as well as by the out-of-plane behaviour of the beam-to-upright connections.

4.3.3 Analysis of Test Results

The test was initially carried out with two cycles of imposed displacement, symmetric with respect to the undeformed position. Hence, the diagonal members of the transversal frames are subjected to cyclic tensile and compressive stresses. Afterwards, the test was continued with loading cycles consisting of imposing maximum displacements up to a desired amplitude, followed by unloading until a

Fig. 4.33 Buckling of the diagonal members

"zero" horizontal force level (Fig. 4.31). In this way, the diagonal members of the transverse frames are subjected to compression forces, and buckle out of plane. Due to non-linear effects, permanent deformation arise in the structure so that, after

Fig. 4.34 Bending of the base plates

unloading, to the "zero" force level (Fig. 4.31b) correspond "non-zero" transversal displacements of the various levels (Fig. 4.31a).

Figure 4.31a, shows the imposed displacements at each level, and highlights how they vary being larger at the third (top) level, and progressively reduce when passing from the third to the first level. The same consideration is valid for the inter-storey drifts shown in Fig. 4.36.

Figure 4.37 shows the forces at the two actuators of the third level.

It is evident that until the second un-symmetric cycle (after the first initial two symmetric ones) the forces in the two actuators are practically equal. During the next cycle, a difference arises between the two actions; this is due to the need to counterbalance the global torsion of the structure, caused by a partial loss of stiffness of the lateral frame where buckling of the diagonal members occurred first.

As already mentioned, the behaviour of the base-plate connections is one of the aspects affecting the global structural response.

Figure 4.38 shows the rotations at the lower part of the uprights and at the base plates, of the central and the south frames for the portion of test after the first two symmetric cycles. It can be noticed that, for both frames, the base plates of the west side practically do not rotate. In fact, in this configuration, the bolts of the base connections are in tension, and prevent large deformation of the base-plate (Fig. 4.39). On the contrary, the base connections on the east side show very large

Fig. 4.35 Upright buckling and failure of the bolted connections

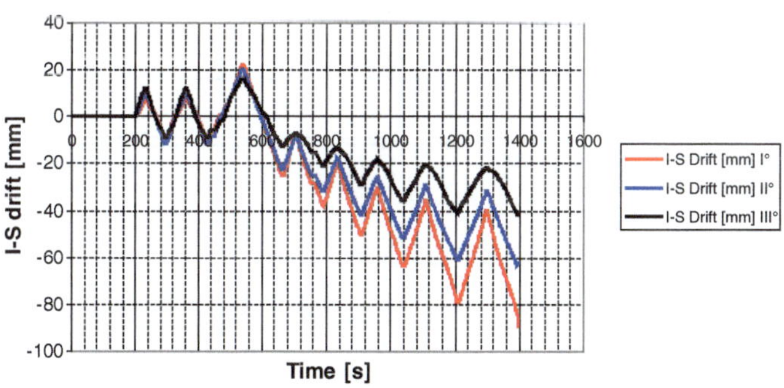

Fig. 4.36 Trend of the inter-storey drifts at the various levels

rotations, because of the deformation of the base-plate, whose bending is not prevented by the bolts in compression. In both cases, however, at unloading the base plates do not show any permanent deformation.

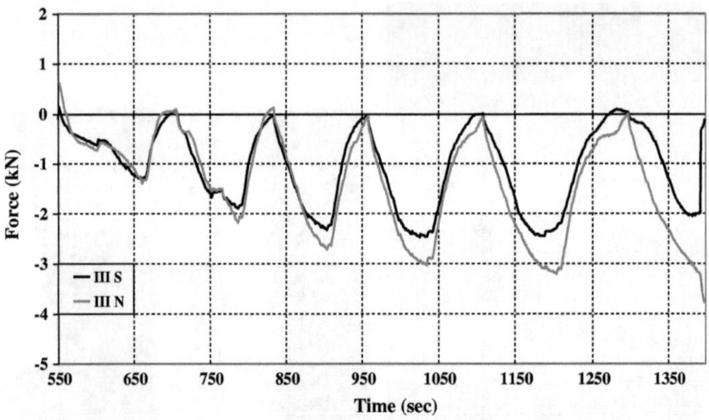

Fig. 4.37 Forces in the two actuators of the third level

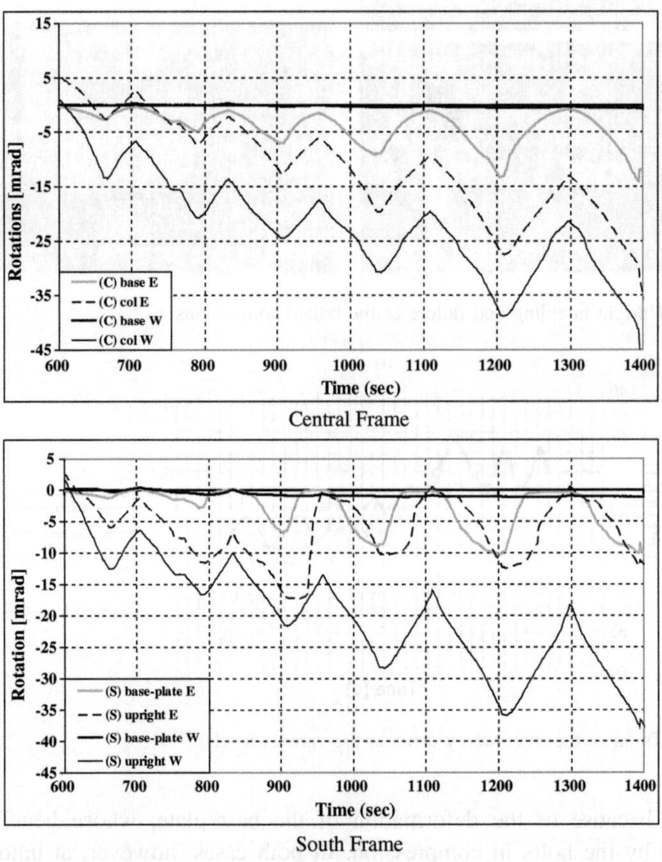

Fig. 4.38 Rotations in uprights and base-plates of the central and the south frames

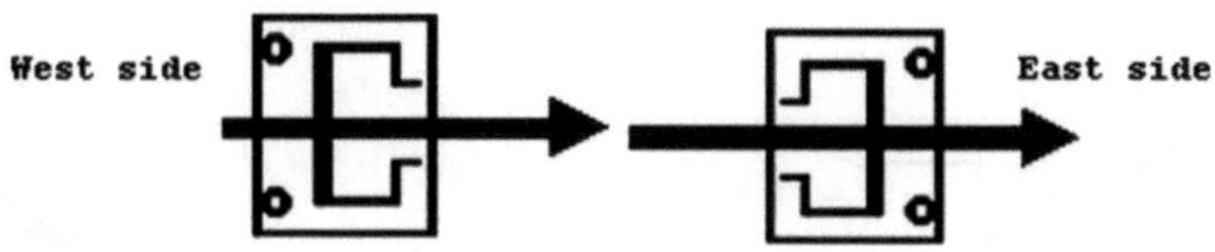

Fig. 4.39 Sketch of the loading conditions of the base-plates

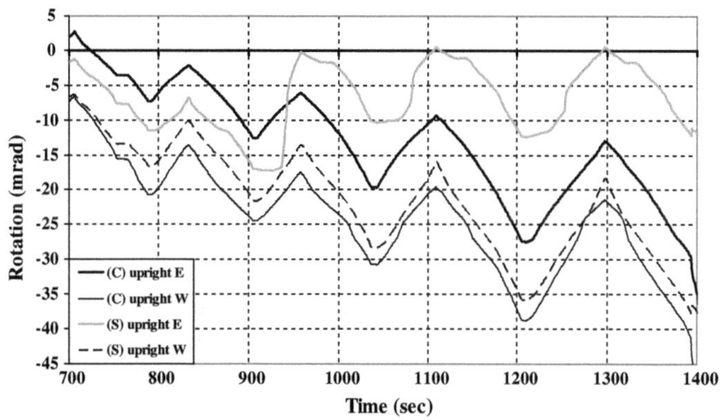

Fig. 4.40 Rotations of the uprights of the central and the south frames

It is also evident that the two frames behave in a similar way until buckling occurs in the diagonal members of the lower level of the south frame (approximately at 950 s). Collapse of the diagonals causes an unloading of the upright of the east side (the one opposite to the actuators); absence of permanent deformations indicates an elastic behaviour of the upright during the first part of the test.

Figure 4.40 shows the rotation at the bottom of the uprights of the central and the south frames. It can be noticed that the two uprights on the west side (the one where the actuators are connected) behave in a very similar way, while the columns on the east side show a difference in the response, already before buckling of the diagonals (which occurs around 950 s).

Difference between the rotations of the uprights of the west and the east side is due to deformation of the bracing system of the transverse frames, as well as to the different behaviour of the base-plate connections and of the beam-to-upright connections.

Figure 4.41 shows a sketched top-view of the beam-to-upright connections on the east and west sides. It is evident that, when loaded by horizontal loads applied transverse to the beam as in the figure, the connections on the west side can transfer the loads of the actuators to the uprights by means of both portions of end-plate in bearing against the upright and of the safety bolt in shear.

On the contrary, the connections on the east side can transfer load only by means of the safety bolt in shear as well as of the hooks in tension; furthermore, bending of the beam in the horizontal plane induces bending in the end-plate. This different

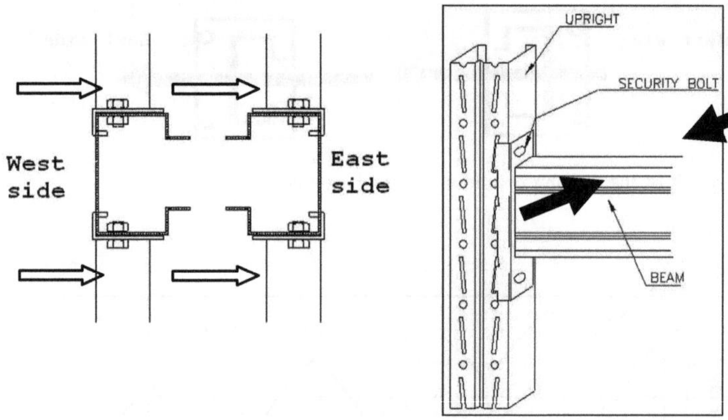

Fig. 4.41 Top-view of the beam-to-upright connections

behaviour of the connections of the east and west side introduces an asymmetry in the load transfer mechanism between beam and uprights, hence contributes to the difference in the global response of the uprights of the two sides.

Figure 4.42 shows the trend of the rotations of the base connections. It is evident the difference between the behaviour of the connections of the east and of the west sides, caused by the already discussed differences in geometry, as shown in Fig. 4.39. In fact, when bolts are in tension they contribute to the global resistance of the connection; such contribution is absent in the case of bolts in the compression zone.

In Fig. 4.43, the hysteresis loops are presented in terms of base shear versus rotation of the base connections and of the bottom of the uprights. It is evident the non linear behaviour of the base connections on the east side, very different from the practically linear one of those on the west side (Fig. 4.43a). From the same figure it is also evident the non linear behaviour of the lower part of the uprights

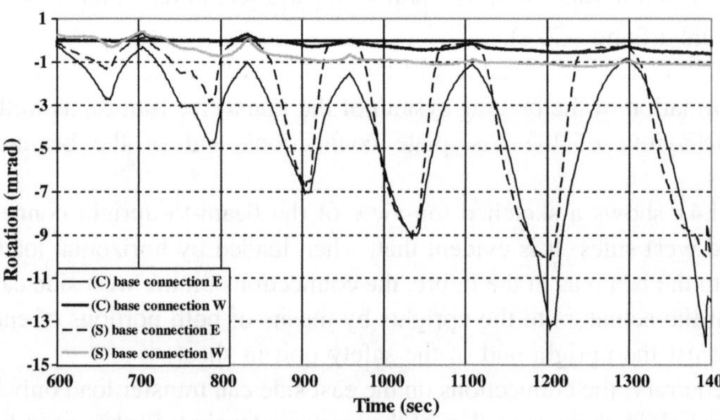

Fig. 4.42 Rotations of the base connections of the central and the south frames

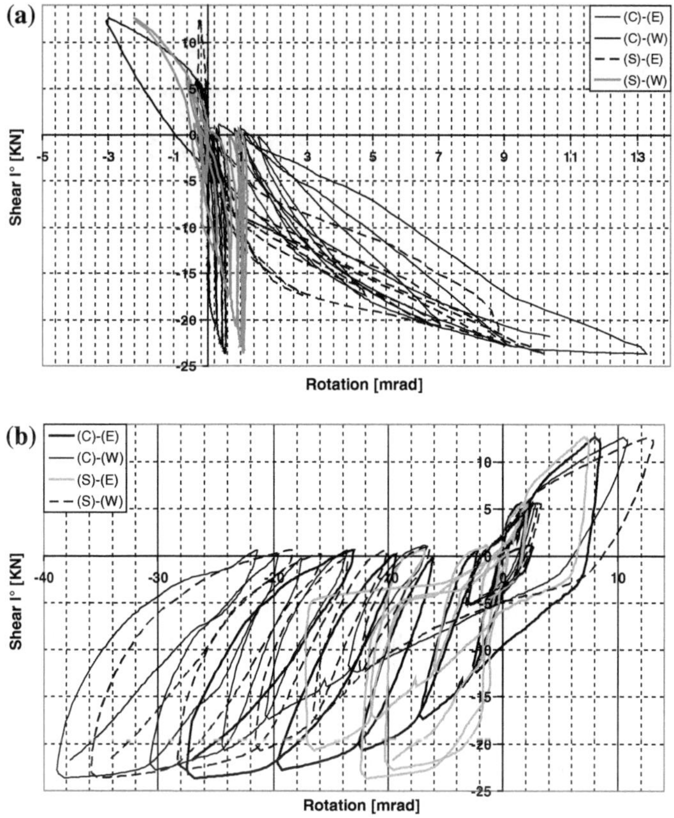

Fig. 4.43 Hysteresis loops of base shear versus rotation of the base connections (**a**) and of the bottom of columns (**b**)

(Fig. 4.43b), that develop rotations as large as 36–38 mrad on the west side. On the east side, due to buckling of the diagonals and the consequent "unloading" of the uprights, smaller rotations are developed; in particular, the east upright of the south frame (where the first buckling of diagonal members occurred) reached a maximum rotation of nearly 18 mrad, while the east upright of the central frame developed a maximum rotation of approximately 28 mrad.

In any case, Fig. 4.43 shows that the energy dissipation in the lower part of the uprights is much larger than that in the base plate connections.

The hysteresis loops of the base shear versus inter-storey drift for the various levels are presented in Fig. 4.44, that shows how the structural behaviour is stable and characterised by a good energy dissipation, larger at the (lower) first level (characterized by the largest inter-storey drifts), and lower at the third one (where the inter-storey drifts are smallest).

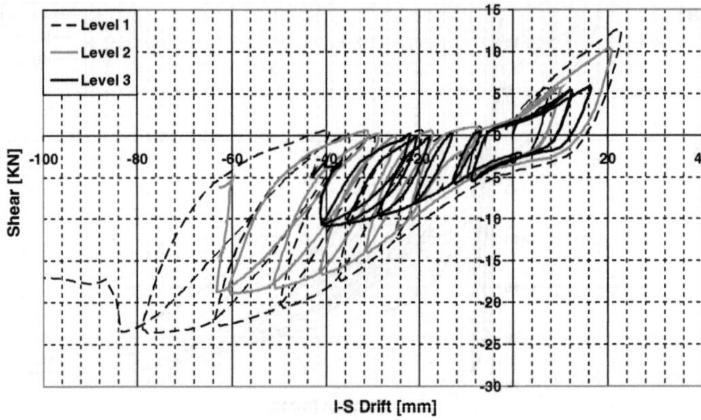

Fig. 4.44 Hysteresis loops of base shear versus inter-storey drift for the various levels

Figure 4.45 shows, for the various levels, the absorbed energy respectively in the first cycles and in the whole test plotted versus the inter-storey drifts. A non linear behaviour is evident already in the first cycles.

By comparing the trend of the total absorbed energy with the trend of the imposed displacements, Fig. 4.46 shows that during the first two symmetric cycles,

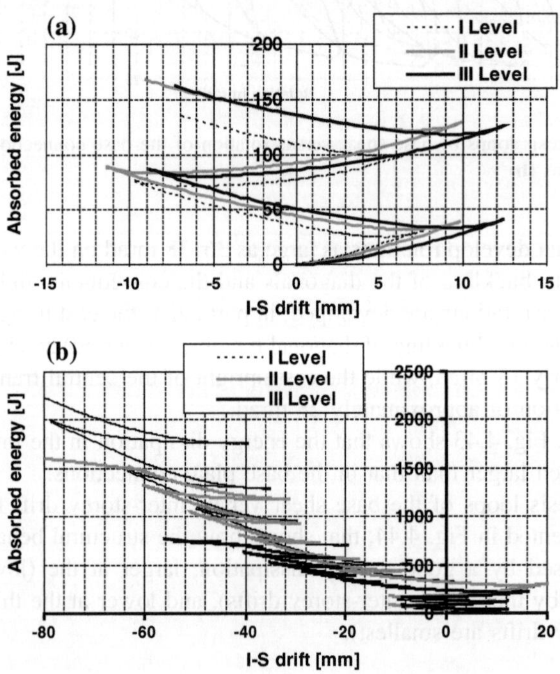

Fig. 4.45 Absorbed energy respectively: **a** in the first cycles; **b** in the whole test

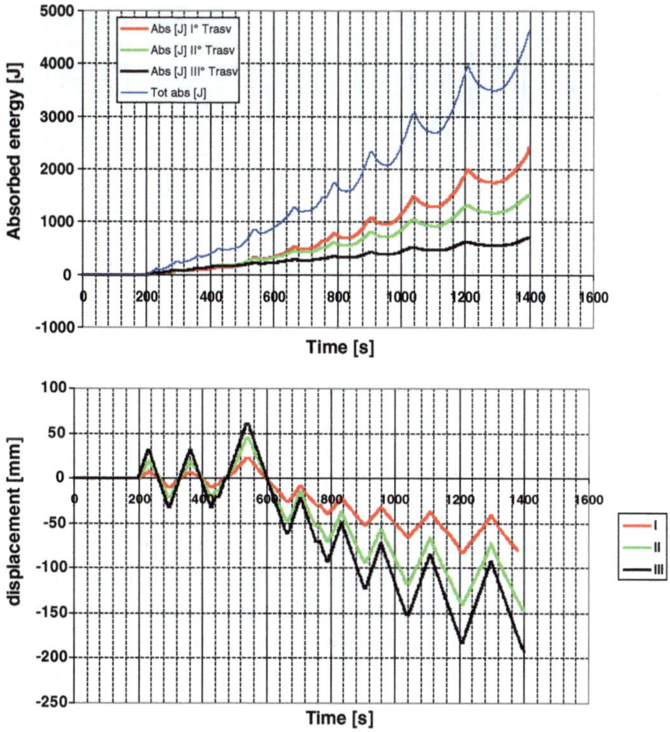

Fig. 4.46 Trend of the absorbed energy compared to the imposed displacements

the behaviour of the three levels is practically identical. Later on, the third level shows a trend of increase of the total absorbed energy that is nearly linear, while the two lower levels show a non linear trend. Each loading and unloading cycle is characterized by oscillations in the total absorbed energy at each level, that decreases during the unloading phase and increases during the loading one, reaching a (relative) maximum value corresponding to the maximum displacement imposed before unloading. A similar trend is shown also by the total absorbed energy of the whole structure (sum of those absorbed at each level).

Deterioration of the mechanical properties of the structure during the test is highlighted in the following Fig. 4.47 where stiffness (evaluated at each level as the ratio of the applied shear to the ratio of the inter-storey drift to the storey height) is plotted versus time, for the test duration.

A clear reduction trend is evidenced, common for all levels, slightly larger for level 1, and smaller for level 3, due to the deformed shape according to the 1st mode.

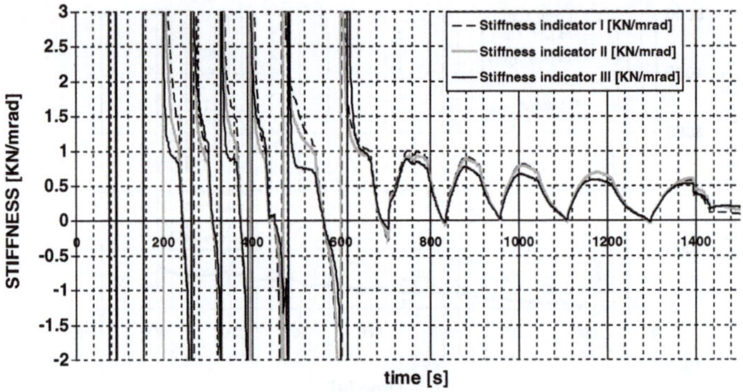

Fig. 4.47 Trend of the stiffness index during the test

4.3.4 Assessment of the q-Factor

Assessment of the q-factor was carried out by analysing the structural response in terms of base shear plotted versus the horizontal top displacement, as shown in Fig. 4.48. The yield force F_y and the yield displacement v_y are conventionally defined according to the ECCS (1986) Recommendations.

One value can be identified for the behaviour factor, based on ductility considerations:

$$q_{\mu_{\max}} = \frac{v_{\max}}{v_y} = \frac{178.5}{75.1} = 2.4$$

defined as the ratio of the displacement v_{max} corresponding to the maximum load carrying capacity of the structure to the yield displacement (v_y).

As the test was stopped before complete structural collapse, the ultimate displacement v_u (the maximum displacement bearable by the structure with a load carrying capacity larger than the yield force F_y) could not be assessed, as well as

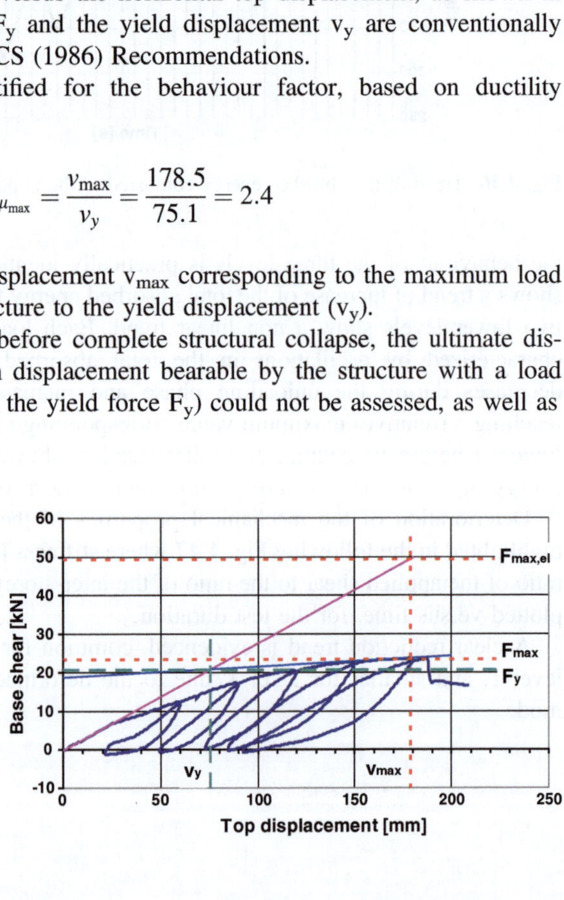

Fig. 4.48 Base shear versus top displacement. Assessment of the q-factor

$$q_{\mu_u} = \frac{v_u}{v_y}$$

With reference to the ductility factor theory, it is the possible to define a value of the q-factor based on strength, as the ratio of the ideal strength $F_{max,el}$ (corresponding to v_{max} and evaluated on the basis of the initial elastic stiffness) to the actual maximum load carrying capacity F_{max}:

$$q_{f_{max}} = \frac{F_{max,el}}{F_{max}} = \frac{49.5}{23.6} = 2.1.$$

4.4 Down-Aisle Cross-Aisle Comparison

The tests carried out in down-aisle and in cross-aisle directions are different in terms of loading histories and force distribution among the vertical levels.

An inverted triangular force distribution was adopted for the pushover test in down-aisle direction, as shown in Fig. 4.4, and the test was performed with steps of loading up to a given top displacement, followed by unloading to the undeformed condition (characterized by zero horizontal displacements at all levels) and by a subsequent reloading up to a larger top displacement value, as shown in the following Fig. 4.7. This procedure didn't allow for cumulative plastic deformation of the structure.

On the contrary, the test in the cross-aisle direction was carried out assuming a horizontal load distribution among the different levels according to the first eigenvector. The loading history adopted is shown in Fig. 4.31. The test was performed with a sequence of loadings and unloadings. In particular, the first two loading-unloading sequences were performed symmetrical to the origin, and with a small absolute top displacement (approximately ±35 mm). This allowed an assessment of the elastic stiffness and of the structural response in the elastic range, when the diagonal members of the upright are respectively in tension or in compression. From the third loading-unloading sequence, however, the test was carried out under increasing values of the maximum displacement, always "pushing" in such a way to have the diagonals of the upright frames in compression. In each loading-unloading sequence, once reached the maximum imposed displacement, the structure was unloaded until zero value of the applied force, and then reloaded up to a larger imposed displacement, resulting in an accumulation of permanent deformations in the structure, as evident in Fig. 4.31a.

Despite these differences, a comparison of the structural response in the two tests is very interesting in order to draw general conclusions about the behaviour of the racking systems in the two directions.

Figure 4.49 shows a comparison of the global response of the structure in cross and down-aisle direction.

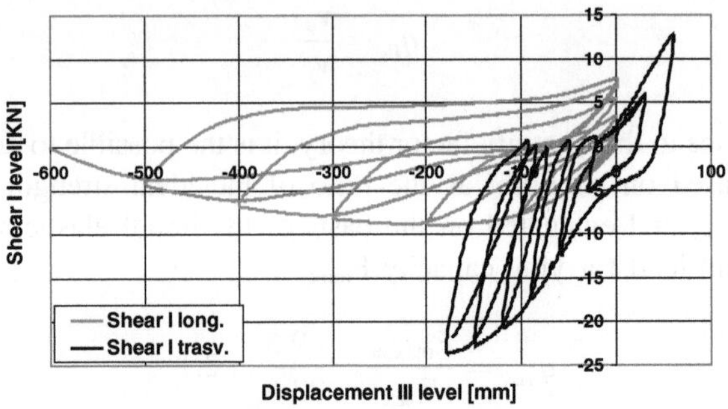

Fig. 4.49 Comparison of the hysteresis loops in terms of base shear versus the top displacement

It can be observed that in cross-aisle direction the structure is stiffer and stronger, but more fragile than in down-aisle direction. Both in down-aisle and in cross-aisle direction, the structural response was satisfactory. In down-aisle direction, formation of "plastic hinges" in all the joints could be observed. In cross-aisle direction, most of the structural components remained in the elastic range, excluding the diagonal members of the bracing system of the upright frames that, beyond a certain level, could not withstand the applied compressive forces and buckled.

Comparison of the absorbed energy in the two tests (Fig. 4.50) shows that the behaviour in the down-aisle direction is more dissipative, thanks to the plastic deformation of the beam-to-upright as well as of the base plate connections. In the cross-aisle direction, on the contrary, dissipation is practically due only to the plastic deformation of the bracing system of the (transversal) upright frames, while most other components remain in the elastic range.

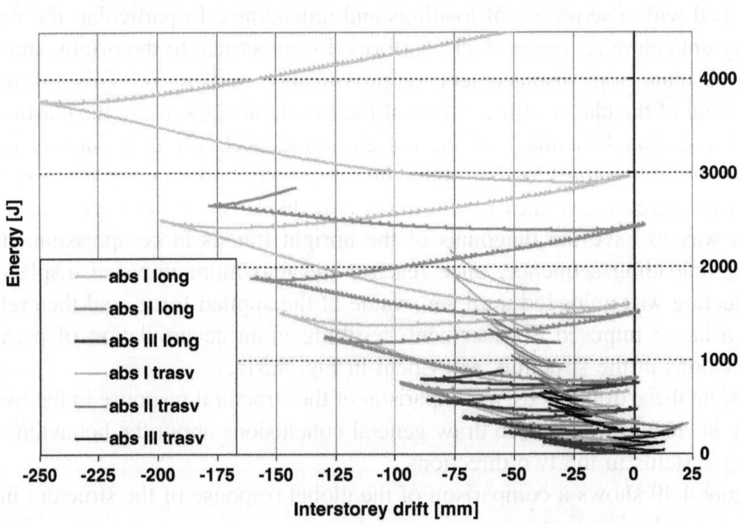

Fig. 4.50 Comparison of the absorbed energy at the various levels

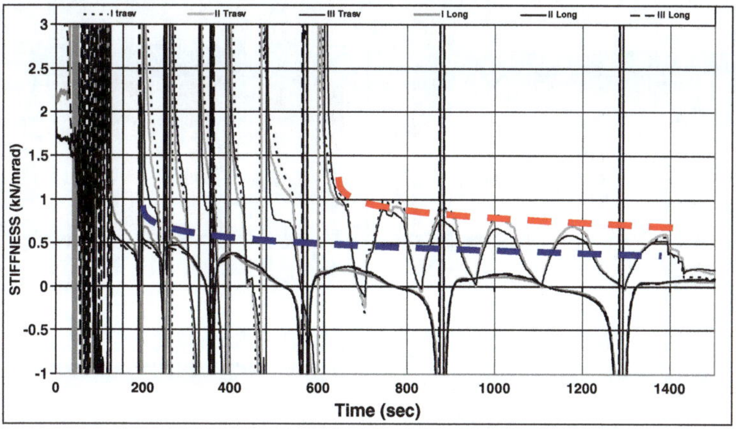

Fig. 4.51 Comparison of the stiffness deterioration at the various levels

Figure 4.51 shows a comparison of the structural stiffness in the two tests. In cross-aisle direction loss of stiffness can be noticed only after the first "pushing" cycle, while the first two "symmetric" cycles do not cause any deterioration. On the contrary, the structure subjected to down-aisle loading shows stiffness deterioration since the beginning of the test. It can also be noticed how the stiffness in the cross-aisle direction is larger than in the down-aisle one.

4.5 Conclusions

The specimen under pushover test in down-aisle direction showed a progressive loss of stiffness associated to accumulation of plastic deformation in the column-base connections and to the large inter-storey drift of the first level. Inter-storey drifts of the upper level are much smaller than the first level one; this is characteristic of a "soft-floor" type of collapse mechanism that may lead to global instability due to second-order effects.

In order to reduce this type of problem, the deformability of the column-base connections should be reduced and, somehow limited. Eventually, adoption of a beam at the ground level might be considered (Fig. 4.52). Despite the increment of cost, this solution could in any case allow use of the space at ground level for storage of goods, while the structure will behave as rotationally restrained at the base.

Due to the bracing systems of the uprights, the specimens show a larger stiffness in cross-aisle direction that in the down-aisle one. Such bracing system is the most stressed structural component. Its failure leads to global collapse, accompanied by flexural-torsional buckling of the columns, consequent to the increment of the

Fig. 4.52 Use of beams at ground level

buckling length of the profile due to failure of the bolted connections with the lattice members.

For this reason, the solution with all diagonals inclined in the same direction should be reconsidered when the structure has to be erected in a seismic zone Fig. 4.53.

Difference between the rotations of the uprights is due to deformation of the bracing system of the transverse frames, as well as to the different behaviour of the base-plate connections and of the beam-to-upright connections.

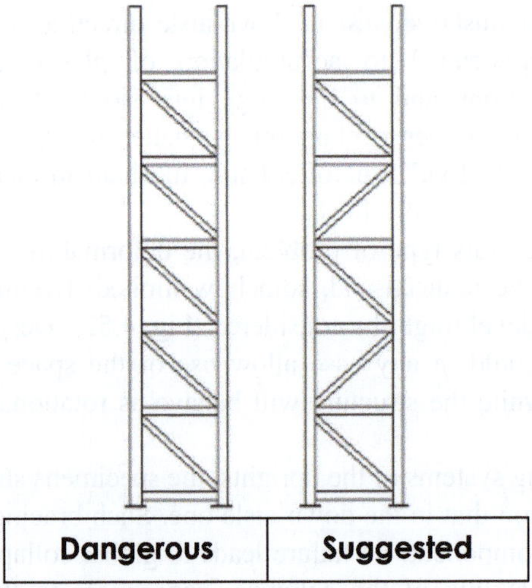

Fig. 4.53 Bracing system of uprights in cross-aisle direction

When loaded by horizontal loads applied transverse to the beam, the connections on one side can transfer the loads by means of both portions of end-plate in bearing against the upright, as well as by the safety bolt in shear. On the contrary, the connections on the other side can transfer load only by means of the safety bolt in shear as well as of the hooks in tension; furthermore, bending of the beam in the horizontal plane induces bending in the end-plate. This different behaviour of the connections contributes to the difference in the global response of the uprights of the two sides.

An evaluation of the behaviour factor has been carried out, with two possible definitions of the q-factor for both down-aisle and cross-aisle directions.

One value can be identified based on ductility considerations as the ratio of the displacement v_{max} corresponding to the maximum load carrying capacity of the structure to the yield displacement (v_y), being $q_{\mu_{max}} = \frac{v_{max}}{v_y} = 3.7$ for the down-aisle direction and $q_{\mu_{max}} = 2.4$ for the cross-aisle direction.

With reference to the ductility factor theory, a value of the q-factor based on strength was also defined as the ratio of the ideal strength $F_{max,el}$ (corresponding to v_{max} and evaluated on the basis of the initial elastic stiffness) to the actual maximum load carrying capacity F_{max}, being $q_{f_{max}} = \frac{F_{max,el}}{F_{max}} = 3.1$ for the down-aisle direction and $q_{f_{max}} = 2.1$ for the cross-aisle direction.

Reference

ECCS. (1986). Recommended testing procedure for assessing the behaviour of structural elements under cyclic loads European Convention for Constructional Steelworks. Publication No. 45.

Further Readings

Brescianini, J. C., & Castiglioni, C. A. (2007). *Caratterizzazione del comportamento dinamico di scaffalature metalliche porta-pallets mediante push-over tests* (pp. 275–282). Catania: Proceedings of XXI C.T.A.

Chintanapakdee, C., & Chopra, A. K. (2003). Evaluation of modal pushover analysis using generic frames. *Earthquake Engineering and Structural Dynamics, 32*, 417–442.

Chopra, A. K., & Goel, R. K. (2002). A modal pushover analysis procedure to estimate seismic demands for buildings. *Earthquake Engineering and Structural Dynamics, 31*, 561–582.

Elnashai, A. S. (2000). Advanced inelastic static (pushover) analysis for seismic design and assessment. In *Proceedings of the G.Penelis international symposium.* Greece: Aristotele University of Thessaloniki, pp. 23–34.

Fajfar, P. (1999). Capacity-spectrum method based on inelastic demand spectra. *Earthquake Engineering and Structural Dynamics, 28*, 979–993.

FEMA. 440. (2005) *Improvement of nonlinear static seismic analysis procedures, prepared by the Applied Technology Council (ATC-55 Project) for the Federal Emergency Management Agency,* Washington D.C.

Goel, R. K., & Chopra, A. K. (2004). Evaluation of modal and FEMA pushover analysis: SAC buildings. *Earthquake Spectra* (Vol. 20(1), pp. 225–254). Oakland CA: Earthquake Engineering Research Institute.

Gupta, A., & Krawinkler, H. (2000a). Estimation of seismic drift demands for frame structures. *Earthquake Engineering and Structural Dynamics*, ***29***, 1287–1305.

Gupta, A., & Krawinkler, H. (2000b). Dynamic P-delta effects for flexible inelastic structures. *Journal of Structural Engineering ASCE*, ***126***(1), 145–154.

Gupta, A., & Krawinkler, H. (2003). Feasibility of pushover analysis for estimation of strength demands. In *Proceedings of STESSA 2003, Behaviour of steel structures in seismic areas*. Naples, Italy, pp. 29–35.

Gupta, B., & Kunnath, S. K. (2000). Adaptive spectra-based pushover procedure for seismic evaluation of structures. *Earthquake Spectra*, *16*(2), pp. 367–391. Oakland CA: Earthquake Engineering Research Institute.

Jan, T. S., Liu, M. W., & Kao, Y. C. (2004). An upper bound pushover analysis procedure for estimating the seismic demands of high rise buildings. *Engineering Structures*, *26*(1), 117–128.

Krawinkler, H., & Seneviratna, G. D. (1998). Pros and cons of a pushover analysis of seismic performance evaluation. *Engineering Structures*, *20*(4–6), 452–464.

Mwafy, A. M., & Elnashai, A. S. (2001). Static pushover versus dynamic collapse analysis of RC buildings. *Engineering Structures*, *23*(4), 407–424.

Sasaki, K. K., Freeman, S. A., & Paret, T. F. (1998). Multi-mode pushover procedure (MMP)—A method to identify the effects of higher modes in a pushover analysis. In *Proceedings of the Sixth U.S. National Conference on Earthquake Engineering, Seattle, WA*. Oakland, CA: Earthquake Engineering Research Institute.

Chapter 5
Pseudodynamic Tests

5.1 Overview

The pseudo-dynamic test method is a hybrid analysis method, combining analytical techniques with experimental methods, aimed at the study of the seismic behaviour of structures. The method was for the first time proposed by Takanashi et al. (1975), with the original name "Computer-Actuator On-Line System", highlighting its analytical-experimental nature.

The method is considered a valid alternative to dynamic shaking table tests not only because of the lower costs, but because it is not affected by the limitations to the size and the weight of the specimens connected with the physical size and capacity of the shaking tables.

The structure under consideration is schematized through a discrete system, so that its equation of motion can be represented through a system of differential equations of second order of the type:

$$[M]\ddot{\underline{x}} + [C]\dot{\underline{x}} + [K]\underline{x} = \underline{f} \tag{5.1}$$

where:

$[M]$	mass matrix;
$[C]$	viscous camping matrix;
$[K]$	stiffness matrix;
$\underline{x}, \dot{\underline{x}}, \ddot{\underline{x}}$	respectively displacement, velocity and acceleration vectors;
$\underline{f}$	external forces vector.

The mass and the stiffness matrix can be obtained by means of FEM (Clough and Penzien 1975).

C.A. Castiglioni, *Seismic Behavior of Steel Storage Pallet Racking Systems*, Research for Development, DOI 10.1007/978-3-319-28466-8_5

The viscous damping matrix can be derived by means either of the modal method (considering proportionality to mass and stiffness) or by small amplitude, free as well as forced vibration tests on the structure (Beck and Jayakumar 1986).

The advantage of considering the structure as a discrete model is to obtain a system with a finite number of DOF. Furthermore if, in the discrete model, the masses can be considered "lumped" at the various floors, the mass matrix is diagonal.

The main difference between the pseudo-dynamic method and the dynamic analysis is that in the former, the forces sustained by the deformed structure are directly measured, hence any problem arising from modelling the non-linear structural behaviour can be avoided.

The equation of motion for the non-linear structural system under any external excitation can be solved by means of a step-by-step integration method. These methods consider the time history divided into a sequence of equal time intervals Δt (steps). The structural response within each step is computed based on the external excitation during the time interval, as well as on the initial conditions (displacement and velocity), considering the mechanical parameters of the structure constant within the step. Hence, the non-linear analysis of the structural behaviour is transformed into a sequence of linear analyses of structures with stiffness different from step to step.

As the pseudo-dynamic method encompasses measurement of the forces with which the structures "reacts" to the external excitation at each step, the equation of motion becomes:

$$[M]\ddot{\underline{x}} + [C]\dot{\underline{x}} + \underline{r} = \underline{f} \tag{5.2}$$

where $\underline{r}$ is the vector of the "restoring" forces of the deformed structure, replacing, in the equation of motion, the stiffness matrix and the displacements vector.

Hence, the test is carried out according to the following procedure (Fig. 5.1):

1. The restoring forces $\underline{r}$ are experimentally measured on the deformed structure;
2. Based on these known values of the restoring forces, the equation of motion is solved, by means of a step-by-step integration method, obtaining, at the end of each integration time step, a vector $\underline{x}$ of displacements corresponding to each DOF defined in the modelling.
3. The obtained solution is experimentally imposed by the actuators to the structures, in a quasi-static way.

Each step of the test can be divided into two phases as shown in Fig. 5.2:

1. "HOLD" phase, when the following actions are performed:
   - Data acquisition (a)
   - Numerical integration of the equation of motion (b)
   - Check of the compatibility of the obtained solution (c)

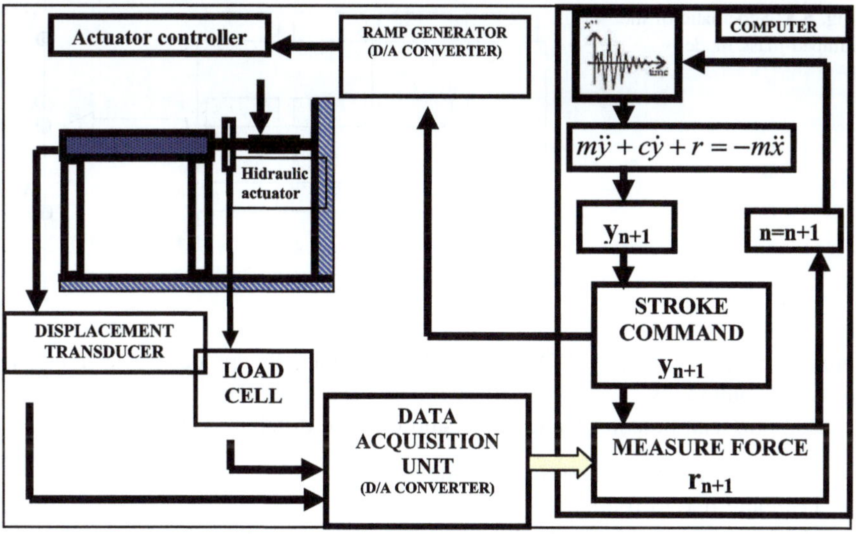

Fig. 5.1 Schematic of the pseudo-dynamic testing procedure

Fig. 5.2 Schematic of the single step of the pseudo-dynamic test

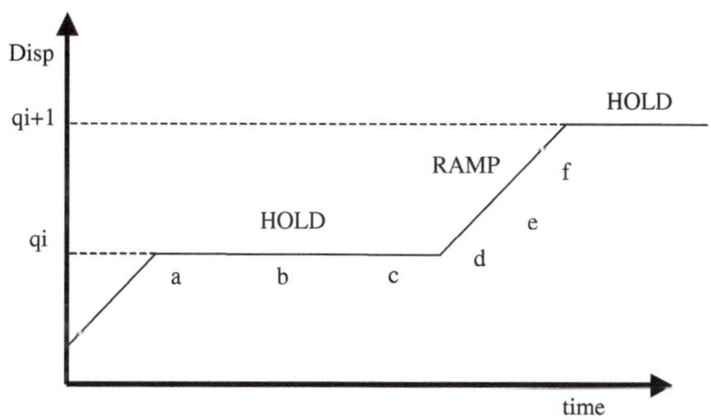

The "HOLD" phase lasts only some milliseconds, in order to avoid structural relaxation, and subsequent loss of load carrying capacity. During this phase, actuators are "still".

2. "RAMP" phase, starting if the solution is compatible, when the following actions are performed:
   - Data storage (on file) (d)
   - Input (from file) of the external excitation (e)
   - Check of the compatibility of all data (f)
   - Plotting of the obtained results.

The "RAMP" phase, during which actuators are "active", should terminate before the target displacement is completely imposed to the structure, in order to

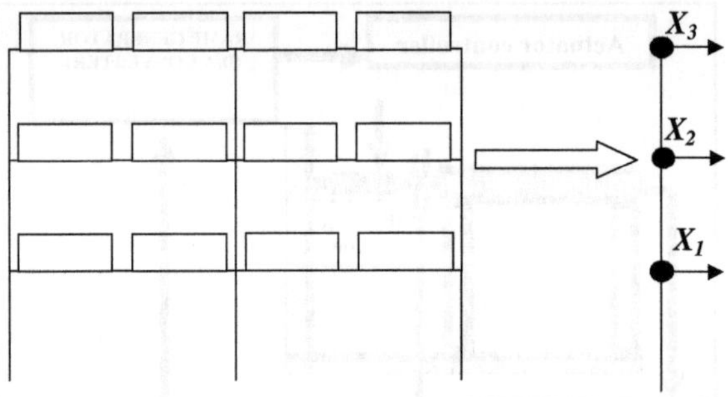

Fig. 5.3 Schematic of the lumped-mass model

minimize the following "HOLD" phase, and make the procedure as much as possible continuous.

Damping can be considered as the sum of three different contributions:

- Viscous damping, proportional to the velocity
- Damping due to friction
- Hysteretic damping, due to the inelastic behaviour of the material.

By defining the damping matrix by means of the modal method or by free and forced small vibrations it is possible to keep into account only the viscous damping contribution.

Both the hysteretic and friction damping contribution affect also the "quasi-static" structural response, hence they are taken into account when measuring the "restoring" forces, that are used in the numerical integration.

Viscous damping changes with the inelastic deformations. However, in case of inelastic deformations, this contribution can be disregarded as very small compared to the hysteretic damping,

In the pseudo-dynamic test on the pallet rack, the coefficients of the damping matrix are considered to be zero, while the mass matrix was obtained considering masses lumped at the three levels, corresponding to the assumed DOF (Fig. 5.3).

Such a model is in good agreement with the reality.

5.2 Test Set-up

The structure tested under pseudo-dynamic conditions is similar to the one adopted for the push-over test. Only tests in the down-aisle direction were performed. Due to the quasi-static loading conditions, no sliding occurs during the test.

Tests were carried out at the European Laboratory for Structural Assessment of the Joint Research Center of Ispra, on three levels, two bays full scale specimens, 6.0 m high and 3.6 m large, with 130 × 45 × 1.5 mm beams and 100/20b uprights. No bracing system is present in down-aisle direction. In cross-aisle direction upright frames are present braced with a system of diagonals, all inclined in the same direction.

Fig. 5.4 "Safety" mass support system

Details of the beam-to-upright as well as of the column-base connections were the same tested at Instituto Superior Tecnico of Lisbon (see Chap. 2).

Four pallets were placed at each level, with an 8.5 kN load on each pallet, for a total vertical load on the structure of 102 kN (34 kN at each floor).

In order to avoid possible fall of the pallets/masses in case of collapse of the specimen, these were hanged to a safety frame by means of steel bars connected to the (steel) masses by means of eye-plates (Fig. 5.4). In order to avoid friction at the eye-plates, in case of large deformation of the specimen, some Teflon was placed around the bars.

Displacements in the down-aisle direction were imposed to the various levels of the specimen by means of three actuators (one per level), as shown in Fig. 5.8.

By means of load distributors, the forces of the actuators were transferred to the uprights directly at the beam-to-upright joints.

In order to avoid local buckling of the uprights due to direct application of the forces, the load distributors were connected to the uprights by means of 200 mm long portions of beams, with the usual endplates and safety bolts (Fig. 5.5).

The actuators were connected to the reaction wall by means of spherical hinges and were hanged to the safety frame, in order to avoid influence of their load on the specimen (Fig. 5.6).

The specimen was instrumented in order to measure:

Forces at actuators (Figs. 5.5 and 5.8)
Displacements at all levels (Figs. 5.7 and 5.8)
Beam-to-upright rotations (Fig. 5.8).

Figure 5.9 summarizes the instrumentation applied to the specimen, while Fig. 5.10 shows the reference system.

Fig. 5.5 Connection detail of the actuator to the uprights

Fig. 5.6 Actuators hanging system

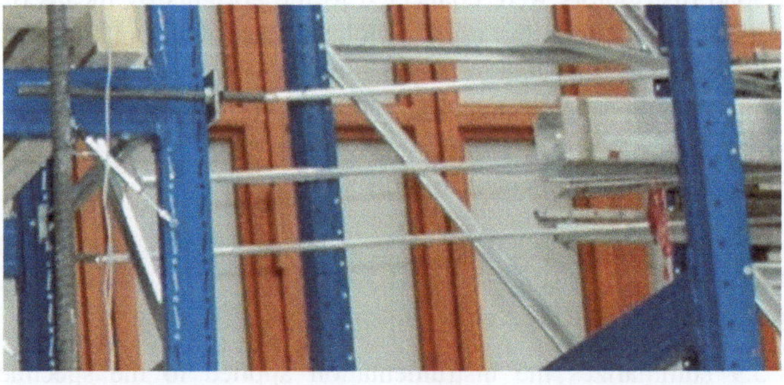

Fig. 5.7 LVDT's for global displacement measurement, at each level

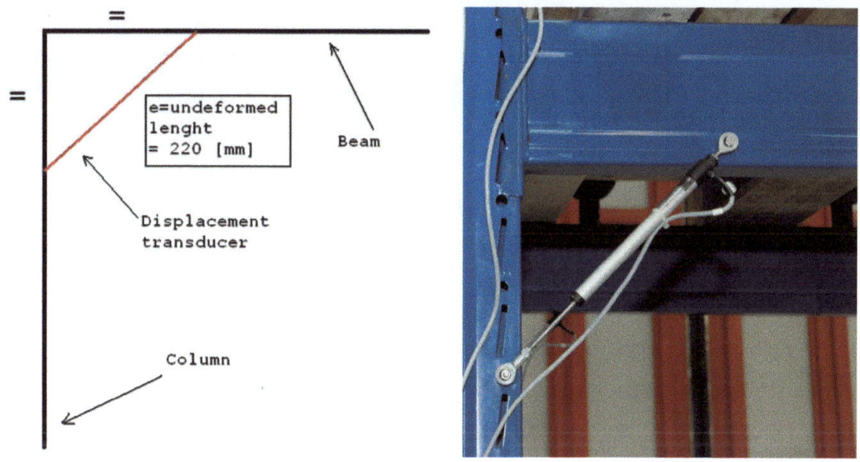

Fig. 5.8 LVDT's for beam-to-upright rotation measurement

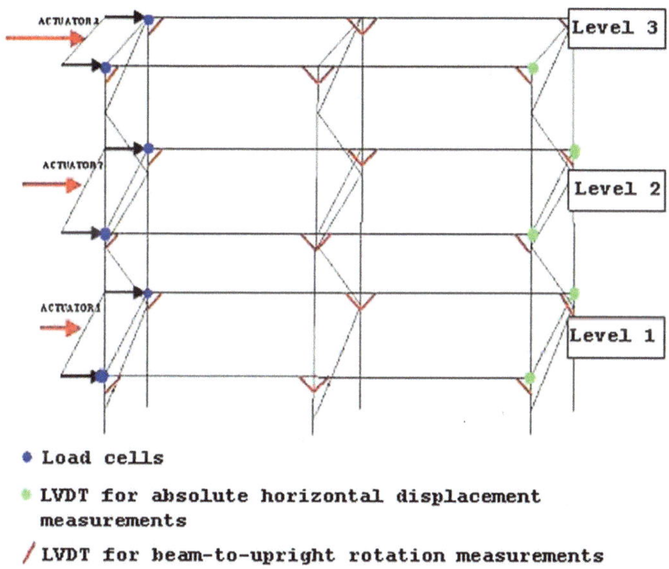

Fig. 5.9 Summary of the instrumentation applied to the specimen

The accelerogram adopted for the test (shown in Fig. 5.11) is an artificial accelerogram, compatible with spectrum type 1, soil D of EC-8. It is the same accelerogram used at the Laboratory of Earthquake Engineering of the National Technical University of Athens to perform, within the SEISRACKS project, shaking table tests on similar rack specimens.

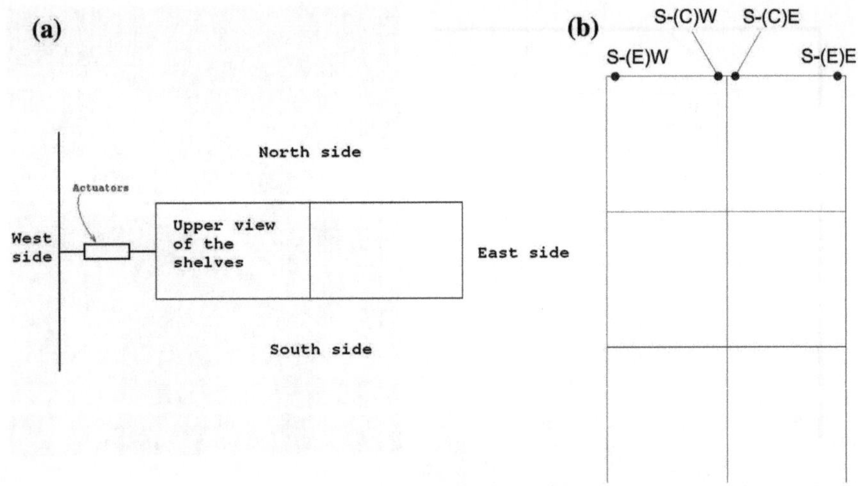

Fig. 5.10 Reference system for the down-aisle pushover test

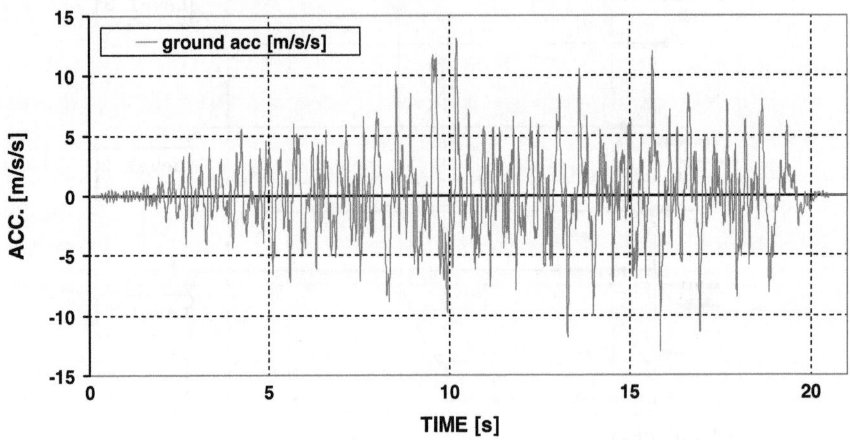

Fig. 5.11 Accelerogram adopted for the pseudo-dynamic test

The tests summarized in Table 5.1 were carried out on the same structure, with different values of the PGA which was increased from one test to the other.

Of course, for all tests after the first one, the initial structural conditions were affected by the damage cumulated during the previous tests. This is introducing an approximation in the analysis; however, as the plastic deformations depend on the maximum deformation reached in the previous tests, the approximation doesn't affects the structural behaviour for oscillations larger than the maximum previously reached.

Table 5.1 Pseudo-dynamic tests in down-aisle direction

SeisRacks—pseudo-dynamic tests down-aisle direction		
Test name	Test type	Date
Y	0.05g EARTHQUAKE	04/07/06
K04	0.35g EARTHQUAKE	05/07/06
K07	0.525g EARTHQUAKE	06/07/06
K08	0.70g EARTHQUAKE	06/07/06
K10	1.05g EARTHQUAKE	07/07/06
K11	1.40g EARTHQUAKE	07/07/06

In any case, the structure resisted an earthquake with PGA of 1.4g, without collapsing. Tests were then stopped, considering further increments of PGA without physical meaning, from an engineering point of view, and in any case beyond the scope of this research.

5.3 Test Results

Hereafter, only the results of the last test are presented.

Damage interested the beam-to-upright connections, where both hooks and openings in the uprights suffered localized deformations as shown in Fig. 5.12.

Figures 5.13, 5.14 and 5.15 show respectively the absolute displacements, the inter-storey drifts and the applied forces at the three levels of the specimen. The structural response didn't evidence any torsional effect.

The absolute displacements are much smaller than those measured at collapse during the push-over test on a similar specimen (see Chap. 4).

The largest displacements and inter-storey drifts were measured at the third level of the specimen; during most part of the test the second floor is characterized by the lowest inter-storey drifts.

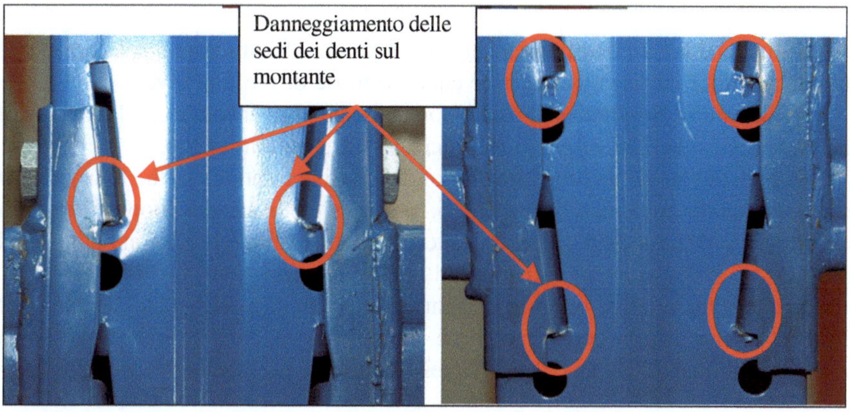

Fig. 5.12 Damage to the beam-to-upright connections

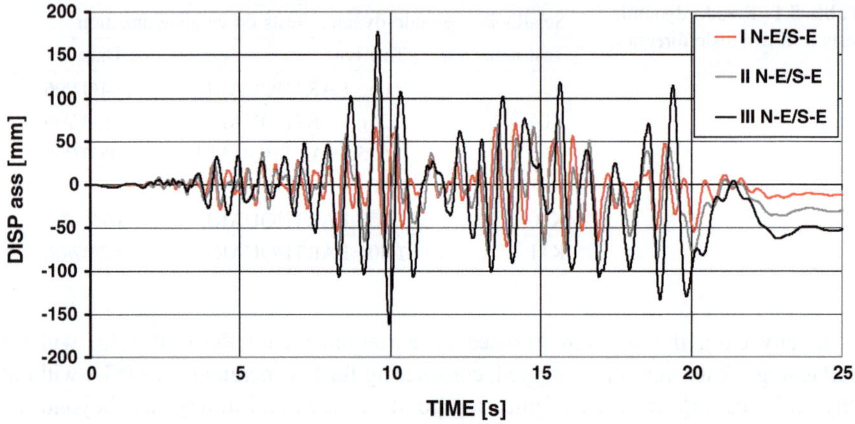

Fig. 5.13 Absolute displacements of the three levels

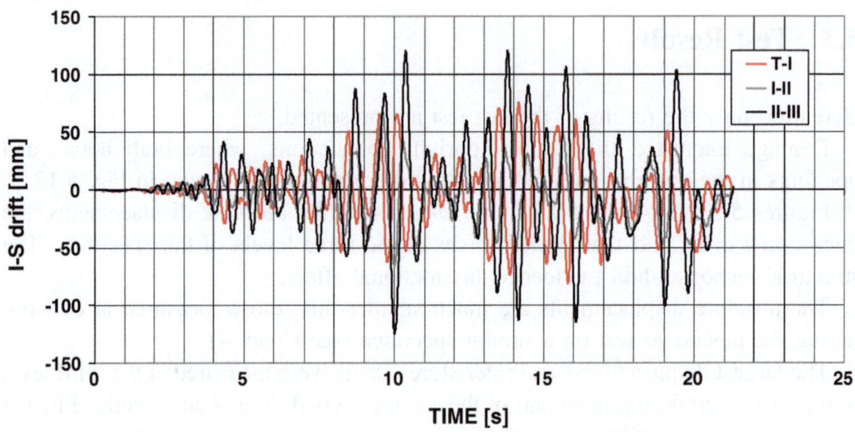

Fig. 5.14 Relative displacements (inter-storey drifts) of the three levels

The first and the second floor are always in phase. The absolute displacements of the second floor are always larger or equal to those of the first level.

The maximum amplitudes of oscillation correspond to absolute displacements of the three levels that are in-phase and increase from bottom to top.

Figure 5.16 shows the time histories of the relative rotations of the four joints at each level of the specimen, while Fig. 5.16 shows the hysteresis loops in terms of shear force versus rotation. It can be noticed that the external joints (on both W and E sides of the specimen) experienced larger rotation than the central joints. Furthermore, the relative beam-to-upright rotations (both positive and negative) of the third level are the largest, while those of the first level are the smallest, as shown in Fig. 5.18 that compares the hysteresis loops at the various levels for each

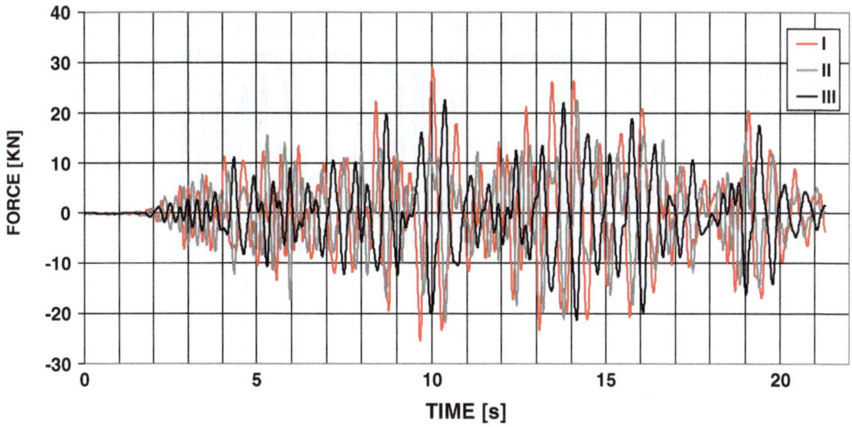

Fig. 5.15 Applied forces at the three levels

corresponding joint of the south frame (namely S_(E)W, S_(C)W, S_(C)E and S_(E)E (see Figs. 5.9 and 5.17).

The absorbed energy was computed as:

$$E_{abs} = \int r^T \cdot du \tag{5.3}$$

where:

- r^T are the restoring forces returned by the structure
- du are the displacements.

As shown in Fig. 5.19, the largest energy amount is absorbed at the first level, and the minimum at the second level.

The largest energy amount absorbed by the 1st level indicates the largest plastic deformation of the components.

Plastic deformation of the base-plate connection, behaving as a semi-rigid connection, contributes to the energy absorption at the 1st level.

The 2nd level, evidencing the smallest relative displacements (hence the smallest plastic deformations), dissipates the smallest energy amount.

Examining Figs. 5.19 and 5.20, showing the energy absorbed at the different levels, it can be noticed the non-linear behaviour of the first level since the earlier cycles. Absence of a linear behaviour since the first cycle can be explained with the kinematics of the beam-to-upright connections allowing energy dissipation due to friction among the components in association with the low testing velocity.

The restoring forces at the different levels are plotted versus the absolute displacements and versus the inter-storey drifts of the various levels respectively in Figs. 5.21 and 5.22. The restoring force, returned by the structure at the 1st level, is

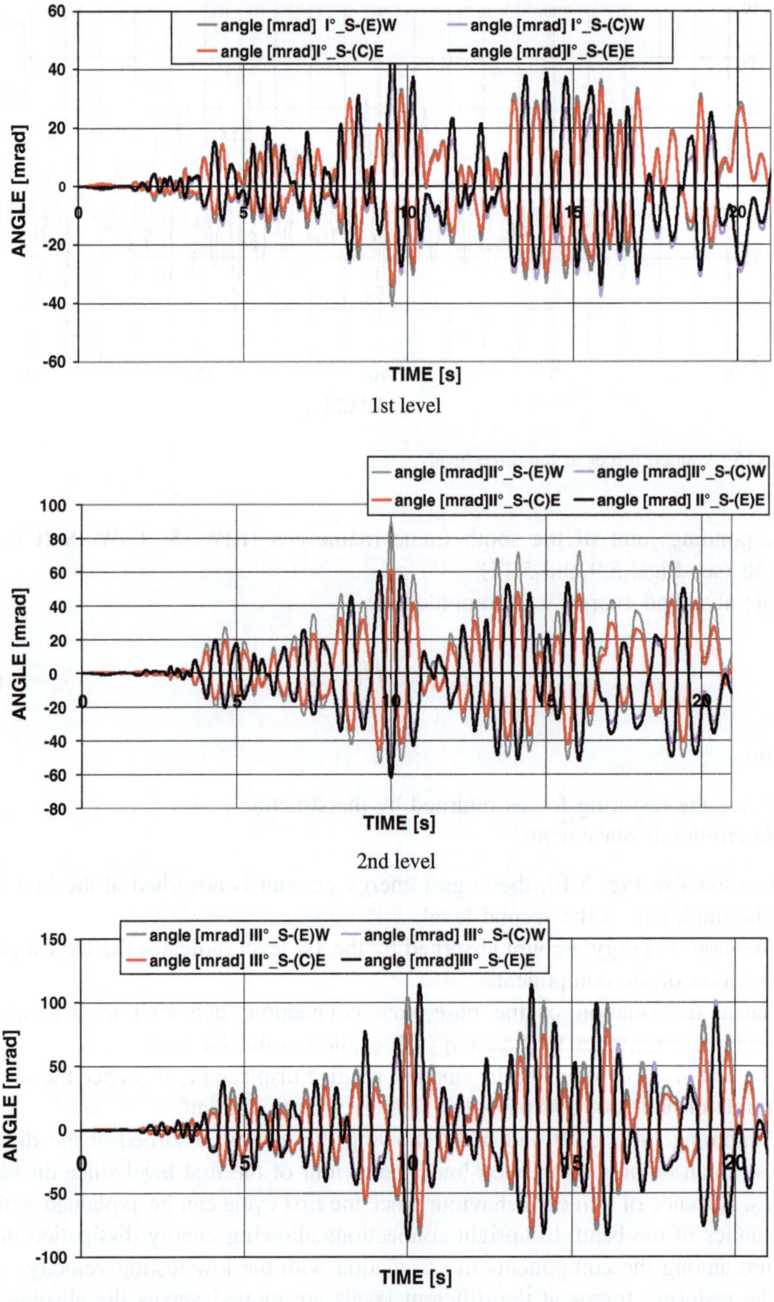

Fig. 5.16 Time histories of the relative beam-to-upright rotation at the various levels

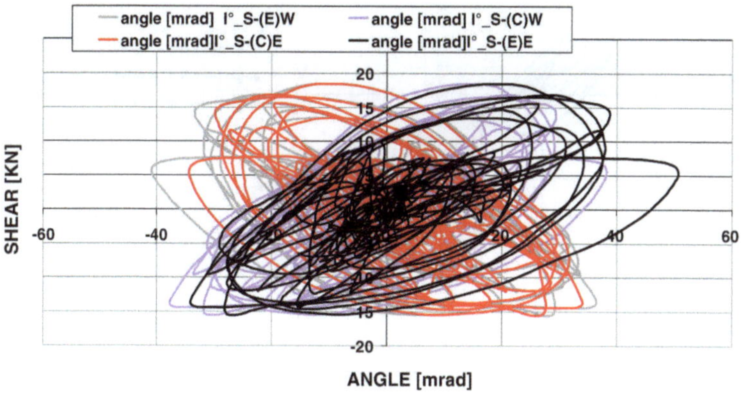

1st level

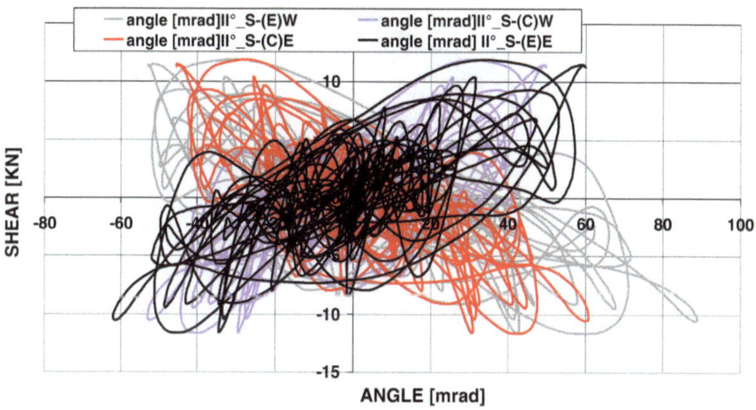

2nd level

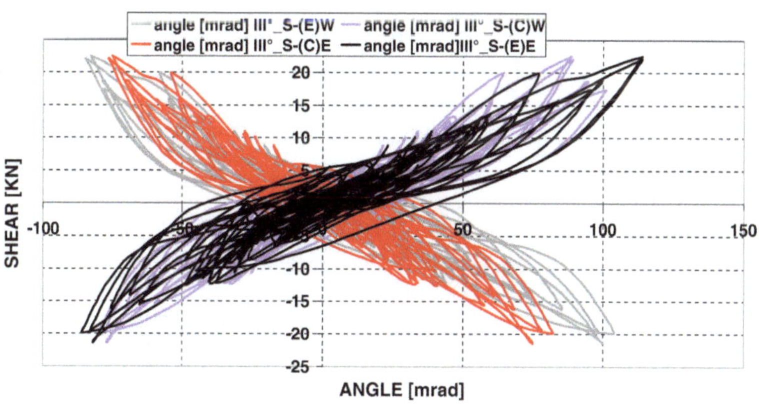

3rd level

Fig. 5.17 Shear versus rotation hysteresis loops at the various levels

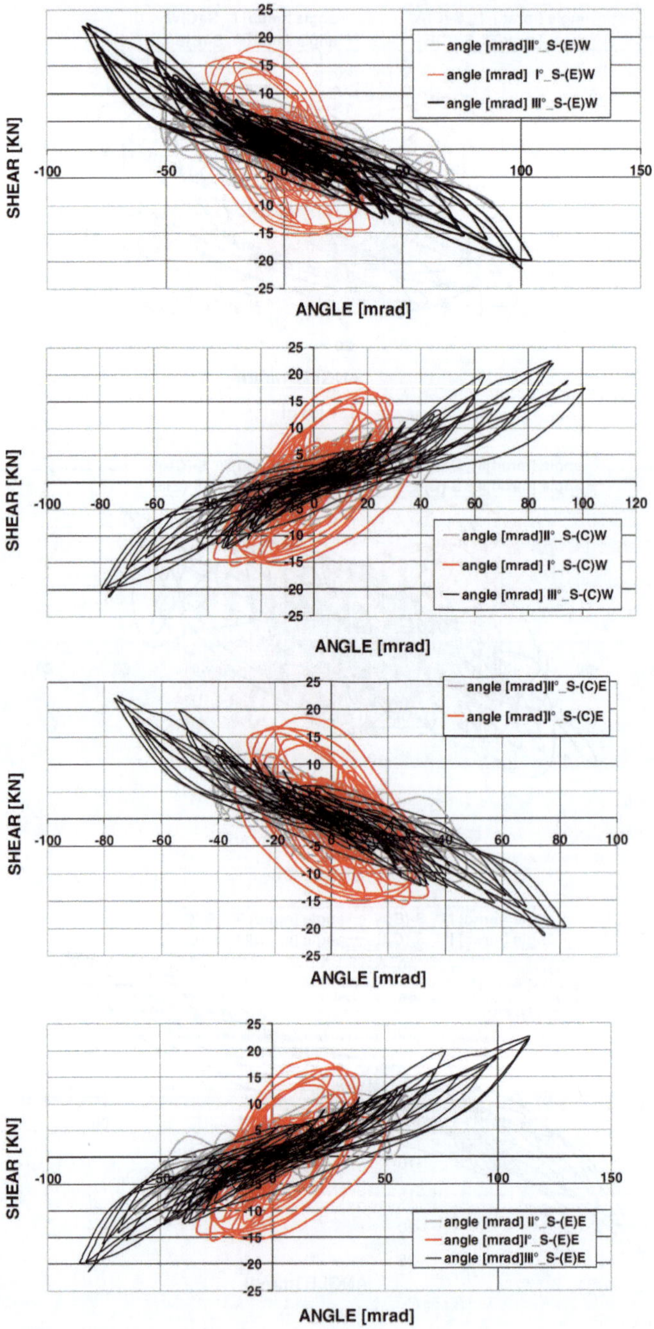

Fig. 5.18 Comparison of shear versus rotation hysteresis loops at the various levels

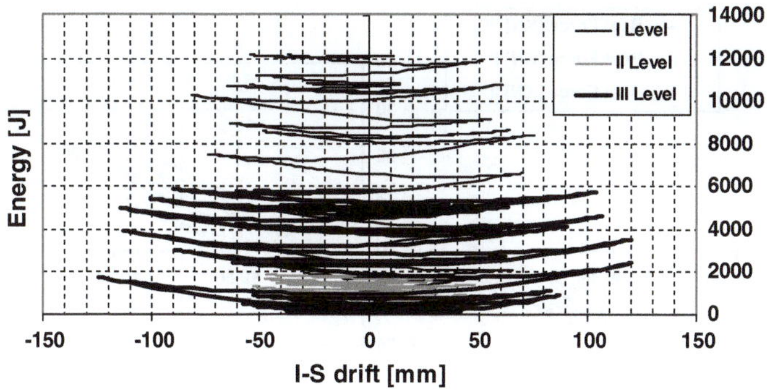

Fig. 5.19 Energy absorbed at the various levels

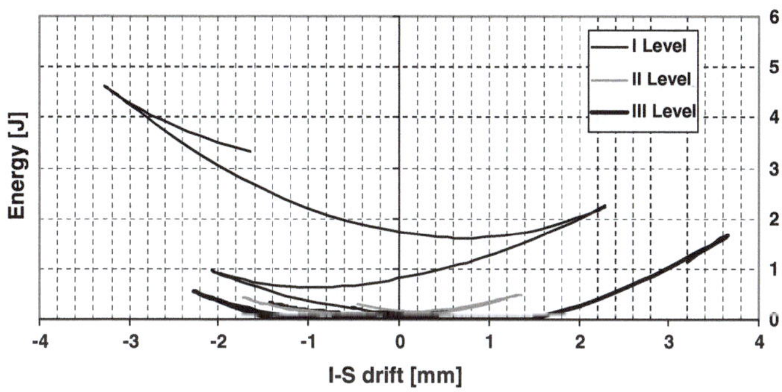

Fig. 5.20 Energy absorbed at the various levels in the 1st cycle

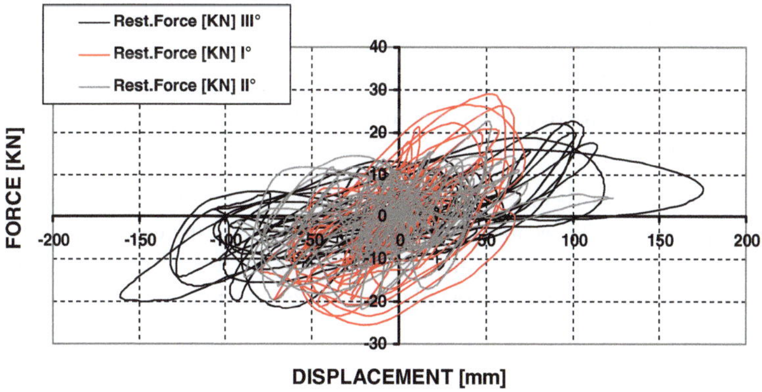

Fig. 5.21 Restoring forces at the different levels versus absolute displacements

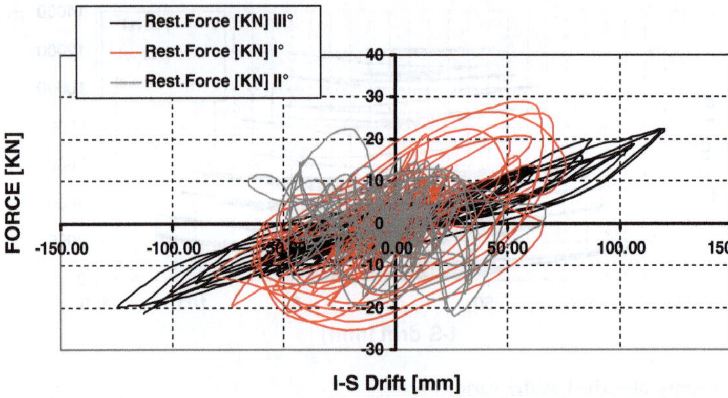

Fig. 5.22 Restoring forces at the different levels versus inter-storey drifts

the highest because is associated with the smallest displacements and it is influenced by the stiffness of the base-plate connection.

The kinetic energy of the structure can be computed as

$$E_c = \frac{1}{2} m v^2 \tag{5.4}$$

The total energy input can be evaluated as the sum of the absorbed and of the kinetic energy.

Figure 5.23 shows that the kinetic energy of the system is very low, with a peak of 5660 J, around 10 s, corresponding to a peak in the input accelerogram. The total energy input strongly depends on the input accelerogram: increasing the acceleration results in an increment of the total energy input.

Figure 5.24 compares the total energy input with the structural response (from an energetic point of view). It can be noticed that there is a perfect balance between energy input and dissipation.

In Fig. 5.25 the first three eigen-frequencies are plotted versus the test duration and compared with the displacements at the various levels plotted versus the test duration. It can be noticed that:

- In the time-period 0–11.5 s, the eigen-frequencies progressively reduce, while the amplitude of the oscillations at the various levels increase.
- In the time-period 11.5–20 s, the eigen-frequencies remain practically constant, while the amplitude of the oscillations never exceeds the maximum values previously attained.

Increasing the amplitude of the oscillations results in plastic deformations and in a reduction of the specimen stiffness, hence in a reduction of the eigen-frequencies. When the oscillations don't exceed the maximum value previously attained, the stiffness of the specimen remains constant as well as its eigen-frequencies.

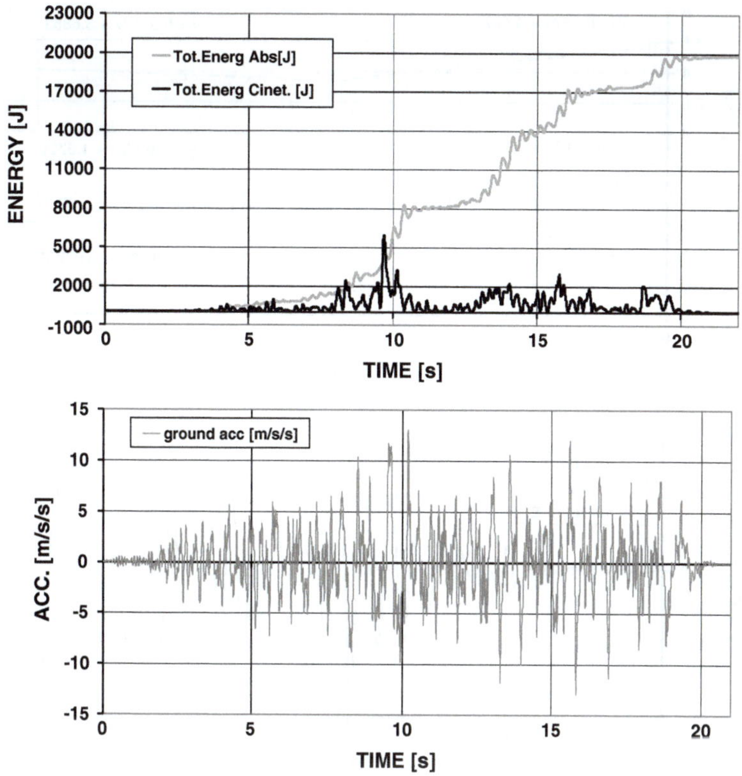

Fig. 5.23 System energy input

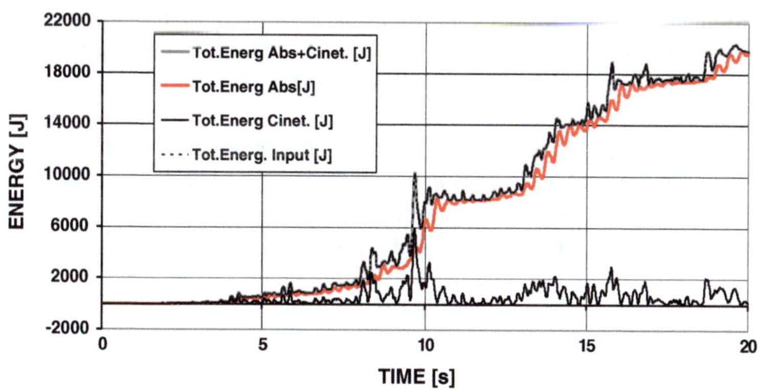

Fig. 5.24 System energy balance

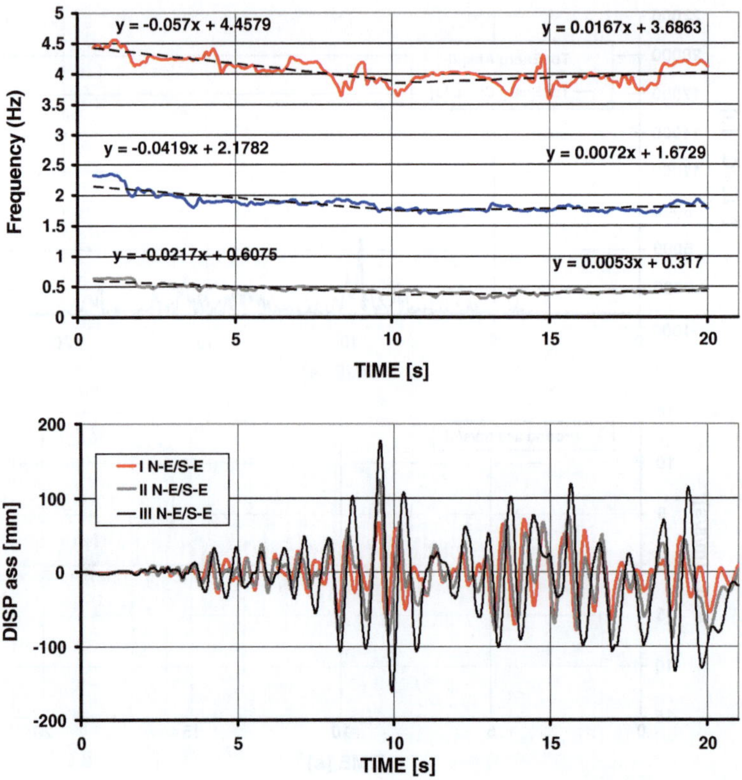

Fig. 5.25 Trend of eigen-frequencies and of displacements at various levels

Additional information about the inelastic behaviour of the specimen can be obtained by examining Fig. 5.26 that shows the damping ratio for the first three eigen-modes as well as the displacements at the various levels of the specimen plotted versus the test duration.

Viscous damping has been disregarded in the analysis (i.e. all coefficients of matrix [C] in the equation of motion are equal to zero; hence, only the hysteretic and friction damping contributions are considered.

In this case, the equation of motion becomes:

$$[M]\ddot{\underline{x}} + Fres = -m\ddot{\underline{y}} \tag{5.5}$$

where F_{res} are the restoring forces returned by the structure during the test, measured by means of the load cells on the jacks. These forces include also the contributions of both hysteretic (which is predominant) and friction damping.

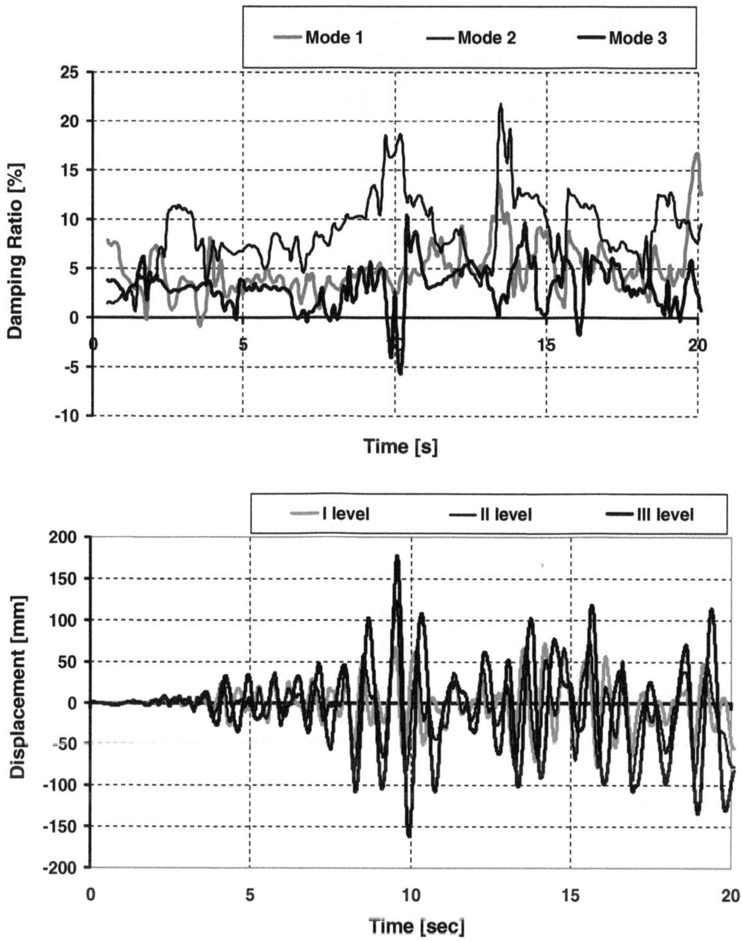

Fig. 5.26 Trend of damping factor and of displacements at various levels

Observing Fig. 5.26 it can be noticed that the damping ratio is strongly influenced by the amplitude of the oscillations of the various levels. In fact, the hysteretic damping contribution increases with the amplitude of the oscillations. This effect is particularly evident by observing the displacements of the third level.

Figure 5.27 shows typical deformed shapes of the structure, during the test. The specimen responded mainly according to the second mode. This fact can be explained by considering the predominant period (T_p = 0.52 s) and the mean period (T_m = 0.40 s) of the input accelerogram; these values correspond respectively to frequencies of 1.92 and 2.5 Hz, that are close to the second eigen-frequency of the (undamaged) specimen, equal to 2.44 Hz.

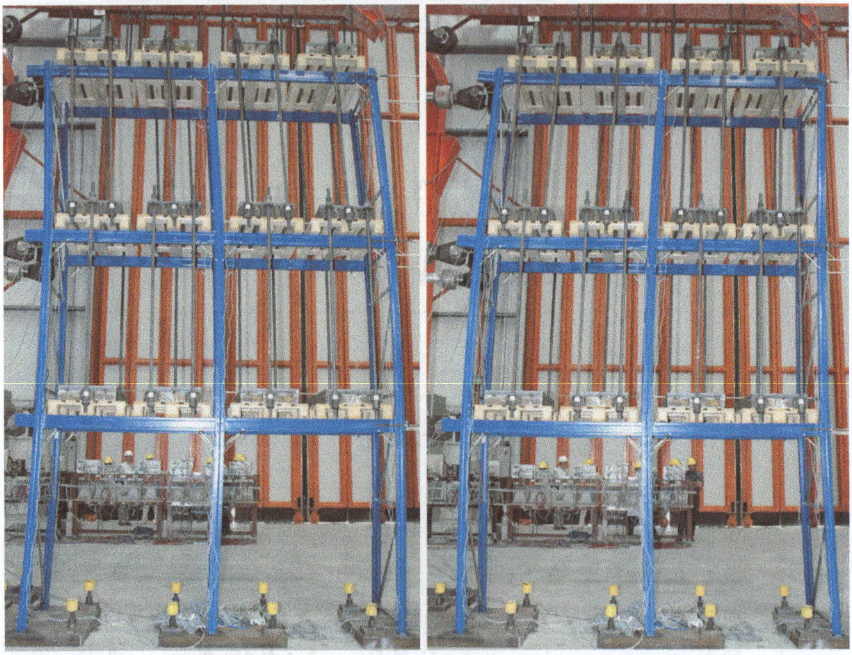

Fig. 5.27 Deformed shapes

5.4 Conclusions

The results of the pseudo-dynamic test on the rack specimen under down-aisle seismic loading previously presented are fully compatible with those obtained on similar specimens, tested under dynamic conditions on the shaking table of the Laboratory for Earthquake Engineering of the national technical University of Athens (see Chap. 6).

Under pseudo-dynamic conditions, in fact, the specimen could sustain the series of earthquake events summarised in Table 5.1 although it didn't collapse during the last test, performed with PGA = 1.4g (ePGA = 1.5g); under dynamic conditions, it will be shown that specimen A1 collapsed under an earthquake with a PGA = 1.46g (ePGA = 1.41g).

Probably, the strain rate effect plays some role in this type of structure; the small movements allowed to the hooks in the holes, in fact, under dynamic conditions result in local impacts that, under increasing number of cycles, may cause cracking either in the hooks or at the edges of the holes.

In any case, comparing the deterioration of the second eigen-frequency of the specimen tested under pseudo-dynamic conditions with the similar one tested in Athens on the shaking table (Fig. 5.28 and Table 5.2) it can be noticed that the trend of reduction of the second eigen-frequency (the most excited one) is similar.

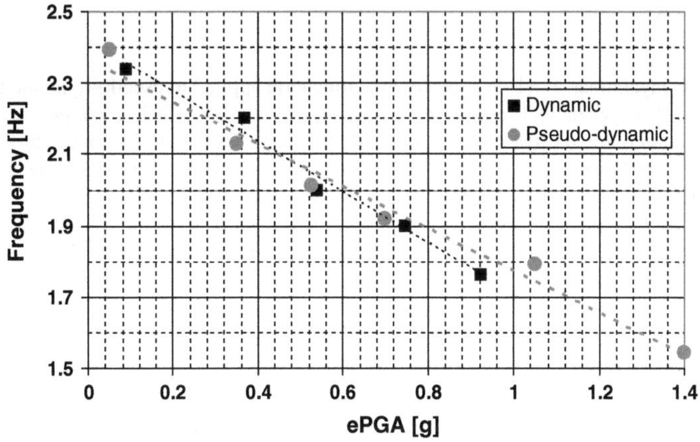

Fig. 5.28 Variation of the second eigen-frequency of specimens tested under dynamic and pseudo-dynamic conditions

Table 5.2 Variation of the second eigen-frequency of specimens tested under dynamic and pseudo-dynamic conditions

	f(Hz) initial	f(Hz) final	Δ %
Dynamic	2.34	1.76	−24.79
Pseudo-dynamic	2.39	1.79	−25

This means that, in general, damage accumulated in the specimen during the two different types of test is similar.

Hence it can be concluded that, from the point of view of the assessment of the seismic resistance and of the damage accumulation of pallet racking systems, pseudo-dynamic tests and shaking table tests are fully compatible, although local damage due to local dynamic effects cannot be reproduced by the pseudo-dynamic testing methodology.

Of course, due to the intrinsic quasi-static nature of the pseudo-dynamic testing procedure, no information can be derived about the effects caused by the sliding of the pallets on the beams during a seismic event.

The values of acceleration that were reached during the pseudo-dynamic tests largely exceed the upper bound of the pallet sliding acceleration that has been derived in Chap. 3.

This means that only full scale dynamic testing will allow a clear assessment of the limit states of pallet racking systems under seismic loading.

References

Beck, J. L., & Jayakumar, P. (1986). System identification applied to pseudo-dynamic test data: A treatment of experimental errors. *Proceedings of the 3rd ASCE engineering mechanics specialty conference on dynamic response of structures*. UCLA, California, April 1986.

Clough, R. W., & Penzien, J. (1975). *Dynamics of structures*. New York: McGraw-Hill.

Takanashi, K., Udagawa, K., Seki, M., Okada, T., & Tanaka, H. (1975). *Non-linear earthquake response analysis of structures by a computer-actuator on-line system*. Tokyo, Japan: Institute for Industrial Science, University of Tokyo (Bulletin of Earthquake Resistant Structure Research Center 8).

Further Readings

Bursi, O. S., Shing, P. B., & Radakovic-Guzina, Z. (1994). Pseudo-dynamic testing of strani-softening systems with adaptive time steps. *Earthquake Engineering & Structural Dynamics, 23*, 745–760.

Calvi, G. M., & Nakashima, M. (1991). Sulla Sperimentazione Pseudodinamica di Strutture n Muratura. *Ingegneria Sismica* (Vol. VIII, No. 3, pp. 3–16). December 1991 (in Italian).

Dermitzakis, S. N., & Mahin, S. A. (1985). *Development of substructuring techniques for on-line computer controlled seismic performance testing, UCB/EERC-85/04*. Berkeley: Earthquake Engineering Research Center, University of California.

Hilber, H. M., Hughes, T. J. R., & Taylor, R. L. (1977). Improved numerical dissipation for time integration algorithms in structural dynamics. *Earthquake Engineering & Structural Dynamics, 5*, 283–292.

Inoue, K., Suita, K., Takeuchi, I., Chusilp, P., Nakashima, M., & Zhou, F. (2006). Seismic-resistant weld-free steel frame buildings with mechanical joints and hysteretic dampers. *Journal of Structural Engineering, 132*(6), 864–872.

Mahin, S. A., & Shing, P. B. (1984). *Pseudo-dynamic test method for seismic performance evaluation: Theory and implementation*. Berkeley: UCB/EERC-84/01 Earthquake Engineering Research Center, University of California.

Mahin, S. A., & Shing, P. B. (1985). Pseudo-dynamic method for seismic testing. *Journal of Structural Engineering, ASCE, 111*(7), 1482–1503.

Nakashima, M. (1995). Strain-hardening behavior of shear panels made of low-yield steel, I: Test. *Journal of Structural Engineering, ASCE, 121*(12), 1742–1749.

Nakashima, M., & Kato H. (1987). *Experimental error growth behaviour and error growth control in on-line computer test control method, Research paper*. Japan: Building Research Institute, Ministry of Construction.

Nakashima, M., et al. (1990). Integration techniques for substructure pseudo-dynamic test. In *Proceedings of 4th U.S. National Conference on Earthquake Engineering* (pp. 515–524). Palm Springs, CA.

Nakashima, M. et al. (1994). Energy dissipation behavior of shear panels made of low yield steel. *International Journal of Earthquake Engineering & Structural Dynamics, 23*(12), 1299–1313.

Nakashima, M., Akazawa, T., & Igarashi, S. (1995a). Pseudodynamic testing using conventional testing devices. *Earthquake Engineering & Structural Dynamics, 24*(10), 1409–1422.

Nakashima, M., Akazawa, T., & Igarashi, S. (1995b), Strain-hardening behaviour of shear panels made of low-yield steel. II: model. *Journal of Structural Engineering, ASCE, 121*(12), 1750–1757.

Nakashima, M., Kato, H., & Takaoka, E. (1992). Development of real-time pseudo-dynamic testing. *Earthquake Engineering & Structural Dynamics, 21*(1), 79–92.

Negro, P., & Verzelletti, G. (1992). Indagini sperimentali nell'ingegneria sismica: il metodo pseudodinamico. In *Proceedings of Seminar "Evoluzione sperimentale per le costruzioni"* (pp. 26–28). JRC Ispra.

Newmark, N. M. (1959). A method of computation for structural dynamics. *Journal of the Engineering Mechanics Division, 85*(EM3), 67–94. (Proc. ASCE).

Pan, P., Nakashima, M., & Tomofuji, H. (2005). Online test using displacement-force mixed control. *Journal of Earthquake Engineering & Structural Dynamics, 34*, 869–888.

Pan, P., Tomofuji, H., Wang, T., Nakashima, M. Makoto Ohsaki, & Khalid, Mosalam. (2006). Development of peer-to-peer (P2P) internet online hybrid test system. *Journal of Earthquake Engineering & Structural Dynamics, 35*(7), 867–890.

Schneider, S. P., & Roeder, C. W. (1994). An inelastic substructure technicque for the pseudo-dynamic test method. *Earthquake Engineering & Structural Dynamics, 23*, 761–775.

Shing, P. B., Bursi, O. S., & Vannan, M. T. (1994). Pseudo-dynamic tests of a concentrically braced frame using substructuring technicques. *Journal of Constructional Steel Research, 29*, 121–148.

Shing, P. B., & Mahin, S. A. (1983). *Experimental error propagation in pseudo-dynamic testing, UCB/EERC-83/12*. Berkeley: Earthquake Engineering Research Center, University of California.

Shing, P. B., & Mahin, S. A. (1985). Computational aspects of a seismic performance test method using on-line computer control. *Earthquake Engineering & Structural Dynamics, 13*, 507–526.

Shing, P. B., & Mahin, S. A. (1987). Cumulative experimental errors in pseudo-dynamic tests. *Earthquake Engineering & Structural Dynamics, 15*, 409–424.

Shing, P. B., & Mahin, S. A. (1988). Rate-of-loading effect on pseudodynamic tests. *Journal of Structural Engineering, ASCE, 114*(11), 2403–2420.

Shing, P. B., & Mahin, S. A. (1990). Experimental error effects in pseudodynamic testing. *Journal of Engineering Mechanics, ASCE, 116*(4), 805–821.

Shing, P. B., Nakashima, M., & Bursi, O. S. (1996). Application of pseudo-dynamic test method to structural research. *Earthquake Spectra, 12*(1), 29–56.

Shing, P. B., & Vannan, M. T. (1991). Implicit algorithm for pseudo-dynamic tests: convergence and energy dissipation. *Earthquake Engineering & Structural Dynamics, 20*, 809–819.

Shing, P. B., Vannan, M. T., & Carter, E. W. (1991). Implicit time integration for pseudo-dynamic tests. *Earthquake Engineering and Structural Dynamics, 20*, 551–576.

Takanashi, K., & Nakashima, M. (1987). Japanese activities on on-line testing. *Journal of Engineering Mechanics, ASCE, 113*(7), 1014–1032.

Takanashi, K., & Nakashima, M. (1988). On-line computer test control method and its application to earthquake response simulation of steel structural models. *Journal of Constructional Steel Research, 11*, 27–40.

Thewalt, C. R., & Mahin, S. A. (1987). *Hybrid solution techniques for generalized pseudo-dynamic testing, UCB/EERC-87/09*. Berkeley: Earthquake Engineering Research Center, University of California.

UNI EN 1998-1:2005. (2005). Eurocode 8—Design of structures for earthquake resistance—Part 1: General rules, seismic actions and rules for buildings.

Wang, T., Nakashima, M., & Pan, P. (2006). On-line hybrid test combining with general-purpose finite element software. *Journal of Earthquake Engineering & Structural Dynamics*, *35*(12), pp. 1471–1488.

Yamazaki, Y., Nakashima, M., & Kaminosono, T. (1989). Reliability of pseudo dynamic test in earthquake response simulation. *Journal of Structural Engineering, ASCE, 115*(8), 2098–2112.

Chapter 6
Dynamic Full-Scale Tests

6.1 Overview

Shake-table testing of full-scale storage rack systems loaded with real merchandise represents the most direct experimental procedure to assess their seismic behaviour. Unfortunately, this type of testing is expensive compared to other methodologies and only a very limited number of shake-table studies on storage racks have been performed to date in the world and are available in the literature.

6.1.1 Previous Studies

6.1.1.1 Research in the U.S.

6.1.1.1.1 Dynamic In Situ Testing of Storage Rack Systems

The first published in situ dynamic investigation of storage racks was performed in the mid 1970s at various distribution centers in the San Francisco Bay Area (John A. Blume & Associates 1973). Ambient and manmade vibration measurements were applied to representative steel industrial storage racks of the standard pallet, drive-in and drive-through, cantilever, and stacker crane types for the purpose of obtaining range of natural periods and damping ratios. The ambient vibration and man-made excitation measurements generated average response accelerations at the top of the racks in the order of $0.005g$ and $0.015g$, respectively. The experimental results indicated that the fundamental translational period of storage racks could not be predicted well by the empirical formula used to characterize building periods provided in the *Uniform Building Code*.

C.A. Castiglioni, *Seismic Behavior of Steel Storage Pallet Racking Systems*,
Research for Development, DOI 10.1007/978-3-319-28466-8_6

Measured fundamental periods over a range of actual merchandise loading conditions averaged 0.6 s in the down-aisle direction, and 0.2 s in the cross-aisle direction.

Torsional periods averaging 0.4 s were identified in many of the rack configurations. It was noted that these period values would increase at least by 20 % under response amplitudes representative of a strong earthquake. Measured structural damping ratios for storage racks averaged 2–3 % of critical at root-mean-square response acceleration levels of 0.01–0.02*g*. It was noted that these damping values would increase at least by a factor of 2 under response amplitude representative of a strong earthquake for which significant energy dissipation would occur due to rocking, slippage and interaction of stored merchandise. Thus, it was concluded that a damping ratio of 5 % of critical would be a reasonable value for storage racks under seismic excitations.

Krawinkler et al. (1979) subjected two full pallet rack assemblies loaded with gravity loads to forced and free vibration tests to obtain information on natural frequencies, mode shapes and damping characteristics in the down-aisle and cross-aisle direction. Measured fundamental periods averaged 0.7 s in the down-aisle direction, and 0.5 s in the cross-aisle direction. The vibration decay obtained from the free-vibration tests in the down-aisle direction exhibited a textbook example of Coulomb-type friction decay. At large amplitudes, the friction between the grip-type connectors and the perforations in the uprights caused significant damping (in the order of 2.5–3.5 % of critical). Once the connectors locked up at smaller amplitudes, the damping dropped drastically to a very small value (in the order of 0.7 % of critical). The damping characteristics in the cross-aisle direction were more constant with vibration amplitudes (in the order of 1–2 % of critical).

6.1.1.1.2 Shake-Table Testing of Storage Rack Systems

The first published shake-table study on storage racks in the United States was performed in the late seventies on the shake-table at the University of California, Berkeley (Chen et al. 1980a, b, 1981). This study provided the background information for the seismic design provisions for storage racks in the U.S.

Four types of full-scale industrial steel storage racks were subjected to scaled ground motions actually recorded during two California earthquakes (1940 El Centro and 1966 Parkfield earthquakes). The ground motions were scaled so that the resulting base shear coefficients nearly equaled the design base shear coefficients given in *Uniform Building Code* for ordinary moment frames ($K = 1.0$) in the down-aisle directions and ordinary braced frames ($K = 1.33$) in the cross-aisle direction.

The types of storage racks tested were: single standard pallet racks, back-to-back pallet racks, drive-in racks, and stacker racks. In general, the storage racks performed well during the tests with the exception of the drive-in stacker racks in the cross-aisle direction, for which considerable buckling was observed in the first storey diagonal members. The fundamental periods of vibration ranged from 2 to 3 s for the standard pallet and drive-in racks in the down-aisle direction and 0.5–1.0 s for the standard pallet, drive-in, and stacker racks in the cross-aisle direction. The first mode damping values were much larger in the down-aisle direction (on the order of 3–9 % of critical) than in the cross-aisle direction (0.5–3 % of critical). It was also observed that the ductility and energy dissipation capacity of the racks were much larger in the down-aisle direction (where the structure behaves as a moment resisting frame) than in the cross-aisle direction (where the structure behaves as a braced frame). Therefore, the racks could undergo significant inelastic deformations without suffering major damage in the down-aisle direction, but could only develop limited amount of inelastic deformations in the cross-aisle direction. Second order ($P - \Delta$) effects contributed significantly to the response of the racks in the down-aisle direction.

More recently (Filiatrault 2001), five different back-to-back pallet racks loaded with real merchandise were tested under a single component of the ground motion recorded at Canoga Park (during the 1994 Northridge earthquake in California) scaled at various amplitudes. Three of the tests were conducted in the cross-aisle direction, while the two other tests were conducted in the down-aisle direction. In general, the racks performed well. Significantly more flexibility, ductility, and energy dissipation capacity were observed in the down-aisle direction than in the cross-aisle direction. The fundamental periods of vibration averaged 1.4 s in the down-aisle direction and 0.6 s in the cross-aisle direction. No structural damage occurred in any of the rack configurations for peak ground motion amplitudes less than $0.42g$.

A shake-table study was recently conducted at the University at Buffalo on four different pallet rack configurations incorporating bolted beam-to-upright connections (Filiatrault and Wanitkorkul 2004a, b). All racks were tested in the down-aisle direction.

The test results indicated that the rotational stiffness of beam-to-upright connections is the main factor influencing the down-aisle seismic response of pallet racks.

The initial values obtained for the rotational stiffness of beam-to-upright connections are two to three times higher than values used in current design. These values reduce with the amplitude of vibration but generally return to their initial values once the pallets rack returns to its undeformed position. Furthermore, very ductile seismic behaviour was observed in the down-aisle direction with peak inter-storey drifts exceeding 7 % without any sign of incipient collapse.

In a recent paper (Filiatrault et al. 2006a), propose a simple displacement-based seismic design procedure to verify the collapse prevention of storage racks in their down-aisle direction under Maximum Considered Earthquake (MCE) ground motions.

The model is aimed at developing simplified equations for the fundamental period, the base shear, and the lateral displacement at the top of storage racks in their down-aisle direction as a function of the beam-to-upright rotational characteristics at a given target design lateral displacement.

Experimental results, obtained from mono-axial shake-table tests performed on two different three-level two-bay pallet-type steel storage racks in their down-aisle direction, are used to assess the predictive capabilities of the model.

Bolted connections were used between beams and uprights. This type of connections differed substantially from the rack specimens investigated in the Blume Report (hook and button grip connectors).

The pallet weight was simulated by standard concrete bricks stacked and plastic wrapped on standard pallets. Each pallet weighed 14 kN. Each beam could accommodate two standard pallets along its span. Therefore, each rack specimen could contain 12 pallets over its three beam levels.

The strongest horizontal component recorded at Canoga Park during the 1994 Northridge earthquake was used for the seismic tests. This record scaled to a peak ground acceleration of 0.5g can be associated with an ordinary ground motion matching the NEHRP spectral accelerations for a probability of exceedance of 10 % in 50 years for a soil type D (Krawinkler et al. 2000). This input motion is significantly more intense than the ground motions used in the Blume Report (1940 El Centro and 1966 Parkfield earthquakes).

Even more recently, Filiatrault and Wanitkorkul (2006a, b), performed a series of Shake Table Tests on Ridg-U-Rak, with and without base-isolation.

6.1.1.1.3 Shake-Table Testing of Merchandise

As part of the shake-table testing conducted by Chen et al. (1980a), the seismic response of a two-story, one bay wide, one bay deep rack was obtained for two different cases: (a) the case in which the merchandise was tied with metal straps to the rack beams, and (b) the case in which the merchandise was not fastened to the rack. For the purpose of this investigation, eight types of merchandise with different geometry and weight were considered.

The tests were conducted with scaled ground motions that resulted in base shear coefficients that were about the same as code level base shear coefficients for buildings. The results of the tests indicated that neither merchandise nor wood pallets fell off the racks during the tests. Only some uppermost paper products and canned goods moved slightly during the tests in the down-aisle direction. It was found, however, that the seismic response of the racks without the merchandise tied up to the beams was smaller compared to that of the same rack with the contents rigidly attached to the beams. This response reduction was particularly obvious during the strongest portion of the response. Low amplitude free vibration tests also indicated that the damping was slightly higher for the case where the merchandise was not tied up to the rack.

The shake-table tests conducted recently by Filiatrault (2001) also included real merchandise. It was clearly demonstrated that the use of plastic wrap was a very efficient mean of restraining the merchandise from falling off the racks. No fallen merchandise was observed for ground motion amplitudes less than 0.30g. Significant slippage of heavy merchandise on pallets was observed but no pallet overturned during any of the tests. Tall merchandise items (e.g., water heaters) were shown to be vulnerable to overturning and need of restraint.

6.1.1.2 Research in Europe

The only European research project (known to this author) on full scale steel pallet racks under dynamic conditions has been carried out with the financial support of E.U. (within the ECOLEADER Research Program for Free Access to Large Scale Testing Facilities), of Section X of the European Federation of Maintenance (F.E.M.) (now European Racking Federation), and of the Italian Ministry of University and Research (M.I.U.R) in 2002.

Castiglioni et al. (2003) performed, at the Laboratory of Earthquake Engineering of the National Technical University of Athens, shake-table tests on four full-scale steel storage pallet racks loaded by concrete blocks mounted on pallets simulating content merchandise. The four specimens were chosen among six structures designed by two different European manufacturers based on the Eurocode 8. The experimental results indicated that sliding of pallets occurred for ground motion intensities less than the design levels. Also, the diagonal bracing configuration in the down-aisle and cross-aisle directions has a significant influence on the seismic response. In particular, eccentric bracing configurations can lead to significant torsional response. The main results of this study will be presented in Sect.6.3.3.

More recently (Seisracks 2008), with the financial support of the EU, Research Programme of the Research Fund for Coal and Steel—Steel RTD, Contract no. RFS-PR-03114, the SEISRACKS research Project, "Storage Racks in Seismic Areas", has been carried out under the co-ordination of the Italian Association of Steel Constructors (ACAI), and the scientific supervision of this author, in coop-eration with the Instituto Superior Technico of Lisbon (Prof. Calado), the Laboratory for Earthquake Engineering of the National Technical University of Athens (Prof. Carydis), Politecnico di Milano (Prof. Castiglioni), University of Liege (Prof. Plumier) and the European Laboratory for Structural Assessment of the Joint Research Center (ELSA-JRC) of the European Union (Dr. Negro).

Within the SEISRACKS project, the following activities were carried out:

Characterization of the component behaviour (the results were presented in Chap. 2)
Assessment of the sliding conditions of pallets on rack beams (presented in Chap. 3)
Push-over tests on two full-scale racks models (presented in Chap. 4)
Pseudo-dynamic tests on one full-scale rack model (presented in Chap. 5)
Assessment of the actual service loading conditions of racks, by means of in situ continuous monitoring of an installation for more than two years
Shake table tests on six full-scale rack models (presented in the following Sect. 6.3.4)
Numerical modelling of the seismic behaviour of racks accounting for pallet sliding.

6.1.2 Research Needs

As presented in this brief review, the current engineering knowledge-base concerning the earthquake safety and vulnerability of storage racks is 20–30 years old and is limited to contents and racks unlike many modern real applications.

The retail industry and the state-of-the-art of the design of storage racks have changed considerably in the meantime.

Large chains of stores now routinely invite the public to shop in physical conditions that formerly were only found in warehouses. Racks have more complex configurations and are taller, and their contents have become heavier. These facts clearly pinpoint to urgent research needs related to the seismic behaviour of storage rack systems.

Literally quoting FEMA 460 (2005),There is an urgent need to increase the experimental database on the seismic response of complete storage rack systems through shake-table testing. The main variables that need to be investigated in such experimental programs are:

The layout and types of storage racks representing current construction practices and innovative systems such as eccentric bracing.

The layout and types of merchandise contents.

The types of seismic restraints (e.g., plastic wraps, screens, ledges, etc.) for contents.

The structural interaction between neighbouring racks.

The direction of the horizontal seismic input, relative to the rack's orientation (transverse, longitudinal, or non-orthogonal).

The characteristics of the input ground motions, including consideration of whether vertical accelerations must be characterized and near-field motions, and relating these input motions to seismic hazard mapping and codes in use in the United States.

Testing needs in the cross-aisle are even more urgent since the understanding of this directions behaviour is even less understood. Failures of racks in earthquakes are most commonly reported as cross-aisle failures.

The information on the seismic response of merchandise contents installed in storage racks is very limited. There is an urgent need to conduct shake-table studies of merchandise. For this purpose, shake-table testing could be used to simulate the motions experienced by various levels of storage racks during earthquakes. A robust numerical model would be required to develop these input motions. Various merchandise items could be mounted on the shake-table via a rigid assembly representative of the level on which they are mounted…

The SEISRACKS Project, whose dynamic tests main results will be summarized hereafter in Sect.6.3.4, was conceived in order to try to give some answers to the previously addressed questions, and to (at least partially) cover knowledge gaps.

6.2 Dynamic Tests on Merchandise

Within the SEISRACKS Research Project, a number of tests were recently carried out on three different types of palletised goods. Seisracks (2008): good type 1 is baby diapers, good type 2 are boxes with powder clothes detergent and good type 3 are boxes with liquid clothes detergent. These palletized goods, presented in Fig. 6.1, were lent, free-of-charge, to the testing laboratory by Procter & Gamble.

For the definition of their frequency and damping ratio, a sine logarithmic sweep signal was applied. The frequency range of sine logarithmic test was 1–32 Hz at an increment rate of one octave per minute. The tests were executed along each one of the global axes X and Y separately with a constant acceleration of 0.50 m/s^2. Since sine sweep rate is one octave per minute and start frequency is 1.0 Hz the exciting frequency versus time is given (in Hz) by the following expression:

$$f = 1.0 * 2^{\frac{Time(s)}{60}}$$

The pallets were fixed directly on the shaking table.

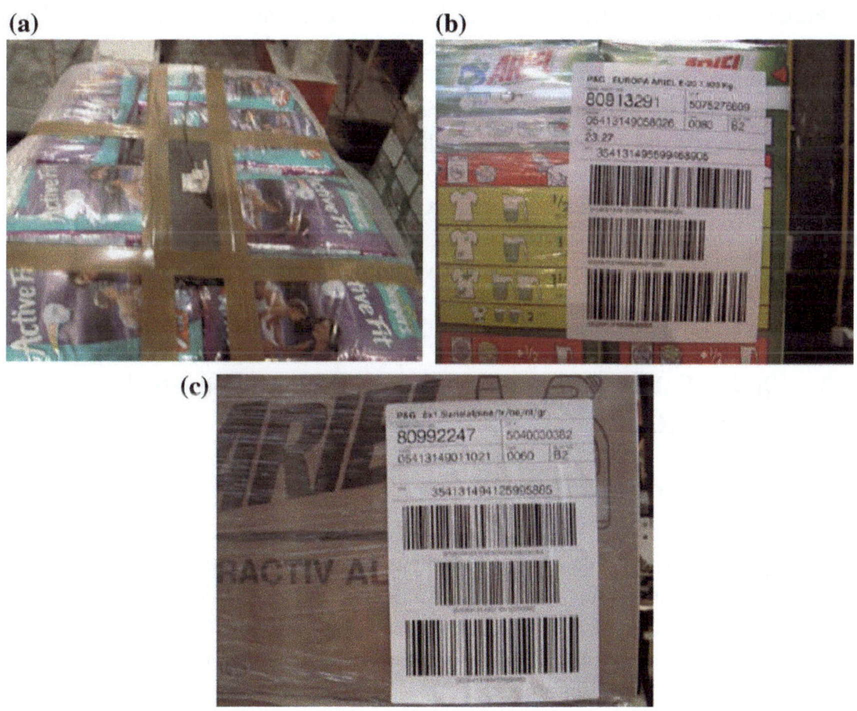

Fig. 6.1 Type of "palletised" goods: **a** baby diapers, **b** powder clothes detergent, **c** boxes with liquid clothes detergent

Table 6.1 Measured natural frequencies and damping ratio

Type of goods	Direction	First mode		Second mode	
		Frequency	Damping	Frequency	Damping
		[Hz]	[%]	[Hz]	[%]
I (A5)	X	5.2	19.5	20.6	10.2
	Y	4.3	16.0	19.2	9.0
II (A4)	X	3.4	19.0	24.1	9.0
	Y	2.3	5.7	21.1	7.9
III (A7)	X	5.7	25.0	18.0	7.3
	Y	1.5	10.5	13.4	15.2

The fundamental natural frequencies and the damping of each type of "palletised" good are shown in Table 6.1. These frequencies were determined from the acceleration time histories of the response of each type of goods to the sine logarithmic sweep excitation. Damping ratio was calculated by using the half power bandwidth method.

Examining the results presented in Table 6.1, it can be stated that the dynamic characteristics of the "palletized" goods depend on the nature of the goods and the way of packing.

Further research is needed in this field for an exhaustive assessment of the problem. During the Seisracks Project, however, it was impossible to identify any other supplier (in addition to Procter & Gamble) available to lend their palletised merchandise and to allow testing on it, because fearing of eventual damages.

For this reason, the obtained results, although interesting, are limited in number, and should be considered as those of a pilot study. Recently (in November 2007) the European Racking Federation has expressed to this author its interest in this research topic, promising support for an extensive testing campaign on palletised merchandise, that should be performed within the present year 2008.

6.3 Shake-Table Tests on Full-Scale Pallet-Type Steel Storage Racks

6.3.1 Test Infrastructure

A mechanical shaking table existed at Athens' National Technical University since 1965, at the Institute for Structural Analysis and Aseismic Research, where Professor Dr. P.G. Carydis was Lecturer since 1964 under the leadership of the late Professor F. Kokkinopoulos. There, a lot work was done, especially in scaled models, with analogical computers.

It is well known that Greece suffers from destructive earthquakes. The best way to mitigate the effects of earthquakes is to improve the knowledge on the subjects and to compile the necessary regulations for the a-seismic design.

The destructive earthquakes that occurred in the major cities of Greece during the period 1978–1981 gave a "support" for the construction of the shaking simulator at the Laboratory of Earthquake Engineering at National technical University of Athens. The establishment was totally funded from Greek national sources. The simulator (Fig. 6.2) consists of a rigid platform with dimension 4.0 × 4.0 m with 6° of freedom and of a system controlling the input motion and the response of the specimen tested on the platform.

The earthquake simulator was calibrated and became fully operational at the beginning of 1987.

The 6 DOF shaking simulator with its control system was unique in the word at that time, and is still one of the largest in Europe. The weight of the reinforced concrete reaction mass amounts to 25 MN. The facility has a full active control along the 6° of freedom. The whole laboratory and the adjacent office building are specially designed and constructed to serve the facility under consideration.

The main characteristics of the earthquake simulator may be summarised as follows:

Dimensions: 4.0 (m) × 4.0 (m) × 0.6 (m)
Weight: 100 (kN)
Material: Steel

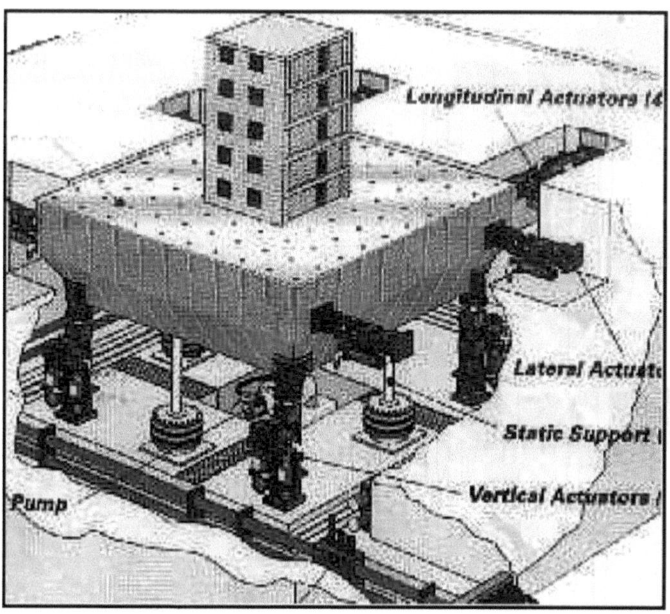

Fig. 6.2 Sketch of the shake table facility

Number of independent degrees of freedom: six (6)
Max Weight of Specimen: 100 (kN), if centre of mass is at 2 (m) above the Simulator's platform
Max Horizontal Force (at each direction X, Y): 320 (kN)
Max Vertical Force (direction Z): 640 (kN)
Max displacement of the platform (each axis): ±10 (cm)
Max rotation (about each axis): $\pm 7 \times 10^{-2}$ (rad)
Max acceleration to each horizontal direction (X, Y): 2.0 (g)
Max acceleration to vertical direction (Z): 4.0 (g)
Max Velocity to each axis (simultaneous performance): 100 (cm s^{-1})
Operating frequencies (for each degree of freedom): 0.1–100 Hz
Electric Power Installed: 1200 kVA.

6.3.2 Test Set-up

All the rack structural models tested on the shake-table consisted in two bays of 1800 mm each, three levels, for a total height of the structure about 6000 mm, width of the upright frames 1100 mm, and were positioned on the table and loaded as shown in Fig. 6.3.

Live loads consisting of concrete blocks (7.7 kN each) were positioned on pallets, two pallets per bay. Hence, at each level, a live load of 30.8 kN was present (15.4 kN for pair of beams), for a total live load of 92.4 kN.

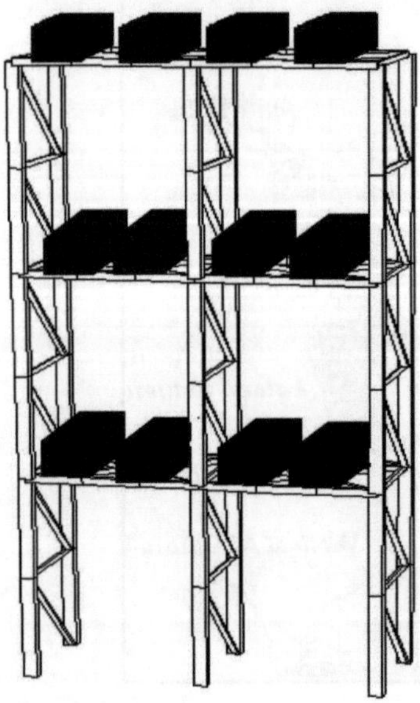

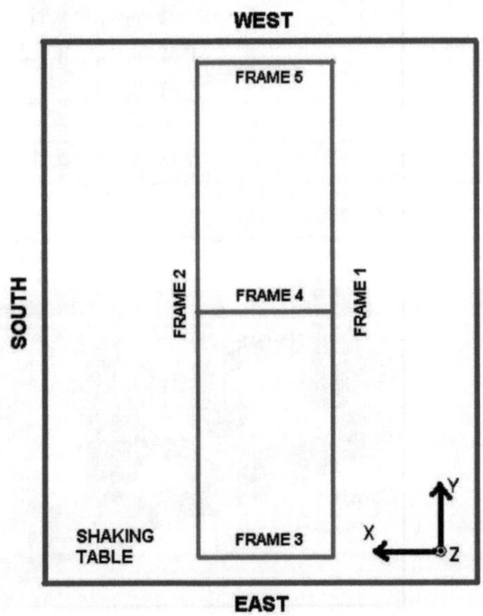

Fig. 6.3 Test set-up

Instrumentation varied from specimen to specimen, according to the direction of the excitation and the type of test.

6.3.3 The ECOLEADER Project on "Seismic Behaviour of Pallet Rack Systems"

Within the ECOLEADER Research Program of the EU, for Free Access to Large Scale Testing Facilities, the research project "Seismic Behaviour of Pallet Rack Systems" has been carried out, planning both a theoretical and experimental study of racks subjected to dynamic forces (Castiglioni 2003a, b; Castiglioni et al. 2003; Brescianini et al. 2003).

Two manufacturers, referenced as manufacturer M1 and manufacturer M2, provided six different specimens to be dynamically tested in full scale on the shaking table facility of the Laboratory for Earthquake Engineering at the National Technical University of Athens. Out of these six structures only four were tested: they are labeled A and B for the manufacturer M1 and C and D for the manufacturer M2.

6.3.3.1 The Specimens

The manufacturers designed their specimens according to the geometric and loading characteristics previously described, and for seismic actions characterized by an acceleration spectrum type 1, subsoil D of Eurocode 8, with PGA respectively of 0.075g, 0.15g and 0.35g.

All the structures complied the requests, except specimen D that was designed for a live load of 8 kN for each pair of beams.

In order to reproduce real service conditions, six concrete blocks (reproducing the concrete slab of a warehouse foundation) were fixed to the shaking table by means of four anchor bolts (Fig. 6.4).

The racks were clamped to these concrete blocks by means of two chemical anchorages for each base-plate. As a consequence, the base of the racks was at about 50 cm above the shaking table level.

Furthermore, like in usual commercial installations in seismic zones, all the beam to upright connections were equipped with a pair of locking pins whose effect is to prevent the unintentional release of the beams.

Hereafter, a brief description is given of the main characteristics of the four specimens.

6.3.3.1.1 Specimen A

Specimen A is designed by Manufacturer M1 to withstand an earthquake with maximum ground acceleration of 0.075g. This model presents only the transversal

Fig. 6.4 Overview of the test set-up

bracing system, therefore the longitudinal forces are absorbed by the beam to upright connections only.

A bi-dimensional view of the structure is reported in Fig. 6.5, while the cross sections of the uprights and beams are shown in Fig. 6.6.

As the design earthquake is of low intensity, it has been possible to use open sections both for the uprights and for the braces, while the beams have a much stiffer rectangular hollow cross section.

The rectangular base-plate (130 × 140 mm) is connected to the uprights by means of two bolts positioned on the external side. Other two bolts, positioned in the internal side, are used to fix the base-plate to the concrete blocks simulating the floor. The eccentricity of the anchorage from the upright axis is of about 80 mm (Fig. 6.7).

6.3.3.1.2 Specimen B

Specimen B (Fig. 6.8) is designed by Manufacturer M1 to withstand an earthquake with maximum ground acceleration of 0.35g.

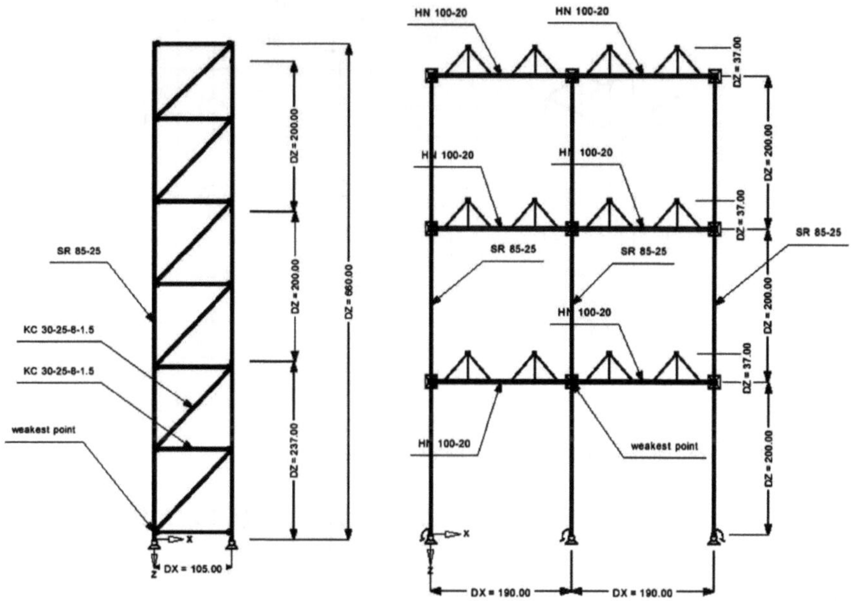

Fig. 6.5 Bi-dimensional view of specimen A

Fig. 6.6 Cross sections of the uprights (**a**) and of the beams (**b**) of specimen A

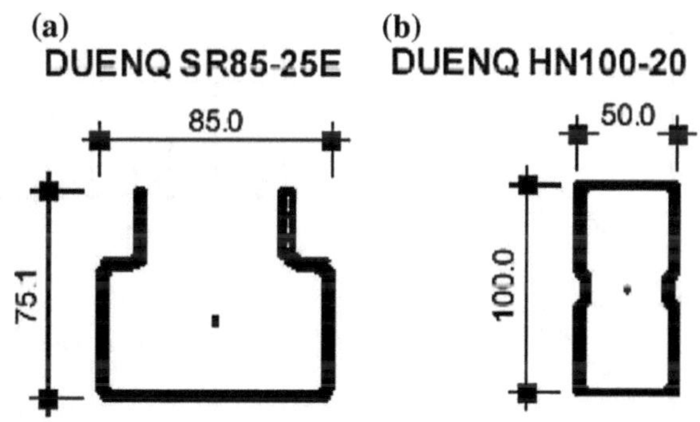

Due to the relevant dynamic design forces acting on the specimen, bracings at each level are provided both in the down-aisle direction and in plane, made of thin bars with a compact rectangular section (Fig. 6.9).

In the cross-aisle direction, bracings are made of rectangular hollow section (Fig. 6.10). In order to allow positioning of the pallets, the down-aisle bracing system has an eccentricity of about 0.20 m from the rear of the rack (Fig. 6.9): this produces a considerable asymmetry in the behaviour of the structure.

The beams have rectangular hollow section 150 × 50 mm, while the uprights are made of an open section with 100 mm width (Fig. 6.11). The base plate detail is the same as for specimen A (Fig. 6.7).

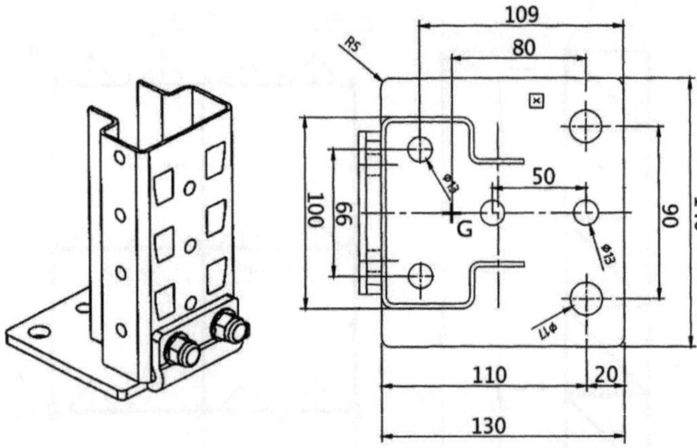

Fig. 6.7 Details of the base-plate connections of specimens A and B

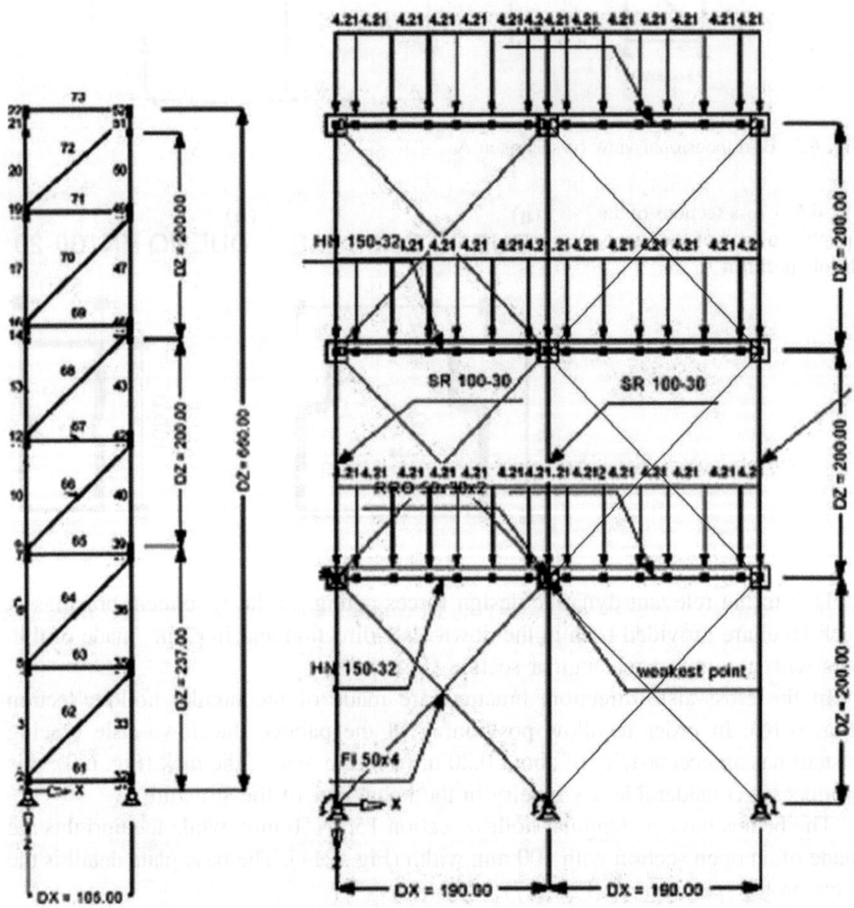

Fig. 6.8 Bi-dimensional view of specimen B

Fig. 6.9 Details of the bracings in down-aisle direction and in plane of spec. B

Fig. 6.10 Details of the bracing in cross-aisle direction of spec. B

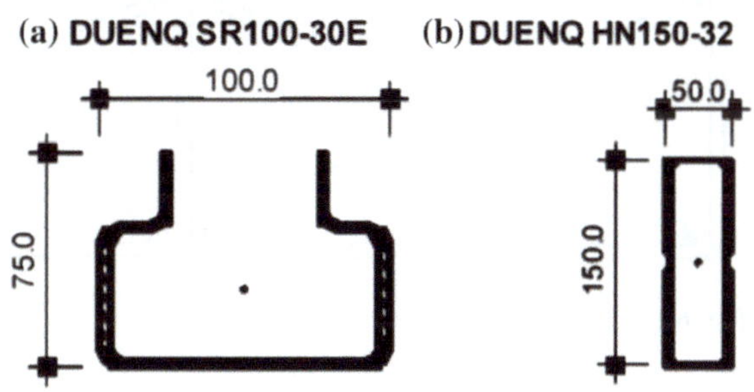

Fig. 6.11 Cross sections of the uprights (**a**) and of the beams (**b**) of specimen B

6.3.3.1.3 Specimen C

Specimen C is designed by manufacturer M2 to withstand an earthquake with maximum ground acceleration of 0.15g. A bi-dimensional view of the structure is reported in Fig. 6.12 while the cross sections of the uprights and beams are reported in Fig. 6.13.

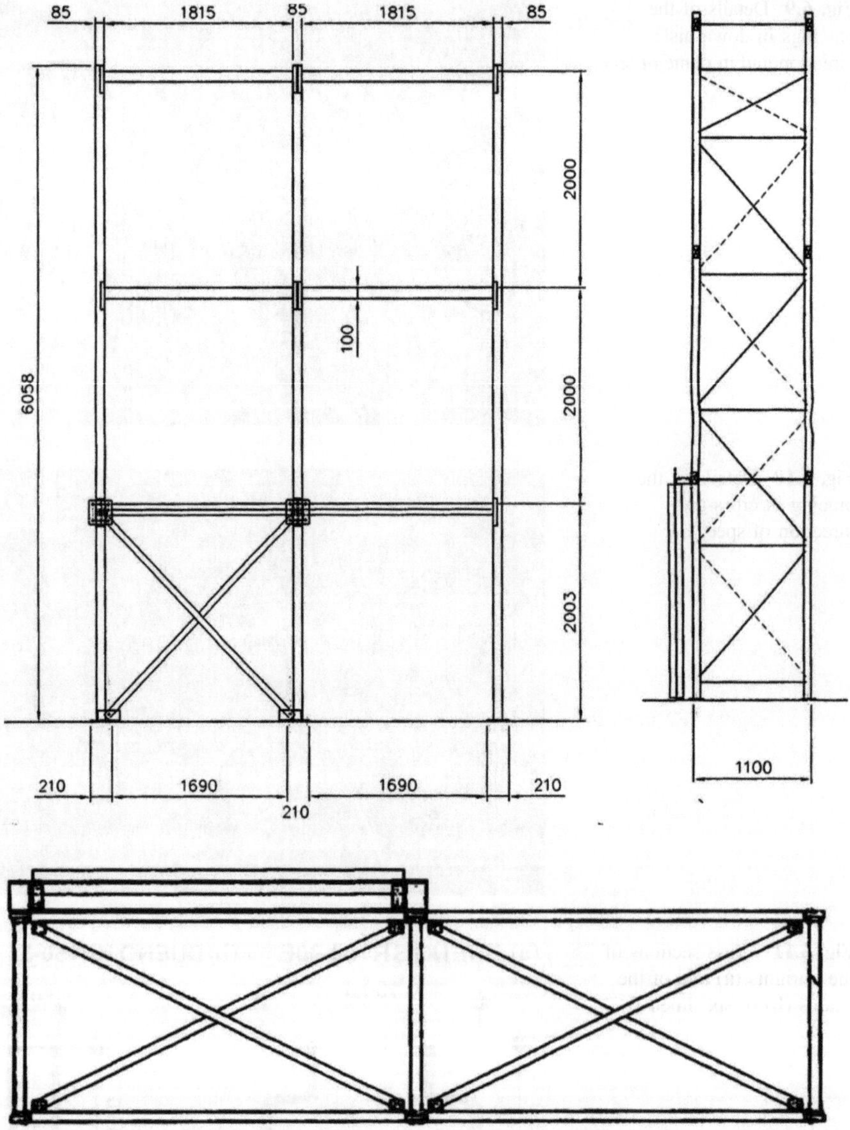

Fig. 6.12 Bi-dimensional view of specimen C

Different solutions were adopted for the bracing systems and the base-plate with respect to the specimens A and B. In this specimen the cross-aisle bracing system is made of thin drilled bars, much weaker than the bars used by manufacturer M1 (Fig. 6.14).

The goal of this different conception of cross-aisle bracing system is to dissipate energy by local plastic deformation that should occur near the holes hence

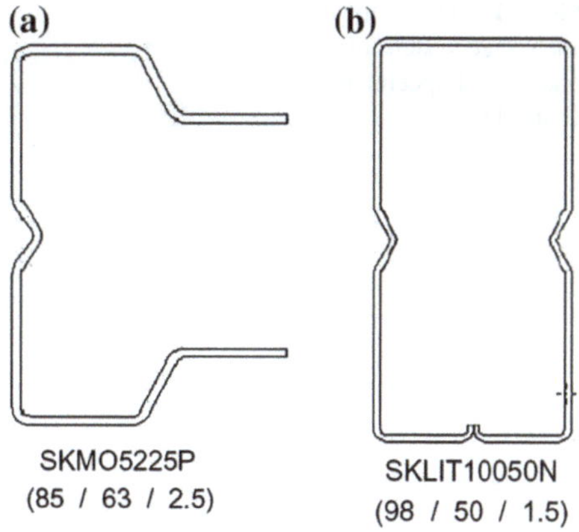

Fig. 6.13 Cross sections of the uprights (**a**) and of the beams (**b**) of specimen C

Fig. 6.14 Details of the bracing systems of specimen C

transforming the "brittle" behaviour, typical of the "braced frame" typology, in a ductile one. On the contrary, the bracing system in down-aisle direction is made of rigid beams and is provided only for the first level of one bay (Fig. 6.12).

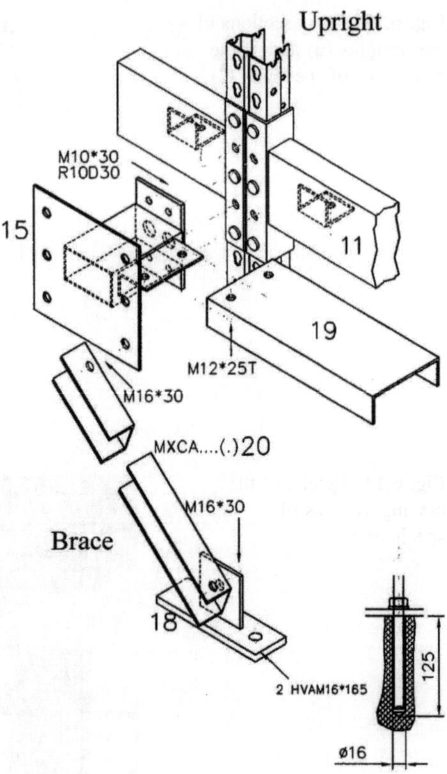

Fig. 6.15 Details of the anchorage system of the bracings of specimens C and D

A detail of the anchoring system of the down-aisle bracing is drawn in Fig. 6.15. Eccentricity of the down-aisle bracing system introduces relevant asymmetries in the behaviour of the structure.

The base plates are rectangular (210 × 130 mm) with four holes, but only two are used in order to fix the rack to the floor by means of bolts M16 × 165. The upright is fixed to the base plate by means of four bolts M16 × 30, so that the eccentricity between its axis and the anchor bolts is negligible (Fig. 6.16).

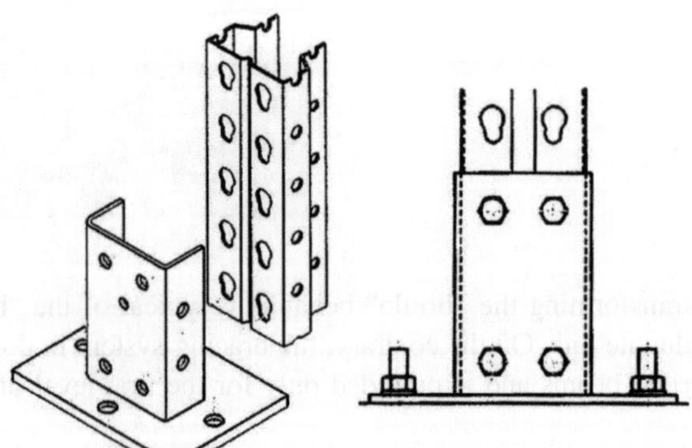

Fig. 6.16 Details of the base plate connections of specimens C and D

6.3.3.1.4 Specimen D

Specimen D is designed to withstand an earthquake with maximum ground acceleration of 0.35g. A bi-dimensional view of the structure is reported in Fig. 6.17 while the cross sections of the uprights and beams are reported in Fig. 6.18.

As in specimen C the cross-aisle bracing system is made of thin drilled bars and extends over all the height of the upright, while the other bracing systems (both in

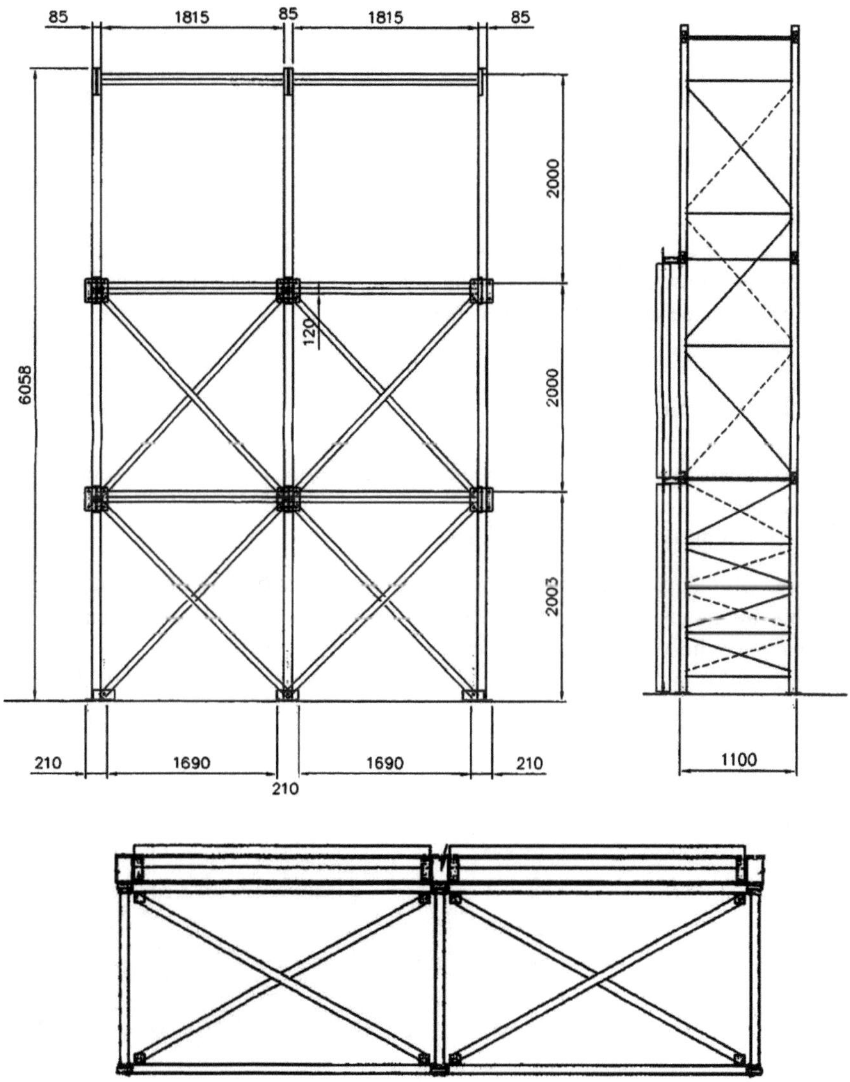

Fig. 6.17 Bi-dimensional view of specimen D

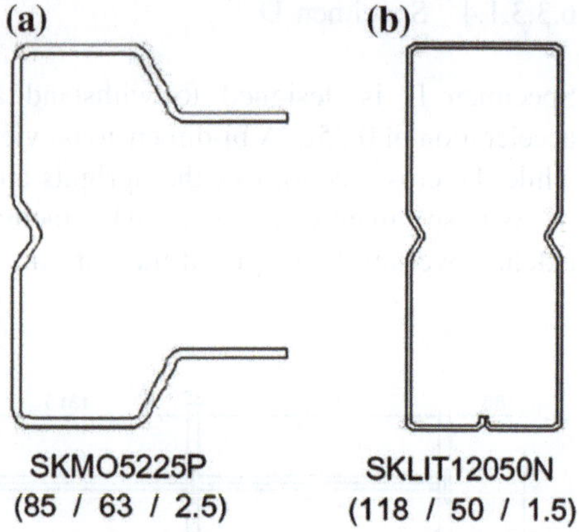

Fig. 6.18 Cross sections of the uprights (**a**) and of the beams (**b**) of specimen D

down-aisle direction and in plane) are made of rigid beams. The down-aisle bracing system extends only for the first two levels of the rack and is connected to the main structure by means of the same detail adopted for specimen C (Fig. 6.15), hence introducing relevant asymmetries in the global structural behaviour. The base plate detail is the same as for specimen C (Fig. 6.16).

6.3.3.2 Dynamic Tests

The specimens were subjected to different excitations in order to identify the eigenfrequencies and, if possible, the level where damage occurs.

The different types of test performed are:

sine sweep: a sinusoidal vibration with a constant acceleration of 0.50 m/s^2 and a variable frequency, in order to obtain the first eigenmodes of the structure. The frequency of the vibration doubles every minute (Fig. 6.19);
single impulse: an initial perturbation that provides free oscillations to the racks (Fig. 6.20);

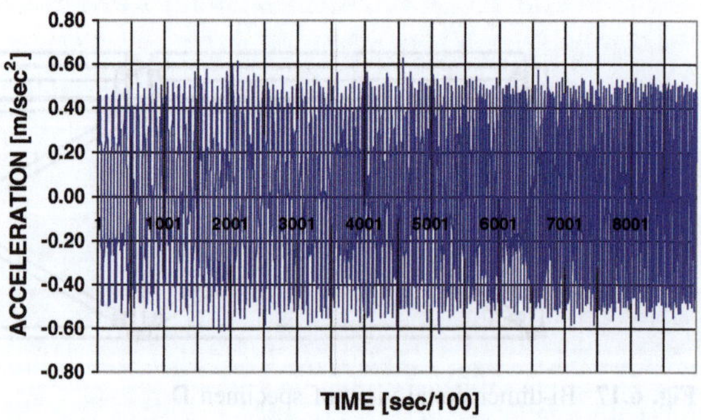

Fig. 6.19 Typical sine sweep time history

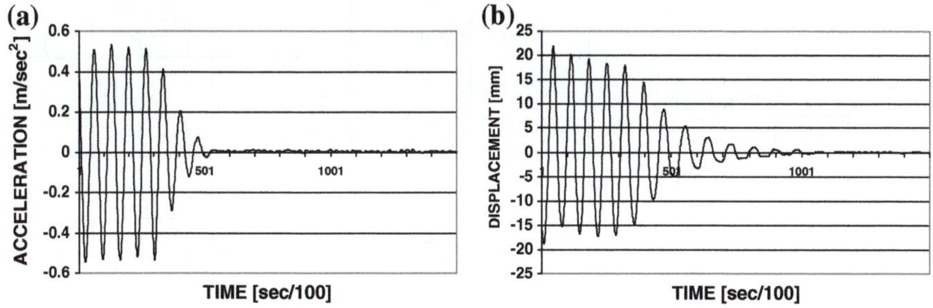

Fig. 6.20 Typical "single impulse" time history: **a** shake table input motion, **b** structure response

earthquake: an artificial earthquake (Fig. 6.21) numerically generated fitting the EC8 acceleration spectrum type 1, for subsoil type D, adequately scaled, from test to test, in order to achieve the desired PGA.

The earthquake tests were performed increasing the maximum ground acceleration until either the design value was reached, or sliding of the pallets occurred. Each earthquake test was followed by either a sine sweep or a single impulse test, in order to monitor the damage suffered by the structure during the earthquake test, through evaluation of the changes in both the eigenfrequencies and the damping factor.

It should be pointed out that, as clearly evident in Fig. 6.21b, in order to cut off the frequencies lower than 1.0 Hz, associated with displacement amplitudes larger than ±100 mm (i.e. beyond the shake table capacity), the input signal had to be filtered. Cutting the low frequencies of the input motion in the case of racks, that are rather flexible structures, may result in excitation of modes higher than the first natural one.

During the tests, accelerations and displacements were recorded in the most significant points. Furthermore, six instruments were applied to the upright nodes in order to evaluate the relative rotations (Fig. 6.22).

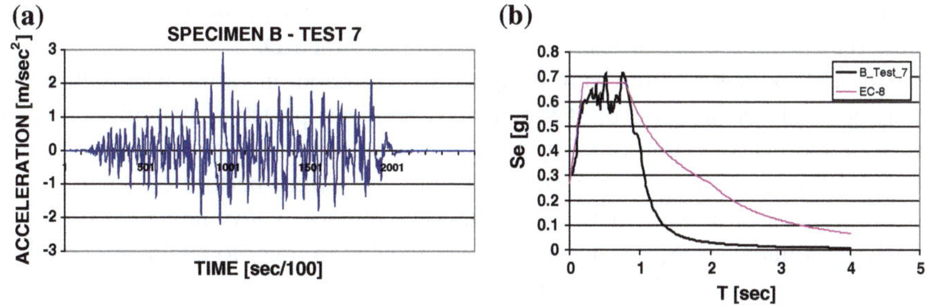

Fig. 6.21 Typical acceleration time history adopted for the earthquake tests: **a** shake table input time-history; **b** elastic response spectrum of input signal

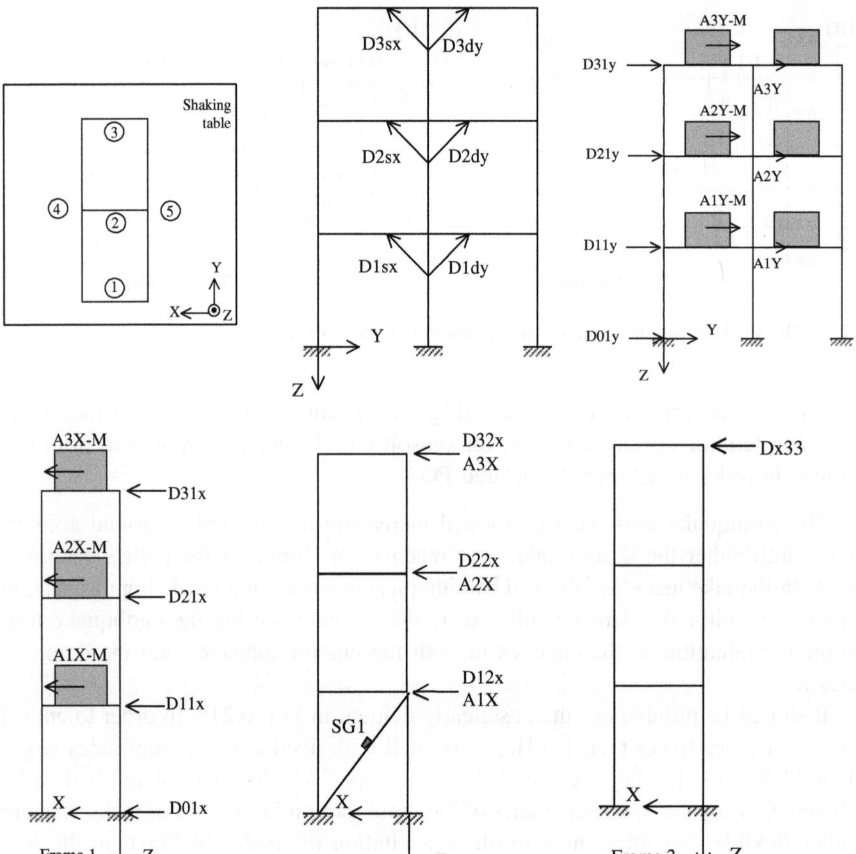

Fig. 6.22 Instrumentation of the specimens

In the following, the test history as well as the experimental results will be reported only for specimens B and D that were designed for an acceleration of 0.35g.

The list of tests performed on specimen B is reported in Table 6.2, while Table 6.3 shows the list of tests performed on specimen D.

Although both structures were designed to withstand an earthquake with maximum ground acceleration of 0.35g, it has not been possible, in both cases, to exceed a base acceleration of 0.20g both in down-aisle and in cross-aisle direction because sliding of the pallets occurred. In fact tests were performed without any "safety frame/scaffolding" that could sustain the masses and/or the whole structure in case of collapse. Hence, tests had to be terminated in order to avoid any possible damage to the shake-table testing facility.

Table 6.2 Test history for specimen B

Test no	Type	Direction	Notes
1	Sine-sweep	Down-aisle	From 1 to 8 Hz
2	Sine-sweep	Cross-aisle	From 1 to 8 Hz
3	Earthquake	Down-aisle	$a_{max} = 0.10g$
4	Sine-sweep	Down-aisle	From 0.5 to 4 Hz
5	Earthquake	Cross-aisle	$a_{max} = 0.10g$
6	Sine-sweep	Cross-aisle	From 0.5 to 4 Hz
7	Earthquake	Down-aisle	$a_{max} = 0.20g$
8	Sine-sweep	Down-aisle	From 0.5 to 4 Hz
9	Earthquake	Cross-aisle	$a_{max} = 0.15g$
10	Sine-sweep	Cross-aisle	From 0.5 to 4 Hz

Table 6.3 Test history for specimen D

Test no	Type	Direction	Notes
1	Sine-sweep	Down-aisle (Y)	From 0.5 to 4 Hz
2	Sine-sweep	Cross-aisle (X)	From 0.5 to 4 Hz
3	Earthquake	Down-aisle (Y)	$a_{max} = 0.10g$
4	Single impulse	Down-aisle (Y)	
5	Earthquake	Cross-aisle (X)	$a_{max} = 0.10g$
6	Single impulse	Cross-aisle (X)	
7	Earthquake	Down-aisle (Y)	$a_{max} = 0.20g$
8	Single impulse	Down-aisle (Y)	
9	Earthquake	Cross-aisle (X)	$a_{max} = 0.20g$
10	Single impulse	Cross-aisle (X)	
11	Earthquake	2-D (X + Y)	$a_{max} = 0.20g$

6.3.3.3 Results

6.3.3.3.1 Specimen B

Various earthquake tests were performed on this specimen, each of them followed by sine-sweep test as summarized in Table 6.2.

Figure 6.23 shows the deformed shapes of an upright, recorded at three different moments of tests no. 1 and 2, sine-sweep tests respectively in down and in cross-aisle direction. Figure 6.23a shows that, although the excitation was in down-aisle direction, the upright shows lateral deflections also in cross-aisle direction that are at every level larger than the one in the down-aisle direction. This is due to the presence of the bracing system at the rear of the rack that, as already mentioned, is eccentric with respect to the centre of gravity of the applied masses and induces torsional effects in the behaviour of the structure.

Figure 6.23b shows that, on the contrary, when the structure is subjected to a sine sweep in cross-aisle direction, the transversal displacements are much larger than the longitudinal ones and no torsion effect is evident.

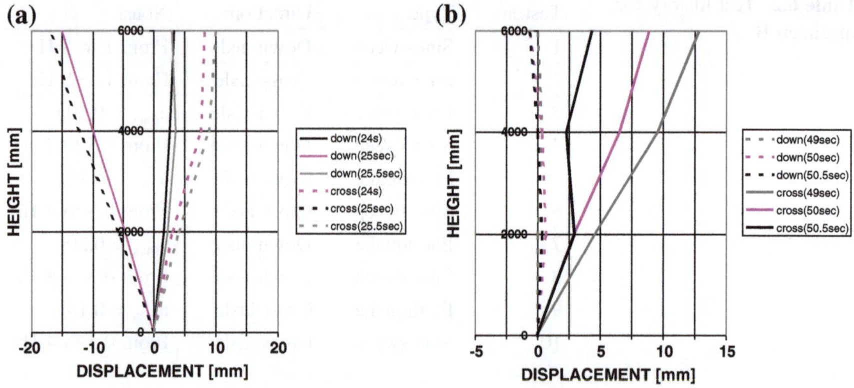

Fig. 6.23 Shapes assumed by the uprights of the specimen B at three different moments: **a** during the test 1: sine-sweep in down-aisle direction and **b** during the test 2: sine-sweep in cross-aisle direction

In general, response of the structure is according to the first eigenmode. Hence the lower level has to resist very large shear loads. In all the tests, the base-plates (having an eccentricity of 80 mm with respect to the upright axis) showed relevant bending due to plastic deformations (Fig. 6.24). The down-aisle bracing system, at the first level, had to be checked after every test and bolts were re-tightened if necessary.

The pallets slide considerably during the earthquake test in down-aisle direction with a_{max} = 0.20g (test 7) and during the earthquake test in cross-aisle with a_{max} = 0.15g (test 9). The bolts of the base-plates loosened after test 9 and had to be re-tightened before performing the next test.

Hereafter, comments are reported only for the two strongest earthquake tests performed in each principal direction, respectively test 7 and test 9, with input motions whose characteristics are summarized in Table 6.4.

Fig. 6.24 Detail of the bending at the base-plates of specimen B during the tests

Table 6.4 Characteristics of the input motion for test 7 and 9 on specimen B

	Test 7	Test 9
Direction	Y	X
Maximum acceleration [g]	0.30 at time t = 9.67 s	0.17 at time t = 12.07 s
Maximum velocity [cm/s]	24.9 at time t = 9.57 s	17.7 at time t = 11.25 s
Maximum displacement [cm]	4.4 at time t = 30.47 s	6.1 at time t = 11.54 s
Acceleration RMS [g]	0.06	0.05
Velocity RMS [cm/s]	4.9	4.9
Displacement RMS [cm]	2.2	2.3
Predominant period [s]	0.76	0.90
Mean period [s]	0.47	0.53
Effective design acceleration [g]	0.28	0.17
Acceleration spectrum intensity [g s]	0.24	0.17
Velocity spectrum intensity [cm]	96.8	85.9
Characteristic intensity	0.08	0.05
Specific energy density [cm^2/s]	726.9	490.3

Test 7 is an earthquake test in down-aisle direction. The acceleration time history, as well as its elastic response spectrum (compared with Spectrum type 1 of Eurocode-8 for sub-soil D), are shown in Fig. 6.25. The structural response under this time history is characterized by longitudinal (down-aisle) displacements of the different levels that are in-phase and of increasing amplitude when passing from the bottom to the top, as shown in Fig. 6.26. The positions of the instruments can be found in Fig. 6.22.

On the contrary, the transversal (cross-aisle) displacements at the top (3rd) level of the two lateral uprights are in counter-phase and of the same order of magnitude of the down-aisle ones, while those of the central one are practically zero, as shown in Fig. 6.27. This confirms the presence of severe global torsional effects in the structure induced by the eccentricity between the longitudinal bracing system and the center-of-gravity of the masses.

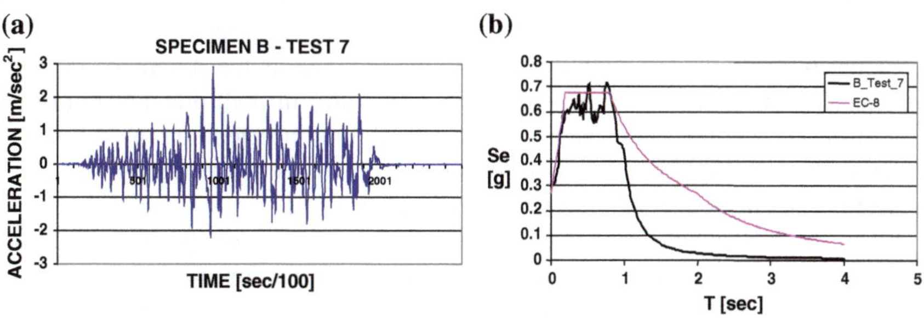

Fig. 6.25 Input motion for test 7 on specimen B; **a** acceleration time history; **b** elastic acceleration spectrum

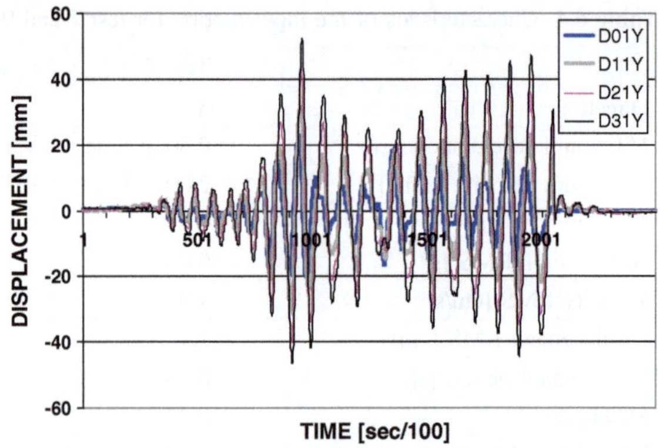

Fig. 6.26 Down-aisle displacements at various levels for specimen B, test 7

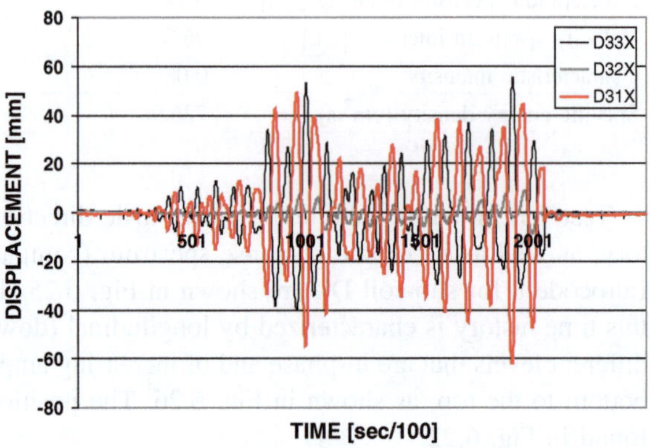

Fig. 6.27 Cross-aisle displacements at various levels for specimen B, test 7

Considerable sliding of the pallets was observed during the test.

Due to the global torsion of the structure, sliding occurred in both longitudinal and transversal directions. This is evident when examining Figs. 6.28 and 6.29. It can be observed that acceleration of the structure in down-aisle direction at all levels is larger than the one in cross-aisle direction at the same level. Acceleration of the masses is similar in both cross-aisle (x) and down-aisle (y) directions, comparable with the acceleration of the structure in down-aisle direction and much larger than the acceleration of the structure in cross-aisle direction.

Figure 6.30 shows the sliding accelerations (defined as the difference between the acceleration of the mass and that of the structure) in both down-aisle (a) and cross-aisle (b). It can be observed that the sliding acceleration in cross-aisle direction (Fig. 6.30b) is larger than the one in down-aisle direction (Fig. 6.30a), despite the input motion is in down-aisle direction.

Figure 6.31 shows the global shear measured at the base of specimen B during test 7 that is, correctly, larger in the down-aisle direction rather than in the cross-aisle one. This parameter, in fact, is not influenced by the torsional effects.

Test 9 is an earthquake test in cross-aisle direction. The acceleration time history, as well as its elastic response spectrum (compared with the type 1 Spectrum of

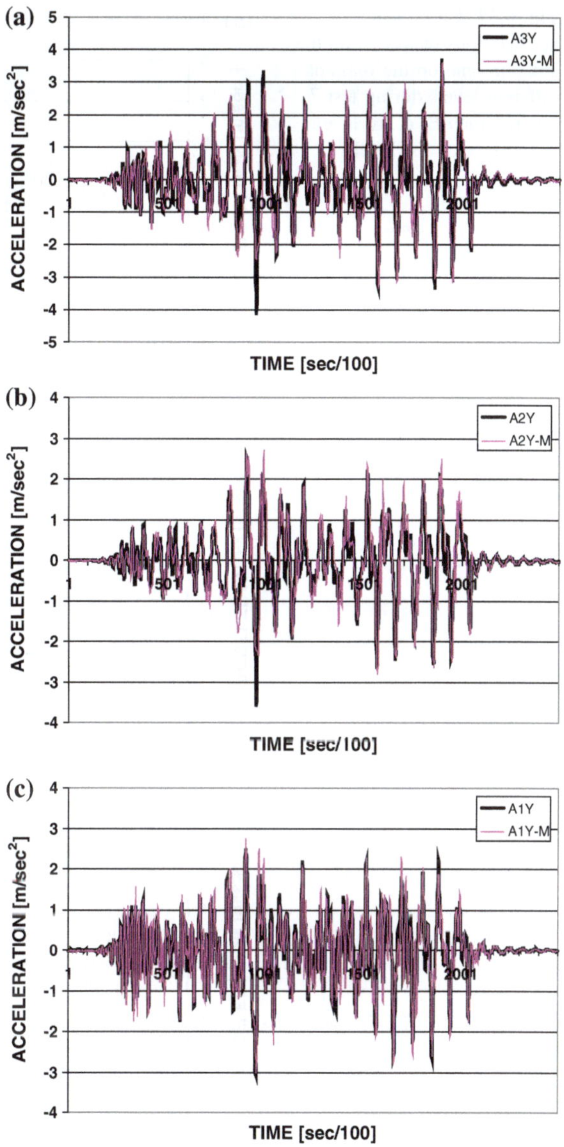

Fig. 6.28 Down-aisle accelerations measured on the structure and on the mass at different levels during test 7 on specimen B: **a** 3rd level, **b** 2nd level, **c** 1st level

Eurocode-8 for sub-soil D), are shown in Fig. 6.32. Geometry of the structure, distribution of the masses and loading condition are such that the central upright frame is, in this case, loaded by horizontal forces that are two times larger than those resisted by the two lateral upright frames. Possible deformation might hence occur in the horizontal plane, at the various levels. However Fig. 6.33, presenting the horizontal displacements in cross-aisle directions at the third level of the various upright frames, shows effectiveness of the in-plane bracing system, provided at each

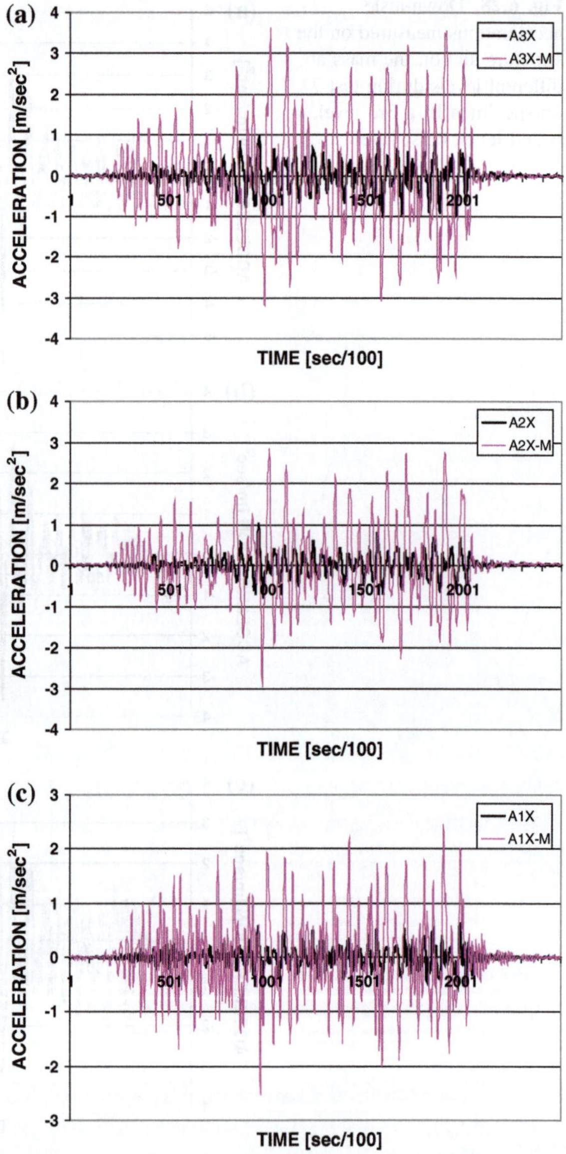

Fig. 6.29 Cross-aisle accelerations measured on the structure and on the mass at different levels during test 7 on specimen B: **a** 3rd level, **b** 2nd level, **c** 1st level

structural level; in fact, it can be observed that all three uprights suffer the same transversal displacements.

Thanks to the effectiveness of the vertical bracing of the upright frames, the structural response is according to the first eigenmode, as already shown in Fig. 6.23b, and as evident in Fig. 6.34 that shows how the cross-aisle displacements at different levels of the central upright frame are increasing (nearly linearly) from bottom to top.

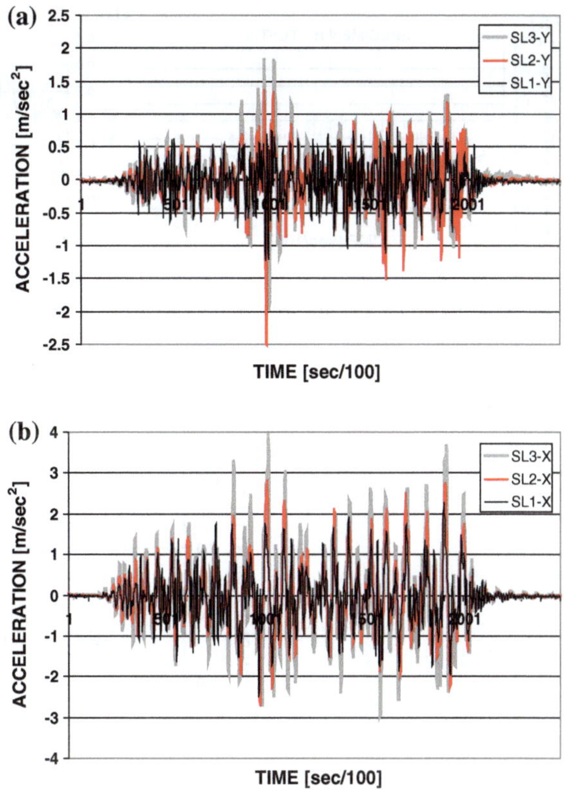

Fig. 6.30 Sliding acceleration of the masses during test 7 on specimen B: **a** down-aisle direction; **b** cross-aisle direction

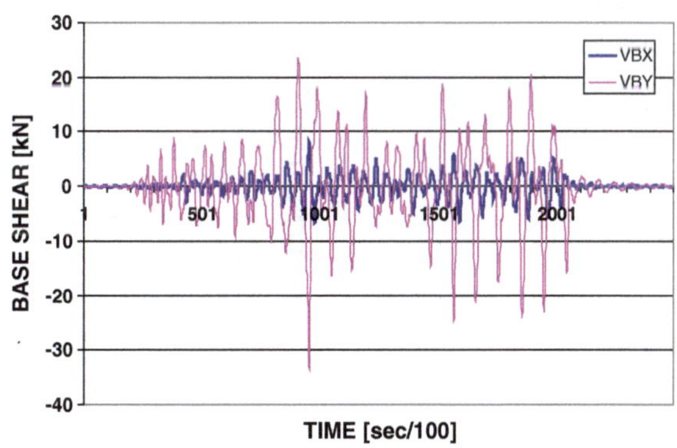

Fig. 6.31 Base shear for test 7 on specimen B

The down-aisle displacements of the structure are shown in Fig. 6.35 and are much lower than those measured in cross-aisle direction. From Fig. 6.35 it can also be noticed that the down-aisle response of all three structural levels, where the masses are applied, presents some peaks, although of limited amplitude.

Also in test 9, evident sliding of the pallets occurred.

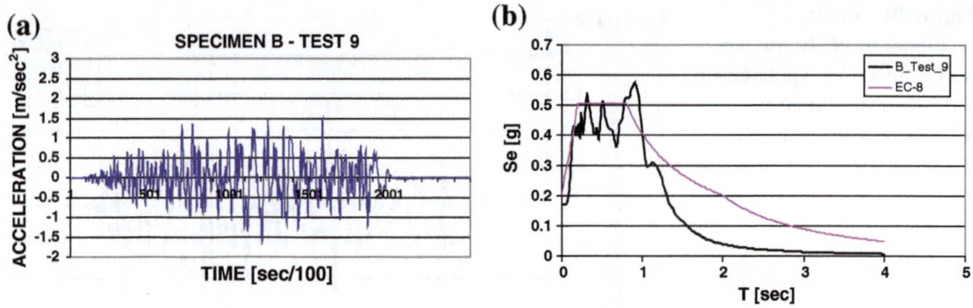

Fig. 6.32 Input motion for test 9 on specimen B; a Acceleration time history; b Elastic acceleration spectrum

Fig. 6.33 Horizontal displacements in cross-aisle directions at the third level of the various upright frames

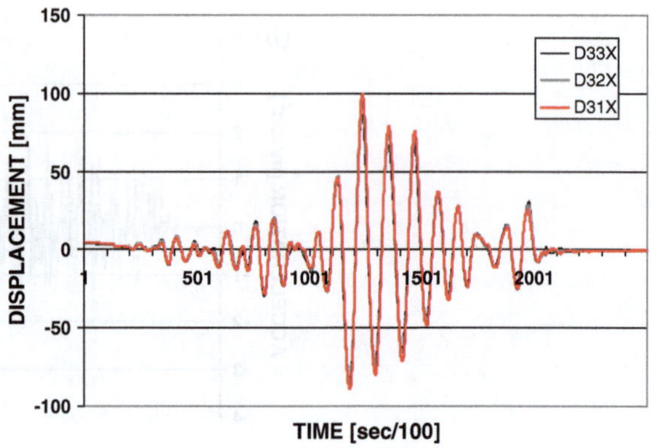

Fig. 6.34 Cross-aisle displacements at various levels of the central upright frame

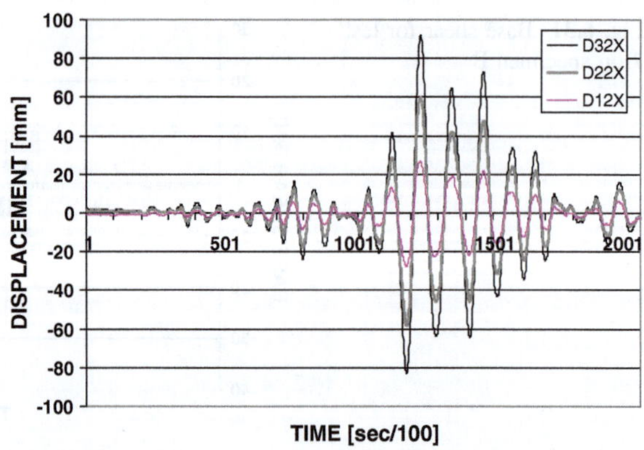

Figures 6.36 and 6.37 show respectively the cross-aisle and down-aisle accelerations measured on the structure and on the masses at the various levels. In this case, accelerations measured in down-aisle direction are much smaller than those measured in cross-aisle.

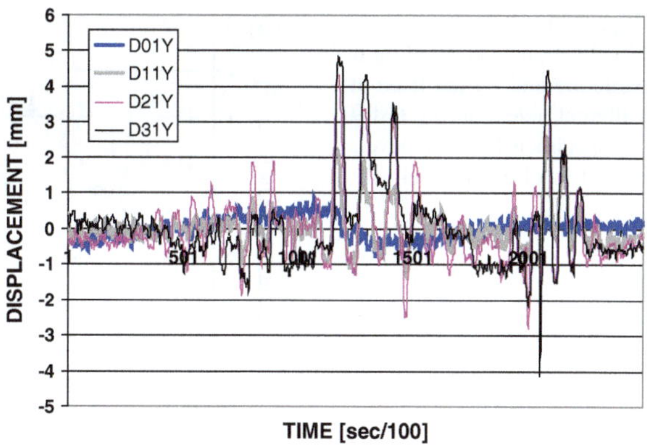

Fig. 6.35 Down-aisle displacements at various levels of the longitudinal frame

Consequently, also the sliding accelerations in down-aisle direction, shown in Fig. 6.38b, are much smaller than those in cross-aisle direction, shown in Fig. 6.38a. It is worth to be noticed that sliding in cross-aisle direction is felt already at the first level, with accelerations (Fig. 6.38a) of the same order of magnitude of the base input motion (Fig. 6.32). It should also be pointed out that the peaks observed in the down-aisle response of the structure (Fig. 6.35) correspond to peaks in the sliding acceleration in down-aisle direction (Fig. 6.38b).

Figure 6.39 shows that the base shear in down-aisle direction (VBY) is practically zero, and however much smaller than the one (VBX) in cross-aisle direction (i.e. in the direction of excitation). This fact is confirming the absence of torsional effects in the structure.

During this test, due to the already discussed eccentricity of the connection (Fig. 6.7) relevant plastic deformation occurred in the base-plates (Fig. 6.24). Anchor bolts loosened and had to be re-tightened before testing activities could be resumed.

It is interesting to observe the variation of the eigenfrequencies and the damping ratios obtained from re-analysis of the acceleration data recorded at the third level during the sine-sweep tests carried out after each earthquake test (Table 6.5).

The sequences of tests 1-4-8 and of tests 2-6-10 show a reduction of the first eigenfrequency due to a possible damage of the rack.

Figure 6.40 shows the residual displacements recorded at the end of each test, and allows an assessment of the accumulation of plastic deformation after each test.

It can be noticed that damage in the down-aisle (y) direction occurred during tests 4, 7 and 8, while in cross-aisle direction it occurred during tests 6, 7 and 9. Test 7 corresponds to the earthquake simulation in down-aisle. The torsional effects, caused by the eccentricity of the bracing system, induce damage accumulation also in cross-aisle direction. Test 9 is an earthquake simulation in cross-aisle direction.

However, tests 4, 6 and 8 are sine sweep tests that were carried out with a constant acceleration of 0.05g. Despite this low acceleration value, passing through resonance results in damage to the structure.

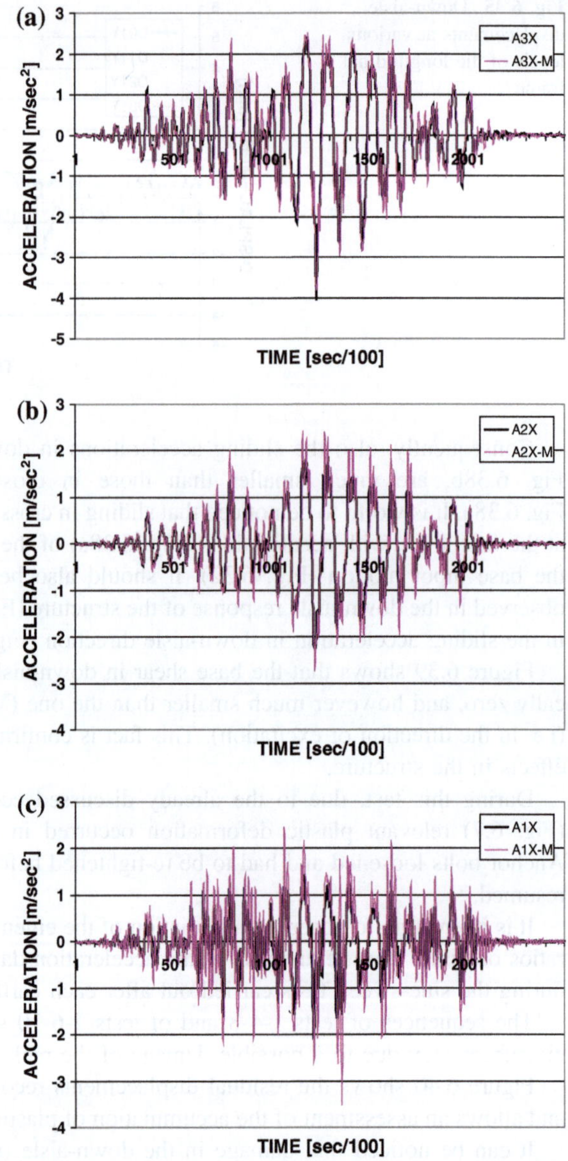

Fig. 6.36 Cross-aisle accelerations measured on the structure and on the mass at different levels during test 9 on specimen B: **a** 3rd level, **b** 2nd level, **c** 1st level

6.3.3.3.2 Specimen D

Also for specimen D various earthquake tests, at increasing PGA, were performed each one followed by a single-impulse test, as summarized in Table 6.3, in order to assess the first natural eigenfrequencies as well as the damping ratio.

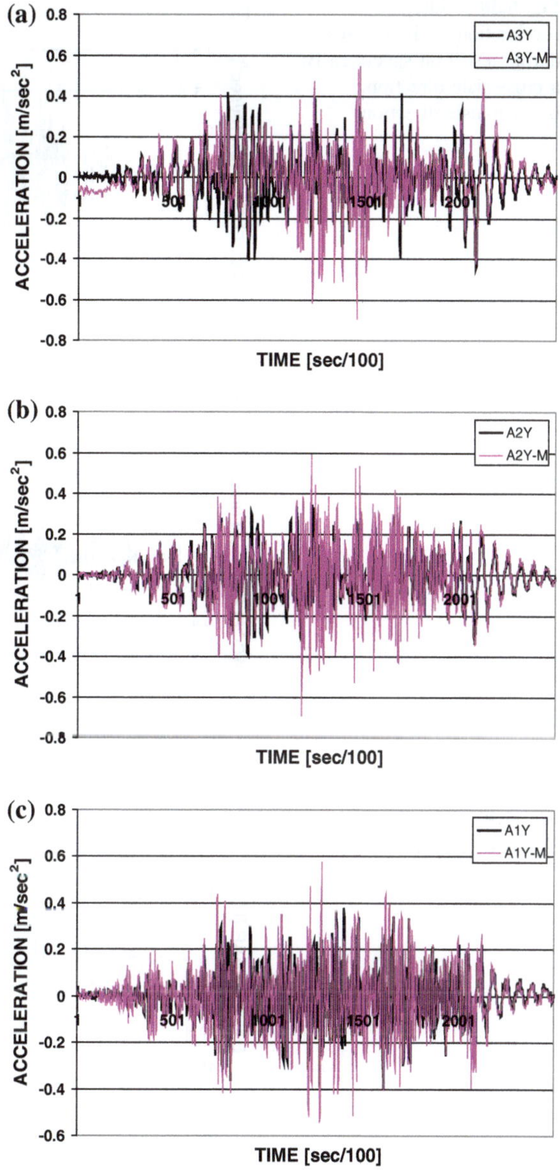

Fig. 6.37 Down-aisle accelerations measured on the structure and on the mass at different levels during test 9 on specimen B: **a** 3rd level, **b** 2nd level, **c** 1st level

Specimen D, however, differs from specimen B because the bracing system is present in the first two levels only, as well as because of the shape of the members of both the cross-aisle and in plane bracings, that are thin drilled bars (Fig. 6.14), aimed at dissipating energy through localized plastic deformation.

Figure 6.41 shows the deformed shapes of an upright, recorded at three different moments of tests no. 1 and 2, sine-sweep tests respectively in down and in

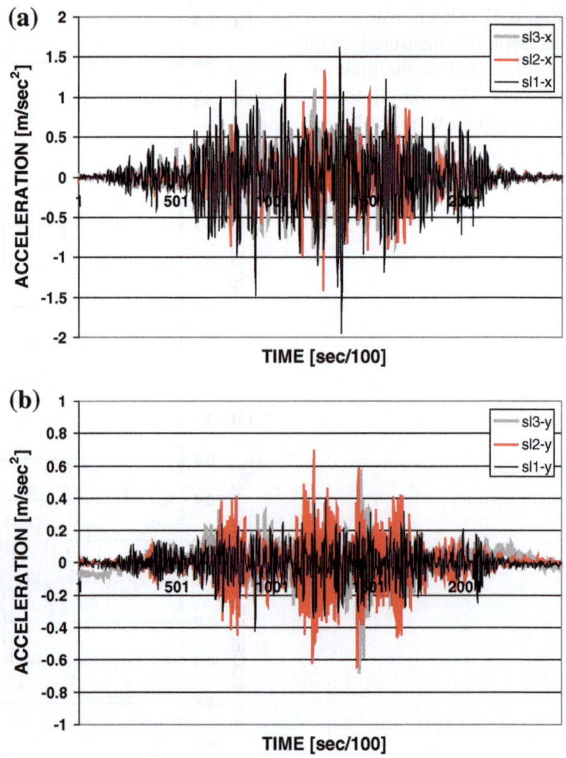

Fig. 6.38 Sliding acceleration of the masses during test 9 on specimen B: **a** cross-aisle direction; **b** down-aisle direction

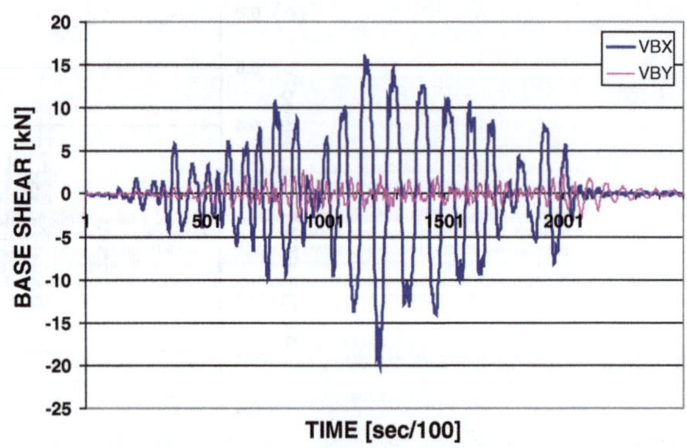

Fig. 6.39 Base shear for test 9 on specimen B

cross-aisle direction. Figure 6.41b shows that, also for specimen D, no torsional effect is present in the case of excitation in cross-aisle direction. Furthermore it can be noticed that presence of bracings in the upright frames reduces the transversal displacements of the 2nd and 3rd level, where base shear is lower. However most deformability is concentrated at the base where shear is higher and, consequently, plastic deformations occur in the drilled bars resulting in large displacements.

Table 6.5 Eigen frequencies and damping ratios of specimen B

Test no	Direction	1st freq. [Hz]	2nd freq. [Hz]	Damping [%]
1	Y	1.34	3.00	0.7
2	X	1.83	6.01	0.8
4	Y	1.29	3.05	0.9
6	X	1.55	–	0.9
8	Y	1.08	3.04	1.1
10	X	1.27	–	0.9

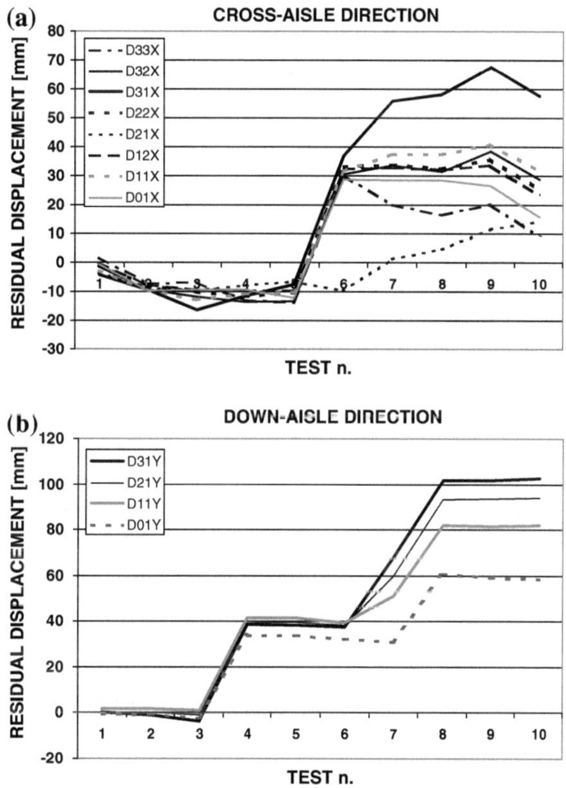

Fig. 6.40 Residual displacements of specimen B recorded after each test. **a** Cross-aisle direction; **b** down-aisle direction

Large out-of-plane deflections of the thin drilled bars in the cross-aisle bracing system could be observed at the end of the tests, index of inelastic elongations due to plastic deformations in the net sections around the holes. Hence, the bracing system lost part of its effectiveness during the sequence of the tests.

Figure 6.41a shows that, although the excitation is in down-aisle direction, the upright shows lateral deflections also in cross-aisle direction, at every level larger than the one in the down-aisle direction. This is due to the presence of the bracing system at the rear of the rack that, as already mentioned, is eccentric with respect to

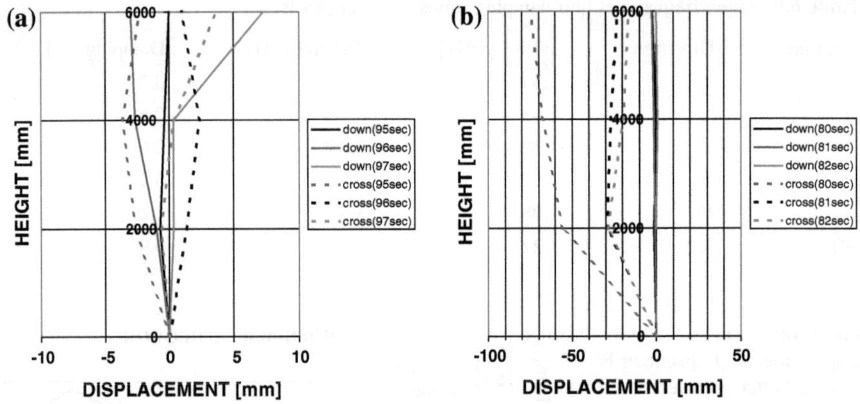

Fig. 6.41 Shapes assumed by the uprights of the specimen D at three different moments: **a** during the test 1: sine-sweep in down-aisle direction and **b** during the test 2: sine-sweep in cross-aisle direction

the centre of gravity of the applied masses and induces torsional effects in the behaviour of the structure. It should be noticed that, presence of cross-bracings in down-aisle direction reduces the longitudinal displacements of the first two levels. However, as the bracing system in the down-aisle direction stops at the second level, the third level of this specimen behaves as a sort of "soft floor", where most of the deformability of the structure is concentrated both in the down-aisle and in the cross-aisle direction (because of the torsional effects induced by eccentricity of the bracing system).

In general, response of the structure is according to the first eigenmode. Hence the lower level has to resist very large shear loads. The down-aisle bracing system, at the first level, had to be checked after every test and bolts were re-tightened if necessary.

Considerable pallet sliding was observed during the earthquake tests with a_{max} = 0.20g both in down-aisle (test 7) and in cross-aisle direction (test 9), and it was considered unsafe to perform further tests with increased base acceleration.

Hereafter, comments are reported only for the two strongest earthquake simulation tests performed in each principal direction (respectively test 7 and test 9) with input motions whose characteristics are summarized in Table 6.6.

Test 7 is an earthquake test in down-aisle direction. The acceleration time history, as well as its elastic response spectrum (compared with Spectrum type 1 of Eurocode-8 for sub-soil D), are shown in Fig. 6.42.

The structural response under this time history is characterized by longitudinal (down-aisle) displacements of the different levels that are in-phase and of increasing amplitude when passing from the bottom to the top, as shown in Fig. 6.43. The positions of the instruments can be found in Fig. 6.22.

On the contrary, the transversal (cross-aisle) displacements at the top (3rd) level of the two lateral uprights are in counter-phase and of the same order of magnitude

Table 6.6 Characteristics of the input motion for test 7 and 9 on specimen D

	Test 7	Test 9
Direction	Y	X
Maximum acceleration [g]	0.31 at time t = 9.60 s	0.23 at time t = 12.07 s
Maximum velocity [cm/s]	25.2 at time t = 8.28 s	25.1 at time t = 11.25 s
Maximum displacement [cm]	46.1 at time t = 9.34 s	6.1 at time t = 11.54 s
Acceleration RMS [g]	0.06	0.06
Velocity RMS [cm/s]	6.2	5.7
Displacement RMS [cm]	18.8	2.1
Predominant period [s]	0.48	0.90
Mean period [s]	0.46	0.53
Effective design acceleration [g]	0.29	0.23
Acceleration spectrum intensity [g s]	0.25	0.25
Velocity spectrum intensity [cm]	91.9	121.6
Characteristic intensity	0.08	0.08
Specific energy density [cm^2/s]	1164.0	978.5

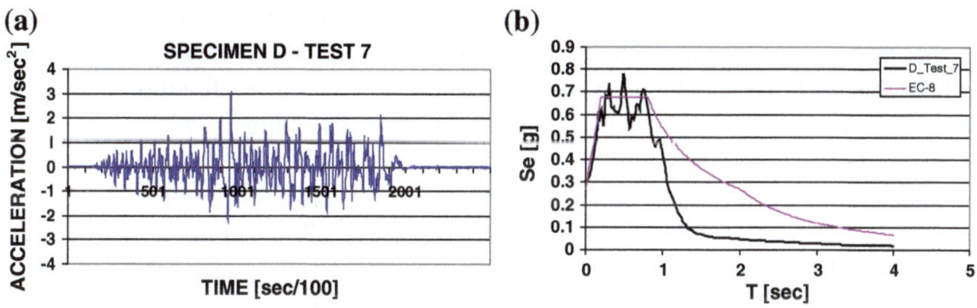

Fig. 6.42 Input motion for test 7 on specimen B; **a** acceleration time history; **b** elastic acceleration spectrum

Fig. 6.43 Down-aisle displacements at various levels for specimen D, test 7

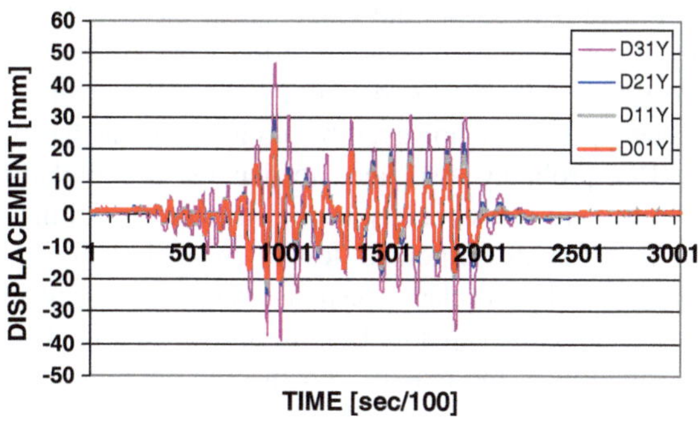

Fig. 6.44 Cross-aisle displacements at various levels for specimen D, test 7

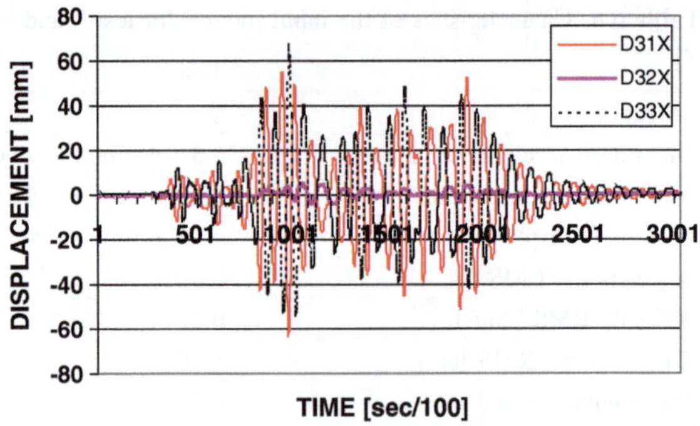

of the down-aisle ones, while those of the central one are practically zero, as shown in Fig. 6.44. This confirms the presence of severe global torsional effects in the structure induced by the eccentricity between the longitudinal bracing system and the center-of-gravity of the masses.

Due to the global torsion of the structure, sliding occurred in both longitudinal and transversal directions. This is evident from examining Figs. 6.45 and 6.46. It can be observed that acceleration of the structure in down-aisle direction at all levels is larger than the one in cross-aisle direction at the same level. Acceleration of the masses is similar in both cross-aisle (x) and down-aisle (y) directions, comparable with the acceleration of the structure in down-aisle direction and much larger than the acceleration of the structure in cross-aisle direction.

Figure 6.47 shows the sliding accelerations (defined as the difference between the acceleration of the mass and that of the structure) in both down-aisle (a) and cross-aisle (b). It can be observed that at the second level the sliding acceleration in down-aisle direction (Fig. 6.47a) is larger than the one in cross-aisle direction (Fig. 6.47b). The contrary is true at the first and third levels. This behaviour is strictly correlated with the presence of the longitudinal bracing system at the first and second levels only. The abrupt change in stiffness (in the down-aisle direction) at the second level results in this high difference between acceleration measured on the structure and on the masses.

Figure 6.48 shows the global shear measured at the base of specimen D during test 7 that, correctly, is larger in the down-aisle direction (VBY) rather than in the cross-aisle one. This parameter, in fact, is not influenced by the torsional effects. Some peaks can however be noticed in the base shear in cross-aisle direction (VBX), probably associated to damage accumulation in the longitudinal bracing system (and consequently in the structure) resulting in an asymmetric response.

Test 9 is an earthquake test in cross-aisle direction. The acceleration time history, as well as its elastic response spectrum (compared with the type 1 Spectrum of Eurocode-8 for sub-soil D), are shown in Fig. 6.49. Geometry of the structure, distribution of the masses and loading condition are such that the central upright frame is, in this case, loaded by horizontal forces that are two times larger than those resisted by the two lateral upright frames. Possible deformation might hence

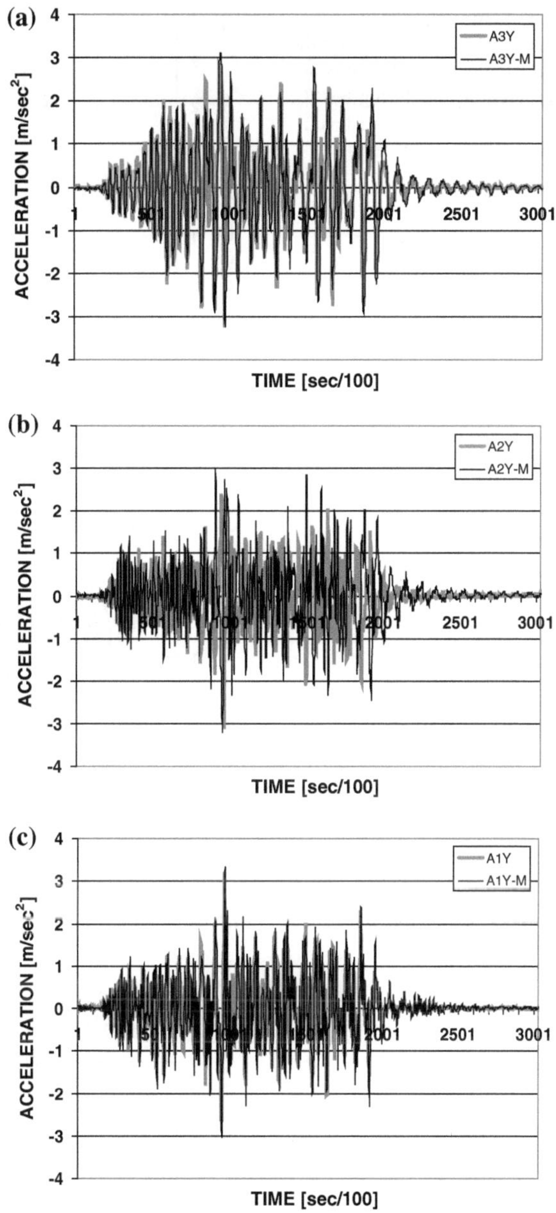

Fig. 6.45 Down-aisle accelerations measured on the structure and on the mass at different levels during test 7 on specimen D: **a** 3rd level, **b** 2nd level, **c** 1st level

occur in the horizontal plane, at the various levels. However Fig. 6.50, presenting the horizontal displacements in cross-aisle directions at the third level of the various upright frames, shows effectiveness of the in-plane bracing system, provided at each structural level; in fact, it can be observed that all three uprights suffer the same transversal displacements. The small differences that can be noticed between the

Fig. 6.46 Cross-aisle accelerations measured on the structure and on the mass at different levels during test 7 on specimen D: **a** 3rd level, **b** 2nd level, **c** 1st level

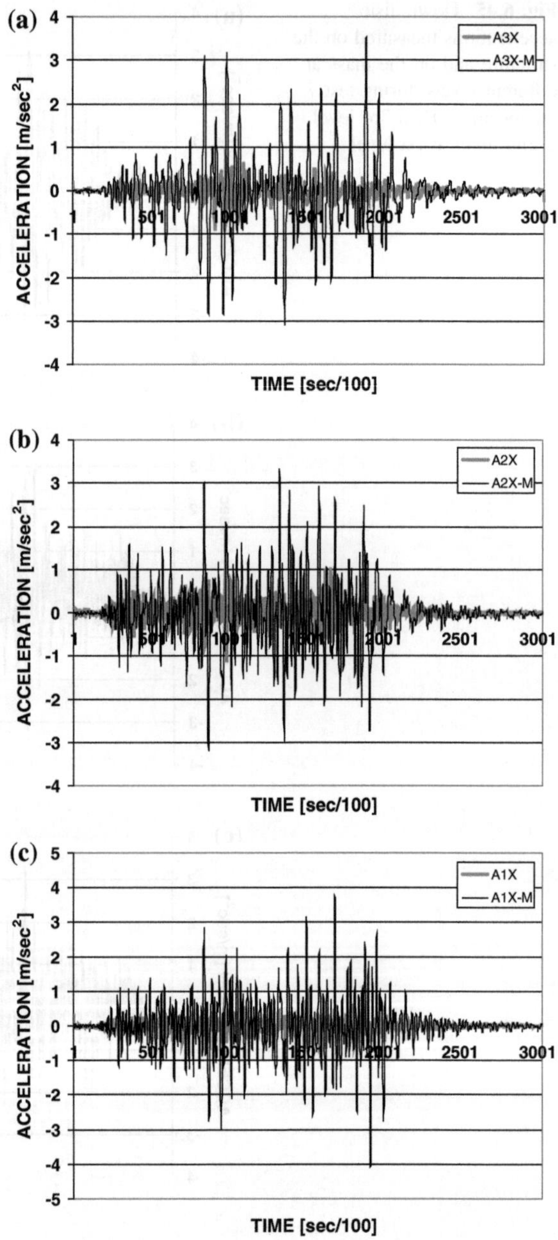

responses of the three upright frames indicate, however, that some plastic deformation occurred in the thin drilled bars of the in-plane bracing system.

Thanks to the effectiveness of the vertical bracing of the upright frames, the structural response is according to the first eigenmode, as already shown in

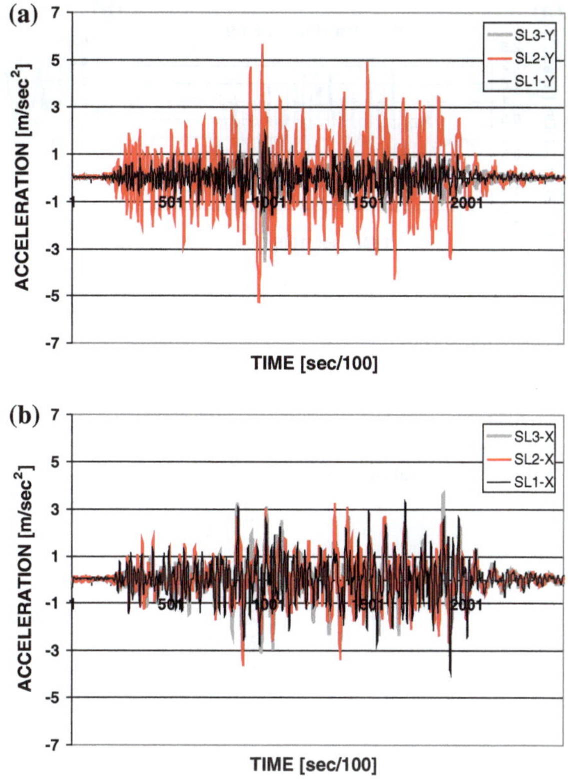

Fig. 6.47 Sliding acceleration of the masses during test 7 on specimen B: **a** down-aisle direction; **b** cross-aisle direction

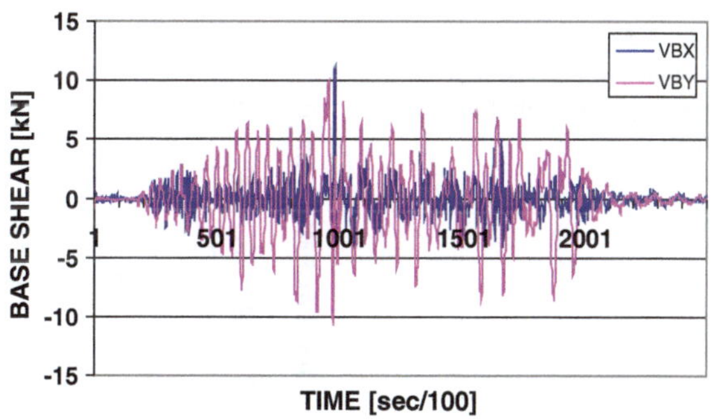

Fig. 6.48 Base shear for test 7 on specimen D

Fig. 6.41b, and as evident in Fig. 6.51 that shows how the cross-aisle displacements at different levels of the central upright frame are increasing (nearly linearly) from bottom to top.

The down-aisle displacements of the structure are shown in Fig. 6.52 and are much lower than those measured in cross-aisle direction. From Fig. 6.52 it can also be noticed that, as already observed in the case of specimen B, the down-aisle

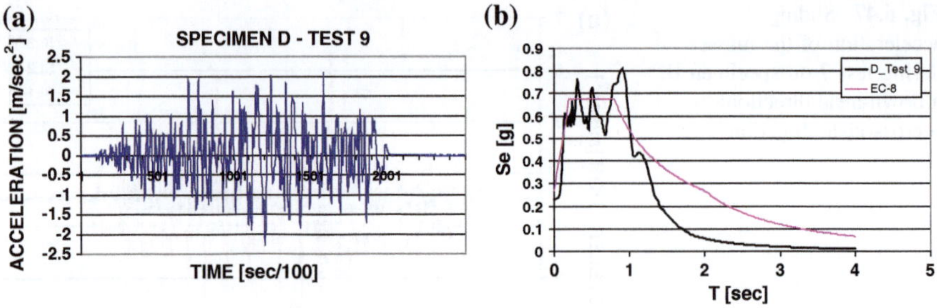

Fig. 6.49 Input motion for test 9 on specimen D; **a** Acceleration time history; **b** Elastic acceleration spectrum

Fig. 6.50 Cross-aisle displacements at various levels of the central upright frame

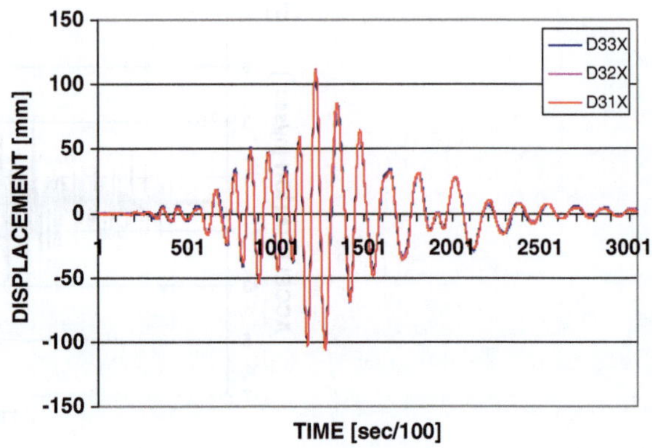

Fig. 6.51 Cross-aisle displacements at various levels of the central upright frame

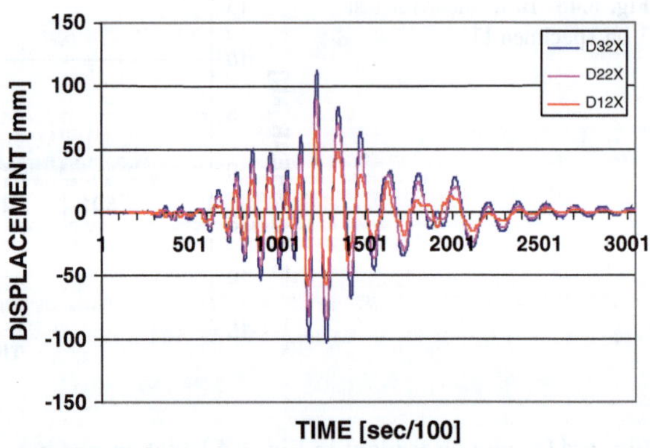

response of all three structural levels, where the masses are applied, presents some peaks, although of limited amplitude.

Also in test 9, evident sliding of the pallets occurred in both directions, although the sliding accelerations in down-aisle direction, shown in Fig. 6.53b, are much

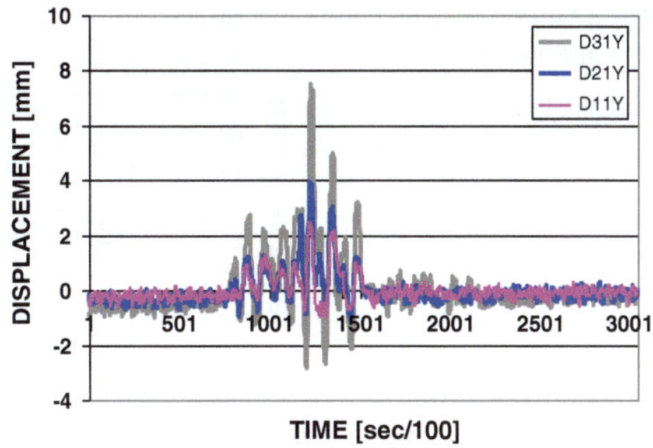

Fig. 6.52 Down-aisle displacements at various levels of the longitudinal frame

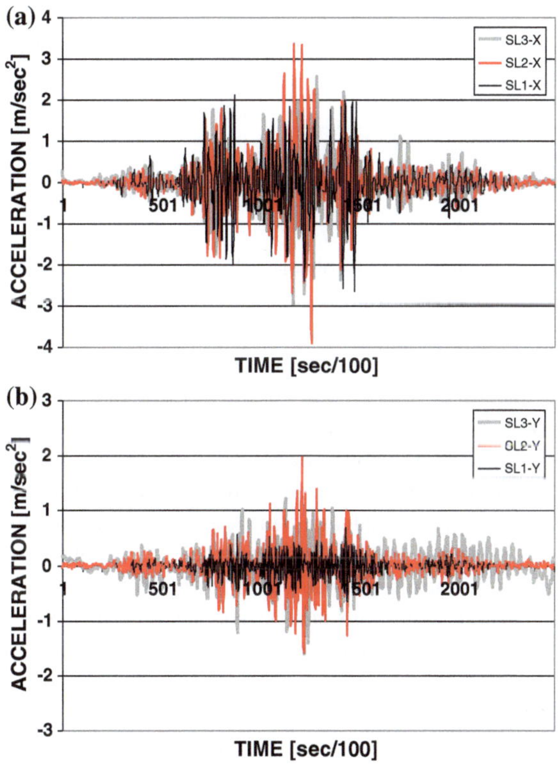

Fig. 6.53 Sliding acceleration of the masses during test 9 on specimen D: **a** cross-aisle direction; **b** down-aisle direction

smaller than those in cross-aisle direction (Fig. 6.53a), as is the base shear (Fig. 6.54). As in the case of specimen B, the peaks observed in the down-aisle response of the structure (Fig. 6.52) correspond to peaks in the sliding acceleration in down-aisle direction (Fig. 6.53b).

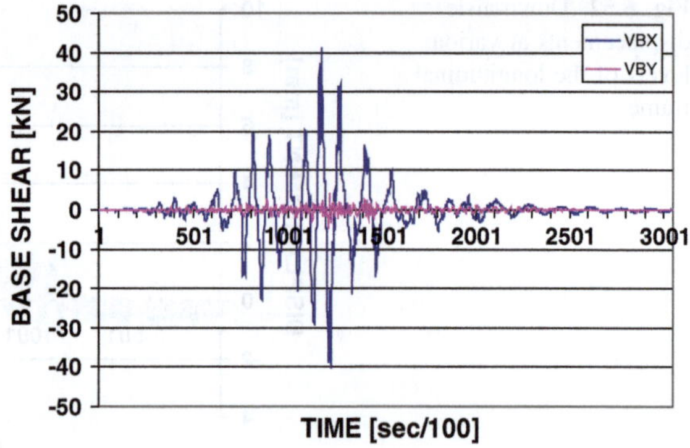

Fig. 6.54 Base shear for test 9 on specimen D

Table 6.7 Eigen frequencies and damping ratios of specimen D

Test no	Direction	1st freq. [Hz]	2nd freq.[Hz]	Damping [%]
1	Y	1.58	2.40	1.4
2	X	1.28	2.62	4.0
4	Y	1.31	–	8.0
8	Y	1.13	–	4.2

It is interesting to observe the variation of the eigenfrequencies and the damping ratios obtained from re-analysis of the acceleration data recorded at the third level during the first two sine-sweep tests and the single-impulse tests carried out after each earthquake simulation test (Table 6.7). The sequence of tests 1-4-8 shows a reduction of the first eigenfrequency due to a possible damage of the rack, although Table 6.7 shows a reduction in the damping ratio corresponding to test 8.

Figure 6.55 shows the residual displacements recorded at the end of each test, and allows an assessment of the accumulation of plastic deformation.

An evident damage (in terms of residual displacements) occurred in the down-aisle (y) direction during tests 1, 4 and 8 (Fig. 6.55b), and in cross-aisle direction during tests 2, 4 and 8 (Fig. 6.55a).

Test 1 and 2 are sine sweep tests performed with a constant acceleration of 0.05g, respectively in down and in cross-aisle direction.

Tests 4 and 8 are single impulse tests that were carried out in down-aisle direction.

The largest residual displacements were recorded in cross-aisle direction, although tests 4 and 8 correspond to free vibration tests in down-aisle direction. This effect is the consequence both of the high eccentricity of the structure in down-aisle direction and of the weak bracing in cross-aisle direction.

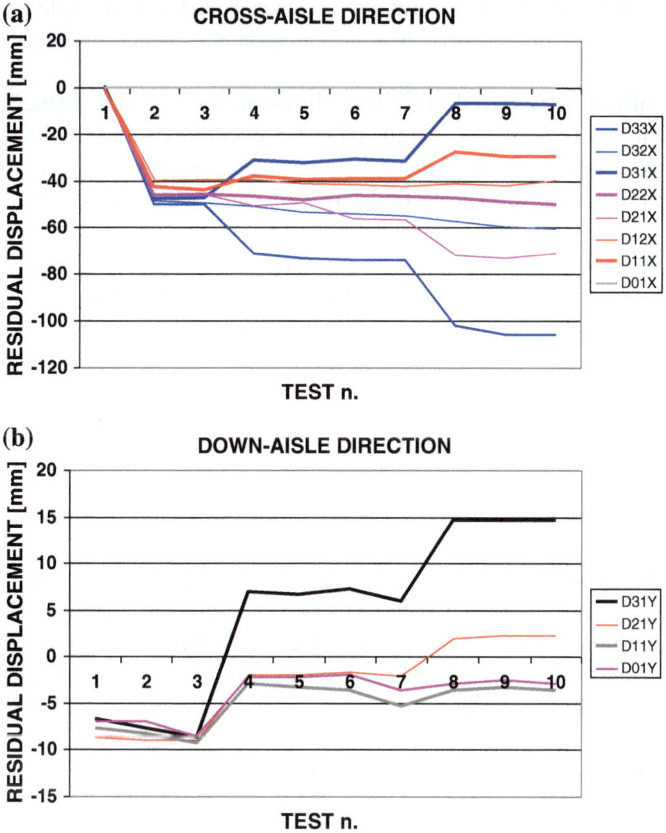

Fig. 6.55 Residual displacements of specimen D recorded after each test. **a** Cross-aisle direction; **b** down-aisle direction

6.3.3.4 Conclusions

From re-analysis of the tests carried out within the ECOLEADER Research Project "Seismic Behaviour of Pallet Rack Systems", the following conclusions could be drawn:

Structural regularity (both in elevation and in plan) is a very important parameter to be considered in the design and enforced in codes. Eccentricity of the vertical down-aisle bracing system may result in severe torsional deformation and in major structural damage.

In-plane bracings at each level where masses are applied allows redistribution of the horizontal forces among all upright frames, avoiding "side-effects", that may be significant, in particular for "short" racking systems.

Due to high flexibility of the structures, a sine sweep test or a single impulse test, even when carried out at very low acceleration can result in major structural damage.

Sliding of pallets occurred in all tests. In absence of a safety frame/scaffolding that could sustain the structure and the pallets, preventing damage to the shake-table facility in case either of fall of the pallets or of global collapse of the structure, it was impossible to perform tests under design value of the acceleration (namely 0.35g).

As no test could be carried out up to complete structural collapse (due to the aforementioned reasons), it was impossible to draw any conclusion about the actual ductility of these structures under seismic loading.

6.3.4 The SEISRACKS Project

6.3.4.1 Overview

Based on the previous experience, and with the aim of deriving information related to the actual ductility of steel racks and to their ultimate response under strong earthquakes, with the financial support of the EU, Research Programme of the Research Fund for Coal and Steel—Steel RTD, a new Research Project (SEISRACKS, "Storage Racks in Seismic Areas", Contract no. RFS-PR-03114), has been carried out under the co-ordination of the Italian Association of Steel Constructors (ACAI), and the scientific supervision of this author, in cooperation with five European Research Institutions (SEISRACKS 2008).

Shake-table tests were carried out on six full-scale rack models, produced by the same manufacturer, affiliated to ACAI, who provided the specimens free-of-charge.

As in the case of the ECOLEADER Project, all the specimens have two bays, 1.8 m wide, and three levels for a total height of 6.0 m. Neither vertical (down-aisle) nor horizontal (in-plane) bracing systems are present in the specimens that were mainly designed for static loading conditions, without any specific requirement for earthquake loading.

The test set-up is the same shown in Fig. 6.3 and described in Sect. 6.3.2. In addition, however, a safety frame was provided capable to sustain both masses and structure in case of collapse or fall of the pallets from the rack as shown in Fig. 6.56. In this way it was possible to perform tests up to collapse, without risk of damaging the testing facility.

The characteristics of each specimen are summarized in Table 6.8 and will be described in detail in the following paragraphs, as well as the instrumentation that varied from specimen to specimen. At the base, specimens A1, A2, A3 and A4 were bolted and welded on stiff steel plates, which were directly fixed on the shaking table, while specimens A5 and A6 were mounted on proprietary base-isolation

Fig. 6.56 Detail of the test set-up with the safety frame

Table 6.8 Summary of characteristics of specimens tested on the shake table

Specimen	Beams	No levels	Testing direction	Sliding	Base
A1	130/15	3	Down-aisle (Y)	No	Fixed
A2	130/15	3	Down-aisle (Y)	Yes	Fixed
A3	70/15	3	Down-aisle (Y)	Yes	Fixed
A4	130/15	3	Cross-aisle (X)	Yes	Fixed
A5	130/15	3	Down-aisle (Y)	Yes	Isolator_1
A6	130/15	3	Down-aisle (Y)	Yes	Isolator_2

systems specifically designed for racking systems in seismic areas (allowing multi-directional in-plane displacements of ±200 mm), directly provided by two different Italian manufacturers.

6.3.4.2 Testing Procedure

The testing procedure consisted in two different types of tests: random vibration test and earthquake tests.

6.3.4.2.1 Random Vibration Test

The specimen was excited by a random acceleration signal along X or Y axes. The frequency range of random test was from 0.5 to 50 Hz. The amplitude of vibration along both X and Y direction was 0.04g (Fig. 6.57a).

These tests were performed, as first test for each specimen, in order to determine its natural frequencies and damping ratios.

6.3.4.2.2 Earthquake Tests

A series of uniaxial earthquake tests along Y or X direction was performed for each specimen.

The time history was an artificial time history which was generated to match the elastic response spectrum Type 1 of EC8 with peak ground acceleration 0.35g, subsoil glass D and damping 2 % (Fig. 6.57b), adequately scaled, from test to test, in order to achieve the desired PGA. In order to adjust to the available displacement capacity of the shaking table, the artificial accelerograms were filtered with a high pass filter 1.0 Hz.

Several earthquake tests were performed for each specimen with the maximum acceleration of the input signal, along the tested axis, being increased from test to test, until failure of the specimen. A summary of the tests performed for each specimen is given in Tables 6.9, 6.10, 6.11, 6.12, 6.13, and 6.14.

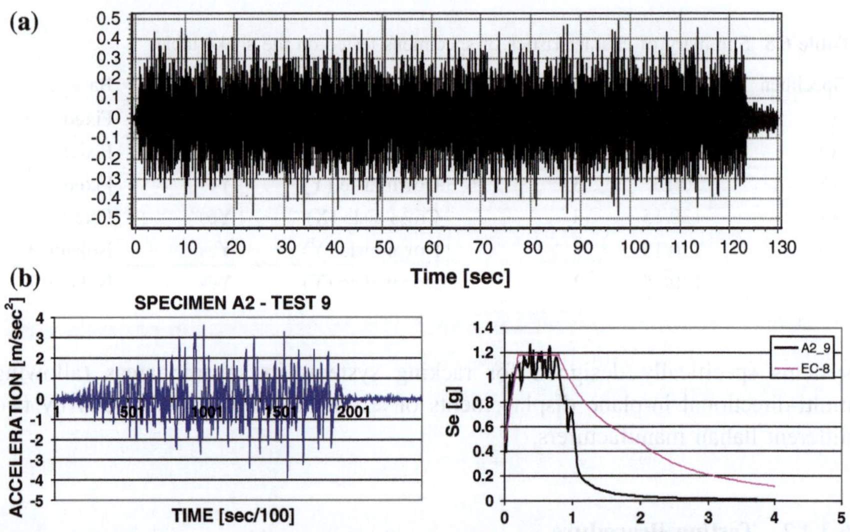

Fig. 6.57 Typical acceleration time histories adopted for the dynamic tests: **a** random vibration tests, **b** earthquake tests: shake table input time-history and elastic response spectrum of input signal (for specimen A2-test 9)

Table 6.9 Tests performed on specimen A1

Test no	Test type	Direction	PGA [g]	EDA [g]
1	Random test	Y		
2	Earthquake	Y	0.09	0.084
3	Earthquake	Y	0.21	0.173
4	Earthquake	Y	0.29	0.263
5	Earthquake	Y	0.40	0.355
6	Earthquake	Y	0.50	0.444
7	Earthquake	Y	0.62	0.539
8	Earthquake	Y	0.76	0.633
9	Earthquake	Y	0.87	0.730
10	Earthquake	Y	1.00	0.830
11	Earthquake	Y	1.07	0.931
12	Earthquake	Y	1.18	1.045
13	Earthquake	Y	1.31	1.175
14	Earthquake	Y	1.46	1.313

Table 6.10 Tests performed on specimen A2

Test no	Test type	Direction	PGA [g]	EDA [g]
1	Random test	Y		
2	Earthquake	Y	0.05	0.043
3	Earthquake	Y	0.09	0.082
4	Earthquake	Y	0.15	0.121
5	Earthquake	Y	0.20	0.167
6	Earthquake	Y	0.25	0.207
7	Earthquake	Y	0.30	0.250
8	Earthquake	Y	0.35	0.289
9	Earthquake	Y	0.40	0.330
10	Earthquake	Y	0.43	0.370
11	Earthquake	Y	0.47	0.411
12	Earthquake	Y	0.46	0.411
13	Earthquake	Y	0.54	0.516
14	Earthquake	Y	0.63	0.600
15	Earthquake	Y	0.75	0.668
15	Earthquake	Y	0.86	0.778
17	Earthquake	Y	0.92	0.881
18	Random test	Y		

Table 6.11 Tests performed on specimen A3

Test no	Test type	Direction	PGA [g]	EDA [g]
1	Random test	Y		
2	Earthquake	Y	0.05	0.043
3	Earthquake	Y	0.10	0.085
4	Earthquake	Y	0.20	0.170
5	Earthquake	Y	0.31	0.260
6	Earthquake	Y	0.42	0.350
7	Earthquake	Y	0.51	0.440
8	Earthquake	Y	0.60	0.530
9	Earthquake	Y	0.73	0.623
10	Earthquake	Y	0.85	0.718
11	Earthquake	Y	0.98	0.820
12	Earthquake	Y	1.10	0.934
13	Earthquake	Y	1.21	1.050

Table 6.12 Tests performed on specimen A4

Test no	Test type	Direction	PGA [g]	EDA [g]
1	Random test	X		
2	Earthquake	X	0.12	0.089
3	Earthquake	X	0.22	0.176
4	Earthquake	X	0.33	0.264
5	Earthquake	X	0.45	0.350
6	Earthquake	X	0.58	0.440
7	Earthquake	X	0.67	0.533
8	Earthquake	X	0.85	0.627
9	Earthquake	X	0.95	0.717
10	Earthquake	X	1.08	0.806
11	Earthquake	X	1.24	0.894

6.3.4.3 Test Specimens

6.3.4.3.1 Specimen A1

Specimen A1 is made with beams type 130 × 45 × 1.5 mm and uprights type 100/20b, i.e. of the same types of profiles adopted when characterizing the behaviour of beam-to-upright as well as base connections, in the testing campaign carried out at Instituto Superior Technico in Lisbon, and described in Chap. 2.

Pallets were rigidly connected to the steel beams, in order to avoid any possible sliding. This was achieved by means of wooden boards, screwed on the beams, on which the wooden pallets were nailed (Fig. 6.58).

Specimen A1 was submitted along the down-aisle direction (the Y main direction of shake table) to the test history summarized in Table 6.9, where for each test

Table 6.13 Tests performed on specimen A5

Test no	Test type	Direction	PGA [g]	EDA [g]
1	Random test	Y		
2	Earthquake	Y	0.05	0.042
3	Earthquake	Y	0.09	0.083
4	Earthquake	Y	0.19	0.172
5	Earthquake	Y	0.30	0.263
6	Earthquake	Y	0.40	0.352
7	Earthquake	Y	0.51	0.441
8	Earthquake	Y	0.62	0.534
9	Earthquake	Y	0.74	0.629
10	Earthquake	Y	0.86	0.722
11	Earthquake	Y	0.99	0.812
12	Earthquake	Y	1.06	0.895
13	Earthquake	Y	1.12	0.991
14	Earthquake	Y	1.15	1.097
15	Earthquake	Y	1.28	1.211
16	Earthquake	Y	1.37	1.284
17	Random test	Y		

Table 6.14 Tests performed on specimen A4

Test no	Test type	Direction	PGA [g]	EDA [g]
1	Random test	Y		
2	Earthquake	Y	0.05	0.042
3	Earthquake	Y	0.10	0.083
4	Earthquake	Y	0.21	0.173
5	Earthquake	Y	0.30	0.263
6	Earthquake	Y	0.41	0.361
7	Earthquake	Y	0.54	0.454
8	Earthquake	Y	0.64	0.533
9	Earthquake	Y	0.75	0.641
10	Earthquake	Y	0.88	0.732
11	Earthquake	Y	1.00	0.832
12	Earthquake	Y	1.11	0.942
13	Earthquake	Y	1.26	1.074
14	Earthquake	Y	1.42	1.200

both the Peak Ground Acceleration (PGA) as well as the Effective Design Acceleration (EDA) are reported. This parameter corresponds to the peak acceleration value found after low-pass filtering the input time history with a cut-off frequency of 9 Hz (Benjamin and Associates 1988). Totally 14 tests were carried out. Instrumentation consisted of 8 accelerometers and 6 LVDT's, positioned as shown in Fig. 6.59.

Fig. 6.58 The pallets were fixed on supporting beams of specimen A1

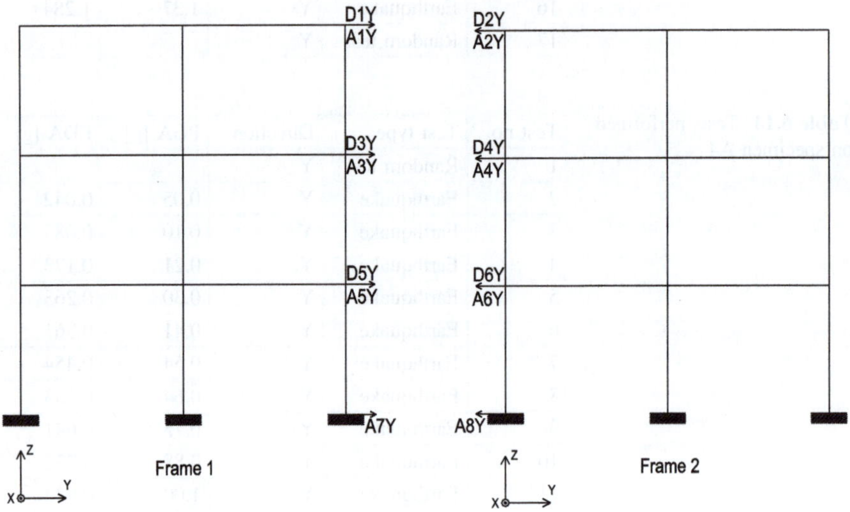

Fig. 6.59 Instrumentation of specimen A1

6.3.4.3.2 Specimen A2

Specimen A2 is similar to specimen A1, but the pallets are free to slide on the steel beams, and was submitted along the down-aisle direction to the test history summarized in Table 6.10.

Totally 18 tests were carried out. After the seismic tests, a random test was executed in order to check any variation of the dynamic characteristics.

Instrumentation consisted of 6 accelerometers and 6 LVDT's, positioned as shown in Fig. 6.60.

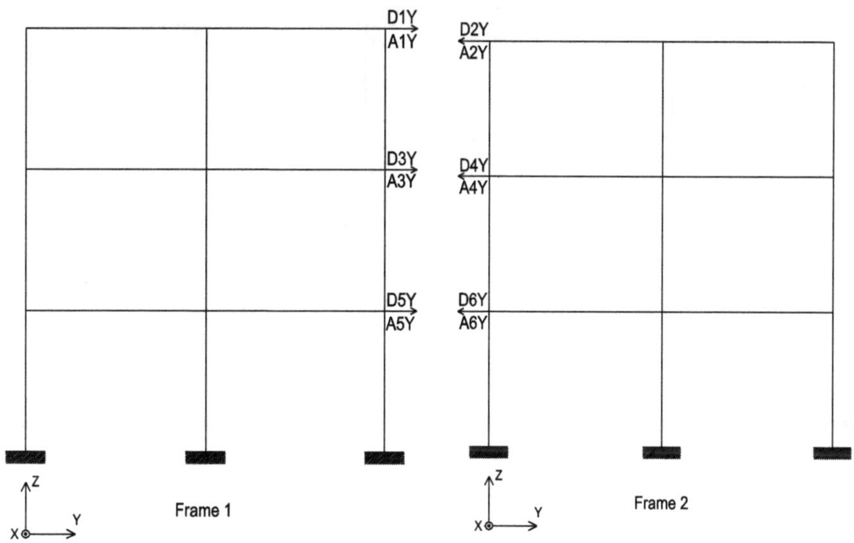

Fig. 6.60 Instrumentation of specimen A2

6.3.4.3.3 Specimen A3

Specimen A3 is similar to specimen A2, but with beams type 70 × 45 × 1.5 mm. This reduced beam size was chosen with the aim of testing a specimen with beams working at stress ratios similar to those usual in design practice.

In fact, beams 130 × 45 × 1.5 mm are frequently adopted in 2.7 m long spans and accommodate three pallets. In order to test two bay specimens on the shake table facility (of dimensions 4.0 × 4.0 m), bay widths of 1.8 m were chosen, supporting two pallets per bay. Hence, the stress ratio is 67 % of the one usually assumed in design practice for the 130 × 45 × 1.5 mm beams. The 70 × 45 × 1.5 mm beams, not only were challenged at the limit allowable stress ratio, but being more flexible and less deep, allowed a "semi-rigid" behaviour of the end-plate connectors of the beam-to-upright connection (as shown in Chap. 2). The pallets are free to slide on the steel beams. The specimen was submitted, along the down-aisle direction to the test history summarized in Table 6.11. A total of 13 tests were carried out. Instrumentation consisted of 8 accelerometers and 6 LVDT's, positioned as shown in Fig. 6.61.

6.3.4.3.4 Specimen A4

Specimen A4 is similar to specimen A2 with the pallets free to slide and was submitted, along the cross-aisle direction, to the test history summarized in Table 6.12.

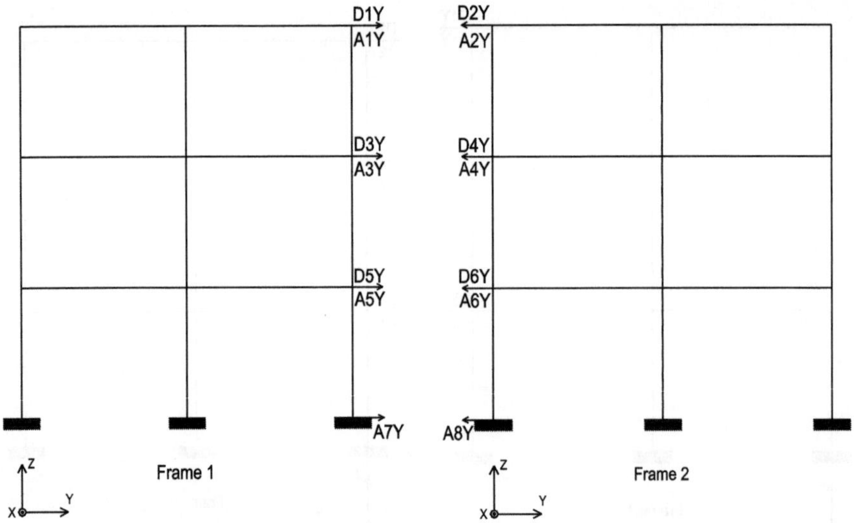

Fig. 6.61 Instrumentation of specimen A3

Totally 11 tests were carried out.

Instrumentation consisted of 9 accelerometers and 9 LVDT's, positioned as shown in Fig. 6.62.

6.3.4.3.5 Specimen A5

Specimen A5 is similar to specimen A2, with the pallets free to slide on the steel beams, but each upright base is mounted on a proprietary isolation system (that will be indicated as type 1) as shown in Fig. 6.63.

As this type of isolator allows independent in-plane movement of each base of the uprights, it was specified by the manufacturer to introduce beams at the base level, as well as an in-plane bracing system at the same level (Fig. 6.63) in order to avoid relative movements of the column bases.

The specimen was submitted, along the down-aisle direction, to the test history summarized in Table 6.13. Totally 17 tests were carried out.

Instrumentation consisted of 8 accelerometers and 8 LVDT's, positioned as shown in Fig. 6.64.

6.3.4.3.6 Specimen A6

Specimen A6 is similar to specimen A2, with the pallets free to slide on the steel beams, but is mounted on a proprietary base isolation system (that will be indicated as type 2), shown in Fig. 6.65.

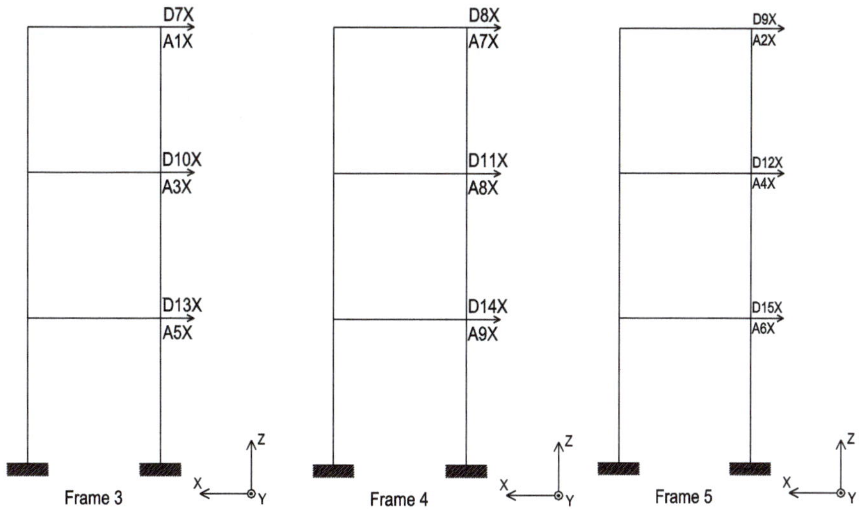

Fig. 6.62 Instrumentation of specimen A4

Fig. 6.63 Specimen A5, with base isolation system type 1

This type of base isolation system allows in-plane movement, but the device comes already with a system of beams and braces, at the base level, so that the specimen can be mounted without any additional "restraining" member.

The specimen was submitted, along the down-aisle direction to the test history summarized in Table 6.14. In total 14 tests were carried out.

Instrumentation consisted of 8 accelerometers and 6 LVDT's, positioned as shown in Fig. 6.66.

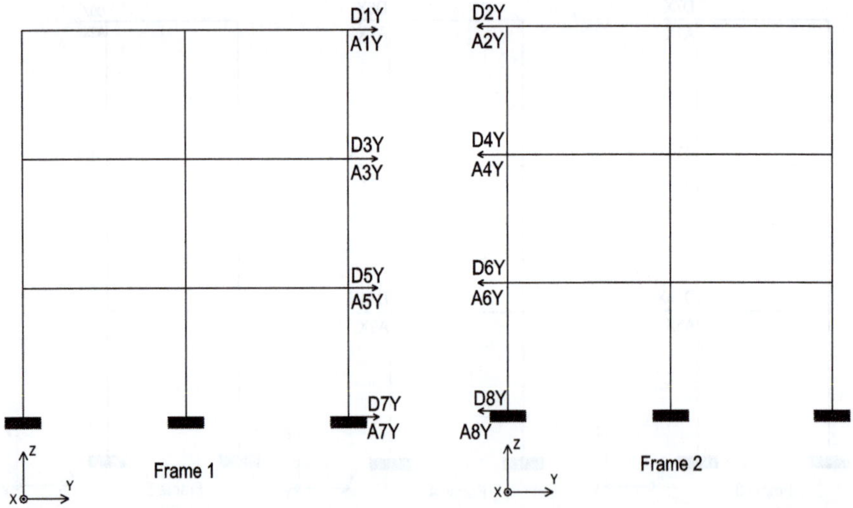

Fig. 6.64 Instrumentation of specimen A5

Fig. 6.65 Base isolation system type 2

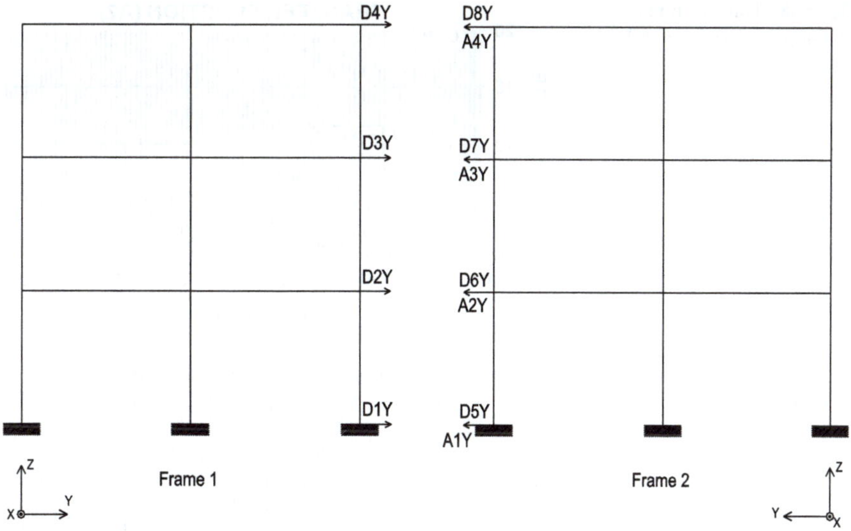

Fig. 6.66 Instrumentation of specimen A6

6.3.4.4 Test Results

A complete and exhaustive re-analysis of test results is currently still underway, as the "SEISRACKS" project terminated just recently, and the final research report to the EU is not ready yet, although is due in three months. In the following, however, a summary of the main results obtained from the experimental campaign is presented with the aim of allowing understanding of the actual dynamic behaviour of steel pallet racks under seismic conditions.

6.3.4.4.1 Specimen Eigen-Frequencies

The natural frequencies were directly measured from the peak values of the transfer functions (shown in Figs. 6.67, 6.68, 6.69, 6.70, 6.71, 6.72, 6.73, and

Fig. 6.67 Transfer function table-top for specimen A1

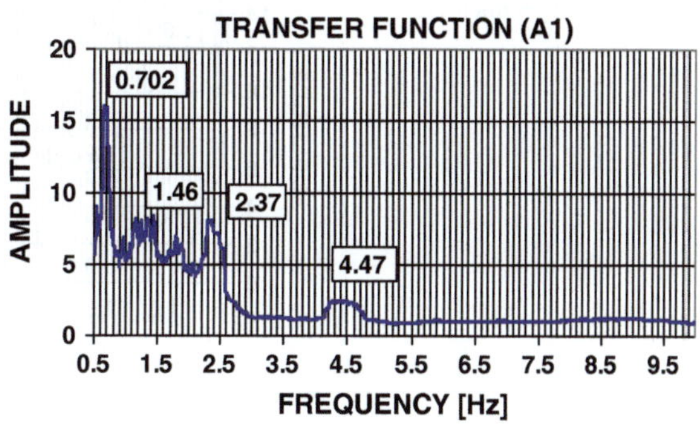

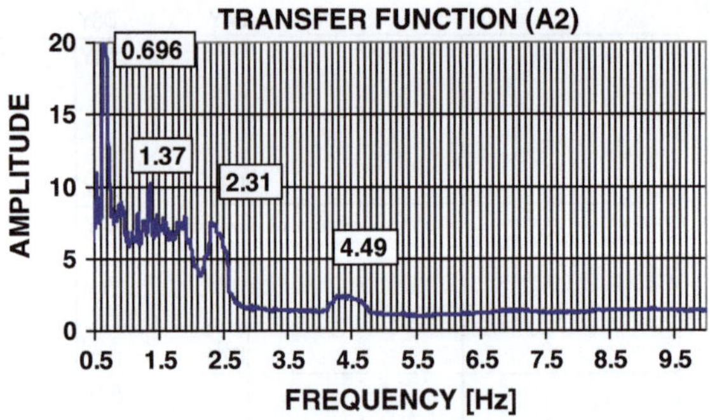

Fig. 6.68 Transfer function table-top for specimen A2

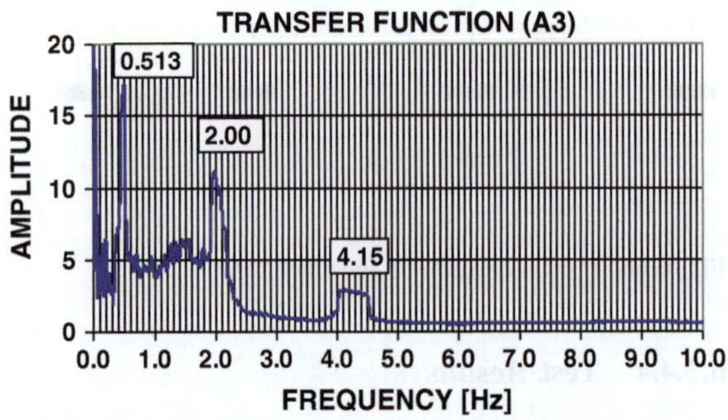

Fig. 6.69 Transfer function table-top for specimen A3

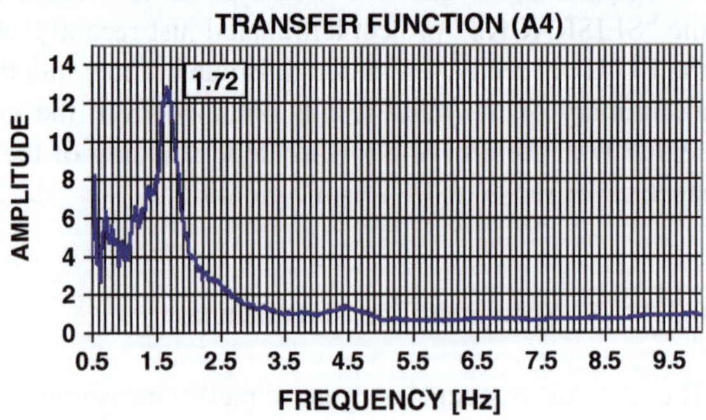

Fig. 6.70 Transfer function table-top for specimen A4

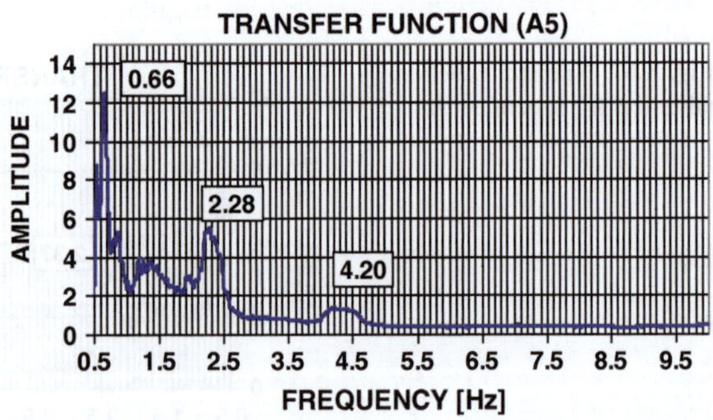

Fig. 6.71 Transfer function table-top for specimen A5

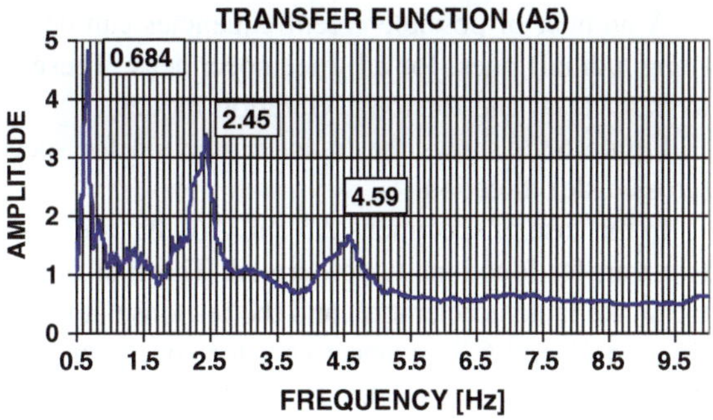

Fig. 6.72 Transfer function column base-top for specimen A5

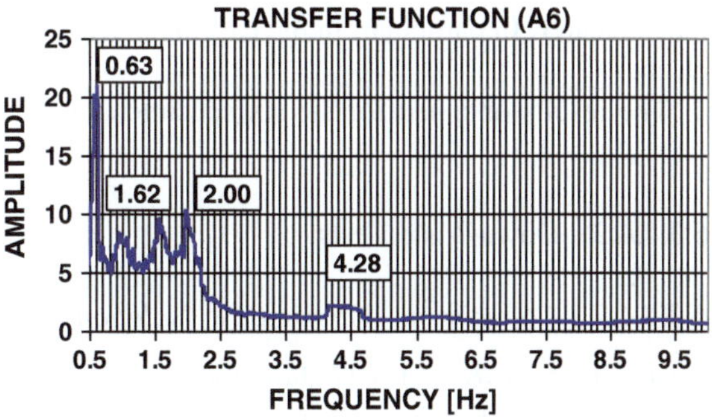

Fig. 6.73 Transfer function table-top for specimen A6

6.74) between the base acceleration and response acceleration of each specimen at the top level.

Examining the transfer functions, it can be noticed how the response of specimens A1 (Fig. 6.67) and A2 (Fig. 6.68) is very similar, although specimen A1 is slightly more rigid, most probably due to the fixed masses. Specimen A3 (Fig. 6.69) is more flexible, and the first eigenfrequency is shorter then those of the previous two specimens.

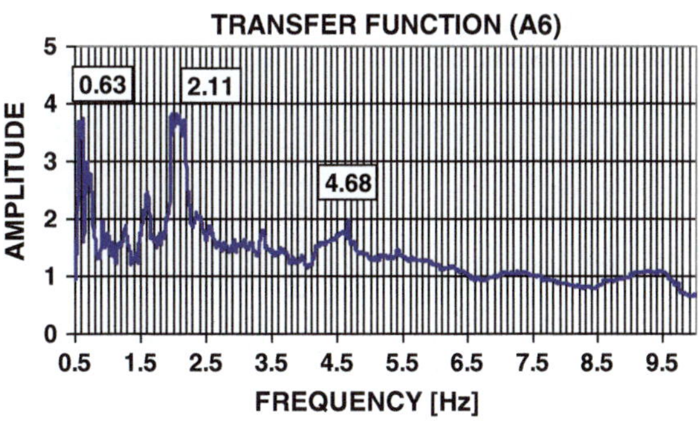

Fig. 6.74 Transfer function column base-top for specimen A6

A number of possible eigenfrequencies can be identified between the two main (longitudinal) ones, for all three specimens. These eigenfrequencies are associated to transversal and torsional vibration modes.

Figure 6.70 shows the transfer function for specimen A4, tested in transversal (cross-aisle) direction.

In the case of specimens A5 and A6, that are mounted on base-isolation devices, two transfer-functions were derived. The first one shows the relationship between the shake table acceleration and the response at the third (top) level of the specimen (Figs. 6.71 and 6.73, respectively for specimens A5 and A6). The second transfer function refers to the relationship between the column base (just above the isolating device) and the third level of the specimen (Figs. 6.72 and 6.74). It can be noticed that assessment of the eigenfrequencies through these latter transfer functions is much clearer. In particular, in the case of specimen A5, the obtained signal is much cleaner.

The results obtained for the six specimens are summarized in Table 6.15.

6.3.4.4.2 Down-Aisle Tests

6.3.4.4.2.1 Specimens Without Base Isolation

Figures 6.75, 6.76, and 6.77 show the maximum deformed shapes assumed by the non base-isolated specimens (namely A1, A2 and A3) tested in down-aisle direction

Table 6.15 Natural frequencies of the tested specimens

Specimen	Direction	Frequency [Hz]
A1	Longitudinal	0.702
A2	Longitudinal	0.696
A3	Longitudinal	0.513
A4	Transversal	1.720
A5	Longitudinal	0.684
A6	Longitudinal	0.630

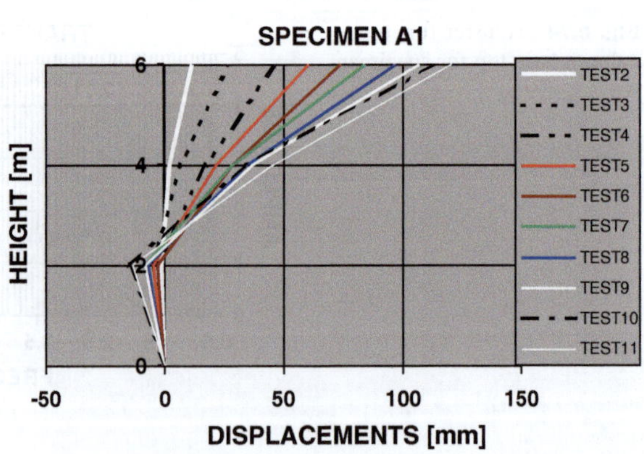

Fig. 6.75 Maximum deformed shapes of specimen A1, during various tests

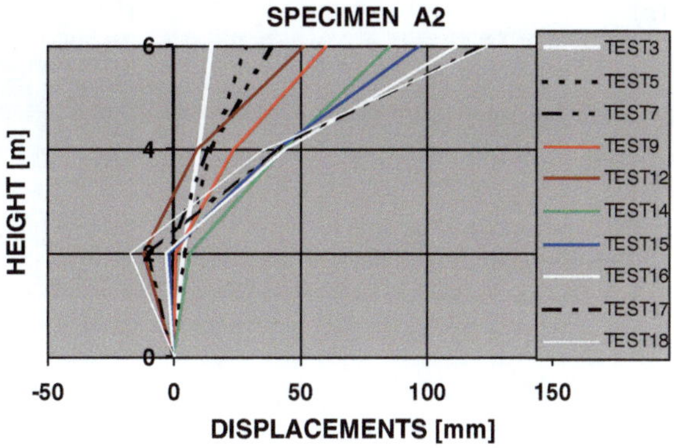

Fig. 6.76 Maximum deformed shapes of specimen A2, during various tests

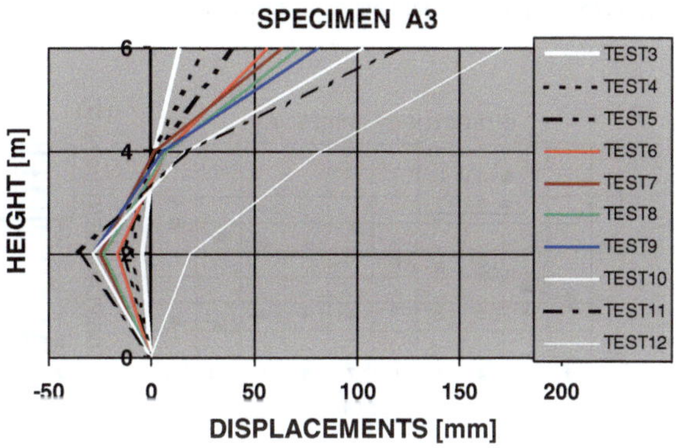

Fig. 6.77 Maximum deformed shapes of specimen A3, during various tests

during the test series at increasing maximum acceleration of the input motion. Actual values of the Peak Ground Acceleration (PGA), as well as of the Effective Design Acceleration (EDA) corresponding to each test were presented for each specimen respectively in Tables 6.9, 6.10, and 6.11.

It can be observed that, in down-aisle (longitudinal) direction, the three specimens respond mainly according to the second mode. First mode type of response is evident only for specimen A1 up to test 3 (performed with PGA of 0.21g) and for specimen A2 up to test 5 (characterized by a PGA of 0.20g). The response of specimen A3, that is more flexible because of the smaller beams (70 × 45 × 1.5 mm type), is according to the second mode already from test 3 (performed with PGA of 0.21g). Typical modal shapes in down-aisle direction, obtainable by numerical analysis, are shown in Fig. 6.78.

The deformed shape of the second mode is such that displacements of the second level are much smaller than those of levels 1 and 3, although the inter-storey drifts are very large, as shown in Figs. 6.79, 6.80, and 6.81. These figures present, for specimens A1, A2 and A3: (a) the trend of the maximum values of the inter-storey drift at the three levels (indicated as Δ I–T for the I level, Δ II–I for the second and Δ III–II for the third) as well as (b) the trend of the maximum absolute

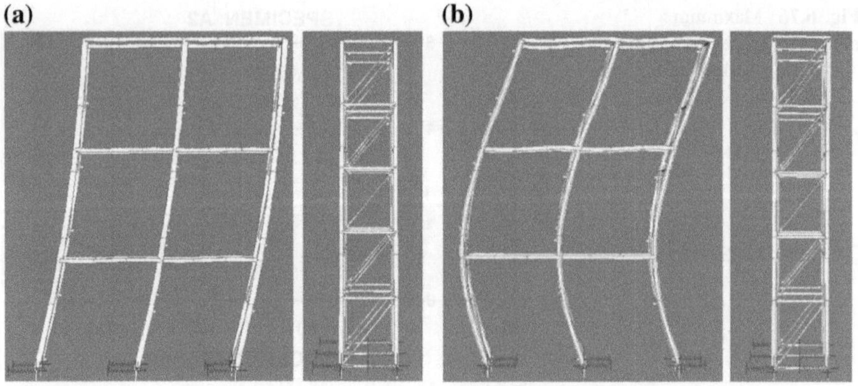

Fig. 6.78 Typical modal shapes in down-aisle direction **a** 1st mode, **b** 2nd mode

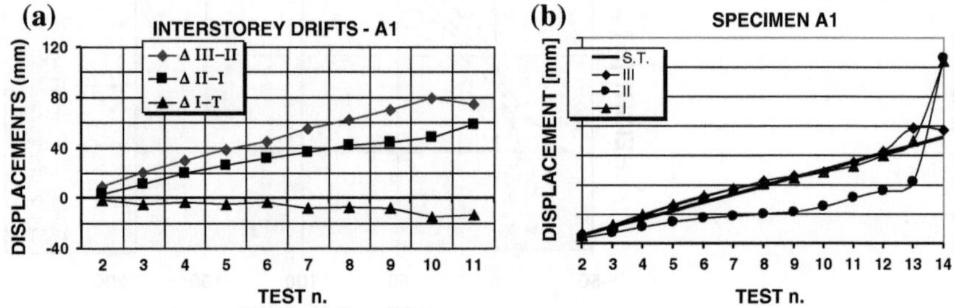

Fig. 6.79 Trend of the maximum inter-storey drifts (**a**) and of the absolute displacements (**b**) at the various levels of specimen A1, during the test sequence

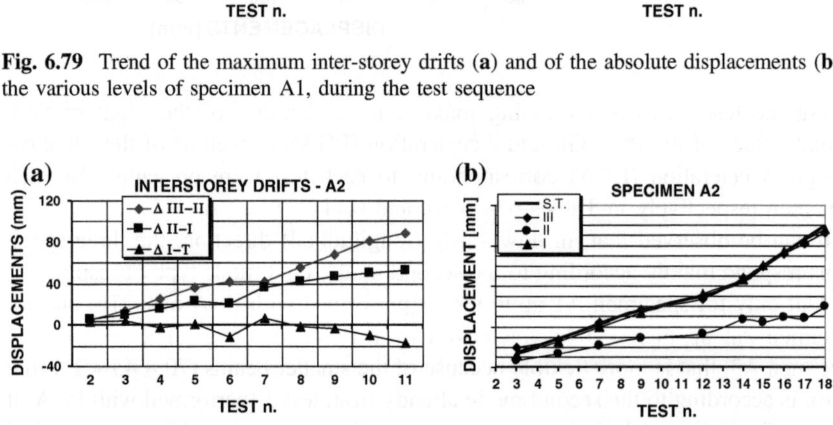

Fig. 6.80 Trend of the maximum inter-storey drifts (**a**) and of the absolute displacements (**b**) at the various levels of specimen A2, during the test sequence

displacements of the same levels together with those of the shaking table, measured during the test sequence.

It is evident that all the three specimens show a similar dynamic behaviour. Inter-storey drifts of the first level are always opposite in sign (i.e. in counter-phase)

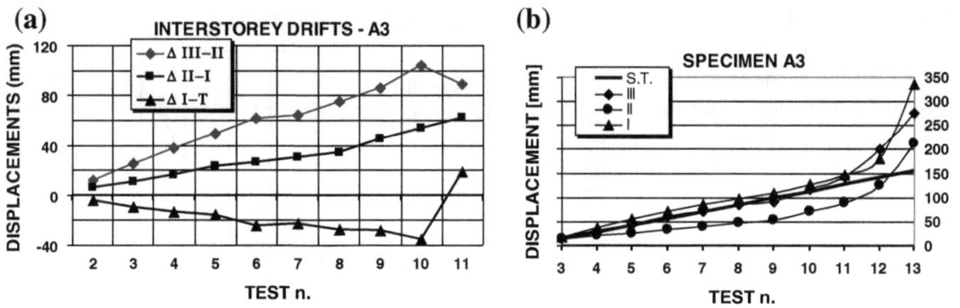

Fig. 6.81 Trend of the maximum inter-storey drifts (**a**) and of the absolute displacements (**b**) at the various levels of specimen A3, during the test sequence

and smaller than those of the upper levels. Specimen A3 is more flexible than specimens A1 and A2. Test 18 was carried out on specimen A2 with a maximum PGA of 0.92g. Specimen A1 was subjected to a maximum PGA of 1.46g at test 14. Comparing the maximum displacements of specimen A2 at test 18 (Fig. 6.80b) with those exhibited by specimen A1 (Fig. 6.79b) during tests 9 or 10 (carried out at PGA of 0.87g and 1.0g respectively) it can be noticed that deformability of the two specimens is similar.

It can also be observed that the maximum displacements of the first and the third levels are similar to those of the shake table, while those of the second level are smaller. In the last test, under sustained values of the PGA, a non-linear trend can be observed in the maximum displacements, clearly indicating incipient collapse.

The trend of accumulation of the residual displacements measured at the various levels of the specimens at the end of each test is plotted vs. the sustained PGA of the test in Figs. 6.82, 6.83, and 6.84.

In the figures, for each level are reported the displacements measured on the two longitudinal frames (Frame 1 and Frame 2, of Fig. 6.3). Differences between measurements at the same level indicate that the specimen suffered a permanent torsional deformation. Despite being designed for gravity loads only (i.e. without

Fig. 6.82 Trend of accumulation of the residual displacements at the various levels of specimen A1, during the test sequence at increasing PGA

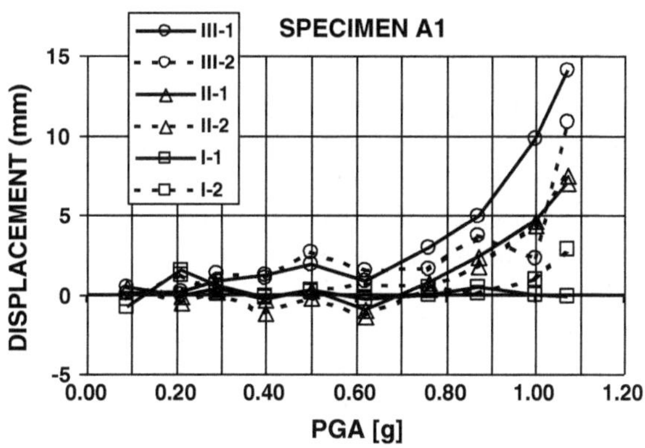

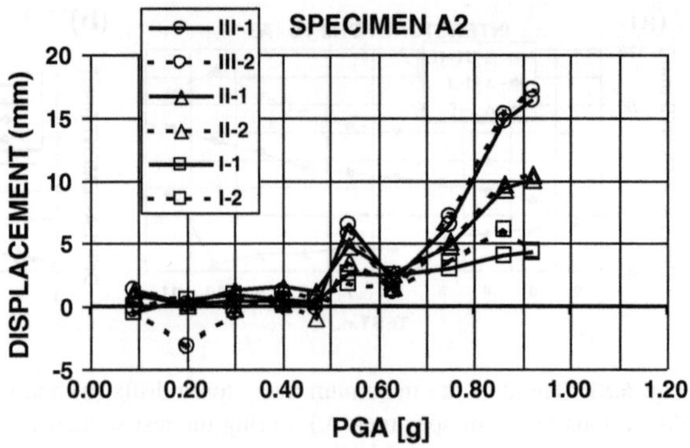

Fig. 6.83 Trend of accumulation of the residual displacements at the various levels of specimen A2, during the test sequence at increasing PGA

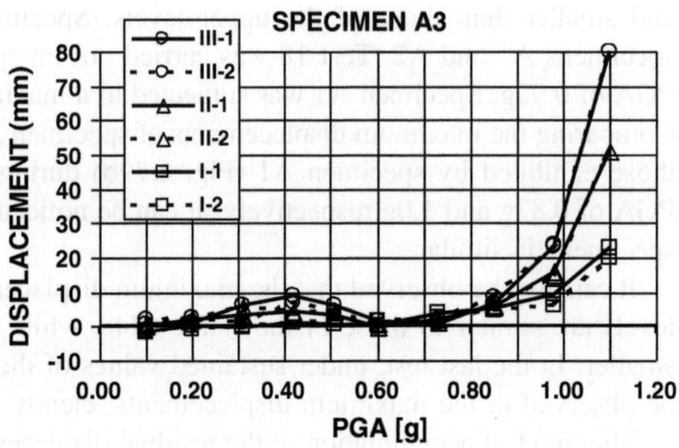

Fig. 6.84 Trend of accumulation of the residual displacements at the various levels of specimen A3, during the test sequence at increasing PGA

any specific a-seismic requirement), the specimens don't show any particular accumulation of residual displacements up to a PGA of 0.4g. Beyond such values of PGA, the specimens start to accumulate permanent plastic deformations, with a non-linear increasing trend. In all three specimens the largest residual displacements occurred at the upper (third) level, and they decrease toward the bottom. No significant transversal residual displacement could be observed.

6.3.4.4.2.2 Specimens with Base Isolation

Although both proprietary base-isolation systems tested on specimens A5 and A6 are bi-dimensional isolating systems, hereafter only the results related to uni-axial (down-aisle) tests are presented.

The presence of a base isolation system strongly influences the structural response. Both the two proprietary systems adopted in this testing campaign, behave as "sliders", allowing in plane movement of the column base. Hence, under seismic input motion, the structure responds, at least for small base displacements, as a rigid body. This preserves the column-base connection from plastic deformations. Under strong input motions, the stiffness of the elastomeric "spring" (that increases non linearly with the displacements), restrains the structure. This effect,

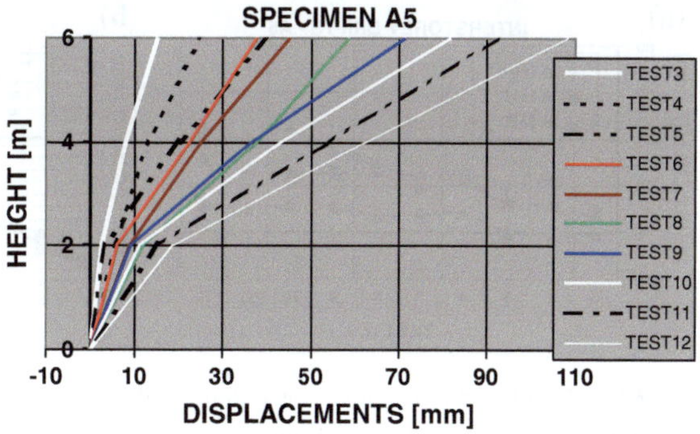

Fig. 6.85 Maximum deformed shapes of specimen A5, during various tests

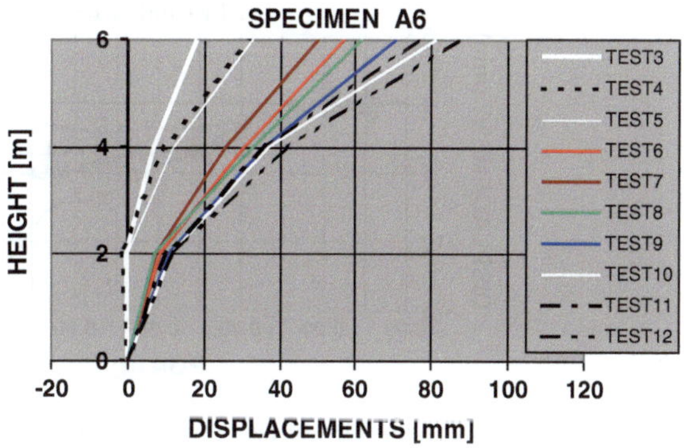

Fig. 6.86 Maximum deformed shapes of specimen A6, during various tests

combined with the inertial forces on the masses, results in deformation of the structure that, in any case, responds according to the first mode. This behaviour is shown in Fig. 6.85 (for specimen A5) and in Fig. 6.86 (specimen A6).

Effects of the presence of base isolation systems on the structural response are well evident in Fig. 6.87 (specimen A5) and Fig. 6.88 (specimen A6) reporting the trend of the maximum inter-storey drifts and of the absolute displacements at the various levels of the specimens, during the test sequence. The corresponding values of the

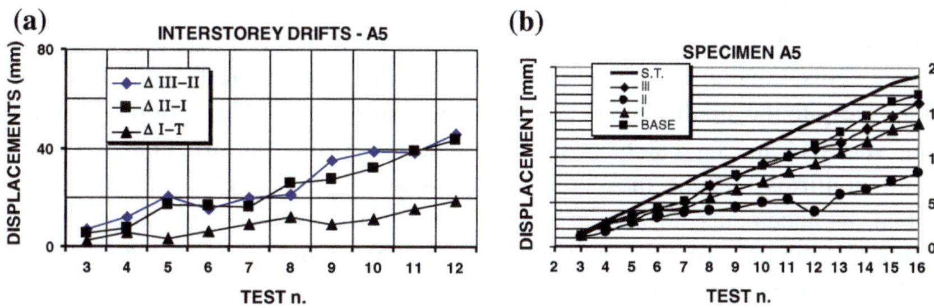

Fig. 6.87 Trend of the maximum inter-storey drifts (**a**) and of the absolute displacements (**b**) at the various levels of specimen A5, during the test sequence

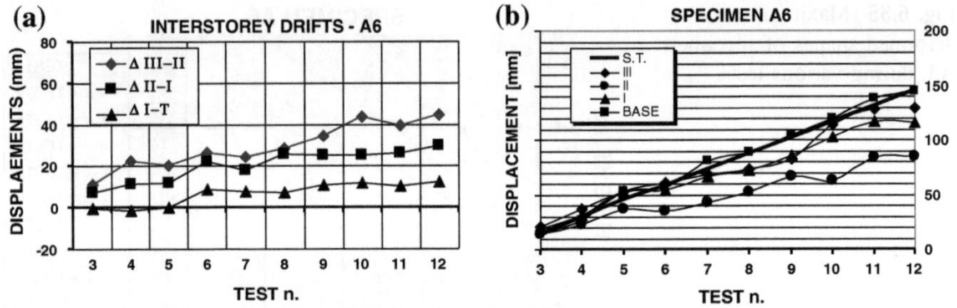

Fig. 6.88 Trend of the maximum inter-storey drifts (**a**) and of the absolute displacements (**b**) at the various levels of specimen A6, during the test sequence

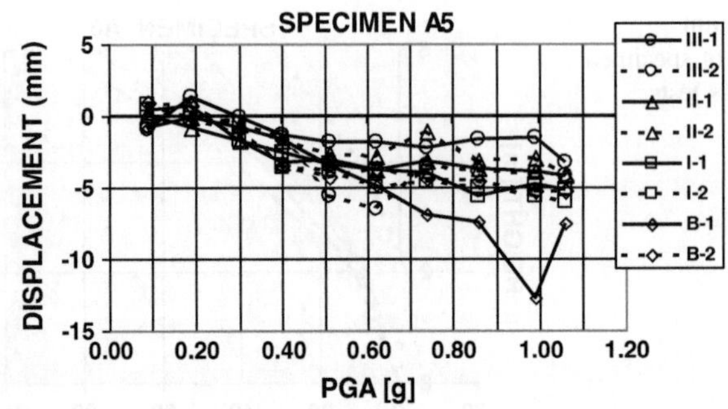

Fig. 6.89 Trend of accumulation of the residual displacements at the various levels of specimen A5, during the test sequence at increasing PGA

PGA can be found in Table 6.13 (for A5) and in Table 6.14 (for A6). The three levels respond in phase, and their absolute displacements are always smaller than those of the shake table (i.e. of the input motion). A comparison with the corresponding values presented for non base-isolated specimens (Fig. 6.79 for specimen A1, Fig. 6.80 for A2 and Fig. 6.81 for A3) shows that both inter-storey drifts and absolute displacements of specimens A5 and A6 are much smaller than those measured on the specimens without base-isolation, under corresponding values of PGA.

Figure 6.89 shows the trend of accumulation of the residual displacements at the various levels of specimen A5, during the test sequence at increasing PGA. Residual displacements, smaller than 5 mm, are evident since the first test, under low PGA values. Despite the small scatters among the measurements at the various levels and on the various frames, it can be observed a general trend for the structure as a whole.

This confirms that the rack is behaving as a rigid body. The accumulation of residual displacements, in this case, is not due to permanent plastic deformations accumulating in the structure, but to "friction" problems in the base-isolators, that do not allow a complete and perfect "re-centering" of the structure after the earthquake.

As already discussed, the possibility of independent movement of the column bases allowed by the base-isolation system tested on specimen A5, required introduction of an additional pair of beams and braces at the base level (Fig. 6.63). Despite this, some torsion occurred at base level, as shown in Fig. 6.89, where it is evident that under an earthquake test with PGA of 1.0g the base of frame 2 suffered permanent displacements larger than those measured on frame 1.

It should also be noticed that, in down-aisle direction, specimen A5 sustained a total of 16 earthquake tests (plus two random tests), the strongest with a PGA of 1.37g. Similarly, specimen A6 sustained a total of 13 earthquake tests (plus one random test), the strongest with a PGA of 1.42g. Under such high accelerations, the base-isolators nearly reached their ultimate working conditions. Hence, it was considered useless to further increase the base acceleration, and testing was terminated, with both specimens practically undamaged.

6.3.4.4.3 Cross-Aisle Tests

Specimen A4, similar to specimen A2, was submitted in cross-aisle direction to the test sequence summarized in Table 6.12.

Figure 6.90 shows the maximum deformed shapes assumed by the central (a) and the lateral (b) upright frames during the test series at increasing maximum acceleration of the input motion. The central upright frame, carrying double forces than the lateral ones, shows larger displacements. The deformed shapes of the lateral upright are similar to those of the central frame, but with smaller displacements.

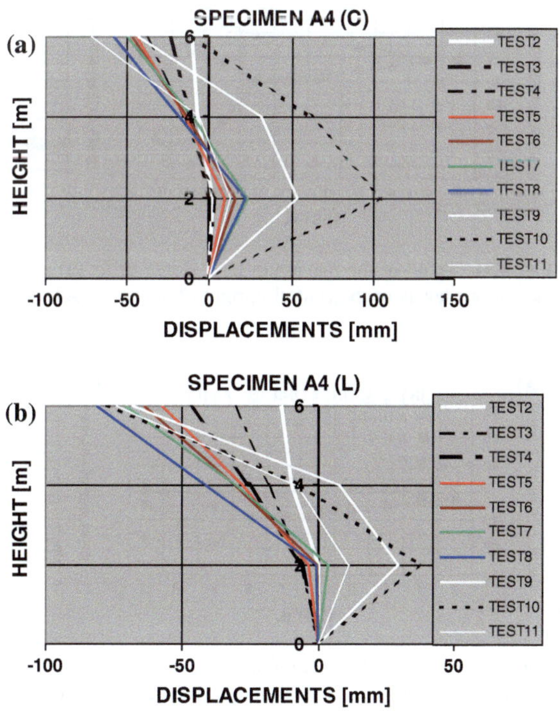

Fig. 6.90 Maximum deformed shapes of specimen A4, during various tests in cross-aisle direction **a** central upright, **b** lateral upright

Initially, the structural response is according to the first mode, and the displacements of all three levels are in phase. Increasing the acceleration, during the test sequence, results in plastic deformations at the base of the central upright. Effect of the reduced stiffness of the base connection is a change in the deformed shape, with the displacements of the first and the third level in counter-phase. Rotation of the column base results in an increment of the transversal displacements at the first level. As the deformations of the lateral frames are smaller, the large transversal deformations of the central frame induce, at the beam-ends, very large rotations in the horizontal plane that cause plastic deformations in the beam-to-upright connections. This behavior is also highlighted in Figs. 6.91 and 6.92 that show, respectively for the central upright frame and for the lateral one, the trend of the maximum inter-storey drifts (a) and of the absolute displacements (b) at the various levels of specimen A4, during the test sequence. In Fig. 6.91a it can be observed that, up to test 8 (i.e. to a PGA of 0.85g), the inter-storey drift at the first level of the central upright are of the same order of magnitude of those at the above levels. In test 9 (PGA = 0.95g, EDA = 0.72g) inter-storey drifts show a very large increment. When considering the absolute displacements, Fig. 6.91b highlights that the transversal displacements at level three of the central upright are always larger than those of the shake table and show a non-linear trend after test 8. The same

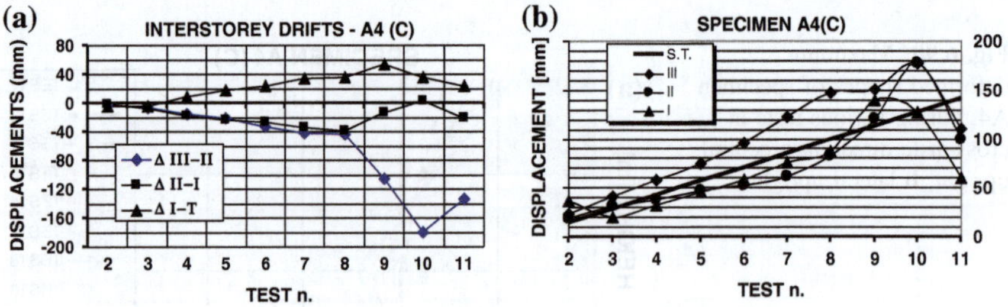

Fig. 6.91 Trend of the maximum inter-storey drifts (**a**) and of the absolute displacements (**b**) at the various levels of the central upright of specimen A4, during the test sequence

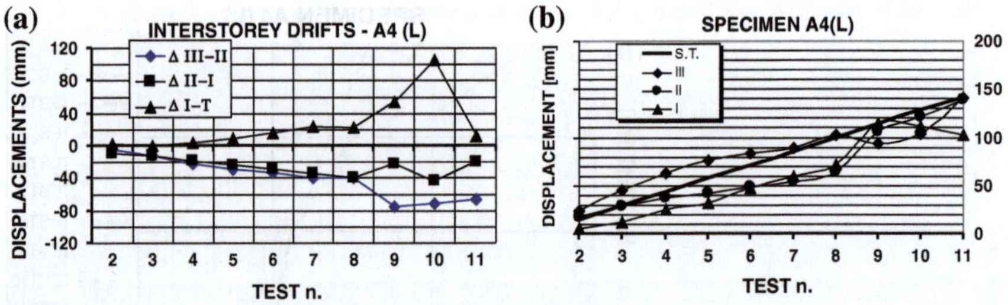

Fig. 6.92 Trend of the maximum inter-storey drifts (**a**) and of the absolute displacements (**b**) at the various levels of the lateral upright of specimen A4, during the test sequence

non-linear increment trend, after test 8, is shown by the displacements at levels one and two that, in the previous tests, were smaller than the shake-table ones.

Similar considerations are valid also in the case of the lateral upright frame (Fig. 6.91). In this case, however, during the first 8 tests the inter-storey drifts at the first level are nearly half of those of the upper levels. While the inter-storey drift at the first level is much smaller than that at the same level of the central upright, at the upper levels the inter-storey drifts of lateral and central uprights are comparable. Comparing the behaviour of the lateral and the central upright frames it can be observed that the displacements of the first two levels are similar. At the third level, on the contrary, the displacements of the lateral uprights do not show the large non-linear increasing trend shown by those of the central one.

Accumulation of the residual displacements at the various levels of specimen A4, during the test sequence at increasing PGA, shown in Fig. 6.93, confirms that plastic deformation at the third level occurs already under values of PGA of the order of magnitude of 0.2–0.3g. Beyond values of the PGA of 0.6g, the central upright frame accumulates large residual displacements at all three levels. When passing from the first to the third level, the amplitude of these residual displacements, as well as their trend of increment with the PGA, increase.

The column-base connections of the central upright frame collapsed during test 9, under a PGA of 0.95g. During the same test, the beam-to-upright connections of the third level collapsed due to excessive bending in the horizontal plane.

Global collapse occurred during test 11, with PGA 1.24g and EDA = 0.89g, with complete failure of the beam-to-upright connections at the lower levels.

6.3.4.4.4 Failure Modes

In the following some images and comments will be presented, related to the failure modes of the three specimens (A1, A3 and A4) that were tested up to complete failure. In fact, as already said, in the case of specimens A5 and A6, base-isolated, testing was ended due to attainment of the ultimate design conditions of the base-isolators, when the structures were still in good conditions, and practically undamaged. In the case of specimen A2, testing was ended after an earthquake

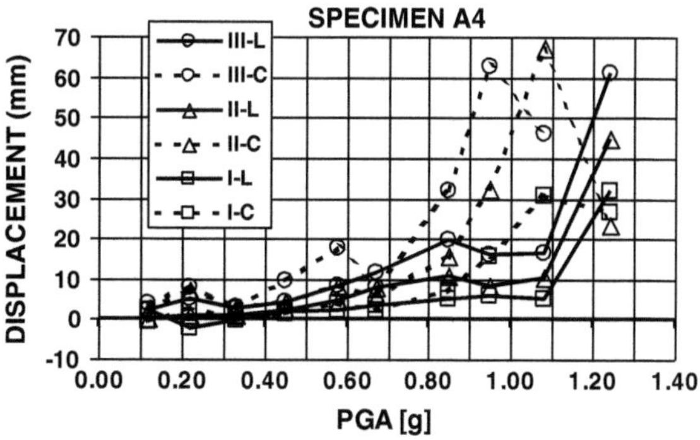

Fig. 6.93 Trend of accumulation of the residual displacements at the various levels of specimen A4, during the test sequence at increasing PGA

simulation test with a PGA of 0.92g, when the structure was starting to accumulate plastic deformations in the down-aisle direction, but was not completely collapsed.

6.3.4.4.4.1 Specimen A1

In order to prevent pallet sliding, in specimen A1 wooden boards were connected to the steel beams and the pallets were then fixed on the wooden boards (Fig. 6.58).

The concrete blocks, representing the live loads, were then rigidly fixed to the wooden pallets by means of some steel angles. This choice turned out to be correct in order to eliminate sliding of the pallets, but created some problems, somehow modifying the structural response of specimen A1.

Figure 6.94 shows the deformed shape of specimen A1 after collapse, occurred during test 14. It can be noticed that the structure is hanging on the safety frame. Evident plastic deformations occurred at the column bases.

A non-symmetry is however evident in the deformed shapes of the uprights, and is highlighted in Fig. 6.95. During the earthquake tests under high PGA values, in order to accommodate the large inter-storey drift demand, the upright deformed so much to enter in contact with the concrete blocks, rigidly fixed on the beams. This most probably occurred during test 11, under a PGA of 1.07g, as it can be deduced by examining Fig. 6.79 that shows a reduction of the inter-storey drift at the first and third levels, despite increasing the PGA.

Figure 6.95 shows the plastic deformations occurred in the uprights, at the first level. The central upright highlights a "piece-wise" deformed shape, in its central portion, near the beam-to-upright connection. The presence of the wooden board, rigidly connected to the steel beams increased the stiffness of the beam. The

Fig. 6.94 Collapse of specimen A1

Fig. 6.95 Detail of the first level of specimen A1 after collapse

structure, hence behaved as a strong-beam weak-column system. This resulted in large moments in the upright profiles that suffered plastic deformations. Complete formation of plastic hinges in the columns was avoided on one side because of the reduced stress ratio of the columns (due to presence of two pallets only per bay, instead of the three that are commonly adopted in practical applications for this types of profiles) and on the other side thanks to the semi-rigid behaviour of the beam-to-upright connections. In any case, at the third level, shear in the central upright could not be resisted and, near the free, unstiffened, edge the cross-section was severed, fracturing at the openings, as shown in Fig. 6.96.

Details of the failure at the column base connection are presented in Fig. 6.97.

Fig. 6.96 Collapse of specimen A1: detail of the failure of the upright at the beam-to-column connection at the third level

Fig. 6.97 Collapse of specimen A1: detail of the column-base failure

6.3.4.4.4.2 Specimen A3

Specimen A3 was erected with beams type 70 × 45 × 1.5 mm, smaller than those of all the other specimens.

Figure 6.98 shows the deformed shape of specimen A3 after collapse occurred during test 13, an earthquake simulation test with PGA of 1.46g

It can be noticed that the structure is hanging on the safety frame.

Also in this case, as for specimen A1, evident plastic deformations occurred at the column bases, together with fracture of the bolts connecting the upright profiles to the gusset plates welded to the base plate. It should be remembered that the base plate was welded to the shake table and, consequently, its bending deformation capacity was completely restrained. Therefore, the weaker components of the connection are the bolts fastening the upright profiles to the gussets and the welds between the gussets and the base plates, which collapsed in all specimens.

Figure 6.98 shows also an evident damage in the beam-to-upright connection, with plastic deformation of the end-plate connector as well as local buckling in the upright profile.

6.3.4.4.4.3 Specimen A4

Specimen A4 was tested in cross-aisle direction. Neither vertical (down-aisle) nor horizontal (in-plane) bracing systems are present in the specimen that is mainly designed for static loading conditions, without any specific seismic requirement.

As already described, collapse is associated with failure of the central upright frame, that is carrying horizontal forces double than the two lateral uprights. In absence of an in-plane bracing system, in fact, such forces could not be re-distributed among the three upright frames. The failure mechanism is shown in Fig. 6.99. Collapse interested the column-base connections (Fig. 6.99c) as well as the beam-to-upright connections that suffered large plastic deformations in the

Fig. 6.98 Collapse of specimen A3: global collapse and details of the column base and of the beam-to-column connection failure

horizontal plane (Fig. 6.100). Failure occurred for accumulation of deformation in the direction in which diagonal members of the upright frames are in compression.

However, no evident buckling deformation of the diagonal members could be observed. On the contrary, failure occurred at the bolted connections between the diagonals and the uprights, as shown in Fig. 6.99b.

Significant sliding could be observed during the tests, as highlighted in Fig. 6.100 (shot at the end of test 11), where relative displacements between pallets and beams are evident.

6.3.4.5 Conclusions

Within the SEISRACKS research project, earthquake simulation tests were carried out on six full scale rack models of similar geometry as those of the first series of tests carried out within the ECOLEADER project. Geometry and member size of all specimens were similar, with exception of specimen A3 that mounted a smaller size type of beam. Specimen A4 was tested in cross-aisle direction, while all other

Fig. 6.99 Collapse of specimen A4: **a** global collapse; **b** failure of the bolted connection between diagonal and upright; **c** failure of the column-base connections

specimens were tested in down-aisle direction (two of which, A5 and A6, with base isolation systems). Specimen A1 was tested, with fixed pallets, in order to avoid sliding, while in all other specimens pallets were free to slide.

Two different base-isolations systems were mounted under specimens A5 and A6.

The models were subjected to a sequence of earthquake simulation tests, progressively increasing the PGA. Failure was attained under high values of PGA, despite the fact that the specimens were mainly designed for static loading conditions, without any specific seismic requirement. In any case, all specimens could

Fig. 6.100 Collapse of specimen A4: failure of the beam-to-upright connections in the horizontal plane, and evident pallet sliding

sustain an earthquake simulation test with PGA of 0.60g without relevant damage. This stresses the good performance of very flexible structures.

Specimens A5 and A6 sustained tests with a maximum PGA as high as 1.4g without collapsing or showing any structural damage.

The importance of structural detailing was also highlighted. Most of the observed failures, in fact, were caused by failure of bolted or welded connections. All these details could be easily improved, at low cost, effectively enhancing the structural performance.

6.4 Assessment of the Behaviour Factor (Q-Factor)

6.4.1 Overview

Eurocode-8 (2004) suggests the reduction of the design seismic forces through the behaviour factor, generally called "q-factor". Other modern codes for the seismic design of structures adopt different formulations although, in each case, a reduction factor can be identified, that can somehow be related to the q-factor (Mazzolani and Piluso 1997). The q-factor takes implicitly into account the global and local ductility resources of the structure (which depend on the structural typology, the ductility of the material, the $P - \Delta$ effects and possible brittle fracture mechanisms) and depends on a wide number of factors related to the structure and the seismic input.

The present codes, such as Eurocode 8, give just a generic estimate of the q-factor, generally related to the typology of the structure (frame building, braced structures etc.), its geometrical regularity and ductility of the elements and connections. This generic q-factor is not capable of representing in detail the real structural response and completely disregards important aspects such as, for example, the duration of the input motion, the number of cycles and their amplitude.

In order to develop simple design rules for steel buildings in seismic zones it is important to characterise the behaviour of steel structures under cyclic reversal loading, and in particular to focus on the damage caused by plastic deformations and low-cycle fatigue.

In general, the q-factor is defined as the ratio of the peak ground acceleration producing collapse of the structure ($a_{\max}$) to the design PGA (a_d):

$$q = \frac{a_{\max}}{a_d}$$

In this definition the peak ground acceleration a_{max} at collapse depends on the type of collapse mechanism that is considered. Hence, in a general prospective, the q-factor can be defined as the minimum value of the q-factors obtained for the same structure, considering n different failure modes and limit states

$$q = \min[q_1, q_2 \ldots q_n].$$

Such a definition of q-factor yields to a safety margin to failure for the structure, if all the collapse mechanisms have been considered.

Due to the importance of taking into account the real inelastic resources of the structure, several methods for the assessment of the q-factor have been proposed in the literature.

An extensive review and comparison of these methods can be found in the state-of-the-art presented by Guerra et al. (1991), by Mazzolani and Piluso (1997) as well as by Gioncu and Mazzolani (2002).

All this methods, however, are based on assessment of the q-factor by means of numerical or theoretical approaches.

6.4.2 Assessment of Behaviour Factor for Pallet Racks

Hereafter it is tried to assess the q-factor for steel pallet racks, based on re-analysis of the experimental shake-table test data previously presented and discussed.

Starting from the definition of the q-factor as "the ratio of the elastic structural response (in terms of accelerations) to the inelastic one", the following procedure was identified. The procedure allows assessment of the q-factor as the ratio of the actual maximum base shear experienced by the structure, to the one that might be derived considering an indefinitely linear elastic structural behaviour.

The actual base shear, at a given time instant during the test, can be easily obtained as the sum of the shear forces at the various levels at that instant. These, in turn, can be computed by multiplying the accelerations measured at each level of the specimens by the actual values of the masses present at the same level. Hence it is possible to correlate the actual maximum value of the base shear with either the

Peak Ground Acceleration or some other acceleration parameter characterising the input motion.

Assessment of the elastic behaviour of the specimen may, on the contrary, be a little tricky. In fact, data relative to the elastic response of the structure are required. The following problems were encountered:

Reduction of stiffness was identified already in tests performed with PGA values of the order of 0.10g–0.15g. Intensity of the "first" test should hence be very low.

Possible differences in the response (i.e. of the stiffness of the specimen) depending on the verse of the excitation.

Extrapolation of results obtained for low values of the acceleration toward large acceleration values might, of course, result in significant errors when assessment of the "initial conditions" is not properly precise.

For example, simply averaging the results obtained for two different verses of the excitation might lead to significant errors, when extrapolated in the high acceleration range.

Figure 6.101 shows a typical application, where the elastic and actual responses of specimens A1 and A2 are presented.

Large differences can be observed in the estimated elastic responses of the two specimens, although the only difference was the fact that the pallets were fixed on specimen A1, and could slide on specimen A2. The actual behaviour of the two specimens, in fact, is rather similar.

The results obtained for specimens A1, A2, A3 and A4 are presented in Table 6.16, where a_{max} (positive or negative) is the maximum (or minimum) acceleration corresponding to the assessment of both the actual base shear (V) and the elastic base shear (V_e). In Table 6.16, the values (q) of the q-factor obtained under maximum (positive) and minimum (negative) values of the acceleration are presented, as well as the mean value of the two ($q_{average}$). This last value is presented with both 1 and (between parentheses) two decimal digits.

Adopting the same procedure, an evaluation of the q-factor was attempted also for the other specimens presented in this study although none of them was tested up to ultimate conditions.

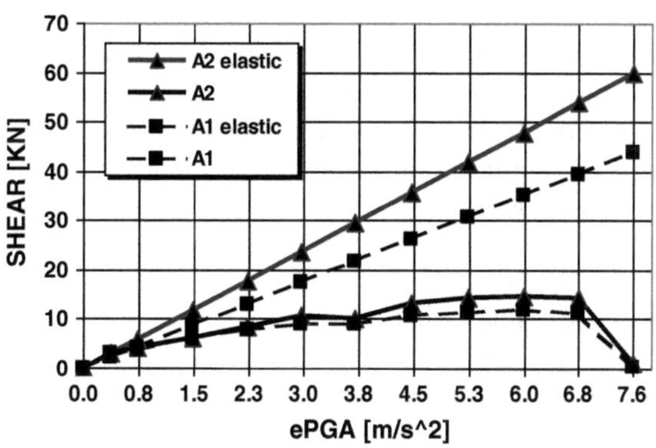

Fig. 6.101 Assessment of the q-factor

Table 6.16 q-factor values of steel racks

SPECIMEN	a_{max} [g]	V_e [kN]	V [kN]	q	$q_{average}$
A1 (down-aisle)	1.31	39.11	14.82	2.64	2.8
	−1.07	−31.95	−11.08	2.89	(2.77)
A2 (down-aisle)	0.92	62.39	17.88	3.49	3.7
	−0.88	−59.67	−15.57	3.83	(3.66)
A3 (down-aisle)	0.82	60.43	16.55	3.65	3.7
	−0.61	−44.96	−12.15	3.70	(3.68)
A4 (cross-aisle)	0.78	36.77	11.40	3.23	2.7
	−0.63	−29.70	−13.38	2.22	(2.73)

In particular, for specimens B, C and D of the ECOLEADER project (presented in Sect. 6.3.3) the procedure did not lead to any reasonable result, as testing was ended because of the sliding of the pallets, when the PGA was still rather low (0.20g or less) and the structures were still in the elastic range. The maximum values returned by application of the procedure for these specimens are in the range 1.3–1.4. Only for specimen A, that was originally designed for a PGA of 0.075g, a value of the q-factor of 2.7 for both the down and cross-aisle direction could be estimated.

On the contrary, it is interesting to underline the result obtained for specimen A5. Although also this specimen could not be tested up to failure, corresponding to a PGA of 1.1g in the down-aisle direction a q-factor of 6.9 could be estimated, nearly two times larger than those obtained for similar specimens, non base-isolated.

6.5 Conclusions

Two series of dynamic tests were performed.

In the first series, within the ECOLEADER Project, four full scale rack models of three levels (total height 6.0 m) and 2 bays (total width 3.6 m) were tested under earthquake simulations in cross-aisle and in down-aisle direction.

Results show the importance of structural regularity, both in elevation and in plan. Eccentricity of the vertical down-aisle bracing system may result in severe torsional deformation and in major structural damage.

In-plane bracings at each level where the masses are applied allows redistribution of the horizontal forces among all upright frames, avoiding “side-effects”, that may be significant, in particular for “short” racking systems.

It was observed that pallet sliding can occur already for PGA values as low as 0.10g.

Specimens could not be tested up to failure, in order to avoid possible damage to the shake table testing facility.

Due to the high flexibility of the structures, a sine sweep test or a single impulse test, even when carried out at very low acceleration, can result in major structural damage. Care should be taken in order to avoid structural damage when performing tests with the only aim to assess eigenfrequencies or damping ratio.

In the second series of tests, performed within the SEISRACKS research project, in addition to some palletised merchandise tests, earthquake simulation tests were carried out on six full scale rack models of similar geometry as those of the first series, but without longitudinal as well as in-plane bracings.

Five specimens were tested in down-aisle direction (two of which with base isolation systems) and one in cross-aisle direction.

Effects such as the beam size, presence/absence of pallet sliding as well as of a base isolation system were investigated.

The importance of small structural detailing was also highlighted. In most cases, in fact, collapse involved failure of bolted or welded connections.

An assessment of the q-factor was also performed, based on the experimental results. q-factor values of 3.7 and 2.7 were obtained respectively for the down-aisle and the cross-aisle directions. In the case of pallets rigidly fixed on the beams (in order to prevent sliding) a q-factor of 2.8 was identified. This value, however, might be affected by the excessive stiffening of the structure, associated with the way in which the pallets were connected to the steel beams, generating a sort of "composite" beam. The estimated values are similar to those obtained by re-analysis of the push-over tests carried out on similar structures.

Positive effects of the presence of the base isolators were also highlighted. The two specimens resisted without any damage earthquakes with a PGA higher than 1.30g. Assessment of the q-factor for specimen A5 lead to a value of 6.9.

References

Benjamin, J. R. and Associates. (1988). *A criterion for determining exceedance of the operating basis earthquake, EPRI Report NP-5930*. Palo Alto, California: Electric Power Research Institute.

Brescianini, J. C., Castiglioni, C. A., & Panzeri, N. (2003). Dynamic experimental tests on steel pallet racks. In *Proceedings of XIX CTA Conference, Genova, Italy* (pp. 107–116).

Castiglioni, C. A. (2003a). Dynamic tests on steel pallet racks. *Costruzioni Metalliche, 55*(3), 35–44.

Castiglioni, C. A. (2003b). Seismic behaviour of steel storage racks. In *Proceedings of the IV Congresso de Construcao Metalica e Mista, Lisbon* (December 4–5, 2003, pp. 41–62)

Castiglioni, C. A., Panzeri, N., Brescianini, J. C., & Carydis, P. (2003). Shaking table tests on steel pallet racks. In *Proceedings of the Conference on Behaviour of Steel Structures in Seismic Areas-Stessa 2003, Naples, Italy* (pp. 775–781).

Chen, C. K., Scholl, R. E., & Blume, J. A. (1981). Seismic-resistant design of industrial storage racks. *Proceedings of the second specialty conference on dynamic response of structures: experimentation, observation and control*, Atlanta, GA, 745–759.

Chen, C. K., Scholl, R. E., & Blume, J. A. (1980a). *Seismic study of industrial storage racks.* Report prepared for the National Science Foundation and for the Rack Manufacturers Institute and Automated Storage and Retrieval Systems (sections of the Material Handling Institute), URS/John A. Blume & Associates, San Francisco, CA, 569 pp.

Chen, C. K., Scholl, R. E., & Blume, J. A. (1980b). Earthquake simulation tests of industrial steel storage racks. In *Proceedings of 7th World Conference on Earthquake Engineering, Istanbul, Turkey* (pp. 379–386).

EN 1998-1 (2004): Eurocode 8: Design of structures for earthquake resistance-Part 1: General rules, seismic actions and rules for buildings.

Federal Emergency Management Agency (FEMA). (2005). Seismic considerations for steel storage racks located in areas accessible to the public. FEMA460, Department of Homeland Security, Washington, D.C., September 2005.

Filiatrault, A. (2001). Shake table tests of storage racks and contents. *Presentation material at the April 12, 2001 Seismic safety commission hearing on industrial storage racks,* San Francisco, CA (available from author).

Filiatrault, A., Bachman, R. E., & Mahoney, M. G. (2006a). Performance-based seismic design of pallet-type steel storage racks. *Earthquake Spectra, 22*(1), 47–64.

Filiatrault, A., & Wanitkorkul, A. (2004a). *Shake table testing of Frazier industrial storage pallet racks*. Report No. CSEE-SEESL-2004-02, University at Buffalo, State University of New York, Buffalo, NY, 83 p.

Filiatrault, A., & Wanitkorkul, A. (2004b). *Shake table testing of Frazier industrial storage racks*, Report No. CSEE-SEESL-2005-02, Structural Engineering and Earthquake Simulation Laboratory, Departmental of Civil, Structural and Environmental Engineering, University at Buffalo, State University of New York, 83 pp.

Filiatrault, A., & Wanitkorkul. A. (2006a). *Shake table testing of Ridg-U-Rak rigid based and Ridg-U-Rak patent pending base isolated industrial storage racks: final production unit testing*, Report No. CSEE-SEESL-2006-19, University at Buffalo, State University of New York, Buffalo, NY, 67 p.

Filiatrault, A., & Wanitkorkul. A. (2006b). *Shake table testing of Ridg-U-Rak rigid based and Ridg-U-Rak patent pending base isolated industrial storage racks: production unit testing*, Report No. CSEE-SEESL-2006-07, University at Buffalo, State University of New York, Buffalo, NY, 43 p.

Gioncu, V., & Mazzolani, F. M. (2002). *Ductility of seismic resistant steel structures*. Prague: Spoon Press.

Guerra, C. A., Mazzolani, F. M. & Piluso, V. (1991). Esame critico sulle metodologie di valutazione del fattore di struttura nelle strutture intelaiate in acciaio, *L'Ingegneria sismica in Italia*, pp. 468–479.

Krawinkler, H., Parisi, F., Ibarra, L., Ayoub, A., & Medina, R. (2000). *Development of a testing protocol for wood frame structures*, CUREE Publication No. W-02, Richmond, CA.

Krawinkler, H., Cofie, N. G., Astiz, M. A., & Kircher, C. A. (1979). *Experimental study on the seismic behavior of industrial storage racks*. Report No. 41, The John A. Blume Earthquake Engineering Center, Department of Civil Engineering, Stanford University, Stanford, CA.

Mazzolani, F. M. & Piluso, V. (1997). *Theory and design of seismic resistant steel frames*. London: E & FN Spon, Champman & Hall.

Storage Racks in Seismic Areas. (2008). Contract no. RFS-PR-03114 of Research Programme of the Research Fund for Coal and Steel—Steel RTD—Final Report (To be published).

Further Readings

Ballio, G., & Castiglioni, C. A. (1994). An approach to the seismic design of steel structures based on cumulative damage criteria. *Earthquake Engineering and Structural Dynamics, 23*, 969–986.

Ballio, G., Castiglioni, C. A., & Perotti, F. (1988). On the assessment of structural design factors for steel structures. In *Proceedings IX World Conference on Earthquake Engineering, Tokyo, August 1988* (Vol. V, pp. 1167–1172).

Bernuzzi, C., & Castiglioni, C. A. (2001). Experimental analysis on the cyclic behaviour of beam-to-column joints in steel storage pallet racks. *Thin-Walled Structures, 39*, 841–859.

Castiglioni, C. A. (1995). Assessment of behaviour factor for steel frames based on linear analysis and cumulative damage criteria. In *7th National Conference L'Ingegneria Sismica in Italia* (pp. 589–598).

Castiglioni, C. A., & Calado, L. (1996). Comparison of the two cumulative damage approaches for the assessment of behaviuor factors for low-rise steel building. *Journal of Constructional Steel Research, 40*, 39–61.

Federal Emergency Management Agency (FEMA). (2004). The 2003 NEHRP recommended provisions for new buildings and other structures—Part 1: Provisions. FEMA450, Department of Homeland Security, Washington, D.C.

Filiatrault, A., Wanitkorkul, A., & Higgins, P. (2006b). Experimental stiffness and seismic response of pallet-type steel storage rack connectors. *ASCE Practice Periodical on Structural Design and Construction, 11*(3), 161–170.

Filiatrault, A., Wanitkorkul, A., & Seo, J.-M. (2005). *Shake table testing of Ridg-U-Rak rigid based and Ridg-U-Rak patent pending base isolated industrial storage racks.* Report No. CSEE-SEESL-2005-03, University at Buffalo, State University of New York, Buffalo, NY, 51 p.

International Code Council (ICC). (2003). *International building code—2003 Edition*. VA: Falls Church.

National Fire Protection Association (NFPA). (2003). NFPA 5000—NFPA Building Construction and Safety Code, Quincy, MA.

Rack Manufacturers Institute (RMI). (2002). *Specification for the design, testing, and utilization of industrial steel storage racks—2002 Edition*. Charlotte, NC: RMI.

Setti, P. (1985). A method to compute the behaviour factor for construction in seismic zones. *Costruzioni Metalliche, 37*(3), 128–139.

Chapter 7
Conclusions

Despite their lightness, racking systems carry very high live load (many times larger than the dead load, opposite of what happens for usual civil engineering structures) and can raise a considerable height. Prediction of the structural behaviour of pallet racks is difficult because is affected by the particular geometry of their structural components: members made by high slenderness thin-walled, open-section profiles (hence prone to global, local and distortional buckling problems), beam-to-upright and base-plate joints exhibiting a non-linear behaviour.

Due to their peculiarities, additional modelling and design rules are required for these non-traditional steel structures (not building structures, but load bearing civil construction work from an engineering point of view) and reference cannot be made to usual Structural Design Recommendations and Standards.

The most recent Design Standards for steel storage racks provide a combined numerical-experimental approach in which the design structural analysis is supported by specific tests to evaluate the performance of the key components (members and joints).

The design needs particular attention for storage racks installed in a seismic zone, where they must be able to withstand dynamic forces. Besides the usual global and local collapse mechanisms, an additional limit state for the system is represented by the fall of the pallets with subsequent damage to goods, people and to the structure itself.

In Europe, no official document is currently available for the seismic design of pallet racks, and the designers are compelled to operate with a total lack of references and of commonly accepted design rules. Very often they make reference to the Rack Manufacturers Institute (R.M.I.) Specifications, while the European Racking Federation (F.E.M.-ERF) is presently working in order to produce an official document.

Racks are widely adopted in warehouses where they are loaded with tons of (more or less) valuable goods. The loss of these goods during an earthquake may represent, for the owner, a very large economic loss, much larger than the cost of the whole rack on which the goods are stored, or of the cost for its seismic upgrade.

Racks are also more and more frequently adopted in supermarkets and shopping centers, in areas open to the public.

C.A. Castiglioni, *Seismic Behavior of Steel Storage Pallet Racking Systems*,
Research for Development, DOI 10.1007/978-3-319-28466-8_7

The falling of the pallets, in this case, may endanger the life of the clients as well as that of the workmen and employees, involving not only Civil and Penal Right considerations about the liability of the owners, but also economic considerations related to the insurance coverage.

Sliding of the pallets on the racks and their consequent fall represents a serviceability limit state i.e. a situation that might occur during a seismic event also in the case of a well designed storage rack, the phenomenon depending only on the dynamic friction coefficient between the pallet and the steel beam of the rack.

Hence, solution of the problems connected with safe and reliable design of steel storage racks in seismic areas has a very large economic impact.

At present, there are technical limitations in the field of safety and design of storage racks in seismic areas: lack of knowledge on actions challenging the structures, lack of knowledge on structural behaviour in terms of ductility and sliding conditions of the pallets on the racks and lack of Standard Design Codes in Europe.

To solve some of these limitations, the EU sponsored through the Research Fund for Coal and Steel an RTD project titled "Storage Racks in Seismic Areas" (acronym SEISRACKS, Contract Number: RFS-PR-03114).

The objectives of this project, initiated in December 2004 and terminated in June 2007, are:

- to increase knowledge on actual service conditions of storage racks,
- to increase knowledge on racks' actual structural behaviour
- to assess design rules for racks under earthquake conditions.

The research team was composed by the following units: ACAI the Italian Association of Steel Constructors (Coordinator), Instituto Superior Tecnico of Lisbon (P), National Technical University of Athens (EL), Politecnico di Milano (I), University of Liege (B) and the European Laboratory for Structural Assessment (ELSA) of the Joint Research Center at Ispra (Subcontractor of Politecnico di Milano). This author acted as Scientific Coordinator of the project.

The research activities carried out in co-operation among the partners, in order to achieve the aforementioned objectives within this project, are subdivided in the following Work Packages:

WP 1—Full scale dynamic tests of storage racks
WP 2—Full scale pseudo-dynamic and pushover tests of storage racks
WP 3—In situ monitoring of storage racks
WP 4—Cyclic testing of components
WP 5—Assessment of seismic design rules for storage racks.

The project focuses on steel selective pallet storage racks located in areas of retail warehouse stores and other facilities, eventually accessible to the general public.

Storage racks are composed of specially designed steel elements that permit easy installation and reconfiguration, consistent with the merchandising needs of a warehouse retail store. Except where adjacent to walls, storage racks normally are configured as two rows of racks that are interconnected. Pallets typically can have plan areas of approximately one square meter and can have a maximum loaded weight of approximately 10–15 kN. Storage rack bays are typically 1.0–1.1 m deep and 1.8–2.7 m wide and can accommodate two or three pallets. The overall height of pallet rack structural frames, found in retail warehouse stores, varies between 5 and 6 m. In industrial warehouse facilities, racking system can reach considerable heights, such as 12–15 m or more.

The rack industry calls the longitudinal direction the down-aisle direction, and the transverse direction the cross-aisle direction. Proprietary moment connections are typically used as the structural system in the down-aisle direction and braced frames are typically used as the structural system in the cross-aisle direction

This work was conceived during the development of the SEISRACKS Project, as the author, in addition of being the "team-leader" for the "Politecnico di Milano" unit, was also acting as "Scientific Coordinator" on behalf of ACAI.

The results of WP 4 ("Cyclic tests of rack components"), of WP2 ("Full scale, Pseudo-dynamic and push-over tests of storage racks") as well as those of sliding tests and full-scale shaking table tests carried out within WP1 ("Full scale dynamic tests of storage racks") have been presented in this report.

Results of the cyclic tests on full scale rack components (namely beam-to-upright connections and column bases) carried out at the Department of Civil Engineering and Architecture of Instituto Superior Tecnico of Lisbon (with the cooperation of the author) were presented and discussed. In particular, an innovative cyclic testing procedure for structural elements and components, alternative to the classic ones recommended by ECCS (1986) and ATC (1992), was presented and discussed in Chap. 2.

The seismic behaviour of beam-to-upright connections of racking systems (and more in general of beam-to-column connections showing an unsymmetrical response and damage accumulation) is determined by a hybrid loading consisting of vertical load and horizontal motion effects, and the beam-to-column connections bending moments are shifted in the negative, hogging, direction as a consequence of the vertical load effects. Failure occurs when the critical sections are no longer able to withstand vertical loads as a consequence of accumulated damage (induced by horizontal motion and vertical forces acting together).

However, general recommended testing procedures, encompassing only displacement controlled conditions, fail to address the unsymmetrical displacement histories experienced by critical beam-to-column connections when subjected to earthquake motion acting simultaneously with vertical (live and dead) loads.

Considering the former limitations of the commonly accepted testing procedures (particularly evident when testing is performed in the inverted T configuration, with the beam standing vertically and the column horizontally), an innovative testing procedure was developed which intends to capture the hybrid nature of loading imposed to beam-to-column connections when subjected to combined vertical and horizontal load effects. This testing procedure consists of "half-cycles" performed partially under "force-controlled" conditions (in order to impose to the specimen the "gravity" design load) and partially under "displacement-controlled" conditions (in order to impose to the specimens the displacements required by the ductility demand).

The innovative testing procedure can be considered a development of the commonly accepted testing procedures as it inherits some of their characteristics, such as the cycle repetition in the post-elastic range and the fact that the controlled-displacement part of the testing cycles is indexed to the yielding displacement (determined through monotonic tests).

In the same Chap. 2, results of 30 tests carried out on two different types of beams (respectively 70 and 130 mm deep), connected to the same type of upright were presented together with those of 40 tests on column bases. Tests were carried out on the same type of column base, but different loading directions and axial load in the column, as well as different type of steel base-to-foundation connections were considered (concrete foundation or steel foundation).

The failure mode for 70 mm (hogging bending) and 130 mm (hogging and sagging bending) deep beams consisted of large deformations in the top zone of the beam end connector.

In 70 mm deep beam subject to sagging bending, the failure mode was the fracture of the fillet weld between the beam and the end-plate connector.

The connection behaviour for all tests was not influenced by safety bolt deformation.

The rotation capacity difference between the loading verse (sagging or hogging bending) was larger for the 70 mm deep beam due to different collapse occurred in the sagging loading.

According to Eurocode 3 (2005) the connection with a 70 mm deep beam can be considered as semi-rigid, for both beam lengths of 1.8 and 2.7 m.

According to Eurocode 3 (2005) the connection with a 130 mm deep beam can be considered as semi-rigid for a beam lengths of 2.7 m, and as flexible for a beam lengths of 1.8 m (as in the case of the structures tested full scale within this research); hence, in this last case, the influence of this connection behaviour may be ignored in structural analysis.

The results of test performed with the innovative cyclic testing procedure were fundamentally different from those obtained through the application of the ECCS recommended testing procedure.

In those tests performed according to the innovative cyclic testing procedure, the imposed displacement history is unsymmetrical. Displacements tend to systematically accumulate in the positive direction as a consequence of the absence of closure of the top part of the connection (the bottom part is generally closed throughout the tests). This connection behaviour also leads to a reduction of the pinching effects.

Imposed forces are shifted in the positive (hogging) direction for the innovative testing procedure whereas, apart from the asymmetry that may result from unsymmetrical connection detailing, positive and negative force amplitudes for ECCS tests are not excessively different.

Failure of the connection is explicitly addressed by the innovative testing procedure since failure occurs when the connection is no longer able to withstand vertical load effects.

In the column base connections under cross-aisle bending, the axial compression load is beneficial when the bolts are in the compression zone. In this case, an increase of the axial force results in an increase of resistance and stiffness but in a decrease of rotation capacity of the specimens.

When the loading direction results in tension of the bolts, the axial force in the upright causes a reduction of resistance and stiffness of the column bases, because induces distortional buckling of the free edges of the cross-section profile.

The tests performed on column base connections in the down-aisle direction proved that presence of an axial force of 17 and 34 % of the yield load (25 and 50 kN respectively) increases the initial stiffness and the resistance but decreases the rotation capacity of the connection.

In the cyclic tests, the higher the axial force, the bigger the difference between the resistance under positive and negative bending moments.

The collapse modes exhibited by the specimens were weld failure, base bending and distortional buckling of the cross section of the upright. In the tests with premature weld failure there was a reduction in the rotation capacity of the connection. Therefore, it is extremely important to control the welding process during the manufacturing of the components.

Distortional buckling of the cross section of the upright occurred in the tests with an axial force equal to 34 or 52 % of the yield load (respectively 50 and 75 kN), except for the monotonic tests when the loading direction is such that bolts are in the compression zone. In the tests on column bases under bending in the cross-aisle direction and axial force of 75 kN (52 % of the yield load) distortional buckling occurred prematurely, drastically reducing the mechanical properties of the connection.

The base welded and bolted to the steel deck exhibited an increase in the resistance and rotation capacity of the connection since there was no base plate bending and in the cyclic tests a higher capacity of energy dissipation was observed, when compared to simply bolted connections.

The results proved that connecting the column base to a concrete slab or to a steel deck does not change the mechanical properties or the failure mode of the connection.

In both cases pre-tension in the bolts should be provided.

Chapter 3 dealt with the assessment of the friction factor between pallets and rack beams, which is governing the "pallet sliding" phenomenon. This turned out to be the most important effect governing the dynamic behavior of racking systems.

Assessment of both the static and the dynamic sliding conditions of pallets stored on steel racking systems was carried out within the SEISRACKS research project, by means of static as well as dynamic tests performed at the Earthquake Engineering Laboratory of the National Technical University of Athens.

More than 180 *Static tests* were carried out in both down and cross-aisle direction, by means of an "inclined plane" device, by slowly increasing the inclination of the plane, and measuring the sliding of the pallet on the rack steel beams.

Influence of the following parameters was investigated:

- Type of beam (namely type of surface finish of the beam)
- Type of pallet
- Geometry and weight of mass resting on the pallet.

Influence of the type of beam was investigated by adopting six different types of beam specimens, produced by different manufactures from 3 different European countries, with different types of surface finish in combination with different types of pallet conditions: wooden and plastic Euro pallets as well as wooden American pallets, in new, old, dry or wet conditions. In particular, hot zinc, hot dip and powder coated steel beams were considered.

In both cross and down-aisle direction, the surface finish influenced very much the static friction factor, with differences as large as 20–30 % from one type to the other, in the case of wooden pallets.

Influence of the type of pallet was investigated by adopting three different types of pallets, namely: wooden Euro pallets, wooden American-pallet and plastic Euro pallet. In both cross and down-aisle direction the plastic Euro pallet showed a very low friction factor (in the order of 0.2), practically being non-influenced by the type of beam surface finish. The wooden pallets show a very similar friction factor (in the order of 0.5), and similarly are influenced by the beam surface finish.

In both cross and down-aisle direction, the mass weight didn't affect much the results. However, its geometry (height of the c.o.g.) and its "placement" on the pallet (centered or eccentric) resulted in small variations of the measured friction factor.

Dynamic tests were carried out on the shaking table facility of the National Technical University of Athens, on a simplified set-up, made of two uprights, connected by two horizontal beams, at approximately 0.30 m from the shaking table. On the beams three wooden Euro pallets were positioned, with concrete blocks rigidly fixed on top. Most tests were carried out with a sinusoidal excitation, with constant frequency and increasing acceleration. Some tests, in down-aisle direction only, were carried out with sinusoidal excitation, with constant acceleration and increasing frequency, in order to verify independence of the obtained results on the type of adopted excitation. A lower bound of the acceleration exists, beyond which pallets start sliding on the steel beams. When acceleration of the mass is lower than such "lower bound", the pallet "sticks" on the beams, and no sliding occurs. When the "lower bound" of acceleration is exceeded, increasing the acceleration of the input motion results in a lower increment in the mass acceleration, until an "upper bound" is reached of the mass acceleration. Any further increase in the acceleration of the input motion doesn't affect the acceleration of the mass that is "free" to slide on the beams. "Stiction" between pallet and beam is not resumed until a reduction of the acceleration occurs. The "upper bound" of the sliding acceleration is, in general, lower than the static friction factor.

In both cross and down-aisle direction lateral pallets slide systematically earlier than the central one.

Dynamic behaviour in cross-aisle direction is completely different to the one in down-aisle direction.

In cross-aisle direction, the torsional stiffness as well as the flexural stiffness in the horizontal plane of the beams influence very much the results. In particular, such stiffnesses are affected by the out-of-plane and torsional behaviour of the beam-to-upright connections, whose stiffness rapidly deteriorates under cycling. Test results show, in general, a dependence of the sliding acceleration on the frequency of the input motion. Both the lower and the upper bound of the sliding acceleration seem to decrease when increasing the frequency of the excitation.

Lower bound sliding acceleration as low as 0.1g was measured, for wooden pallets on hot dip coated steel beams. Upper bound values of the acceleration ranging from 0.3 to 0.5g were measured depending on the type of beam surface finish as well as on the position of the pallet (laterals or central one).

In down-aisle direction, the sliding acceleration is in general higher than the one measured in cross-aisle direction, under the same testing conditions, with a lower bound of the measured sliding acceleration of nearly 0.3g, and an upper bound of nearly 0.6g. Also in down-aisle direction, test results show, in general, a dependence of the sliding acceleration on the frequency of the input motion. However, in this case, both the lower and the upper bound of the sliding acceleration seem to increase when increasing the frequency of the excitation.

Results of tests carried out with constant acceleration and increasing frequency are fully compatible with those obtained in tests with constant frequency and increasing acceleration.

Test results confirm that "sliding" is, under severe dynamic conditions, the main factor influencing the rack response. Hysteresis loops were obtained, showing the presence of energy dissipation through sliding.

A few *seismic tests* were carried out, adopting three different input motions recorder in Greece during recent earthquakes, and characterized by different durations and frequency contents. Both mono-directional and bi-directional tests were carried out. The obtained results were compared with those of tests carried out with a sinusoidal excitation, showing full compatibility. Measured sliding accelerations range from 0.15 to 0.35g in cross-aisle direction and from 0.45 to 0.6g in the down-aisle direction. Similar compatibility was also obtained for bi-directional tests, when comparing the resultants of the vector-compositions of the components of the sliding accelerations in the two orthogonal directions.

Chapters 4 and 5 deal respectively with the full scale pushover and pseudo-dynamic tests that were carried out (under the coordination of the author) at the European Laboratory for Structural Assessment of the Joint Research Center of Ispra. In particular, one pseudo-dynamic test and a pushover test in the down-aisle direction, and one pushover test in the cross-aisle direction were carried out on two bays, three storeys full scale rack models. Re-analysis of the results allowed drawing interesting conclusions on the seismic behaviour of racking systems.

The specimen under pushover test in down-aisle direction showed a progressive loss of stiffness associated to accumulation of plastic deformation in the column-base connections and to the large inter-storey drift of the first level. Inter-storey drifts of the upper level are much smaller than those of the first level; this is characteristic of a "soft-floor" type of collapse mechanism that may lead to global instability due to second-order effects.

In order to reduce this type of problem, the deformability of the column-base connections should be reduced and, somehow, limited. Eventually, adoption of a beam at the ground level might be considered. Despite the increment of cost, this solution would in any case allow use of the space at ground level for storage of goods, while the structure will behave as rotationally restrained at the base.

Due to the bracing systems of the uprights, the specimens show a higher stiffness in cross-aisle direction than in the down-aisle one. Such bracing system is the most stressed structural component, and its failure leads to global collapse, accompanied by flexural-torsional buckling of the columns, consequent to the increment of the buckling length of the profile due to failure of the bolted connections with the lattice members.

For this reason, the solution with all diagonals inclined in the same direction should be reconsidered when the structure has to be erected in a seismic zone.

Difference between the rotations of the uprights is due to deformation of the bracing system of the transverse frames, as well as to the different behaviour of the base-plate connections and of the beam-to-upright connections.

When loaded by horizontal loads applied transverse to the beam, the connections on one side can transfer the loads by means of both portions of the end-plate in bearing against the upright. On the contrary, the connections on the other side can transfer load only by means of the safety bolt in shear as well as of the hooks in tension; furthermore, bending of the beam in the horizontal plane induces bending in the end-plate. This different behaviour of the connections contributes to the difference in the global response of the uprights of the two sides.

An evaluation of the behaviour factor has been carried out for both down-aisle and cross-aisle directions, with two possible definitions of the q-factor.

One value can be identified based on ductility considerations as the ratio of the displacement v_{max} corresponding to the maximum load carrying capacity of the structure to the yield displacement (v_y), being $q_{\mu_{max}} = \frac{v_{max}}{v_y} = 3.7$ for the down-aisle direction and $q_{\mu_{max}} = 2.4$ for the cross-aisle direction.

With reference to the ductility factor theory, a value of the q-factor based on strength was also defined as the ratio of the ideal strength $F_{max,el}$ (corresponding to v_{max} and evaluated on the basis of the initial elastic stiffness) to the maximum load carrying capacity F_{max}, being $q_{f_{max}} = \frac{F_{max,el}}{F_{max}} = 3.1$ for the down-aisle direction and $q_{f_{max}} = 2.1$ for the cross-aisle direction.

The results of the pseudo-dynamic test on the rack specimen under down-aisle seismic loading previously presented are fully compatible with those obtained on similar specimens, tested under dynamic conditions on the shaking table of the

Laboratory for Earthquake Engineering of the national technical University of Athens.

Under pseudo-dynamic conditions the specimen could sustain the series of earthquake events although it didn't collapse during the last test, performed with PGA = 1.4g (ePGA = 1.5g). Under dynamic conditions specimen A1 (having the masses fixed on the beams in order to prevent sliding, simulating the "quasi-static" conditions of the pseudo-dynamic tests) collapsed under an earthquake with a PGA = 1.46g (ePGA = 1.41g).

Probably, the strain rate effect plays some role in this type of structure; the small movements allowed to the hooks in the holes, in fact, under dynamic conditions result in local impacts that, under increasing number of cycles, may cause cracking either in the hooks or at the edges of the holes.

In any case, comparing the deterioration of the second eigen-frequency (the most excited one) of the specimen tested under pseudo-dynamic conditions with the similar one tested in Athens on the shaking table it can be noticed that their trend of reduction is similar.

This means that, in general, damage accumulated in the specimen during the two different types of test is similar.

Hence it can be concluded that, from the point of view of the assessment of the seismic resistance and of the damage accumulation of pallet racking systems, pseudo-dynamic tests and shaking table tests are fully compatible, although local damage due to local dynamic effects cannot be reproduced by the pseudo-dynamic testing methodology.

Of course, due to the intrinsic quasi-static nature of the pseudo-dynamic testing procedure, no information can be derived about the effects caused by the sliding of the pallets on the beams during a seismic event.

The values of acceleration that were reached during the pseudo-dynamic tests largely exceed the upper bound of the pallet sliding acceleration.

This means that only full scale dynamic testing will allow a clear assessment of the limit states of pallet racking systems under seismic loading.

Chapter 6 presents the results of the full scale dynamic tests carried out within the SEISRACKS project as well as those carried out previously within the research project on the "Seismic behaviour of steel storage pallet racks" sponsored by the EU within the ECOLEADER Research Program for Free Access to Large Testing Facilities.

Within the ECOLEADER Project, four full scale rack models of three levels (total height 6.0 m) and 2 bays (total width 3.6 m) were tested under earthquake simulations in cross-aisle and in down-aisle direction.

Results show the importance of structural regularity, both in elevation and in plan. Eccentricity of the vertical down-aisle bracing system may result in severe torsional deformation and in major structural damage.

In-plane bracings at each level where the masses are applied allows redistribution of the horizontal forces among all upright frames, avoiding "side-effects", that may be significant, in particular for "short" racking systems.

It was observed that pallet sliding can occur already for PGA values as low as 0.10g.

Specimens could not be tested up to failure, in order to avoid possible damage to the shake table testing facility, in absence of a safety frame supporting both masses and structure in case of collapse.

Due to high flexibility of the structures, a sine sweep test or a single impulse test, even when carried out at very low acceleration, can result in major structural damage. Care should be taken in order to avoid structural damage when performing tests with the only aim of assessing eigen-frequencies or damping ratio.

In the second series of tests, performed within the SEISRACKS research project, in addition to some palletised merchandise tests, earthquake simulation tests were carried out on six full scale rack models of similar geometry as those of the first series. Five specimens were tested in down-aisle direction (two of which with base isolation systems) and one in cross-aisle direction.

Effects such as the beam size, presence/absence of pallet sliding as well as of a base isolation system were investigated.

The importance of small structural detailing, to be taken into account when designing pallet racks in seismic areas, was highlighted. Most of the observed failure, in fact, involved failure of bolted or welded connections.

An assessment of the q-factor was performed, based on the experimental results. q-factor values of 3.7 and 2.7 were obtained respectively for the down-aisle and the cross-aisle directions. In the case of pallets rigidly fixed on the beams (in order to prevent sliding) a q-factor of 2.8 was identified. This value, however, might be affected by the excessive stiffening of the structure, associated with the way in which the pallets were connected to the steel beams, generating a sort of "composite" beam. The estimated values are similar to those obtained by re-analysis of the push-over tests carried out on similar structures.

Positive effects of the presence of the base isolators were also highlighted. The two specimens with base-isolation systems resisted earthquakes with a PGA higher than 1.30g without any damage. Assessment of the q-factor for specimen A5 lead to a value of 6.9.

Different aspects of the numerical modelling and of the analysis of rack structures have also been investigated within the SEISRACKS research project, without being presented in this report.

In particular, two new features of prime importance for an efficient analysis of racks subjected to seismic action have been included in the FEM software FineLg: springs with hysteretic energy dissipation and sliding point-mass with coulomb friction law. These tools are fully operational, even if some future improvements have already been identified (convergence of the sliding model in the stick phase, test of other friction laws, sliding mass model with numerous contact points…).

Models have been validated versus a selection of test results obtained during the SEISRACKS research and during previous research programs. The validation has been carried out for cross- and down-aisle seismic excitation and for braced as well as non braced structures.

Some particular aspects have been emphasized during the calibration procedure: the need for a precise knowledge of the stiffness and resistance of the column bases, the horizontal bracing role played by the pallets as long as they are not sliding or the need for a future calibration of the behaviour of beam-to-column joints regarding rotation around the longitudinal axis of the upright.

Two complementary studies have also been performed in the perspective of normative prescriptions.

The first one is a parameter study about the consequences of pallet sliding on the structural response. The study evidences clearly that the horizontal force reduction coefficient is depending on the intensity of the ground motion, on the value of the friction coefficient and on the structural typology (i.e. the structural natural period and the number of loaded levels). The reduction coefficient ranges roughly from 0.2 to 1.0. Additional studies would be necessary to refine these conclusions and calibrate properly the reduction factor.

The second complementary study compared different types of seismic analyses (lateral force method, response spectrum analysis and pushover analysis) applied to a same structure exhibiting significant second order effects. The main conclusions are that, for the considered structure, all analyses provide similar results, provided that second order effects are really accounted for. To this purpose, the use of the approximate amplification factor $1/(1-\theta)$ is found efficient. Moreover, a verification taking into account second order effects by using a sway buckling length appears strongly over-conservative.

7.1 Comments on FEM 10-2-08

7.1.1 General Introduction

The whole research project SEISRACKS has been an opportunity to analyse the current draft of the normative document pr FEM 10-2-08 "Recommendations for the design of static steel pallet racks under seismic conditions". In particular, a series of items have been identified as questionable and are listed here with the corresponding sections of pr FEM 10-2-08 in its version of December 2005.

- Determination of the period of the structure and of the seismic action, and in particular:
  - Regularity criteria and consequences on the behaviour factor (2.2 5),
  - Effect of the actual position of the gravity centre of the masses, vertical eccentricity with respect to the beams (2.3.6),
  - Methods of analysis (2.4),
  - Definition of regularity criteria (3.1.4),
  - Modelling assumptions in the perspective of the structural analysis (3.3),

- Account for the different sources of energy dissipation (Viscous damping, friction of pallets, energy dissipation within the stored goods)
  - Definition and values of parameters $E_{D,1}$, $E_{D,2}$ and R_F (2.3.1, 2.3.2, 2.3.3, 2.3.4, 4.2.2, 4.2.3)
- Assessment of the structural ductility and associated behaviour factor
  - Definition of ductility classes (3.1.1)
  - Material properties and overstrength coefficient (3.1.2)
  - Definition of the q-factor according to the structural typology (3.1.3, 3.4)
  - Impact of (ir-)regularity (3.1.4, 3.4)
  - Design rules for dissipative and non-dissipative structures (3.1.5)
  - Identification of the resisting system (3.2)
- Detailing of dissipative elements and overstrength criteria (5)

On the base of the knowledge gained during the research project and on engineering judgment, many of these items can be addressed. The final output will be a revised version of pr FEM 10-2-08, but some background comments are given in the following.

7.1.2 Regularity Criteria

In the current draft of pr FEM 10-2-08, all rules are taken directly from EN 1998-1. No contradictory information has been developed on this aspect within the project SEISRACKS and it is thus proposed to confirm the current draft.

7.1.3 Position and Height of the Masses

Pr FEM 10-2-08 proposes rules to account for the fact that the gravity centre of the pallets may be at a certain level above the beams, which induces an overloading of the pallet beams due to the overturning of the palletised goods subjected to horizontal inertial forces acting in the cross-aisle direction. This effect is particularly important for palletized goods having a significant height. These rules are based on a reasonable engineering judgment and no contradictory information has been developed within the project SEISRACKS, essentially because the considered structural test specimens were loaded by concrete blocks or steel plates with a very low position of the gravity centre with respect to the pallet beams. It is proposed to confirm the current draft.

7.1.4 *Methods of Analysis*

Methods of analysis proposed in pr FEM 10-2-08 are derived from EN 1998. The reference method is modal response spectrum analysis and can be replaced by a lateral force method analysis if the structure can be classified as regular. Section 2.4 of the pre-normative document is confirmed.

A comment must however be made about the way to deal with second order effects (Sect. 4.1.2.4). In EN 1998-1, the rule is the following.

If the coefficient θ measuring the sensitivity of the structure to second order effects is lower than 0.1, these second order effects can be disregarded.

If θ is between 0.1 and 0.2, second order effects can be accounted for in an approximate way by multiplying the results of a linear analysis by an amplification factor equal to $1/(1-\theta)$.

If θ is between 0.2 and 0.3, second order effects must be accounted for explicitly.

θ higher than 0.3 is not allowed.

In the case of racks, on the base of the conclusions drawn within the SEISRACKS Project, the proposal of pr FEM 10-2-08 is confirmed:

If the coefficient θ is lower than 0.1, second order effects can be disregarded.

If θ is between 0.1 and 0.3, second order effects can be accounted for in an approximate way by multiplying the results of a linear analysis (obtained either modal response spectrum or lateral force method according to the regularity of the structure) by an amplification factor equal to $1/(1-\theta)$.

If θ is between 0.3 and 0.5, second order effects must be accounted for explicitly, which can be done either through a non linear dynamic time-history analysis or through a push-over analysis.

θ higher than 0.5 is not allowed.

In addition, as soon as θ is higher than 0.1, it is recommended to consider a stiffness of the structure reduced by the pre-compression of the uprights (by using the geometric matrix) for the calculation of the period. However, the use of the initial non reduced stiffness would normally lead to safe results, since the corresponding periods are higher.

7.1.5 *Pallet Weight Modification Factor E_{D2}*

In pr FEM 10-2-08, a pallet weight modification factor is proposed to account for damping inside the palletized goods. This coefficient might probably be removed, since such a source of energy dissipation would be possible only if the frequencies of the stored goods are tuned on the frequency of the structure. This is very unlikely according to the respective range of frequencies observed during the SEISRACKS project. Natural frequencies of the racks are rather low (between 0.25 and 1 Hz), while the few values of frequencies obtained for merchandise are significantly higher (between 1.5 and 24 Hz). As long as extensive measurements are not performed to assess the actual dynamic behaviour of a wide range of merchandise, it is recommended not to account for any dissipation in the goods and thus to set E_{D2} = 1.0 in Sect. 2.3.3 of pr FEM 10-2-08.

7.1.6 Design Spectrum Modification Factor E_{D1}

Design spectrum modification factor E_{D1} is introduced in the design code to account for possible energy dissipation by friction of the pallets on their supporting beams. A constant value equal to 0.8 is proposed in the current draft if no specific devices are present to avoid pallet sliding.

Within the SEISRACKS Project it has been shown that this reduction factor is actually depending on the design acceleration, on the friction coefficient and on the structural typology. With this perspective, the most advantageous situation would be a structure with only one loaded level, with high design acceleration and a very low friction coefficient. For example, if the ground acceleration a_g is equal to 0.5g, the design acceleration of the structure may rise up to 1.25g. If only one level is loaded and if the friction coefficient μ is equal to 0.25, the force acting on the structure is exactly limited to the friction force, which corresponds to an equivalent acceleration of 0.25g. The reduction factor is thus equal to 0.25g/1.25g = 0.2. On the other hand, for structures with many loaded levels and for low values of a_g, it can happen that no sliding occurs. The reduction factor is consequently equal to 1.0. The practical range of E_D is therefore rather wide, between 0.2 and 1.0. It is interesting to note that the value proposed by the American RMI is equal to 0.67, which is somewhere in the middle of this interval.

Some additional studies are suggested to calibrate an expression of E_D that would depend on the structural typology and the ratio a_g/μ. These additional investigations should be a priority.

Complementary comments can also be done on Sects. 2.3.4, 4.2.2 and 4.2.3 dealing with the "no sliding limit state". Indeed, as soon as sliding is used to justify a reduction of forces acting on the structure, it is contradictory to check that pallets are not sliding. Sections 2.3.4, 4.2.2 and 4.2.3 should consequently be removed. Moreover, it has been shown during the research that the amplitude of the sliding motion is highly random and can go up to values like 30 cm. Therefore, clauses 2.3.4 (4) and (5) imposing specific devices to avoid the fall of goods (like for example a 3rd beam) should be mandatory.

In short, the situation regarding sliding can be summarized in two points.

If specific devices are used to avoid sliding, a value of E_D equal to 1.0 should be used.

Otherwise, value of E_D smaller than 1.0 can be used but provisions must be taken to avoid the fall of pallets. Reduced values of E_D should be varying between 0.2 and 1.0 according to the structural typology and to the ratio a_g/μ.

7.1.7 Rack Filling Factor R_F

Even if a long-term monitoring of a typical structure has been realised during the research project, this was obviously a particular case. The proposal of pr FEM 10-2-08 regarding the rack filling ratio R_F to be considered for the seismic assessment is confirmed, namely to use a value equal to 1.0 (structure fully loaded) if nothing else is specified by the Specifier/User.

7.1.8 Ductility and Behaviour Factor q

Various comments can be made on the behaviour factor to be used for the analysis.

- For down-aisle direction,
  - A q-factor equal to 1.5 can be considered in any case;
  - A q-factor equal to 2.0 can be considered for non braced structures, provided that members working in compression and/or bending are at least of class 3. This is supported by the pushover, pseudo-dynamic and dynamic tests realised during the SEISRACKS project (q-factor of about 3.5), affected by a reasonable safety factor.
  - Higher values can be used if properly demonstrated and if the rules of capacity design are followed.
- For cross-aisle direction,
  - For configurations in which the horizontal stability involves at least one bar in compression (configurations partially braced or configurations in Z, D or K), no dissipation should normally be accounted for. This implies that the q-factor must be taken equal to 1.0. A value of 1.5 could however be considered if properly demonstrated by appropriate test results.
  - For concentric X-bracing configurations, a minimum value of 1.5 can be used. If the structure is designed as dissipative following all principles of capacity design, a behaviour factor equal to 4.0 can be used provided that the condition of uniform dissipation all over the height of the upright frame is checked. If this condition is not fulfilled, the structure is actually behaving as an inverted pendulum with an equivalent plastic hinge at the bottom of the frame (only one cell of the upright frame is behaving in the plastic range at ULS). In that case, the behaviour factor should be limited to 2.0.

7.1.9 Characterization of Joints for the Analysis

The proposal of pr FEM 10-2-08 consisting in using the stiffness of beam-to-column connections obtained from static tests according to FEM 10-2-02 for the structural analysis under seismic conditions is confirmed (Sect. 3.3.3). A similar proposal can also be made for column bases, provided that the actual level of compression is duly accounted for.

7.1.10 Detailing Rules for Moment Resisting Frames

In the case of down-aisle non braced configurations resisting by frame effect and designed with dissipative semi-rigid connections, the energy dissipation in the beam-to-column connections and in column bases should be properly assessed. To this purpose, the following procedure is proposed. A non linear pushover analysis is carried out considering joint properties obtained from a monotonic static test. From this analysis, the seismic rotation demand of the joints can be evaluated. A cyclic component test must then be realised following the new testing procedure proposed in the corresponding section of the present report with cycles up to the value of the rotation obtained from the pushover analysis. The joint should be able to keep its strength during a sufficient number of cycles, this number still being to be defined.

7.1.11 Ductility Classes

Even if a distinction needs obviously to be kept between "low dissipative" and "dissipative" structures, the additional sub-classification made within the "dissipative" category between DCM and DCH (ductility classes medium and high) could be removed (see for example Table 3.2 of pr FEM 10-2-08). This would have a very limited impact on the code and could make it much clearer. At the few locations where a distinction is nevertheless needed, the criterion could be based on the section class.

7.1.12 Vertical Component of the Seismic Action

Section 2.2.4 of pr FEM 10-2-08 dealing with the vertical component of the seismic action is confirmed. However, it is felt that additional investigations are necessary to clarify its actual impact on the joint behaviour (beam-to-column joints and column bases) and on the sliding.

7.2 Future Developments

The results presented in this report, as well as those obtained within the whole SEISRACKS Research Project, can give answers only to some of the "unanswered questions" related to the seismic behaviour of Adjustable Steel Pallet Storage Racks.

Also based on the experience gathered within this research, further research topics were identified such as (but not only):

- Dynamic characterization of palletized goods
- Characterization of the static and cyclic behaviour of beam-to-upright connections under bending in the horizontal plane.
- Assessment of the dynamic behaviour (and of the q-factor) in cross-aisle direction (eventually also in presence of a base isolation system).
- Assessment of the dynamic behaviour of storage racks under bi-directional or three-directional ground motions. In fact, the vertical component might play a major role in the case of these structures made by high slenderness profiles, and can also influence the pallet sliding phenomenon.

Globalization of the markets and competitiveness are forcing engineers of one continent to confront themselves with design regulations as well as performance and quality requirements of the others, and vice versa.

Harmonization is needed, in particular between the European and the U.S. design approaches and testing procedures. In this latter field, in particular, a synergy should be found among the research groups. Research funds are more-and-more limited and difficult to be allocated, and in the meanwhile research is "duplicated".

Just recently, in a joint ERF-RMI meeting, held in November 2007 at Politecnico di Milano, some bases were created toward this co-operation. The first actions in this sense should be taken in the current year 2008, and should involve a comparison, discussion and harmonization of the component testing procedures, needed for qualification of products, as well as for assessment of the component behavioural parameters to be adopted in modelling and design.

This should be the topics of the next joint RMI-ERF meeting to be held in April 2008 in the U.S. with the participation of this author. I hope that the content of this report might be helpful in the discussion

Appendix A—Emilia Earthquake

In May 2012, two major earthquakes occurred in Northern Italy, causing 27 deaths and widespread damage. The most affected municipalities are 39. The provinces concerned are Modena, Ferrara and to a lesser extent Bologna, Mantova, Reggio Emilia and Rovigo. The epicenter located between the towns of Finale Emilia and San Felice sul Panaro (Modena) and Sermide (Mantova) (Fig. A. 1).

The maximum registered ground acceleration was about 0.31g and the magnitude Mw = 5, 9. [International Journal of Earthquake Engineering, 2–3 2012]

The earthquake has shown the structural weakness of the racks. During surveys different critical issues that contributes to total or partial collapse mechanisms have been observed:

Fig. A. 1 Earthquake epicenter

C.A. Castiglioni, *Seismic Behavior of Steel Storage Pallet Racking Systems*, Research for Development, DOI 10.1007/978-3-319-28466-8

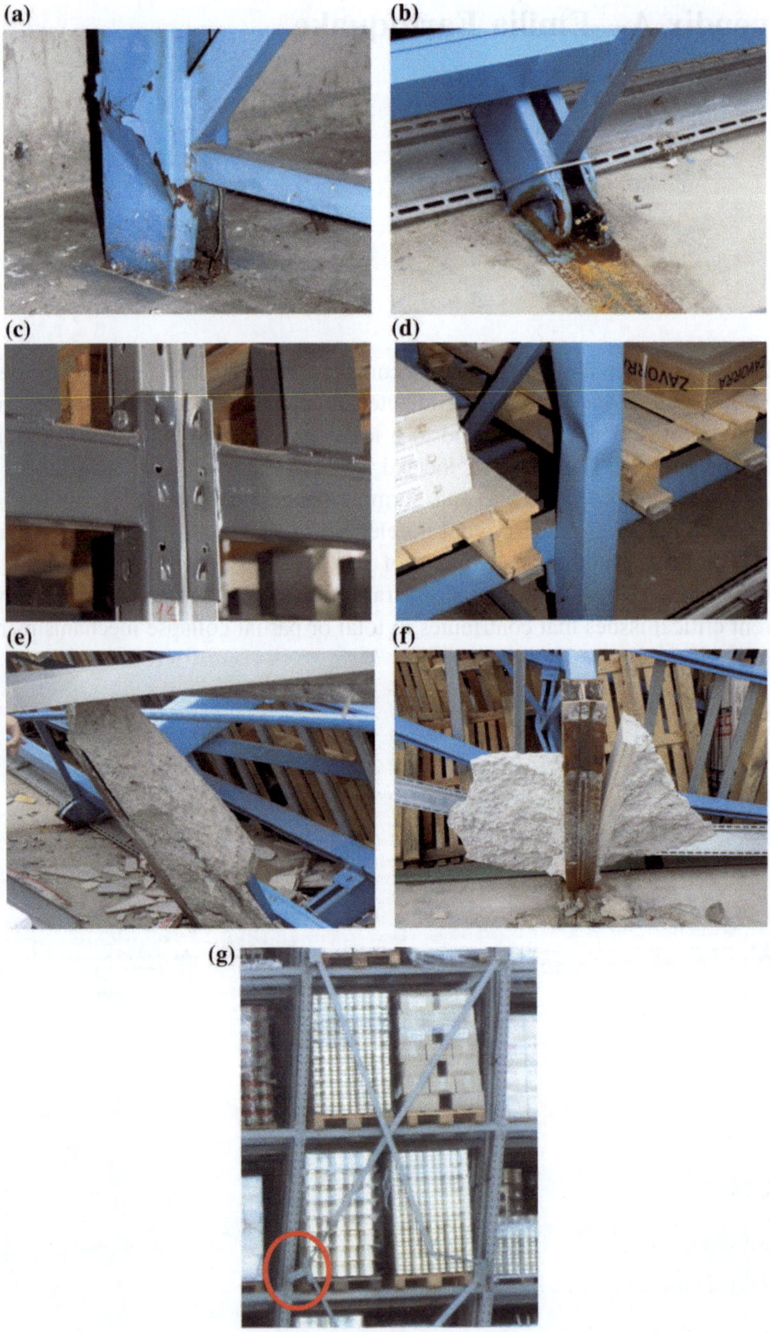

Fig. A. 2 Critical issues observed on **a**, **b** base connection, **c** beam-to-upright connections, **d** upright, **e**, **f** base support, **g** bracing system

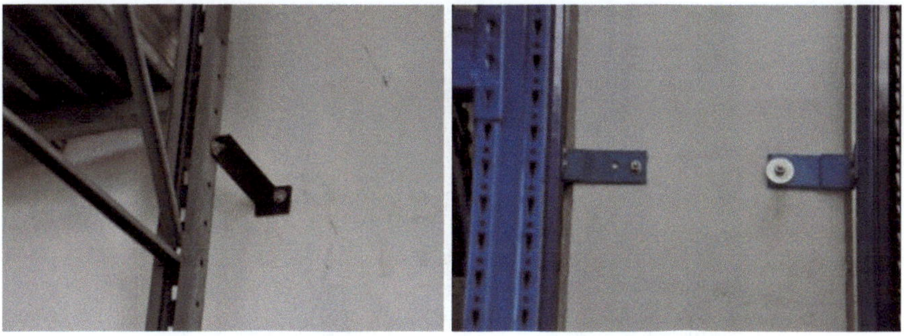

Fig. A. 3 Detail of anchored rack

- Plastic deformations of base connections (Fig. A. 2a, b) and beam end connectors (Fig. A. 2c)
- Local buckling of the uprights (Fig. A. 2d)
- Concrete failure of floor base anchoring (Fig. A. 2e, f)
- Failure of vertical bracing system connections (Fig. A. 2g).

These phenomena underline the importance of structural detailing. Most of the observed failure modes can be easily avoided with small improvement of connections that leads to great enhancement of structural performance.

Other issues linked to the racking systems highlighted by the earthquake are:

- Interaction with the main structure in case of collapse.

Often racking systems are anchored both to non structural (e.g. concrete cladding) or structural (e.g. concrete slabs and foundations) parts of the warehouse. Seismic actions will be transmitted mutually between the two structural systems (rack and warehouse) and can bring to unexpected consequences if not correctly taken into account during design (Fig. A. 3).

- Safety of workers.

The collapse of the rack and the falling/sliding of pallets can cause injuries or even death of the nearby workers. In fact The May 29th earthquake occurred during the daytime, when many people were working. As in the mainshock of May 20th, many of the causalities in the May 29th earthquake were workers inside huge warehouses and factories.

- Obstruction of escape routes (Fig. A. 4).

The fall from shelves and the damage of valuable Parmesan cheese inside storage warehouses, became the symbol of the economic loss related to Emilia Romagna Earthquake (Fig. A. 5).

The following picture shows the beginning of a global collapse mechanism of a fully loaded racking system. In this case part of the lateral walls and of the roof, in precast r.c. elements collapse during the earthquake. The fully loaded racking

Fig. A. 4 Serious calamity at the passage way

Fig. A. 5 Example of collapse with damage of parmesan cheese

system survived and was able to sustain the roof slabs that collapsed during the shock. Collapse of a warehouse where r.c. precast panels of the lateral cladding and of the roof collapsed while the steel rack survived, despite damages, and was able to sustain the roof (Fig. A. 6).

The recent experience of the earthquake in Emily Romagna with its casualties also indicates that the vertical component of the seismic action was larger than expected and should be taken into account ("the main reasons of the damage is the extremely high vertical ground shaking (of the order of 1.0g) in combination with the moderate horizontal motions" [Carydis P., Castiglioni C. A. et al., The Emilia Romagna, May 2012 earthquake sequence. The influence of the vertical earthquake component and related geoscientific and engineering aspects, International Journal of Earthquake Engineering, 2–3 2012, 31–57]).

Fig. A. 6 Onset of global collapse mechanism

Also some clad-rack warehouses were damaged by the earthquake. The clad-rack warehouses are complex storage systems in which the shelving facility is part of the building structure, thereby avoiding the need for the civil works of a conventional building. In clad rack warehouses, the shelving facility not only supports the load of the stored goods, but also the load of the roof and lateral cladding of the building, as well as external actions, such as wind, snow and eventually the earthquake. Most clad rack buildings are automatic systems (AS/RS) using robotic equipment for handling loads. In this case, the collapse of a warehouse or the falling of stored merchandise, is not a risk for the safety of workers, but can cause a large economical loss.

An example of the complexity of the behavior of this type of structures is the ceramic warehouse in Sant'Agostino. As shown in the Fig. A. 7, part of the structure totally collapsed while part of the same warehouse did not move.

Another aspect underlined by the image above is the layout of the stored pallets. As it is known, the distribution of masses along the height of the structure can highly influence the seismic response of the rack.

Fig. A. 7 Collapse of warehouse in Sant'Agostino

MIX
Papier aus verantwortungsvollen Quellen
Paper from responsible sources
FSC® C105338

If you have any concerns about our products,
you can contact us on
ProductSafety@springernature.com

In case Publisher is established outside the EU,
the EU authorized representative is:
Springer Nature Customer Service Center GmbH
Europaplatz 3, 69115 Heidelberg, Germany

Printed by Libri Plureos GmbH
in Hamburg, Germany